The Physical Oceanography of Sea Straits

NATO ASI Series

Advanced Science Institutes Series

A Series presenting the results of activities sponsored by the NATO Science Committee,
which aims at the dissemination of advanced scientific and technological knowledge,
with a view to strengthening links between scientific communities.

The Series is published by an international board of publishers in conjunction with the
NATO Scientific Affairs Division

A Life Sciences	Plenum Publishing Corporation
B Physics	London and New York
C Mathematical	Kluwer Academic Publishers
and Physical Sciences	Dordrecht, Boston and London
D Behavioural and Social Sciences	
E Applied Sciences	
F Computer and Systems Sciences	Springer-Verlag
G Ecological Sciences	Berlin, Heidelberg, New York, London,
H Cell Biology	Paris and Tokyo

The Physical Oceanography of Sea Straits

edited by

L. J. Pratt

Woods Hole Oceanographic Institution,
Woods Hole, Massachusetts, U.S.A.

Kluwer Academic Publishers

Dordrecht / Boston / London

Published in cooperation with NATO Scientific Affairs Division

Proceedings of the NATO Advanced Research Workshop on
The Physical Oceanography of Sea Straits
Les Arcs, France
July 5–9, 1989

Library of Congress Cataloging-in-Publication Data

The physical oceanography of sea straits   edited by L.J. Pratt.
      p.   cm. -- (NATO advanced study institutes series. Series C,
  Mathematical and physical sciences ; v. 318)
    Includes index.
    ISBN-13:978-94-010-6789-8   e-ISBN-13:978-94-009-0677-8
    DOI: 10.1007/978-94-009-0677-8

    1. Straits.   I. Pratt, L. J., 1952-   . II. Series.
  GC99.P47 1990
  551.46'09--dc20   90-44192

ISBN-13:978-94-010-6789-8

Published by Kluwer Academic Publishers,
P.O. Box 17, 3300 AA Dordrecht, The Netherlands.

Kluwer Academic Publishers incorporates the publishing programmes of
D. Reidel, Martinus Nijhoff, Dr W. Junk and MTP Press.

Sold and distributed in the U.S.A. and Canada
by Kluwer Academic Publishers,
101 Philip Drive, Norwell, MA 02061, U.S.A.

In all other countries, sold and distributed
by Kluwer Academic Publishers Group,
P.O. Box 322, 3300 AH Dordrecht, The Netherlands.

Printed on acid-free paper

Table of Contents

*Indicates peer-review paper

Professor Henri Lacombe (Photograph: M. Fieux)

Professor Henri LACOMBE

Professor Henri LACOMBE may be considered the "Father of Physical Oceanography" in France. Born in 1914 he graduated from the prestigious "Ecole Polytechnique" in 1935 at the head of his class and then began his career at the Hydrographic Service of the French Navy. Physical Oceanography was almost non-existent in France at that time. He studied acoustic propagation (1943–1951) and tides and swell diffraction (1946–1953).

In 1955, a Professorship in Physical Oceanography was created at the Museum National d'Histoire Naturelle, and he was elected to this position. The "Laboratoire d'Océanographie Physique" with Professor H. Lacombe at its head and Paul Tchernia as the only member was born. Some time later Bernard Saint-Guily and an ex-Marine officer, Pierre Guibout, joined the Lab. The Laboratory's first cruise was to the Eastern Mediterranean, on board the *Calypso*, rented from Cdt Cousteau. The circulation in the Mediterranean Sea was not well known at that time, and Henri Lacombe decided that this "natural reduced model of the ocean" had to be investigated first by his new-born Laboratory.

In 1957 and 1958, the Laboratory participated in the International Geophysical Year (IGY), and Henri Lacombe conducted two cruises on board *Calypso* to study the outflow of Mediterranean water into the Atlantic Ocean. His study of the Strait of Gibraltar continued in 1958 when he carried out the first cruise in the Strait, on board the *Winnaretta-Singer*.

At the same time, growing interest in the ocean was expressed by the French government and a "Committee for the Exploration of the Ocean" (COMEXO) was created. The formation of the committee and the availability of grants marked the beginning of intense new activity at the Laboratory and the arrival of technicians, engineers and young researchers. Courses in Physical Oceanography were organized and Professor H. Lacombe, in charge of Dynamical Oceanography, supervised the research and theses of most of the French physical oceanography scientists and students. Professor Lacombe always encouraged young researchers of his Lab to spend time abroad, especially at the Woods Hole Oceanographic Institution, where they learnt new techniques and developed contacts with famous scientists, some of whom participated in the Lab's cruises.

Under the sponsorship of NATO, he organized the first international survey of the Strait of Gibraltar in May–June 1961, a project involving Belgium, France, Italy, Norway and Spain. The large data set, obtained using three to five ships in the Strait simultaneously, allowed the publication of an Atlas showing the evolution of the flow during spring and neap tides. The advancement of oceanic techniques lead Henri Lacombe to return in 1967 with a CTD in order to sample internal waves. In the meantime, the remarkable and now classical anticyclonic gyre of the Alboran Sea was discovered (1962) and the "Mediterranean Outflow" project (1965–1967) was carried out to study the detailed trajectory of Mediterranean water entering the Atlantic Ocean. During the following years (1969 to 1975), Henri Lacombe's Laboratory was a leader in deep water formation studies, with the series of cruises

(MEDOC '69, '70, '72, '73, '74 and '75) in the Northwestern Mediterranean. These cruises followed up Lacombe and Tchernia's earlier studies of 1963 winter data collected by the French Navy.

Ocean–Atmosphere interaction studies began in 1962 with the launching in the North-western Mediterranean of the first "laboratory buoy," on which oceanic and atmospheric data were collected for years. Holding fast to his opinion that the Mediterranean Sea is a reduced model of the world ocean, he campaigned for a polygon of buoys in the Mediter-ranean, an ambition which was unfortunately never realized.

When the time came in France to create an Institution devoted to the Ocean, he actively participated in the birth of the CNEXO (National Center for Exploiting the Ocean), which became later the Institut Français de Recherche pour l'Exploitation de la Mer (IFREMER). He was also involved in the launching of the first French oceanographic research ships, *Jean-Charcot* and *Noroit*.

Besides the activities related to his functions as Director of the Laboratory, Professor La-combe occupied important positions in most international committees and organizations: President of the International Association for the Physical Sciences of the Oceans (IAPSO), President of the UNESCO Intergovernmental Oceanographic Commission, President of the Physical Oceanography Committee of CIESM (International Commission for the Scientific Exploration of the Mediterranean Sea), member of the NATO Sub-Committee for Oceano-graphic Research (President of the Gibraltar Working Group). In April 1973, he was elected to the French Academy of Sciences.

Retired in 1983, he continues to be interested in all oceanic activities, working every day at his Lab and participating in seminars and working groups. He was an active presence at the NATO/ONR workshop which lead to this volume.

Claude Richez

Preface

Suppose one were given the task of mapping the general circulation in an unfamiliar ocean. The ocean, like our own, is subdivided into basins and marginal seas interconnected by sea straits. Assuming a limited budget for this undertaking, one would do well to choose the straits as observational starting points. To begin with, the currents flowing from one basin to the next, over possibly wide and time-varying paths, are confined to narrow and stable routes within the straits. Mass, heat and chemical budgets for individual basins can be formulated in terms of the fluxes measured across the straits using a relatively small number of instruments.

The confinement of the flow by a strait can also give rise to profound dynamical consequences including choking or hydraulic control, a process similar to that by which a dam regulates the flow from a reservoir. The funneling geometry can lead to enhanced tidal modulation and increased velocities, giving rise to local instabilities, mixing, internal bores, jumps, and other striking hydraulic and fine scale phenomena. In short, sea straits represent choke points which are observationally and dynamically strategic and which contain a full range of fascinating physical processes.

In July, 1989, a group of scientists gathered in Les Arcs, France, to discuss the physical oceanography of sea straits. The five-day meeting was sponsored by the North Atlantic Treaty Organization (NATO) and the United States Office of Naval Research (ONR). The purpose was to assess the present state of scientific knowledge with respect to strait and sill flow and to identify important and tractable problems for consideration in the near future. This volume is a compilation of the scientific work presented and the discussions held at Les Arcs.

The papers appearing herein fall into two categories: peer- and non-peer-review contributions. In the latter, short or medium length summaries of personal research or overviews of field programs received editorial and technical review. The peer-review papers, on the other hand, were refereed by two anonymous conference participants. In addition to meeting certain quality standards, these contributions were required to devote some space to historical review of the subject. The purpose was to provide up-to-date overviews accessible to students and other new investigators wishing to learn more about the field. The peer-review papers are indicated with asterisks in the table of contents.

Regrettably, practical considerations made it necessary to limit the size and scope of the meeting. For example, discussion had to be restricted largely to straits containing flows with significant hydraulic processes, ruling out larger passages such as the Florida Straits. Also field work prevented a number of investigators of important deep passages such as the Vema Channel and Denmark Strait from attending. However, the overviews of other field programs described here attempt to give a general indication of the observational and instrumentational problems facing investigators today.

One of the most useful discussions took place near the end of the meeting when each participant was asked to write down one research problem he or she would like to see solved within the next 5–10 years. These suggestions are summarized in the book's final paper entitled "Current Research Problems."

The editor wishes to express his gratitude to a number of people for their help and encouragement in the organization of the meeting. They include David Farmer, Larry Armi, Tom Kinder, Harry Bryden, Chris Garrett and Alan Brandt, all of whom were involved in the initial planning. Also, Claude Richez was instrumental in the organization and logistics on the French side. Her biography of Professor Henri Lacombe, to whom this volume is dedicated, appears next. Special thanks are also due to Lisa Garner Wolfe for her technical assistance. Above all, I wish to thank Barbara Gaffron for her careful and thoughtful technical editing, without which the book would never have been completed.

Lastly, the editor, on behalf of the participants, wishes to express sincere thanks to NATO (under grant 713/88) and ONR (under grant N00014-89-J-1182) for their financial support.

Lawrence J. Pratt

List of Participants

Yalçin Arisoy, Dokuz Eyül University, Faculty of Engineering and Architecture, Department of Civil Engineering, Bornova-İzmir, Turkey

Laurence Armi, Scripps Institution of Oceanography, Mail Code A-030, La Jolla, California 92093

Peter Baines, CSIRO Division of Atmospheric Research, P.B. No. 1, Mordialloc, Victoria 3195, Australia

Molly Baringer, Clark 3, Woods Hole Oceanographic Institution, Woods Hole, Massachusetts 02543. Tel: 508 548-1400, x2796

Eric Barthelemy, Institut de Mécanique de Grenoble, B.P. 53X, F-38041 Grenoble-Cedex, France. FAX: 76-82-50-01

Francesco Bignami, c/o Salusti, Dipartimento di Fisica, Università degli Studi di Roma "La Sapienza", Piazzale Aldo Moro, 2; 00185 Roma, Italia. Tel: 396/49914291

Karin Borenäs, Department of Oceanography, University of Gothenburg, Stigbergstorget 8, Box 4038, S-400 40 Göteborg, Sweden. Tel: 031-422800, FAX: 031-421988

Alan Brandt, Small-Scale Physical Oceanography Program, Office of Naval Research, Code 1122 SS, 800 North Quincy Street, Arlington, Virginia 22217-5000. Tel: 202 696-4025, Telemail: OMNET/A.BRANDT

Nancy Bray, Mail Code A-009, University of California at San Diego, Scripps Institution of Oceanography, La Jolla, California 92093. Tel: 619 534-2193, Telemail: COASTAL.SIO

Harry Bryden, Clark 3, Woods Hole Oceanographic Institution, Woods Hole, Massachusetts 02543. Tel: 1-508-548-1400 x2806, Telemail: H.BRYDEN

Julio Candela, Clark 3, Woods Hole Oceanographic Institution, Woods Hole, Massachusetts 02543. Tel: (508) 548-1400, ext. 2907

Stuart Dalziel, DAMTP University of Cambridge, Silver Street, Cambridge, England, CB3 9EW. Tel: (0223) 337840, FAX: (0223) 337918

David Farmer, Institute of Ocean Sciences, P.O. Box 6000, 9860 West Saanich Road, Sidney, British Columbia, Canada V8L 4B2

Christine Gailliard, Florida State University, Department of Oceanography, Tallahassee, Florida 32306-3048. Tel: 904 644-3492

Chris Garrett, Department of Oceanography, Dalhousie University, Halifax, Nova Scotia, Canada B3H 4J1. Tel: 902 424-3674

Gian Pietro Gasparini, Stazione Oceanografica, c/o ENEA, 19036 Pozzuolo di Leriei, Italy. FAX: 39-187-536213

Jean-Pierre Germain, Institut de Mécanique de Grenoble, B.P. 53 X, F-38041 Grenoble - Cedex, France. FAX: 76-82-50.01

Rocky Geyer, Bigelow 106, Woods Hole Oceanographic Institution, Woods Hole, Massachusetts 02543. Tel: 508 548-1400, x2868, Telemail: R.GEYER

Niels Erik Ottesen Hansen, LIC Engineering A/S, Ehlersvej 24, DK-2900 Hellerup, Denmark. Tel: 45 1 621642, Telex: 21437 LICENG.DK

Karl Helfrich, Clark 3, Woods Hole Oceanographic Institution, Woods Hole, Massachusetts 02543. Tel: 508 548-1400, ext. 2870, Telemail: K.HELFRICH

Ken Hunkins, Lamont-Doherty Geological Observatory, Palisades, New York 10964

Jorg Imberger, Department of Math and Mechanical Engineering, The University of Western Australia, Nedlands, Western Australia 6009, Australia

Thomas Kinder, Office of Naval Research Code 1121 CS, Room 532, Ballston Towers 1, 800 North Quincy Street, Arlington, Virginia 22217-5000. Tel: 202 696-4441, Omnet/Telemail: T.KINDER

Henri Lacombe, Laboratoire d'Océanographie Physique, Museum National d'Histoire Naturelle, 43–45 Rue Cuvier, 75231 Paris, Cedex 05, France

Greg Lawrence, Department of Civil Engineering, 2324 Main Mall, U. of British Columbia, Vancouver, B.C., Canada, V6T 1W5. Tel: 604 228-5371, FAX: 604 228-6901

Eric Lindstrom, CSIRO Division of Oceanography, G.P.O. Box 1538, Hobart, Tasmania 7001, Australia

Peter Lundberg, Department of Oceanography, University of Gothenburg, Box 4038, S-400 40 Göteborg, Sweden. Tel: 031/126867

Tony Maxworthy, Dept. of Mechanical Engineering, University of Southern California at Los Angeles, University Park, Los Angeles, California 90089-1453. Tel: 213 743-6240, FAX: 213 749-1271

Tom McClimans, Norsk Hydroteknisk Laboratorium, Norwegian Institute of Technology, N-7034 Trondheim, Norway

Ken Melville, 48-331, Massachusetts Institute of Technology, Cambridge, Massachusetts 02139

Antonio Michelato, O.G.S., Laboratori Marini, P.O. Box 2011, 34016 Trieste, Italy

Jacob Steen Møller, Danish Hydraulic Institute, Agern Alle 5, DK-2970 Hørsholm, Denmark. Tel: + 42868033, TeleFAX: + 42867951, Telex: 37402dhicph.dk

Steve Murray, Coastal Studies Institute, Louisiana State University, Baton Rouge, Louisiana 70803. Tel: 504 388-2954, FAX: 504 388-2520, Telemail: S.MURRAY/OMNET

Doron Nof, Department of Oceanography, Florida State University, Tallahassee, Florida 32306. Tel: 904–644-6700

James O'Donnell, Marine Sciences Department, University of Connecticut, Groton, Connecticut 06340. BITNET: ODONNELL @UCONNUM, Tel: 203 445-3471

Gregorio Parrilla, Instituto Español de Oceanografia, Corazon de Maria 8, 1ª Pa., 28020 Madrid, Spain. Tel: 1 4138013 x-299, FAX: 4156268

Flemming Bo Pedersen, Institute of Hydrodynamics and Hydraulic Engineering, Technical University of Denmark, Building 115, DK-2800 Lyngby, Denmark. Tel: + 45 2 88 22 22, Telex: 37529 DTHDIA DK, Telegram: HYDROENG

Neal Pettigrew, Science & Eng. Research Building, University of New Hampshire, Durham, New Hampshire 03824. Tel: 603 862-3159, FAX: 603 862-1915, Telemail: UNH.OCEAN

Larry Pratt, Clark 3, Woods Hole Oceanographic Institution, Woods Hole, Massachusetts 02543. (508) 548-1400 X2540, Telemail: L.PRATT

Dominique Renouard, Institut de Mécanique de Grenoble, B.P. 53X, F-38 041 Grenoble, Cedex, France. FAX: 76-82-50-01

Claude Richez, Laboratoire d' Océanographie Dynamique et de Climatologie, LODYC Université de Paris 6, 4 Place Jussieu, 75252 Paris Cedex 05, France. Tel: 33 1 46 33 21 31, Telex: 206317, Telemail: J.GASCARD

Ettore Salusti, Dipartimento di Fisica, Università degli Studi di Roma "La Sapienza" Roma, Piazzale Aldo Moro, 2; 00185 Roma, Italia

Takashige Sugimoto, Ocean Research Institute, University of Tokyo, 1-15-1 Minamidai, Nakano-ku, Tokyo 164, Japan. Tel: 03-376-1251, FAX: 03-375-6716, Telex: J-25607 (ORIUT), Telemail: ORI.TOKYO

Ümit Ünlüata, Institute of Marine Sciences, Middle East Technical University, P.K. 28 Erdemli, Icel, Turkey

I. CASE STUDIES

CHARACTERISTICS OF CIRCULATION IN AN INDONESIAN
ARCHIPELAGO STRAIT FROM HYDROGRAPHY, CURRENT
MEASUREMENTS AND MODELING RESULTS

Stephen P. Murray and Dharma Arief
Coastal Studies Institute
Louisiana State University
Baton Rouge, Louisiana   70803-7527

John C. Kindle and Harley E. Hurlburt
Naval Oceanographic Research and
    Development Activity
Code 323
Bay St. Louis, Mississippi   39529

ABSTRACT.   The Lombok Strait, a gap in the lower Indonesian Archipelago
second in cross sectional area only to the Timor passages, provides a
major pathway for the Pacific to Indian throughflow.   A global reduced
gravity model, corroborated by dynamic height climatology from the
Generalized Digital Environmental Model, predicts annual mean sea levels
15-20 cm higher at the Pacific entrance to the Indonesian Seas than in
the Indian Ocean south of the archipelago straits.   Consistent with this
regional pressure gradient, Pacific core layers of the Northern
Subtropical Central Water and the North Pacific Intermediate Water are
traced southward from the Makassar Strait into the Lombok Strait.   Maps
of temperature, salinity, and density distributions and sea surface
dynamic heights in the Lombok Strait from January, June, and September
1985 also indicate a persistent southward flow of appreciable magnitude.
Geostrophic speeds, however, are clearly too large by a factor of two or
more.   Current meter arrays in the north strait (January 1985 - March
1986) provide direct measurements of southward currents which persist
through most of the year and are concentrated in the upper few hundred
meters consistent with Wyrtki's (1987) analysis of the regional pressure
gradient.   Maximum sustained speeds of over 70 cm/sec occur from July to
September with a long period of weak currents from mid-October 1985
through January 1986.   Tropical cyclones in the Timor Sea (December-
April) force strong northward flow reversals which can persist for ten
days.   The wind-forced numerical model identifies the strong westward
wind stresses in the Timor Sea during the southeast monsoon as the major
cause of the annual cycle of current in the Strait.

3

L. J. Pratt (ed.), The Physical Oceanography of Sea Straits, 3–23.

1.   Introduction

Recent research on sea straits emphasizes the variability of forcing mechanisms that control the circulation in many of the major sea straits around the world.   For example, we note the dominant role of evaporation-induced pressure gradients in the Mediterranean and Middle East straits (Bryden and Stommel, 1984), the inertial effect of the Gulf Stream in the Florida Strait (Lee et al., 1985), the cross-stream geostrophically induced surface slope of a major western boundary current driving the flow in the Tsugaru Strait (Conlon, 1981), and the role of large scale meteorological forcing in the Strait of Belle Isle (Garrett and Petrie, 1981).
        From another point of view, sea straits are the systematic result of large-scale geophysical and tectonic processes and often cluster in environmental regions that impose similar dynamical constraints on members of the cluster.   A notable example is the Bab El Mandeb/Hormus/ Tiran/Gibraltar cluster, where the low-frequency circulation is dominated by the effects of regional evaporation.   The island arcs that rim subduction zones are a prominent feature of plate margin tectonics, and the characteristic breaches in these arcs have evolved into important sea straits.   The Indonesian archipelago (Figure 1), containing the Sunda/Malaka/Makassar/Lombok/Timor straits is another critical cluster that provides the only connection between ocean basins in tropical latitudes.   The potential importance to the global circulation of the net transport (the Indonesian throughflow) between the Pacific and Indian Oceans through these straits is now well recognized (Gordon, 1986).   Pacific Ocean, South China Sea and Java Sea waters are of significantly lower salinity and higher temperature than the adjacent Indian Ocean.   Potential inter-basin fluxes of these properties appear critical to the heat and salt balances in the Indian Ocean (Toole and Raymer, 1985).
        Despite numerous breaches in the archipelago, the deep passages (2,000 m depth) bracketing Timor were considered (Wyrtki, 1961) the sole pathways for any significant transport into the Indian Ocean.   In terms of cross-sectional area available for transport, the second most important channel through the lower archipelago is the Lombok Strait between the island of Bali and Lombok (Figure 1).   Lying at the end of a deep bathymetric trough linking it to the Makassar Strait, it provides a direct conduit for Pacific Ocean water into the Indian Ocean.   Depths in the Lombok Strait are typically 800-1,000 m except at the south end where a small island Nusa Penida divides the channel.   The western channel (Badung Strait) has a cross sectional area less than one-fourth that of the main channel between Nusa Penida and Lombok Islands.   An extremely irregular sill with maximal depths of about 300 m connects Nusa Penida to Lombok Island.
        The objectives of this paper are to present both: (a) a general review of observational data taken in the Lombok Strait area during 1985-1986; and (b) the present status of our understanding of the forces controlling the low-frequency circulation in the strait, interpreted mainly from numerical model results.   These results not only shed considerable light on the dynamics controlling the circulation through

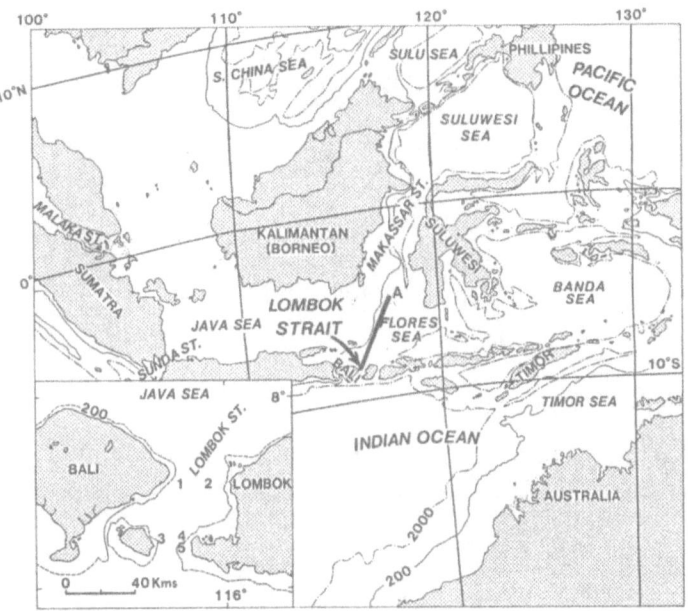

Figure 1. Regional map showing the location of the Lombok
           Strait with an inset showing current meter
           mooring locations in the Strait.

this particular strait, but also provide a general framework for
understanding transport and fluctuations in the other straits of the
archipelago.

2.   Observations and Model

Field work and data collection were conducted from January 1985 to March
1986.   Seven current moorings were deployed, five in the Lombok Strait
(see Figure 1) and two in the bathymetric trough (in the western Flores
Sea) linking the Makassar Strait outflow to the Lombok Strait.   More
than 120 current meter months of data were collected.   A total of 234
CTD casts were taken in the Lombok Strait, its Indian Ocean approaches,
and the western Flores Sea extending as far north as the Makassar Strait
and east into the Flores Sea to 119°E.   The CTD station grid occupied
with some variations in January, June, and September 1985 is shown in
Figure 2.   Thirty-four meter months of sea level (pressure gauge) data
taken in the Strait and at the Makassar trough mooring site will be
discussed separately.   Data on the regional sea level during our
observations were collected from tide gauges in the Philippines and
northwest Australia courtesy of Klaus Wyrtki, University of Hawaii, and
the Flinders Tidal Observatory, Australia.

6

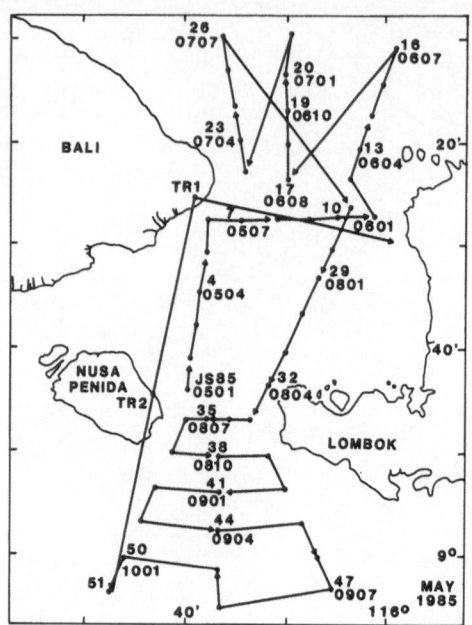

Figure 2.   The basic CTD grid occupied in January, June, and
             September, 1985.

        To examine the mean and seasonal variability of the throughflow in
the Indonesian passages and the forcing mechanisms responsible for the
variability, we use the most recent version of the NORDA Global Model,
forced by monthly averages of the European Centre for Medium-Range
Weather Forecasts (ECMWF) 1,000-mb winds.

        This model is a multi-layer, primitive equation formulation which
incorporates a free surface, arbitrary coastline geometry, full scale
bottom topography in the lowest layer and a semi-implicit time scheme.
The model equations are formulated in spherical geometry over a
latitudinal extent ranging from 71°N to 72°S.   Lateral boundaries are
located at the 200 m contour using bathymetric data from the Synthetic
Bathymetric Profiling Systems (SYNBAPS) data base.   The walls are rigid
and the no-slip condition is prescribed on the tangential flow.

        The simulations described in this paper use a one active layer
reduced gravity version of the model which includes the effects of
mixing and mean thermodynamics.   Density gradients within the upper
layer are permitted and are modified by horizontal advection,
entrainment, eddy diffusion and relaxation to a mean density climatology
based on Levitus (1982).   The relaxation time constant is a function of
layer depth and ranges from three months for a 50 m thick layer to 1.5
years for a layer of 550 m.   Except for very shallow layers, the
relaxation does not produce a strong constraint on the model density
field.   Entrainment from the lower layer is initiated when the layer

thickness reaches a minimum value, mass is conserved through uniform detrainment. The model formulation permits the specification of surface heat fluxes, but that option was not utilized in these simulations. A list of model parameters is given in Table 1. Also see Kindle et al. (1989).

TABLE 1

PARAMETER	DESCRIPTION	VALUE
A(m)	Horizontal eddy viscosity (momentum)	1500 m^2/sec
A(d)	Horizontal eddy viscosity (density)	5000 m^2/sec
g	Gravity	9.8 m/sec^2
dx	Grid spacing in longitude	.7 deg dy
dy	Grid spacing in latitude	.5 deg
dt	Time step	60 min.
H	Initial layer depth	250 m
hm	Maximum layer thickness for which mixing is initiated	60 m

3. Regional Forcing

The Indonesian Seas lie at the western edge of the Pacific trade wind belt where the persistent westerly directed wind stresses pile water up against the boundary. As a result, annual mean sea-surface elevations are as much as 20 cm higher at the Pacific entrances to the archipelago south of the Philippines than in the Indian Ocean south of Java. This pattern is seen clearly in our model results in Figure 3, where the average sea-surface deviations from the initial state over the two-year period 1984 through 1985 are presented. The long-term average climatology of sea surface dynamic height from the Generalized Digital Environmental Model (GDEM) data set referenced to 1,000 db (Figure 4) shows a markedly similar pattern.

In addition to trade wind forcing of mean annual sea-level differences across the archipelago, it is well known that the Asian monsoon drives a strong seasonal cycle of meteorological forcing over the Indonesian Seas (Wyrtki, 1961). The southeast Asian monsoon brings strong east winds across the archipelago from May to early September and west winds from November to March. Transition periods are characterized

8

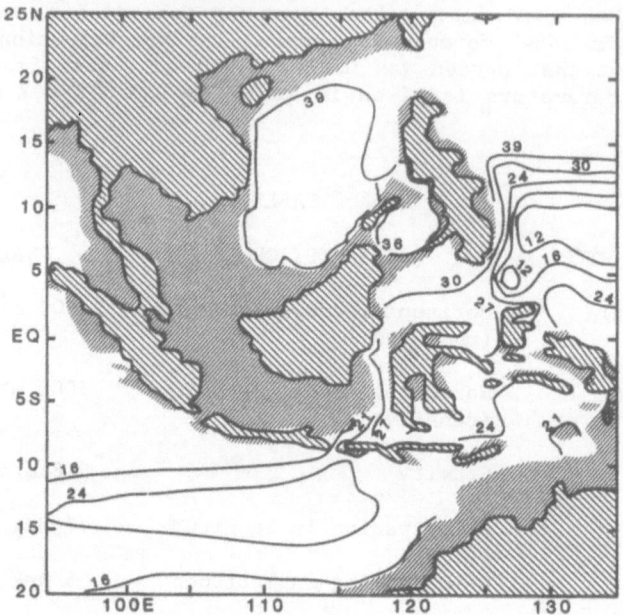

Figure 3. Mean sea surface deviations (cm) from the initial
state of the reduced gravity model for the two-
year period 1984-1985.

by weak, variable winds.   Wyrtki (1961) has shown the step-like
reversals of the monsoon dominate the surface circulation along the Java
Sea - Banda Sea axis.

4.   Thermohaline Characteristics

The general characteristics of the vertical structure of temperature and
salinity in the region are illustrated in Figure 5 from a CTD cast taken
in the north Strait.   Surface isothermal and isohaline layers usually
30-50 m thick overlay a strong thermocline extending to about 400 m
below which the temperature decreases slowly to 4°C.   The low salinity
in the surface layer reflects the regionally intense rainfall.   A
salinity maximum and a salinity minimum occur near 150 m and 300 m,
respectively.
       The presence of Pacific Ocean core water in the Indonesian Seas
was documented by Wyrtki (1961) from historical data widely spaced in
time and space.   Our CTD transect from the southern end of the Makassar
Strait 475 km south through the Lombok Strait distinctly shows (Figure
6) the salinity maximum of the core layer of the Northern Subtropical
Central Water (NSCW) at 150 m $\pm$ 25 m, in early June 1985.   There is no
appreciable change in the salinity maximum over this distance until

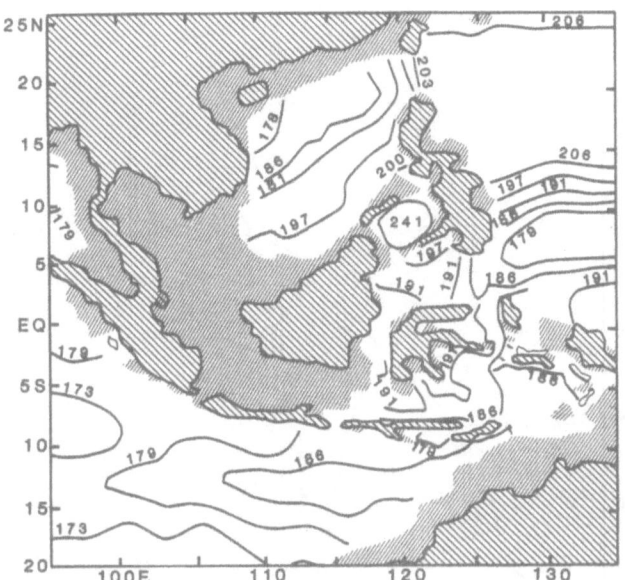

Figure 4.   Dynamic height (dyn. cm) from the long-term GDEM
                climatology.

encountering intensified mixing in the Lombok Strait.   Similarly, the
salinity minimum diagnostic of the North Pacific Intermediate Water
(NPIW) is tracked in this figure from the Makassar Strait across the
Flores Sea with little change until reaching the Lombok Strait.
        The Pacific Ocean core layers penetrate into the Lombok Strait in
all three months we observed.   As shown in Figure 7, there was
considerable seasonal variability in the NSCW.   A much stronger and more
extensive salinity maximum occurred in September than did in June.   The
NPIW core layer, on the other hand, showed no discernible seasonal
variability.   The NSCW salinity maximum core layer at 150 m depth
clearly survived mixing over the sill only in the September data when it
penetrated only about 25 km into the Indian Ocean.   Patches of NPIW
water were occasionally found south of the sill, apparently brought up
over the sill by Bernoulli suction (Bryden and Stommel, 1984; Kinder and
Parrilla, 1987).   It is notable that these core layers, having travelled
thousands of kilometers from their Pacific origin, are destroyed by
locally intense mixing in the Lombok Strait.
        The distribution of thermohaline properties in the surface layer
in the Strait reflect the southward motion indicated by the core layers.
Salinity, temperature, and density distributions on the 10 db surface
are shown in Figure 8a from early June 1985.   The isotherms clearly show
the penetration of the warm (29°C) surface isothermal layer into the
north Strait with very little temperature change along the axis of the
strait until encountering the sill.   South of the sill an extremely

10

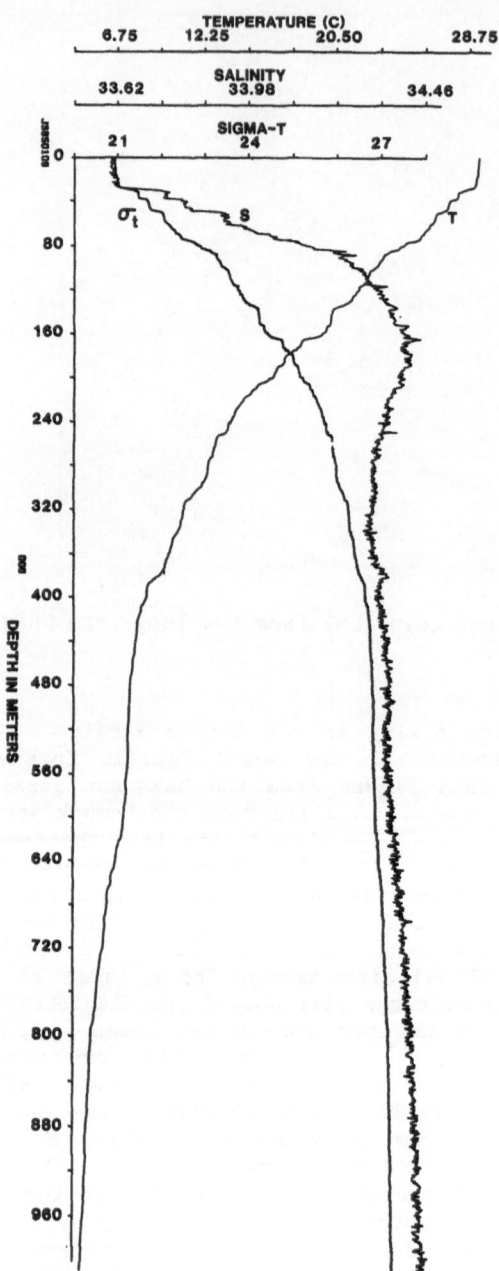

Figure 5. An example of a CTD cast from the north Lombok
Strait showing the well-mixed surface layer and
the salinity maximum and minimum core layers in
the thermocline.

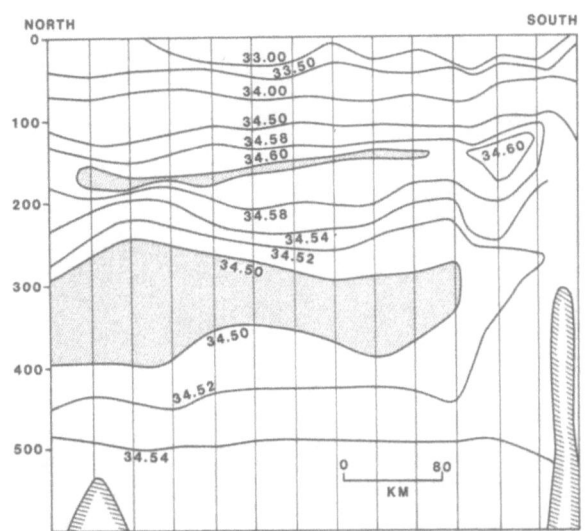

Figure 6.   Vertical section of salinity from the southern
end of the Makassar Strait to the sill at the
southern end of Lombok Strait along the line
labelled A in Figure 1.

well-developed thermal plume intrudes over 30 km into the Indian Ocean.
Temperature gradients are steepest on the western side of the plume
(over 3° drop in 15 km) suggesting the presence of the eastward directed
Java Coastal Current which runs at high velocities along the south coast
of the archipelago from December to June (Wyrtki, 1961).
    The salinity distribution (Figure 8b) on the 10 db surface shows a
similar but less dramatic evolution from north to south in the Strait.
Low salinity coastal boundary layer waters originating from Javanese
rivers enter the Strait from the northwest to combine with higher
salinity water from the Flores Sea.   Salinities at 10 db gradually
increase to the south of the sill. A haline plume penetrates into the
Indian Ocean and deflects to the east as a result, we believe, of
impacting the Java Coastal Current.
    The density distribution (Figure 8c) on the 10 db surface nicely
summarizes the southward increased mixing of homogeneous, warm, less
saline water above 10 db with the cold, saltier water below.   The
penetration of the density plume, with a front-like western limb and an
eastward deflection are also all present in the density field.
    The dynamic topography of the sea surface shown in Figure 9 also
adds important information on the flow pattern in the Strait.   Wyrtki
(1987, 1961) has shown that most of the pressure gradient in this region
is contained in the upper 200 m and thus we choose 500 db as a suitable
reference level.   In January there is a 15 dyn. cm gradient across the
mid-section of the Strait indicating a substantial surface layer flow to

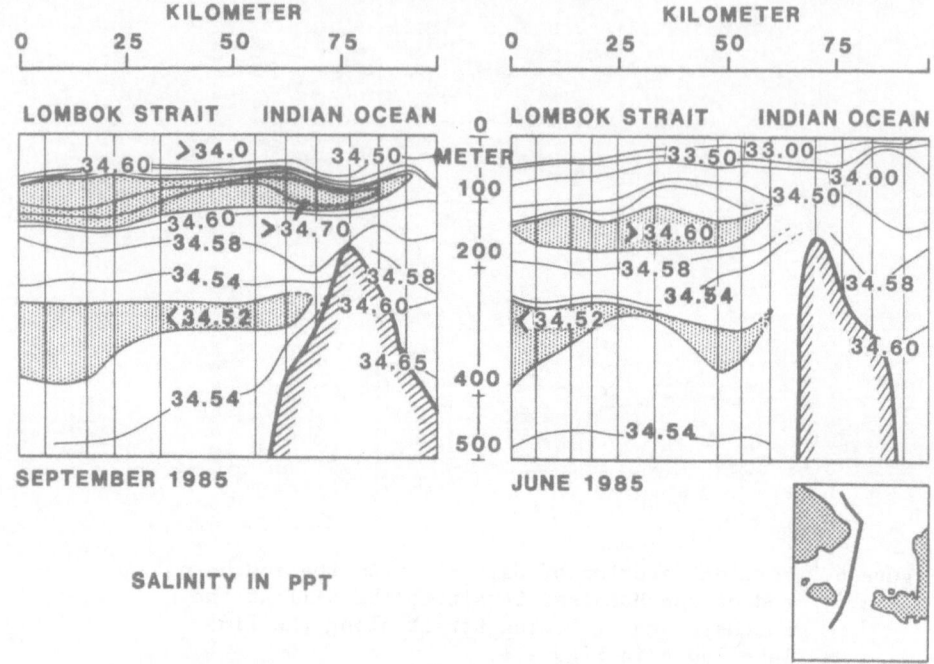

Figure 7. Vertical sections of salinity in September and
           June 1985 along the west side of the Lombok
           Strait showing the location of NSCW and NPIW
           core layers.

the south.   The dynamic height contours fan out radially into the Indian
Ocean, but they are skewed to the southeast reflecting the direction of
movement indicated by the haline and thermal plumes.   Unfortunately,
equipment failure in January did not allow measurements at the CTD
stations in the northern approaches.   In the September map of surface
dynamic height (0/500 db), a similar surface slope exists across the
mid-section of the Strait with a strong southward flow indicated.
Additionally, these dynamic height data show a 10-15 cm drop in the sea
surface along the center line of the strait from the west Flores Sea
inflow region in the north to the Indian Ocean outflow region in the
south.   The internal radius of deformation of the Strait is 120 km, much
larger than the width of the Strait.   Therefore a geostrophic balance is
not expected either across or along the Strait.   In fact, geostrophic
surface layer speeds indicated by the cross strait slopes are over 300
cm/sec.   Such unrealistically high speeds suggest a more complex across-
strait momentum balance for such low latitude straits.   Quantitative
knowledge of the current field must come from direct observations.

13

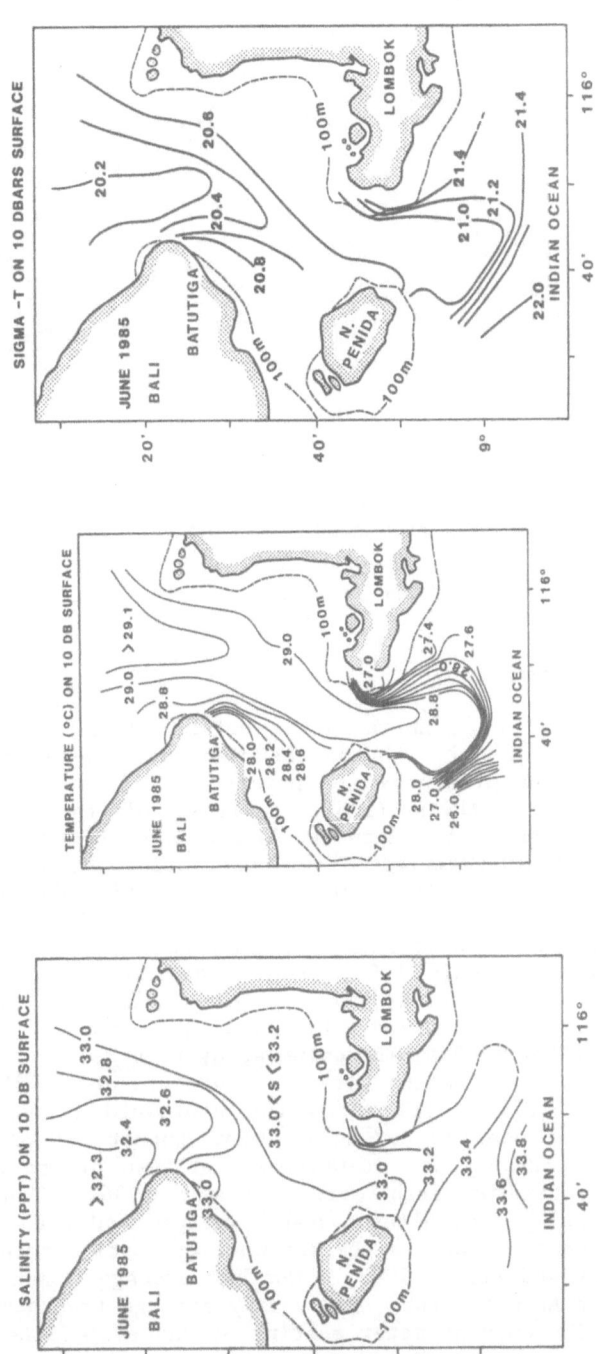

Figure 8. Salinity (A), temperature (B), and density (C) on the 10db surface, June 1985.

14

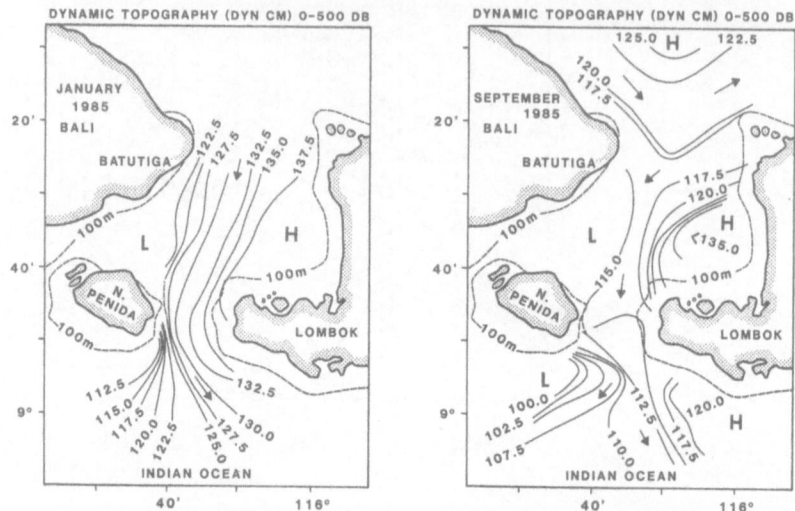

Figure 9.   Dynamic topography of the sea surface in January
            and September 1985.

5.   Currents

The annual mean north-south surface slope across the Indonesian Seas
(Figures 1 and 2) and the thermohaline properties (Figures 7 through 9)
suggest the presence of a persistent southward flow through the Lombok
Strait.   Wyrtki's (1987) analysis of the regional pressure field
indicates such a transport should be concentrated in the upper 200 m.
In Figure 10 examples of our observations of currents at four levels in
the north Strait from January to May 1985 document the presence of a
very strong and persistent southward flow in at least the upper 100 m.
This southward mean flow is also present at the 300 m level, but with a
magnitude of only 3-5 cm/sec.   Dramatic flow reversals to the north
reaching 75 cm/sec at the 35 m level interrupt the persistent southward
transport.   The period February 15 to March 15, as well as the last ten
days of April 1985, is dominated by reversal events.
        A 13-month time series of currents at the 35 m level at the Site 2
mooring (Figure 11) shows the southward transport reaches sustained
maxima of over 70 cm/sec in July, August and September.   Deceleration
occurs abruptly in mid-October and then begins a long period of weak
flow extending into early January 1986.   The current meter data
principally from Sites 1 and 2 were binned into monthly block averages
and used to calculate the seasonal cycle of transport through the Strait
(Murray and Arief, 1988).   Monthly average transport reached maximum of
4 Sv in August with a 1985 yearly average transport of 1.8 Sv.
        The current meter mooring at the sill (Site 5, Figure 1) operated
only during the first deployment as tidal currents (Figure 12) were
extremely rigorous.   Daily maximum speeds reached nearly 300 cm/sec at

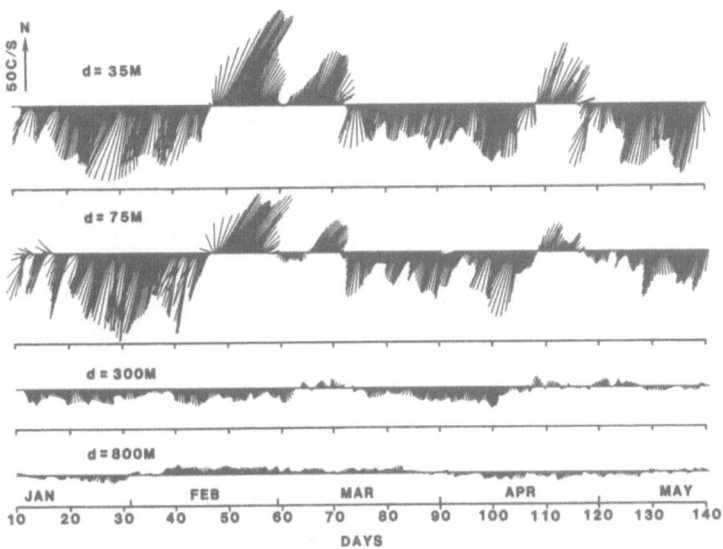

Figure 10. Current velocity vectors (60 hrs. low pass) from
the four levels of the north Strait mooring at
Site 2, January-May 1985.

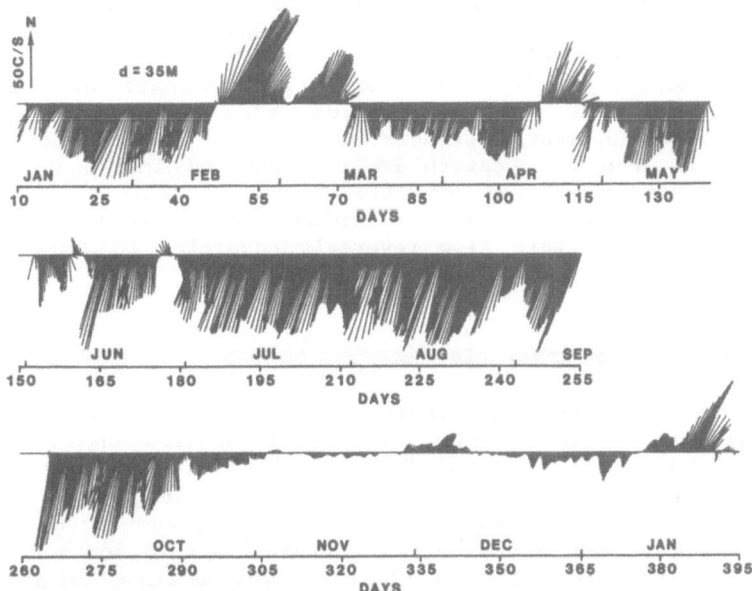

Figure 11. Thirteen-month time series of current from the
35 m level, Site 2 mooring, January 1985-
January 1986.

16

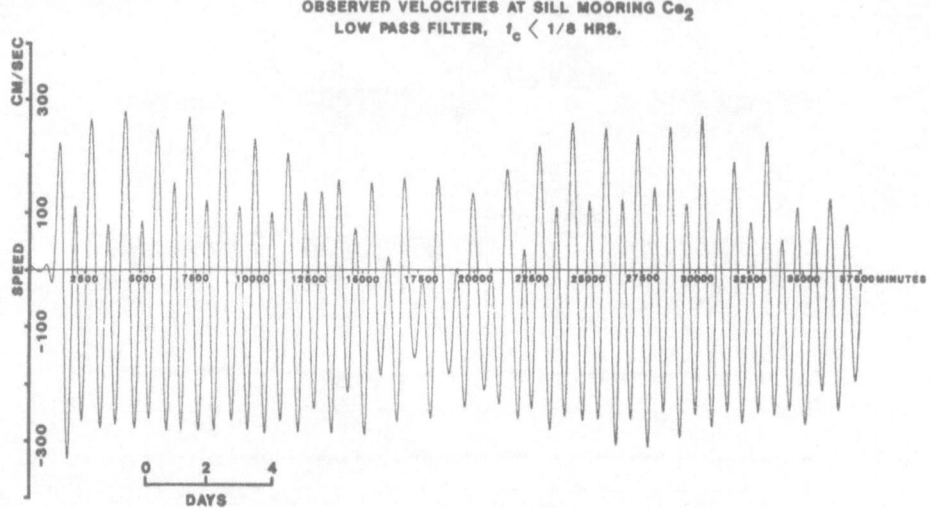

OBSERVED VELOCITIES AT SILL MOORING Cₒ₂
LOW PASS FILTER, $f_c < 1/8$ HRS.

Figure 12. North-South current component (north - positive)
observed at the sill mooring January 1985.

this site on the shallow (150 m) east flank of sill. The low pass data
from the sill mooring mirrors that of the north strait (Figure 10), but,
of course, with greater speeds.

In summary, the current meter data indicate at least three
important time scales in the sub-tidal flow regime: (1) a steady
southward flow apparently associated with an annual mean sea level
difference between the western Pacific and the Indian Ocean; (2) a
seasonal modulation of the mean flow producing maximum speeds in the
months of the northern summer and minima in the late fall and winter;
and (3) strong northward flow reversals occurring in January through
April.

6.   Northward Flow Reversals

December through April is the usual tropical cyclone season in the Timor
Sea.   Although their effects are well recognized on the northwest
Australian coast (McBride and Keenan, 1982) to our knowledge their
influence on Indonesian waters is unreported.   In Figure 13 we plot the
tracks of three tropical storms (Hubert, Isobel, and Jacob) transiting
the Timor Sea in February 1985.   The coastal waters south of Java and
Lombok were under the influence of these cyclone winds for most of the
latter half of February.   The capability of these storms to affect the
sea surface is clear from Figure 14 as the colinear positions of Hubert,
Isobel, and Jacob all bring strong (35-45 knots) sustained westerly
winds along the south coast of the archipelago.   In fact, all other
northward current reversals in the Strait occur in conjunction with

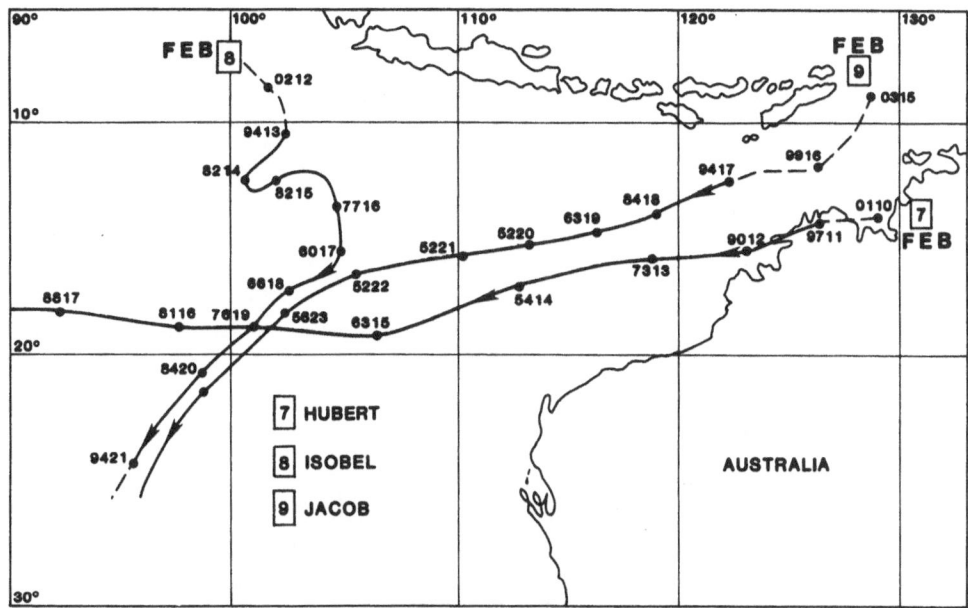

Figure 13. Tracks of three tropical cyclones in the Timor
            Sea in February 1985. The storm center at noon
            is plotted each day with the first two digits
            being central pressure and the second two digits
            the day of the month (from data in Kuuse, 1985).

Timor Sea cyclones. A set-up from the Ekman transport toward the coast
of sufficient magnitude to reverse the pressure gradient along the
Strait appears responsible for the flow reversal. We are pursuing this
quantitatively in cooperation with the Australian Bureau of
Meteorological Research using their operational Timor Sea storm surge
model.

7.   Numerical Simulations

Considerable insight into the forcing controlling the circulation
through the Lombok Strait is gained from the results of the reduced
gravity model of the region. Figure 15 shows the upper layer velocity
field on May 28, 1985. Note the presence of the Mindanao Current off
the southeast coast of the Philippines and the impressive scale of the
Mindanao eddy in the northeast corner of the figure, all in agreement
with recent observations (Lukas, 1988; Richardson and Collins, 1988).
The southward flow seen in the Makassar Strait continues throughout the
year in agreement with the ship drift climatology and continually shunts
water into the Lombok Strait.

18

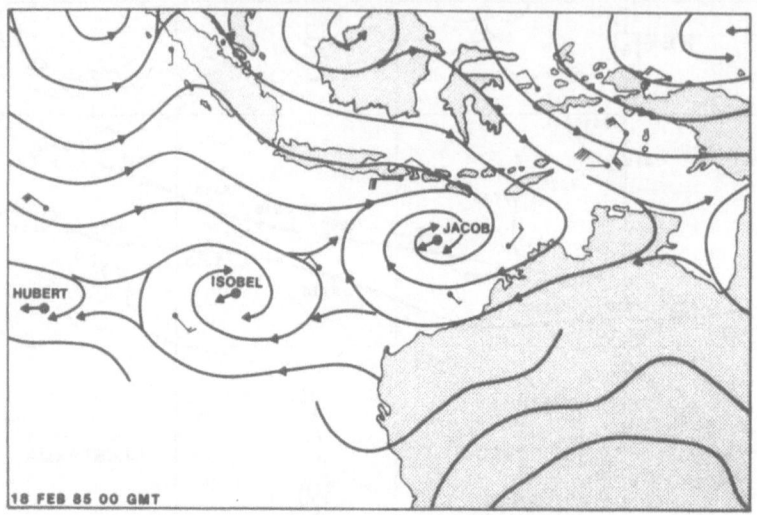

Figure 14.   Sea surface streamline map in the Timor Sea for
             February 18, 1985 from the Singapore Meteoro-
             logical Service.

         A 13-month time series of low frequency transport through Lombok
Strait predicted by the model is compared to our observation in Figure
16.   Due to present grid limitations, the active upper layer in the
model is larger by a factor of 2.2 than the cross sectional area of the
Strait above 200 m where the transport is concentrated.   For the first
approximation we assume the model transport is directly proportional to
cross-sectional area and a corrected transport scale is added to Figure
16.   Monthly block averages of transport utilizing all current meter
data were presented in Murray and Arief (1988).   The observed transport
shown in Figure 16 is calculated somewhat differently by first
establishing the best fit regression relation between these monthly
average transports and individual 40-hour low pass current meter time
series.   The data from the 35 m level instrument at Site 2 provided the
best predictor (R^2 = .78) of the total monthly averaged transport.   The
regression coefficients so determined were then used with the complete
time series from the 35 m level Site 2 meter to estimate the higher
temporal resolution transport seen in Figure 16.   We note the general
agreement between model and observations in the phase of the major
seasonal pulse of southward transport.   There is also good agreement in
magnitude between the adjusted model transport and the observations.
The distinct pulse of low southward transport in February due to intense
cyclonic activity in the Timor Sea is present in both the observations
and the model.   Clearly the February 1985 cyclonic winds were of
sufficiently large time and length scale (Figure 14) to affect the ECMWF
winds driving the model, while other tropical cyclones also driving flow

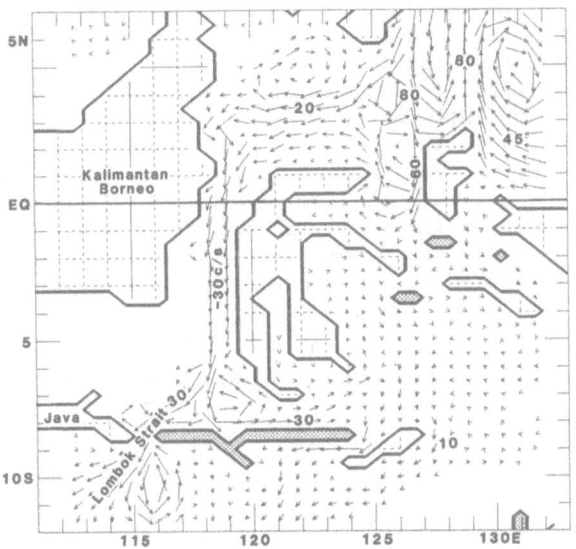

Figure 15. Velocity field from the numerical simulation of
May 28, 1985. Representative current speeds
(cm/sec) are shown at a few locations.

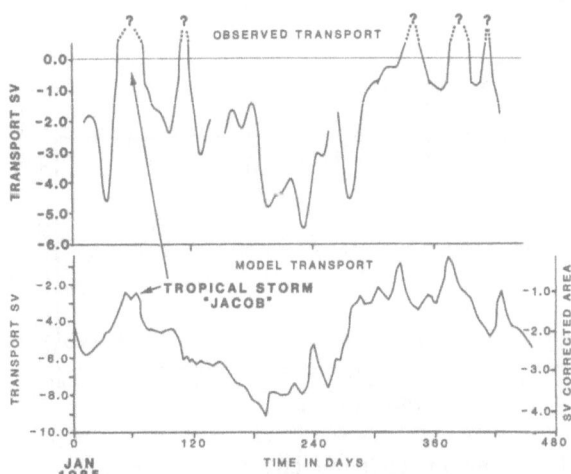

Figure 16. The transport through the Lombok Strait predicted by the model compared to the transport
observed in the north strait. The right side of
the lower panel is a scale corrected for excess
cross-sectional area mandated by model
limitations.

reversals (such as the one in late April) were not. We note the period of low transport in the observations from November 1985 to February 1986 appears to be well modeled also.

Finally, to explain the large seasonal signal in the transport we compare (Figure 17) the ECMWF observed zonal wind stress averaged over the Timor Sea and the observed sea level difference between Davao and Darwin to transport through the Strait during our period of observation. The southeast monsoon produces a large pulse of westward zonal wind stress during the months of the northern summer. The sea surface south of the archipelago is apparently depressed as a result of offshore Ekman transport which is reflected in a maximum in the Davao minus Darwin sea level difference.

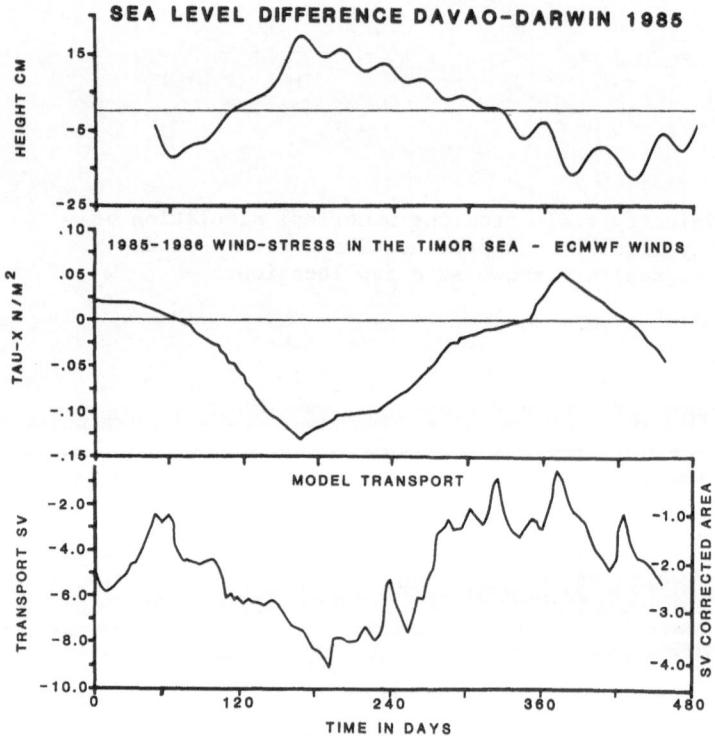

Figure 17.   The model transport through the Lombok Strait compared to the average ECMWF zonal wind stress in the Timor Sea and Davao-Darwin sea level difference for the period January 1985-April 1986.

The net result is the July through September pulse of southward transport figuratively being pulled through the Lombok Strait by the depressed sea surface to the south. Our quantitative results from the numerical simulations are in essential agreement with the throughflow mechanism advanced by Wyrtki (1987). The model simulations also suggest that the high currents observed in the Strait in January 1985 (Figure 11) result from winds east of the Philippines pushing water into the Sulawesi Sea increasing the transport through the Makassar Strait and then in turn through the Lombok Strait.

We have also investigated the north-south component of the wind stress north of the Strait as a possible local driving force, but its seasonal cycle varies from only $-.01$ m/m^2 in February to $.035$ m/m^2 in August. This local northward wind stress maximum in August is 180° out of phase with the observed southward maximum in the transport emphasizing its lack of importance in the mechanics of the throughflow.

8.   Summary and Conclusions

Observation of thermohaline properties in the Lombok Strait and adjacent waters in the Flores Sea and Indian Ocean are combined with current observations and numerical modeling results to obtain an understanding of the low frequency circulation in the Strait. A global reduced gravity model with an active upper layer predicts mean annual sea levels 15-20 cm higher at the northern (Pacific) entrances to the Indonesian archipelago than in the Indian Ocean south of the archipelago. This provides a mechanism for an annual mean transport through the archipelago straits. Consistent with this idea we traced core layers of the Northern Subtropical Central Water and the North Pacific Intermediate Water moving at the 150 m and 300 m depth levels, respectively, southward from the Makassar Strait across the Flores Sea into the Lombok Strait. Their identity is destroyed by intense turbulent mixing associated with 3-4 m/sec tidal currents over the steep sill at the south end of the Strait. Distributions of temperature, salinity, and density on isobaric surfaces indicate a strong persistent flow through the Lombok Strait which forms distinct thermal and haline plumes upon outflow into the Indian Ocean. Maps of surface dynamic height (0/500 db) in the Strait also indicate a strong southward flow with a 10-15 dyn. cm drop between the north and south entrances. Geostrophic velocities in the Strait are far too large suggesting more complex dynamics in the cross-strait momentum balance of such low latitude straits that are much smaller than the internal radius of deformation.

Current measurements on the sill exceeded 3-4 m/sec rendering monitoring there extremely difficult. Moorings in the north Strait, however, allowed identification of the important variability in the low frequency currents. A southward flow concentrated in the upper 200 m of water persists throughout the year. There is a sustained maximum in July through September when speeds exceed 70 cm/sec and a period of weak currents from mid-October through January 1986. During the Timor Sea cyclone season strong westerly winds apparently elevate the sea surface south of the archipelago and force strong northward flow reversals which can persist for ten days.

Simulations from the wind forced numerical model, although of a coarser scale than our observations, reproduce the major cyclone-produced flow events. Additionally, the model simulations clearly identify the strong persistent westward wind stresses in the Timor Sea during the southeast monsoon as the primary cause for the maximum flow phase in the Strait during the months of the northern summer.

We will conclude with a brief discussion relating the Lombok transport to the long-term net throughflow. The combinations of observations and numerical simulations suggest that the Lombok Strait is a major passage of Pacific water into the Indian Ocean. The long-term (1980-1987) net Indonesian throughflow for the model simulations forced by the ECMWF winds is 4.5 Sv. There is now evidence indicating that the present parameterization of wind stress from the ECMWF winds leads to low wind stress estimates. Numerical simulation forced by the Hellerman and Rosenstein (1983) climatological winds suggest that the mean net throughflow may be 7-8 Sv. Extrapolating the Lombok Strait transport observations to the Timor Passages based only on cross-sectional area above 200 m suggests a net throughflow of 10-12 Sv. Hence, in the mean, the flow through Lombok Strait may be approximately 25 percent of the Pacific and Indian Ocean transport. Seasonal and interannual variations may increase this percentage considerably.

Acknowledgement: This work was supported by the Office of Naval Research, Coastal Sciences Program.

References

Bryden, H. L. and Stommel, H. L. (1984) Limiting processes that determine basic features of the circulation in the Mediterranean Sea, Oceanologica Acta, v. 7, pp. 289-296.

Conlon, D. M. (1981) Dynamics of flow in the region of the Tsugaru Strait, Tech. Rept. No. 312, Coastal Studies Institute, Louisiana State University, Baton Rouge, LA.

Garrett, C. and Petrie, B. (1981) Dynamical aspects of the flow through the Strait of Belle Isle, J. Phys. Oceanogr. v. 11, pp. 376-393.

Gordon, A. L. (1986) Interocean exchange of thermocline water, J. Geophys. Res., v. 91, pp. 5037-5046.

Hellerman, S. and Rosenstein, M. (1983) Normal monthly wind stress over the ocean with error estimates, J. Phys. Oceanogr., v. 13, (7), pp. 1093-1104.

Kinder, T. H. and Parrilla, G. (1987) Yes, some of the Mediterranean outflow does come from great depth, J. Geophys. Res., v. 92, pp. 2901-2906.

Kindle, J. C., Hurlburt, H. E., and Metzger, E. J. (1989) On the seasonal and interannual variability of the Pacific to Indian throughflow, Proc. TOGA-COARE Symposium, Noumea, New Caledonia, ORSTOM, Noumea.

Kuuse, J. (1985) The Australian tropical cyclone season 1984-1985, Aust. Met. Mag., v. 33, pp. 129-143.

Lee, T. N., F. Schott, and R. Zantropp (1985) Florida Current: Low frequency variability as observed with moored current meters during April 1982 to June 1983, Science, v. 227, pp. 298-301.

Levitus, S. (1982) Climatological atlas of the world ocean, NOAA Prof. Paper 13, U.S. Department of Commerce, Washington, D.C.

Lukas, R. (1988) Hydrographic observations of the Mindanao Current during WEPOCS III, EOS, v. 69, pp. 1227 (abs.).

McBride, J. L. and Keenan, T. D. (1982) Climatology of the tropical cyclone genesis in the Australian region, J. Climatology, v. 2, pp. 13-33.

Murray, S. P. and Arief, D. (1988) Throughflow into the Indian Ocean through the Lombok Strait, January 1985-January 1986, Nature, v. 333, pp. 444-447.

Richardson, P. L. and Collins, C. A. (1988) Preliminary results from WEPOCS drifters, EOS, v. 69, p. 1227 (abs.).

Toole, J. M. and Raymer, M. E. (1985) Heat and fresh water budget of the Indian Ocean-revisited, Deep Sea Research, v. 32, pp. 917-928.

Wyrtki, K. (1961) Physical oceanography of the southeast Asian waters, NAGA Report Volume 2, Scripps Institution of Oceanography, La Jolla.

Wyrtki, K. (1987) Indonesian throughflow and the associated pressure gradient, J. Geophys. Res., v. 92, pp. 12941-2946.

ON THE PHYSICAL OCEANOGRAPHY OF THE TURKISH STRAITS

Ü. Ünlüata, T. Oğuz, M.A. Latif, E. Özsoy
Institute of Marine Sciences
Middle East Technical University
P.O. Box 28, 33731, Erdemli-Içel, Turkey

ABSTRACT

The Bosphorus and the Dardanelles Straits and the Sea of Marmara constitute a system through which exchange of Mediterranean and the Black Sea waters takes place. The two layer flow regime displays temporal and spatial variability on a wealth of scales. An assessment of the volume fluxes for the various elements of the system, based on recent hydrographic investigations, shows that a major portion of the Mediterranean flow entering through the Dardanelles is transported back to the Aegean Sea due to upward mixing induced by internal hydraulic adjustments of the exchange flow in the straits and by wind in the Sea of Marmara proper. The jet-like Bosphorus outflow in the exit region of the Marmara Sea also has a substantial contribution to the overall upward mixing. A mesoscale anticyclonic eddy to the right of the outflow off the Thracian coast is a quasi-permanent feature of the system. Hydraulic controls in the Bosphorus strait result in a maximal exchange, while a submaximal exchange exists in the Dardanelles. The Mediterranean inflow enters the Black Sea on an essentially continuous basis, with only few, short interruptions.

1. INTRODUCTION

The Turkish Straits, formed by the Bosphorus and Dardanelles Straits and the Sea of Marmara, constitute an oceanographic system through which the exchanges between waters of the Aegean Basin of the Eastern Mediterranean and the Black Seas take place. The low salinity waters of the Black Sea, formed as a result of excess of precipitation and run-off over evaporation, are transported to the Mediterranean through the Turkish Straits System (TSS) as a surface flow. In return, the saltier and heavier waters of the Mediterranean Sea, generated by the excess of evaporation over fresh water input, flow as an undercurrent to the Black Sea to seek their density level.

L. J. Pratt (ed.), The Physical Oceanography of Sea Straits, 25–60.

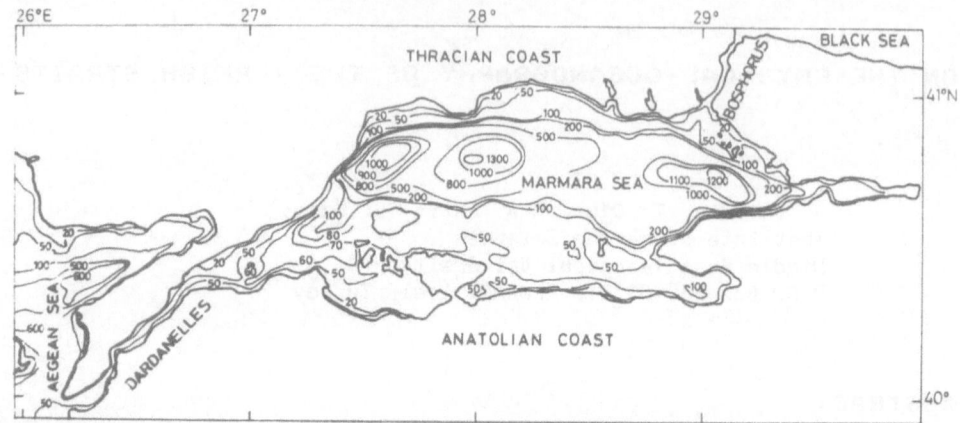

Figure 1. Location map for the Turkish Straits System composed by the Bosphorus and Dardanelles Straits and the Sea of Marmara.

In their evolution through TSS, the traversing water masses generate a unique oceanographic environment, representing a transitory state between the two extremes found in the adjoining seas, with constraints imposed by the hydrodynamical controls of the two straits as well as the interactions of the straits with the Sea of Marmara and the interactions of the entire system with the atmosphere.

The present review is concerned with the physical oceanography of the Turkish Straits. Sections 2 and 3 summarize the salient aspects of the morphometric and atmospheric setting, respectively. Hydrographic characteristics are considered in Section 4. Exchanges between the various components of TSS and with the adjacent seas, and mixing characteristics are covered in Section 5. The dynamics of the Bosphorus and Dardanelles Straits constitute the topics of Section 6. Sea level and current variability are discussed in Section 7. The interaction of the Bosphorus with the Black Sea is given in Section 8 which is followed by an overall summary and conclusions in Section 9. An evaluation of the volume fluxes described in Section 5 is provided in the Appendix.

It is worth mentioning that the present review is largely based on the investigations carried out in recent years by the Institute of Marine Sciences of the Middle East Technical University, Turkey. These continuing investigations commenced in 1986 and are primarily concerned with the evolution of the present state of health and the oceanography of TSS and the regions surrounding the junctions of the system with the adjacent seas. The research program, an exhaustive literature survey and the results to date (including chemical aspects) are given in detail in a series of reports published by the above mentioned Instute of Marine Sciences (Özsoy et al., 1986, 1988; Ünlüata and Özsoy, 1986;

Latif *et al.*, 1989a; Baştürk *et al.*, 1986, 1988). The review by Ünlüata
and Oğuz (1983) provides a summary of results prior to 1983.

2. MORPHOMETRIC CHARACTERISTICS

The Sea of Marmara (Fig. 1) is a relatively small inter-continental
basin with a surface area of 11500 km^2 and a volume of 3378 km^3. It is
connected to the Black Sea and the Aegean Sea through the straits of
Bosphorus and the Dardanelles, respectively. The east-west length of
the basin is roughly 240 km and the north-south width is approximately
70 km. The North Anatolian Fault crosses the region in the east-west
direction, and "pull-apart" basins associated with this fault are
located on the northern side of the sea. Here, three sub-basins with
depths in excess of 1000 m (maximum depth 1300 m) have been formed,
oriented also in the east-west direction. The southern half of the
Marmara Sea is characterised by a relatively shallow shelf region with
an average depth of 100 m. The length of the European shore-line is 264
km. The shore-line on the Asian side is longer (663 km).

The Bosphorus and the Dardanelles are narrow elongated Straits. The
Bosphorus is nearly 31 km in length, and its width varies between 0.7-
3.5 km, with a mean depth of 35 m and a maximum depth of 110 m. The
narrowest width occurs at about 12 km north of the southern end. A sill
of about 33 m depth is located approximately 3 km north of the southern
end. while a sill of 60 m is located 4 km north of its northern end.
The Dardanelles has a length of nearly 62 km and its width varies
between 1.2-7 km, the average width being 4 km. The average depth of
the strait is 55 m. The narrowest section occurs at the Nara Passage
about 25 km east of its junction with the Aegean Sea.

3. ATMOSPHERIC SETTING

The region is affected by two distinct seasonal climatic regimes.
During the winter, the weather is dominated by an almost continuous
passage of cyclonic systems. During the summer, northerly winds from
the Black Sea are dominant. Between 30-40 cyclonic systems, occurring
during October through April, follow three main tracks (Trewartha,
1968). In the southern part of the straits system, the cyclones move
eastward over the Aegean towards the eastern Levantine basin; in the
north, the tracks are from the Balkans towards the eastern Black Sea,
and the third route is northeasterly from the Aegean Sea towards the
Black Sea. The systems affect the region for periods between three to
ten days, and often result in winds of 8 m/s to 10 m/s (hourly average
speeds) sustained over one to two days. Maximum speeds of 35 m/s have
been observed as gusts.

On an annual basis, northerly winds (from the NW-NE sector) are
dominant with a frequency of 60%, with southerlies (SW-SE sector)
occurring 20% of the time (De Filippi *et al.*, 1986). However, during

the winter, winds from either sector are equal in both frequency and strength. For the three year period 1985-1987, in winter (December-March) the average frequency was 35% for each of these directions (Büyükay, 1989).

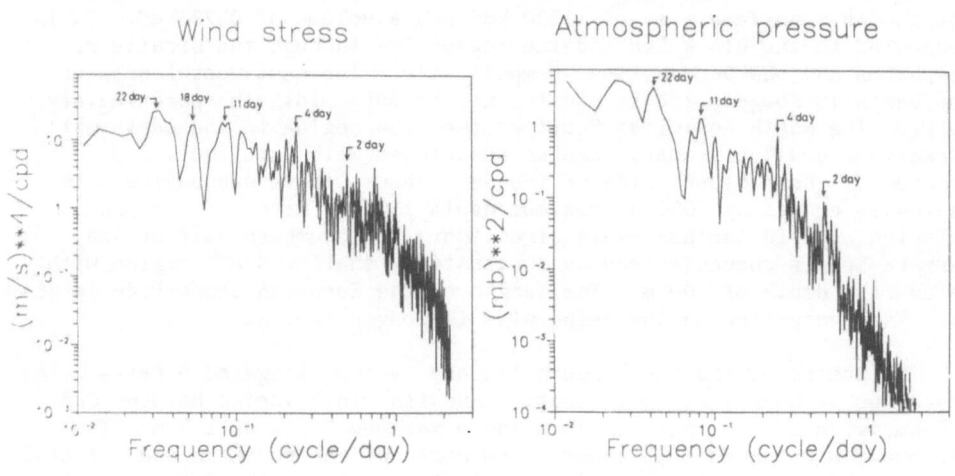

Figure 2. Wind stress and atmospheric pressure spectrum at Kumköy meteorological station, located at the Black Sea entrance of the Bosphorus Strait, for 1985.

Power spectra of wind stress and atmospheric pressure at the Kumköy meteorological station located near the Black Sea entrance of the Bosphorus shows peaks near 3, 4, 5, 7, 11, 15 and 20 days (Fig. 2). These spectra are based on the time series for the entire year of 1985. The observed spectral peaks correspond to the passage of cyclonic systems (Büyükay, 1989). Similar time scales have also been inferred by Gunnerson and Özturgut (1974).

Rainfall occurs mostly during October through March, leading to a precipitation of 7 km³/yr over the Marmara Basin. The mean evaporation is estimated as 11 km³/yr (Özsoy, *et al.*, 1988) . These fluxes are negligible in comparision to the exchange flows with the adjacent basins (Section 5).

The mean air temperatures range from 5 °C in winters to 25 °C in summers, with a diurnal range of 2-10 °C. The solar radiation, the net long wave, evaporative heat and the sensible heat fluxes are estimated as 108, 50, 52. and 2.8 kcal per square centimeter per year (Özsoy, *et al.*, 1986).

4. HYDROGRAPHIC CHARACTERISTICS

The hydrographic characteristics of the Turkish Straits and their variability in space and time are discussed in detail by Özsoy *et al.* (1986, 1988) by utilizing all available data. A wealth of time scales, extending from the inertial period (18 hrs), to days, to months, to seasonal, to interannual and longer periods, exists. Variability on the seasonal or longer time scales has been shown to be related to the response of the region to local climatic changes as well as those occurring in the adjacent Aegean and Black Seas. The seasonal changes are mostly reflected in the upper layer and may further be modulated by the long-term or interannual variations in the climatology. Internal hydraulics of the Straits, jets in the exit regions, the general circulation of the Sea of Marmara, deep water renewals and double diffusive processes also generate a rich variety of spatial scales.

In this section we will be primarily concerned with the seasonal and the sub-basin scale variability in salinity and temperature including the effects of transient wind episodes of few days duration.

In general, TSS is stratified in two layers with a sharp pycnocline whose depth changes from 50 m at the Black Sea entrance to 10 m at the Aegean exit, with a depth of 20-25 m within the Marmara proper. The density changes across the pycnocline are mainly accounted by the differences of salinity between the layers, with the lower salinity waters of Black Sea origin overlying the high salinity Mediterranean waters below the halocline. The degree of stratification changes seasonally depending mainly on the conditions in the adjacent deep basins.

4.1 Local and Seasonal Variability

Seasonal variability in the TSS is presented in Fig. 3 using the profiles of temperature and salinity at two stations located within the central parts of the Bosphorus and Dardanelles Straits. These figures demonstrate that the upper layer has considerable seasonal variations together with additional shorter term changes in response to local meteorological conditions. Specifically, the upper layer deepens occasionally during the periods of increased surface inflow caused by strong northerly winds in the winter months (as indicated by the profiles numbered 2 and 3 in Fig. 3). In the opposite case, the upper layer structure is lost due to short term effects of southwesterly winds as shown by the profiles numbered 1 at the Bosphorus station.

In general terms, the temperature of the upper layer undergoes greater changes in response to the surface heating/cooling. In winter, from November till May, the surface layer becomes colder than the lower layer with temperatures decreasing to about 4 °C. After May, the temperature of near-surface levels increases up to 24 °C as a result of radiative heating. A remnant of cold water is however observed between

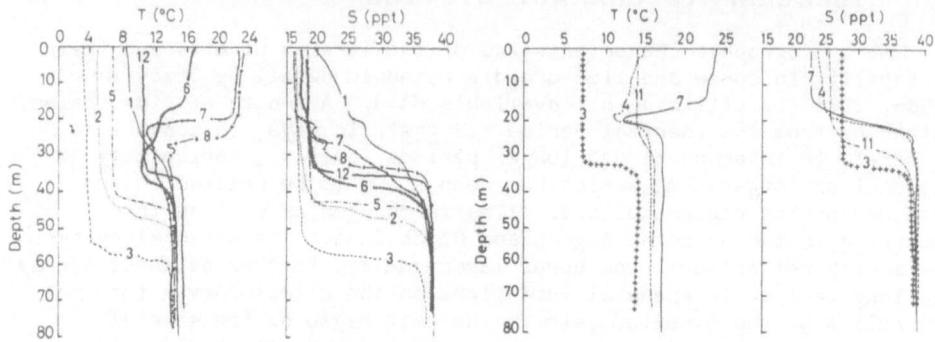

*Figure 3.   Profiles of temperature and salinity at the
Bosphorus Strait (left) and the Dardanelles Strait (right) during 1986.
(Numbers indicate the month).*

the warm surface layer and the interface.   This subsurface cold layer is
in fact partially formed within the TSS and is partially advected from
the Black Sea.   The upper layer salinity varies within the range of 16-
18 ppt in the Bosphorus station and 23-28 ppt in the Dardanelles station
through the year.   Relatively larger surface salinities of the winter
period generally decrease somewhat in the late spring and summer as a
result of increased fresh water inflow from the Black Sea.

The lower layer properties display their most pronounced seasonal
changes in the Dardanelles Strait in response to the seasonal changes in
the properties of the inflowing Aegean waters (Fig. 3).   The lower layer
average temperature in the Dardanelles, for a total of 7 surveys during
1986-1987, indicates the lowest temperature (13.1 °C) in March and the
highest temperatures of about 16.5 °C in July-August.   The variability
in the lower layer average salinity occurs within the range of 38.5-38.7
ppt.   As compared to the variability observed in the Dardanelles, the
subhalocline waters of the Sea of Marmara possess much more stable
properties having average temperature and salinity of 14.48 °C and 38.52
ppt with variations of ±0.04 °C and ±0.03 ppt.   The seasonal variability
of the lower layer properties along the Bosphorus occurs within the
temperature and salinity ranges of 12.5-14.5 °C and 35-37.5 ppt
depending on the intensity of local vertical mixing across the interface
(Fig. 3).

4.2 Spatial Variability

Apart from the temporal variability, the water masses also change
their identities while they are traversing the system in both
directions.   Many of the changes in the lower layer waters of
Mediterranean origin take place in the Dardanelles Strait before they

exit into the deep Marmara basin and within the Bosphorus as they
eventually join into the Black Sea. In contrast, the most significant
changes of the upper layer waters of the Black Sea origin take place in
the southernmost reaches of the Bosphorus, its exit region to the
Marmara Sea, and the southwestern part of the Dardanelles. The upper
layer water masses are particularly modified in localized regions of the
system where the flow is hydraulically controlled and is, therefore,
subject to significant vertical mixing in the regions of supercritical
flow and the subsequent internal hydraulic jumps.

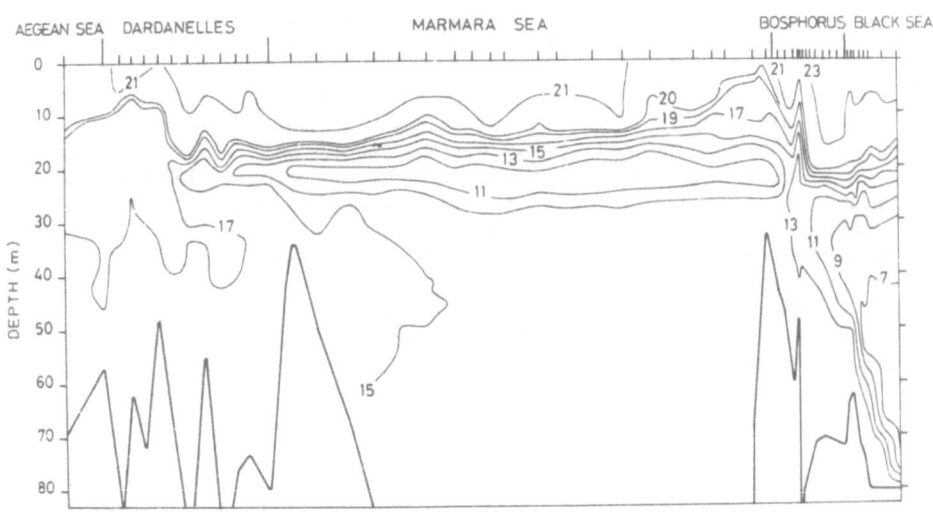

*Figure 4a. The longitudinal variation of temperature in the Turkish
Straits System during July 1986.*

    The above aspects of the TSS are displayed in Figs. 4a,b where
typical longitudinal transects of temperature and salinity in TSS,
extending from the pre-Bosphorus area of the Black Sea to the Aegean
exit, are given. It is seen that near the Black Sea entrance of the
Bosphorus, the upper layer is deep and the interface level is only
slightly above the sill depth (60 m). The transport from the Black Sea
of the cold intermediate layer (Tolmazin, 1985) is seen in the
temperature transect. Towards the southern end of the Bosphorus,
significant changes take place with respect to the position of
interface, and stratification. The interface tilts almost linearly
towards the free surface and the interfacial layer becomes much thicker,
with a typical value of about 20m.

    Relatively uniform conditions prevail within the Marmara Sea proper
up to the central part of the Dardanelles Strait as implied by the
horizontal distributions of the isotherms and the isohalines in Figs.

32

4a,b. The surface layer attains salinity values of about 22-23 ppt
which is an increase of almost 3-4 ppt as compared with further upstream
in the vicinity of the Bosphorus southern entrance. The temperature
structure presents a radiatively heated warm layer of about 15-20 m with
temperatures of 20-21 °C below which a layer of colder water resides
immediately above the halocline level, with a minimum temperature of
approximately 10 °C. The uniformity of the properties within the
Dardanelles is destroyed as the flow passes through the elbow-shaped
Nara Burnu section after which, up to the Aegean exit of the Strait, the
temperature and salinity display asymmetric distributions similar to
those observed in the southern Bosphorus. The upper layer flow joins
the Aegean Sea with 27-28 ppt surface salinity which indicates almost 10
ppt surface salinity difference between two extreme ends of the TSS.

The underflow entering into the TSS from its Aegean end also attains
most of its significant characteristics in the Dardanelles and the
Bosphorus Straits. The relatively denser Aegean waters enter below the
depth of 15-20 m and undergo gradual changes through the Dardanelles
Strait and its transition region to the western Marmara basin. Along
the topographic slope adjoining the wide channel region to the Marmara
Sea, they sink towards the density level where they reside in the form
of a dense plume. The sinking plume subsequently takes part in the
renewal of the subhalocline waters of the Marmara Sea by spreading
horizontally in the form of intrusive layers at the corresponding depths
to which it sinks (Ünlüata and Özsoy, 1986; Özsoy, et al., 1988).

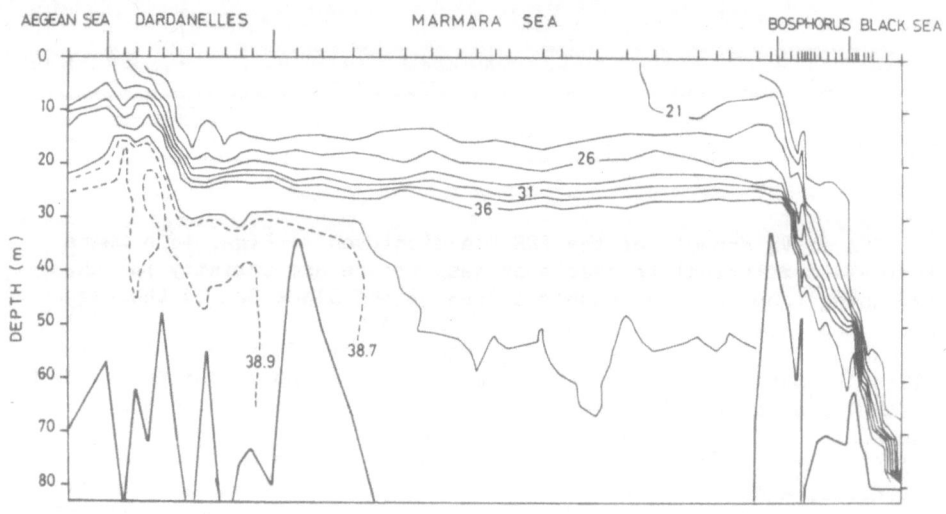

Figure 4b. The longitudinal variation of salinity in the Turkish Straits
System during July 1986.

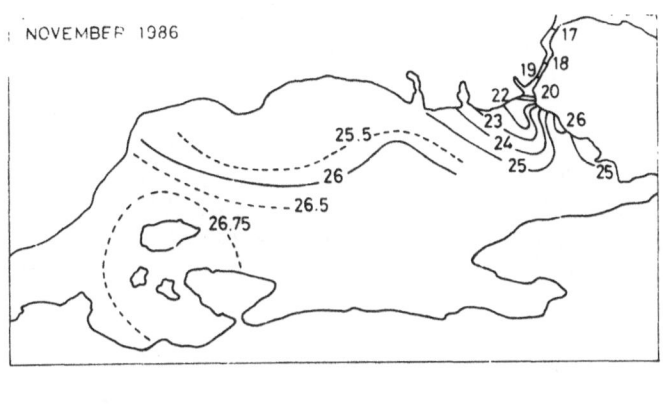

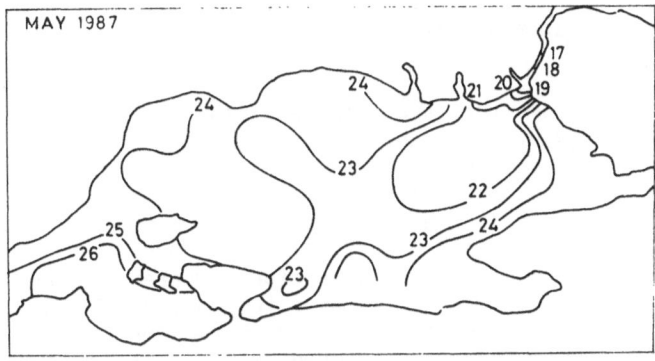

Figure 5a. Distribution of surface salinity in the Turkish Straits System for November 1986 and May 1987.

Upon reaching the Bosphorus-Marmara junction region, the lower layer waters flow into the Bosphorus through the submarine canyon. Thereafter, the underflow interacts with local topography at the southern and northern sill regions of the Bosphorus, becomes progressively diluted and enters the Black Sea shelf as a thin plume. Further characteristics of the Mediterranean effluent exiting from the northern end of the Bosphorus are described separately in Section 8.

The surface flow from the Bosphorus enters the triangular-shaped Marmara junction region in the form of a shallow and narrow turbulent buoyant jet with currents in excess of 2 m/s at the surface. It can be inferred from the surface salinity distributions (Fig. 5a) and from relevant measurements (Ünlüata and Oğuz, 1983; De Filippi, et al., 1986; Özsoy et al., 1986, 1988) that, immediately south of the Bosphorus exit section, the surface jet tends to spread asymmetrically, with more intense flow on the side of the Anatolian coast. The surface outflow proceeding in the southerly direction then bifurcates into two branches.

As one branch continues to flow in the south-southwesterly direction, the other branch of the jet turns anticyclonically and proceeds north-northwest towards the Thracian coast.

Figure 5b. Conceptual sketch of the upper and lower layer circulations in the Bosphorus-Marmara junction region.

The surface waters that are inshore of the northwestward curling main flow form a quasi-permanent anticyclonic mesoscale eddy (Ünlüata and Oğuz, 1983). The size of the eddy and the intensity of the flow dispersing within the region may however vary depending on the atmospheric conditions and the strength of the Bosphorus surface outflow. Most notable changes evidently occur during winter months when the Bosphorus surface outflow is appreciably weaker and during southerly and southwesterly winds prevailing over the region. Under these conditions, the jet core exiting from the Bosphorus does not deflect to the east and concentrates directly towards the northern (Thracian) coast. The presence of anticyclonic circulation within the Bosphorus-Marmara junction (BMJ) region is also supported by recent current measurements (De Filippi *et al.*, 1986; cf. Section 7).

It is worth mentioning that there is evidence to the effect that a secondary cyclonic circulation exists in the lower layer below the anticyclonic eddy at the surface (Ünlüata and Oğuz, 1983; De Filippi *et al.*, 1986). These features are illustrated conceptually in Fig. 5b.

The eastward intensification of the flow emanating from the Bosphorus as well as the subsequent anticyclonic eddy will be reconsidered in Sec. 6.

5. VOLUME FLUXES AND MIXING CHARACTERISTICS

Upon modelling the Turkish Straits as a two-layer system and decomposing it into three coupled compartments, corresponding to the Bosphorus, the Sea of Marmara and the Dardanelles. the horizontal and vertical volume fluxes defining the time-averaged exchanges between the elements of the system as well as the exchanges of the entire system with the adjacent seas have been estimated by Özsoy et al. (1986, 1988).

The fluxes are computed by making use of the steady state salt and mass
conservation equations. The recent measurements carried out during
1986-1987 are utilized for the average salinity values at the junctions
of the system. The results are summarized in Fig. 6.

The present estimates of the Bosphorus flows differ from most
previous estimates by nearly 40 %. This is due to the utilization of
different net fresh water input values for the Black Sea. The present
estimates are, however, consistent with the separate and independent
numerical computations that are based on the sea level differences at
the two ends of the Bosphorus. These points are discussed in the
Appendix.

It is seen from Fig. 6 that, in the mean, the Black Sea water with
salinity ≈17.8 ppt flowing through the Bosphorus as a surface layer
enters the Sea of Marmara with ≈19.4 ppt salinity. While crossing the
Sea of Marmara its salinity increases by nearly 6 ppt. After increasing
by another 4 ppt, it exits from the Dardanelles with a salinity of 29.62
ppt. These changes are in reasonable accord with those reported by
Defant (1961, p.523). On the other hand, the Aegean water with a
salinity of 38.9 ppt entering the Dardanelles traverses the strait with
little changes in its salt content. Within the Marmara basin a
reduction of nearly 2 ppt is observed in the salinity of this water.
After getting diluted by another 2 ppt in transit through the Bosphorus,
the Mediterranean waters enter the Black Sea with nearly 35 ppt
salinity.

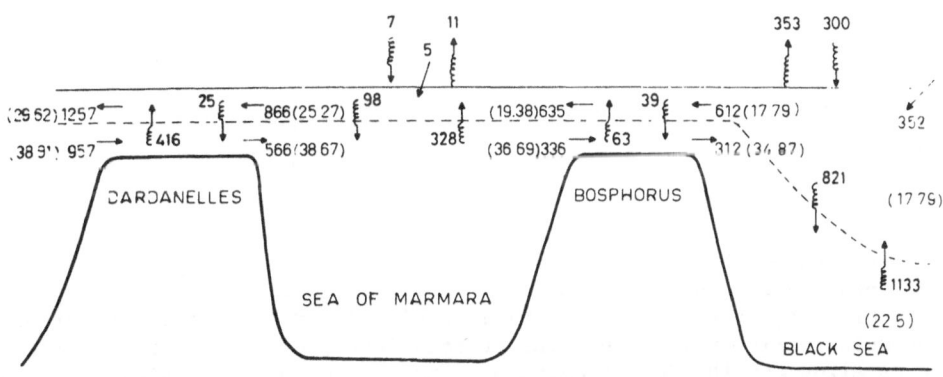

*Figure 6. Volume fluxes across the compartments of the Turkish Straits
System. The numbers are in km³/yr. The numbers in parentheses indicate
salinities in ppt.*

The flow rate of 612 km³/yr of the surface layer at the Black Sea-
Bosphorus junction does not differ much from that of 636 km³/yr at the
Bosphorus-Marmara junction. The surface flow entering the Dardanelles
at its junction with the Sea of Marmara is larger than the flow exiting

36

the Bosphorus by an amount of 230 km³/yr, and increases further by 391
km³/yr at the Dardanelles-Aegean junction. Mediterranean waters
entering the Dardanelles at a rate of 957 km³/yr exit into the Sea of
Marmara and enter into the Bosphorus at the rates of 566 km³/yr and 336
km³/yr, respectively. Such substantial reductions in flow rates do not
occur along the Bosphorus.

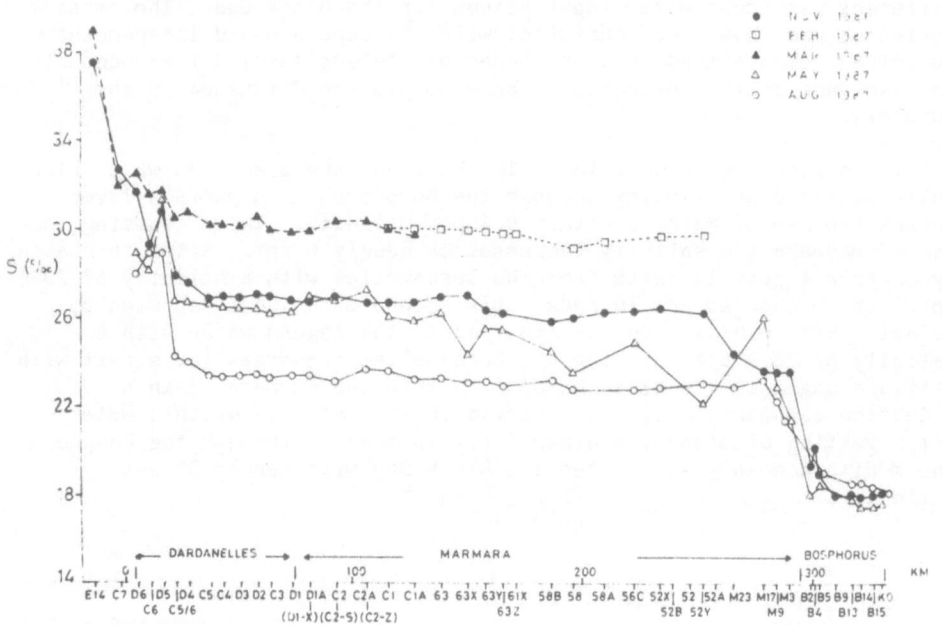

*Figure 7. Upper layer averages of salinity along the Turkish Straits
System.*

In view of the relatively small precipitation and evaporation fluxes
(cf. Section 3), the increases in the flow rates of the surface layer,
together with increases in salinities, clearly indicate entrainment of
the subhalocline waters into the surface layer, especially within the
Dardanelles and the Sea of Marmara. Indeed, the most striking aspect of
the water budget of the system is that nearly seventy percent of the
underflow entering from the Aegean Sea is returned back before reaching
the Black Sea, the values of the inflow being recirculated back to the
Aegean within the Dardanelles, Marmara and the Bosphorus being 41, 24,
and 3 percent, respectively.

With the exception of the Bosphorus, the upward mixing dominates in
the system. The upward and downward mixing in the Bosphorus do not
differ much from each other, the net upward flux being 24 km³/yr which
is 7 % of the inflow from the Marmara basin.

Previous assessments of the exchanges of the Black Sea with the Mediterranean have been based on the fluxes through Bosphorus (Sverdrup et al., 1946; Tixeront, 1970; Ovchinnikov, 1974). A comparison of the Bosphorus flows with the flows through the Dardanelles, especially at the junction with the Aegean, reveals significant differences. The outflow into the Aegean is found to be nearly twice the Bosphorus outflow into the Sea of Marmara, while the inflow from the Aegean is three times larger than the Bosphorus underflow.

Utilizing the fluxes reported in Fig. 6, the residence times are estimated as ≃3 months for the surface layer of volume 230 km³ and ≃5 years for the lower layer waters of volume 3148 km³. It may be of interest to note that the estimated residence times for the Mediterranean and the Black Sea deep waters are 70 and 500-2000 years, respectively (Lacombe et al., 1981; Östlund, 1969, 1986).

It is important to note that even though the budgetary calculations indicate significant upward mixing in the Dardanelles and the Sea of Marmara, the physical mechanisms leading to the entrainment as well as the region of their predominance cannot be inferred from them.

The upward mixing within the Sea of Marmara is due to 3 mechanisms. An internal hydraulic jump (see Section 6) initiated at the southern entrance of the Bosphorus as well as the surface jet emanating from the Bosphorus inject the saltier subhalocline waters into the surface layer. These two mechanisms operate on a continuous basis and induce intense and rapidly varying upward mixing and are effective within the Bosphorus exit region adjoining the Sea of Marmara. Their influence can be inferred from the longitudinal variations of the upper layer salinity across the Turkish Straits displayed in Fig. 7 where it is seen that the salinity of the surface layers rapidly change by as much as 6 ppt within

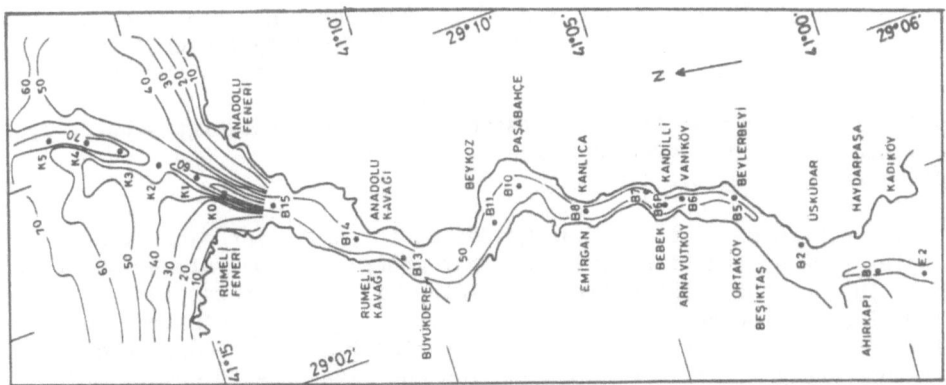

Figure 8a. Plan view of the Bosphorus geometry and locations of hydrographic stations.

20 km of the Marmara junction of the Bosphorus (Özsoy *et al.* 1986, 1988). The third mechanism of vertical entrainment involves the influence of winds down to the shallow halocline (20 m).

The upward mixing within the Dardanelles is induced by internal hydraulic adjustments of the flow initiated at the narrowest section of the Strait (Nara Passage) located at station D4 in Fig. 7 and at the abruptly widening Aegean exit region (Özsoy *et al.* 1986, 1988; Oğuz and Sur, 1989).

6. DYNAMICAL CHARACTERISTICS

The dynamics of the two layer water exchange through the Turkish Straits constitute one of the classical examples in oceanography. The differential surface elevation across TSS due to the net fresh water input to the Black Sea provides a barotropic pressure gradient, leading to the southerly flow of brackish waters. The opposite underflow of denser waters, on the other hand, arises due to the baroclinic pressure gradient established in response to marked salinity differences between the Aegean and the Black Seas. A simple dynamical account of this system of exchange flow was first studied in a two layer model by Defant (1961) in which the hydrostatically derived pressure gradient is balanced by friction. Recent related works are reviewed by Tolmazin (1985) and Özsoy *et al.*,(1986).

An important dynamical feature regarding the overall structure of the water exchange through TSS is the presence of internal hydraulic transitions of the two way exchange flow within the Bosphorus and the Dardanelles Strait. Observations carried out monthly in the Bosphorus and seasonally in the Dardanelles during 1986-1989, indicate the existence of a series of internal hydraulic adjustments of the exchange flow at various sections, similar to those encountered in the Strait of Gibraltar (Armi and Farmer, 1985, 1988). The internal hydraulic controls in the Bosphorus and the Dardanelles are inferred from the asymmetrical characteristics of the measured property fields, rapid transitions at the interface depth, and the associated intense vertical mixing at certain locations (Figs. 8a,b and Fig. 10a; cf. Figs. 4a,b) as well as from the quantitative evidence (Figs. 9a,b and Fig. 10b) provided by the two layer numerical models developed by Oğuz and Sur (1989) and Oğuz *et al.* (1989).

The TSS in general, and the Bosphorus in particular exhibit a complex flow system with considerable temporal and spatial variability. While the seasonal variability is related with the changes in the conditions in the adjacent basins, the low frequency variations on the time scales of few days, associated with the wind and atmospheric pressure changes, may dominate the flow and give rise to substantial modification of the regional flow structure. This is particularly observed in winter at the times of pronounced northerlies and southerlies. In the Bosphorus, when they are sufficiently intense, they

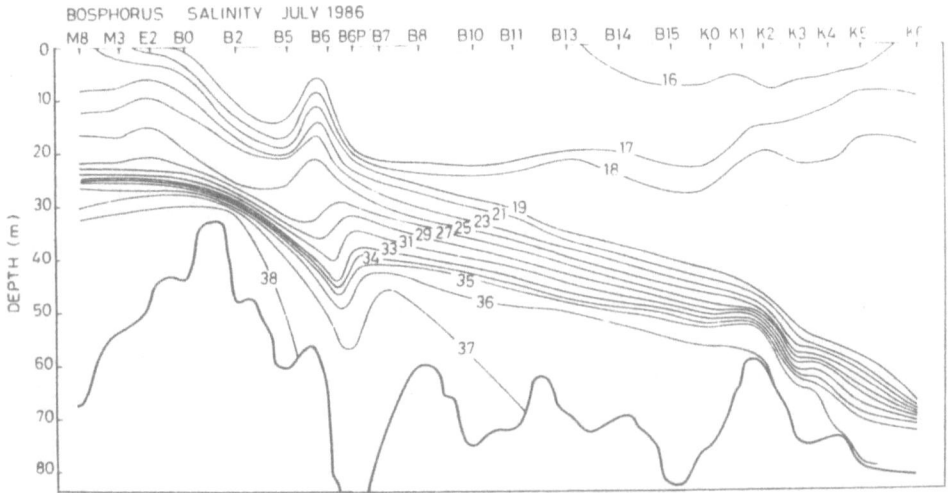

Figure 8b. Salinity transect in the Bosphorus and its exit regions during July 1986, corrsponding to Fig. 4b in expanded scale.

lead to the blockage of either the lower or upper layer depending on the direction of the wind. These cases resemble the solutions with the so-called "intermediate and strong barotropic forcing" given by Farmer and Armi (1986). They further reflect the significance of the time dependent adjustment process during such events. The tidal signal is weak, on the order of 10 cm, and does not have any important contribution to the flow field.

Hydrographic observations carried out recently in the Bosphorus on a monthly basis with a closely spaced station network consistently reveal considerable nonlinearity in the position of the interface and increased vertical mixing, which may possibly be associated with the internal hydraulic jumps and/or lee waves. These features are particularly pronounced in the southern half of the strait, to the south of station B8, where there are certain morphological features (horizontal and vertical constrictions and abrupt expansion of the width) which may lead to the internal hydraulic adjustment of the exhange flow (Fig. 8b). The observations indicate that the Bosphorus Strait possesses interesting internal hydraulic characteristics whose satisfactory description requires current velocity profiles of sufficient resolution and duration, in addition to the existing closely spaced CTD casts.

The two-layer model results (cf. Oğuz *et al.*, 1989) indicate that the Bosphorus possesses distinct regions of supercritical flows (Figs. 9a,b). The upper layer flow is first controlled at the constricted region (between stations B8 and B6). The interface, which is located at

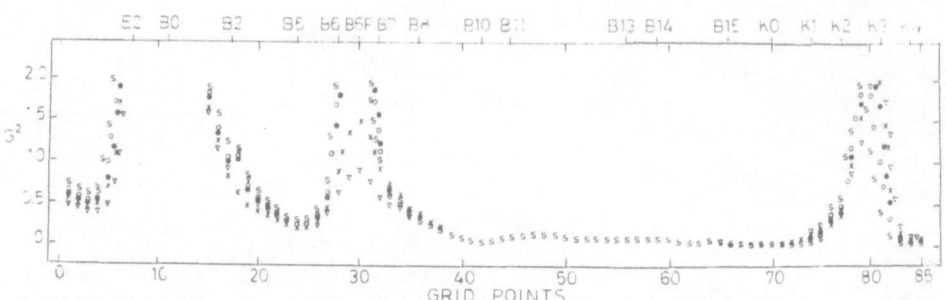

Figure 9a.   The computed composite Froude number distributions along
the Bosphorus for various net barotropic flow (after Oğuz et al., 1989).

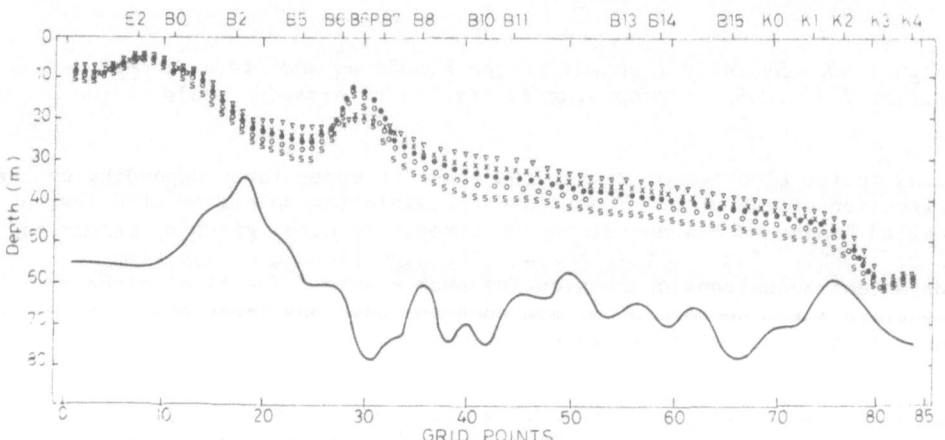

Figure 9b.   The computed variations of interface along the Bosphorus for
various values of net barotropic flow (after Oğuz et al., 1989).

deeper levels to the north of this region, is sharply elevated. In this
way, the flow adjusts itself to the critical hydraulic condition and the
upper layer flow becomes supercritical immediately to the south of the
constriction.   Thereafter, the interface depth declines and the surface
layer undergoes an internal hydraulic jump so that it passes through
another critical section at the southern exit region (generally, between
stations B2 and E2).   The model computations show the supercritical
region at the abruptly widening Marmara exit region comprises a distance
of about 3 km covering a recirculation zone in the surface layer
immediately upstream of the exit.   The supercritical exit flow is
subsequently matched farther downstream (near station M3 in the

Bosphorus-Marmara junction region) to the equilibrium conditions of the
Marmara Sea through another internal hydraulic jump.

Further hydraulic controls in the Bosphorus Strait occur due to the
effect of the sill on the lower layer flow. The underflow traverses
the Dardanelles Strait and the Marmara Sea in its subcritical state.
After it enters into the Bosphorus Strait, it proceeds northward in a
progressively thinner layer towards the Black Sea, and controlled over
the northern sill. The absence of controlled flow at the southern sill
could be either a genuine feature of the model or could be associated
with the computed position of interface which leads to an inadequate
representation of the bottom layer in the model. The effect of southern
sill on the flow structure remains to be explored by future
computational and observational studies.

The underflow may be blocked by the northern sill under extreme
conditions (cf. Section 8). The two layer model by Oğuz et al. (1989)
indicates that the blockage may occur when the net barotropic flow
exceeds about 27000 m^3/s, corresponding to the case of 45 cm maximum
surface elevation difference between the ends of the Bosphorus. Similar
results have been obtained earlier by Sümer and Bakioğlu (1981). A
continuous flow of the Mediterranean effluent into the Black Sea is also
supported by the salt wedge analysis given by Bogdanova and Stepanov
(1974) implying that the blockage can only occur when the upper layer
average current exceeds a value of 60 cm/s, which occurs only in cases
of very strong northerly wind conditions.

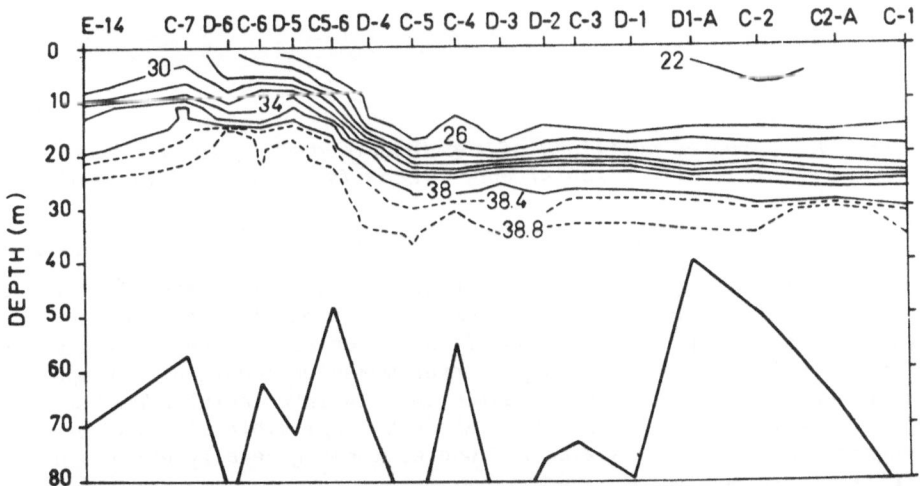

*Figure 10a. Salinity transect in the Dardanelles Strait during July
1986, corresponding to Fig. 4b in expanded scale.*

The upper layer flow passes through the Marmara Sea subcritically
without much change in its character. However, features similar to
those observed in the Bosphorus are also found in the Dardanelles (Figs.
10a,b). The Marmara inflow, which is subcritical at the northeastern
part of the strait, reaches the critical state at the Nara contraction
located just south of station D4 (corresponding to the point 58 in the
computational grid of the model). The flow becomes supercritical
downstream of the narrowest section and is then followed by an internal
hydraulic jump. This sequence is repeated at the Aegean termination of
the strait when the upper layer flow passes through the abruptly
expanding exit section.

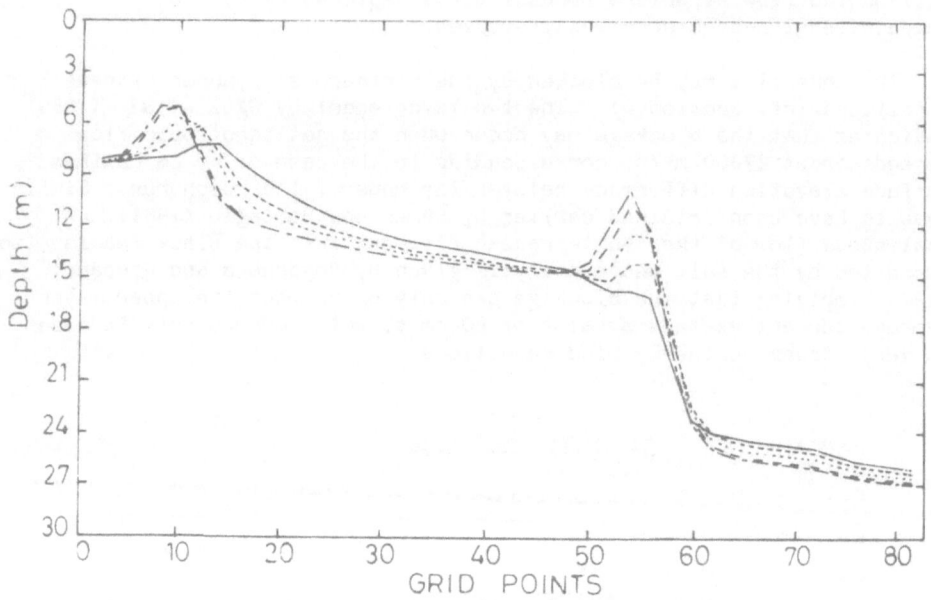

*Figure 10b. Computed positions of the interface along the Dardanelles
for various values of horizontal eddy viscosity coefficent (After Oğuz
and Sur, 1989).*

It is seen in Fig. 10c that the internal hydraulic adjustment of the
flow is basically a nonlinear process; the horizontal advection term
together with partial contribution of the interfacial stress balance the
pressure gradient term in the upper layer momentum equation. It appears
that the horizontal momentum diffusion and the interfacial momentum
transfer due to the entrainment process have negligible effects
throughout the channel. A weak balance of terms generally exists in the
subcritical regions.

The overall picture of internal hydraulics of the TSS is consistent
with the theoretical analyses of Farmer and Armi (1986), Armi and Farmer

(1987) who outline the conditions imposed by the hydraulic controls for
maximal and submaximal exchanges in steady, frictionless and immiscible
flows through a gradually varying channel between two basins. In
accordance with the results of Farmer and Armi (1986), Armi and Farmer
(1987), the Bosphorus possesses maximal exchange because of the presence
of the combination of the northern sill and the contraction and also the
combination of the southern sill and the abrupt expansion to the Sea of
Marmara. The latter combination may not be crucial to the issue because
of the close proximity of the two controls which, at times, appear to
merge and generate a complex situation.

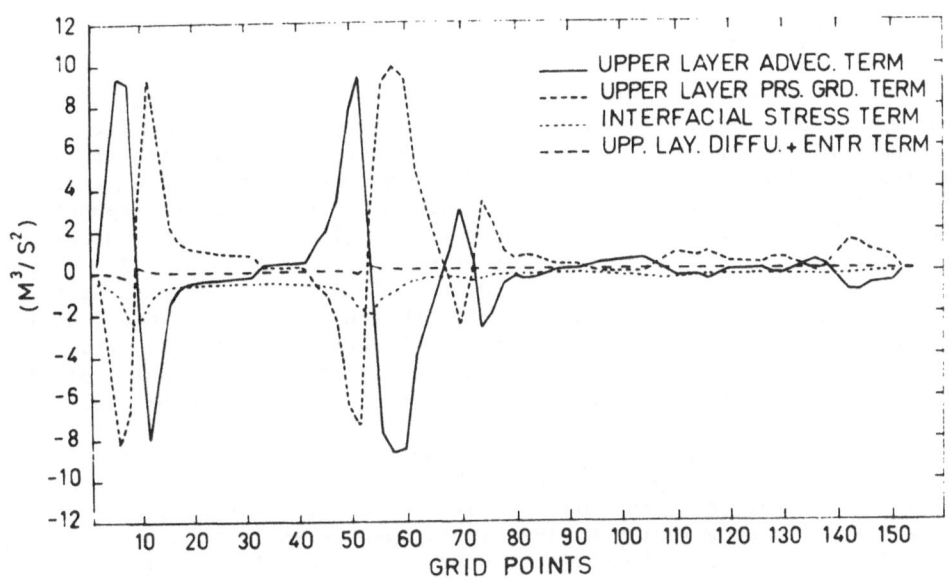

*Figure 10c. The variations of the terms in the upper layer momentum
equation for the two layer Dardanelles model (after Oğuz and Sur, 1989).*

Because of the maximal exchange conditions in the Bosphorus, the
exchange between the Marmara and the Black Seas will be determined by
the conditions within the Bosphorus itself, and not by the conditions at
the adjacent basins. On the other hand, the Dardanelles possesses a
submaximal exchange since the lower layer flow is not subject to an
internal hydraulic control near its Marmara exit. In this case, the
conditions within the Marmara Sea also contribute to the exchange along
the strait, and the flow is therefore no longer fully determined by the
conditions within the strait.

In addition to the two layer model studies, the qualitative features
of the baroclinic flow structure in the Bosphorus have also been studied
by continuously stratified, two dimensional numerical models (Tolmazin,
1981; Johns and Oğuz, 1989). The model given by Johns and Oğuz (1989)

is not able to produce observed penetration of sufficiently high salinity lower layer water into the Black Sea, and does not predict the layer transports in consistency with the results of two-layer models (Çeçen et al., 1981; Oğuz et al., 1989) as well as of the box model (cf. Section 5). However, it simulates a two layer system of exchange separated from each other by an intermediate entrainment layer. The numerical model identifies quantitatively the effect of the southern sill on the resulting flow structure. It is shown that an important part of the incoming lower layer flow from the Marmara Sea (about 34%) is returned back as a part of the surface layer transport to the south of the sill. This rate of upward transport appears to be higher than the two-layer model estimates cited above and arises due to the underprediction of upper and lower transports in the model.

The Johns and Oğuz (1989) model indicates that at some height above the sill, both horizontal and vertical components of the mean flow become vanishingly small, giving rise to a complete stagnation of a part of the mean flow, and predominance of turbulence and associated intense vertical mixing in the exchange flow system. Prediction of stagnant flow and important vertical mixing above the sill seem to be consistent with a feature of the temperature transect shown in Fig. 4a. A gap is observed between the subsurface cold layer temperature contours at the sill region, coinciding approximately with the location of intermediate layer simulated by the model. Significant vertical mixing is evident in Fig. 4a since the 7 °C core temperature of the cold layer on the Black Sea side of the sill is increased to about 10 °C on the Marmara side.

Turning our attention now to the dynamical aspects of surface outflows from straits pertaining to the Bosphorus outflow in the Marmara exit region, Whitehead and Miller (1979) describe several laboratory experiments to show that when the strait opening is less than the baroclinic radius of deformation, buoyant jets tend to form an anticyclonic gyre. Beardsley and Hart (1978), Nof (1978), Preller (1985) and Wang (1987) all relate the left hand attachment of an exit flow and resulting generation of an anticyclonic eddy outside a channel-open sea junction region to the presence of sufficiently large negative relative vorticity (comparable to the Coriolis parameter) of the exit flow. For surface outflows having an internal hydraulic control at the exit of channel, the negative relative vorticity is generated by strong upward vertical velocity induced during the critical transition of the outflow (Wang, 1987). Apart from the three dimensional, continuously stratified model of Wang (1987), the other models are based on the two-layer dynamics. They however do not incorporate the energy loss associated with the entrainment mechanism taking place between the surface jet and the ambient waters.

7. SEA LEVEL AND CURRENTS

The sea level and current variability of the Turkish Straits have been only partially studied in the past. Möller (1928) estimated

average sea level differences of 6 cm and 7 cm, respectively, between
the two ends of the Bosphorus and of the Dardanelles. More recent
estimates yield higher elevation differences. Bogdanova (1965) (quoted
in Yüce, 1986) has found a mean sea level difference across the entire
TSS of 42 cm with considerable seasonal variations ranging between a
minimum value of 35 cm in October and a maximum value of 57 cm in June.
Bogdanova's (1965) estimates are the only available recent sea level
information for the entire TSS. The others are related to the sea level
measurements in the Bosphorus Strait alone (Gunnerson and Özturgut,
1974; De Filippi et al., 1986; Büyükay, 1989).

Based on tide gauge measurements during July 1966–February 1968,
Gunnerson and Özturgut (1974) estimated the average sea level difference
between the two ends of the Bosphorus as 35 cm. Çeçen et al. (1981)
report a value of 33 cm with a standard deviation of 13 cm. De Filippi
et al. (1986) determined an average sea level difference of 37 cm for
April-August, 1984 period. Analysing the sea level data for 1985 and
1986, Büyükay (1989) finds the seasonally-mean sea level differences and
their standard deviations (the latters are shown in parantheses) as:

	(1985)	(1986)
Winter (Dec.-Feb.)	18 (12)	26 (13)
Spring (Mar.-May)	26 (7)	34 (8)
Summer (Jun.-Aug.)	34 (10)	28 (4)
Autumn (Sep.-Nov.)	35 (10)	--
Annual average	28 (10)	29 (8)

While the average sea level difference between the ends of the
Bosphorus is typically of the order of 30-40 cm, the slope of free
surface is found to be nonlinear by both Gunnerson and Özturgut (1974)
and De Filippi et al, (1986) who have indicated that the surface slope
in the southern half is much steeper than in the northern half.

The sea level spectra (Fig. 11) indicate significant variability in
the 3-14 day period range which appears to be related to the variations
in the barometric pressure and winds (cf. Section 3). Similar spectral
bands have also been reported by Gunnerson and Özturgut (1974) and De
Filippi et al, (1986). The high frequency tidal oscillations of 24 hr
and 12 hr periods are also dominant in the sea level signals. The wind
influence on the sea level variations and, particularly, notable effects
of strong southwesterlies in diminishing or even reversing the sea
surface slope are emphasized by Gunnerson and Özturgut (1974) and
Büyükay (1989).

As compared with the sea level, currents and their variability
within the Turkish Straits are less known. The classical study of
currents in the Bosphorus and Dardanelles Straits by Merz (Möller, 1928;
cf. Defant, 1961) reveals the upper layer average current increasing
from 50 cm/s in the northern end to about 200 cm/s at the southern end

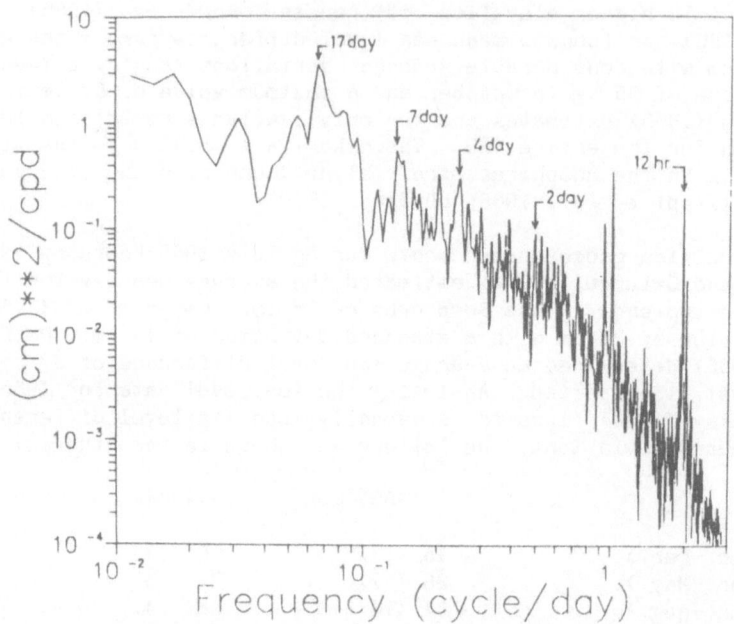

Figure 11. *Sea level spectrum at Anadolu Kava&gi, located near the Black sea end of the Bosphorus for 1985.*

of the Bosphorus Strait. The speed of the undercurrent decreases from 40 cm/s to 20 cm/s in the opposite direction. The upper layer current in the Dardanelles Strait is weaker and varies from about 80 cm/s near the northern entrance to 100 cm/s at the Aegean exit section. The northerly flowing lower layer current speed attains typically a value of 30 cm/s. In both straits, regions of intensification of the upper and lower layer currents are present associated with the internal hydraulic structure of the flow (cf. Section 6). More recent current measurements described in DAMOC (1971) and De Filippi *et al.* (1986) yield similar values of currents in the layers within the Bosphorus Strait.

The measurements carried out by De Filippi *et al.* (1986) in the Marmara exit region of the Bosphorus indicate that the surface currents possess significant variability due to forcing by winds, inertial and tidal oscillations.

8. MEDITERRANEAN INFLOW INTO THE BLACK SEA

The Mediterranean waters emanating from the Bosphorus play a crucial role in the Black Sea salinity and water budgets. The nature of the flow, i.e, whether continuous or sporadic, has been a controversial and poorly understood matter since the early studies of the subject (Ullyott and Ilgaz, 1943, 1946; Pektaş, 1953; Bogdanova and Stepanov,1974 and the references cited therein). A summary of the various conflicting views is given in Ünlüata and Oğuz (1983). Uncertainties in more recent results do not allow a firm settlement of the issue because of incomplete information on the bottom topography of (Büyüközden et al., 1985) and the lack of hydrographic data in (Tolmazin, 1985) the near-field region of the discharge.

An extensive oceanographic investigation has been conducted recently in the Bosphorus-Black Sea junction region, with a primary purpose of studying the bottom topographical features and determining the path of the Mediterranean effluent on the adjacent Black Sea shelf (Latif, et al., 1989a, 1989b). It is found that the Mediterranean flow entering the Black Sea is initially confined in a channel that is a natural extension of the Bosphorus. For the first 8 km, the channel is directed towards the northeast, which is the same orientation as the Strait, and subsequently turns towards the northwest, eventually joining the shelf topography in a formation similar to the morphology of a river delta (Fig. 12).

The initial confinement of the Mediterranean flow can be inferred from Fig. 13) where the distribution of bottom salinity values greater than 24 ppt measured during various cruises are displayed together with the depth contours defining the channel (Latif, et al., 1989a, 1989b). The higher salinity values are found only in the channel and the correspondence between the flow track and the channel is clearly seen.

A sill at 60 m depth in the channel is seen in Fig. 12. Under sufficiently strong and persistent winds, the interface may be lowered below 60 m depth, resulting in blockage of the lower layer flow. Three such cases, in March 1986, April 1987, and Jan. 1989 have been observed (Latif, et al., 1989a, 1989b). The salinity transect during the blocking on 13 March, 1986, is shown in Fig. 14. The Mediterranean inflow, delineated by the 20 ppt isohaline, is seen to be compressed close to the bottom in the northern half of the Strait. The flow resumes quickly after the northerly winds cease or change direction; this was documented in the April 1987 episode; the time scale for the blockage appears to be of the order of two days (Latif, et al., 1989a).

The blocking of the Mediterranean inflow at the sill had been earlier believed, by some researchers, to be a frequent occurrence (Ünlüata and Oğuz, 1983). The reason for this appears to be that the sill depth was incorrectly known as 50 m, after Scholten (1974). Since the interface between the two layers is often located at 50 m in the northern end of the Bosphorus, a blocking situation may be inferred if the Black Sea water is found in the 50 m depth region, when in fact, such is most probably not the case. The fact that the blocking of the Mediterranean inflow is a rare occurrence can be seen from Fig. 15, where the depth of the 20 ppt salinity defining the lower limit of the surface layer, measured during a series of cruises between 1986-88, is displayed (Latif, *et al.*, 1989a, 1989b).

After leaving the channel, the Mediterranean effluent spreads out as a thin gravity current over a relatively flat shelf area and maintains a generally northwesterly track towards the shelf break. It is seen from the bottom salinity values measured during a cruise in June, 1988

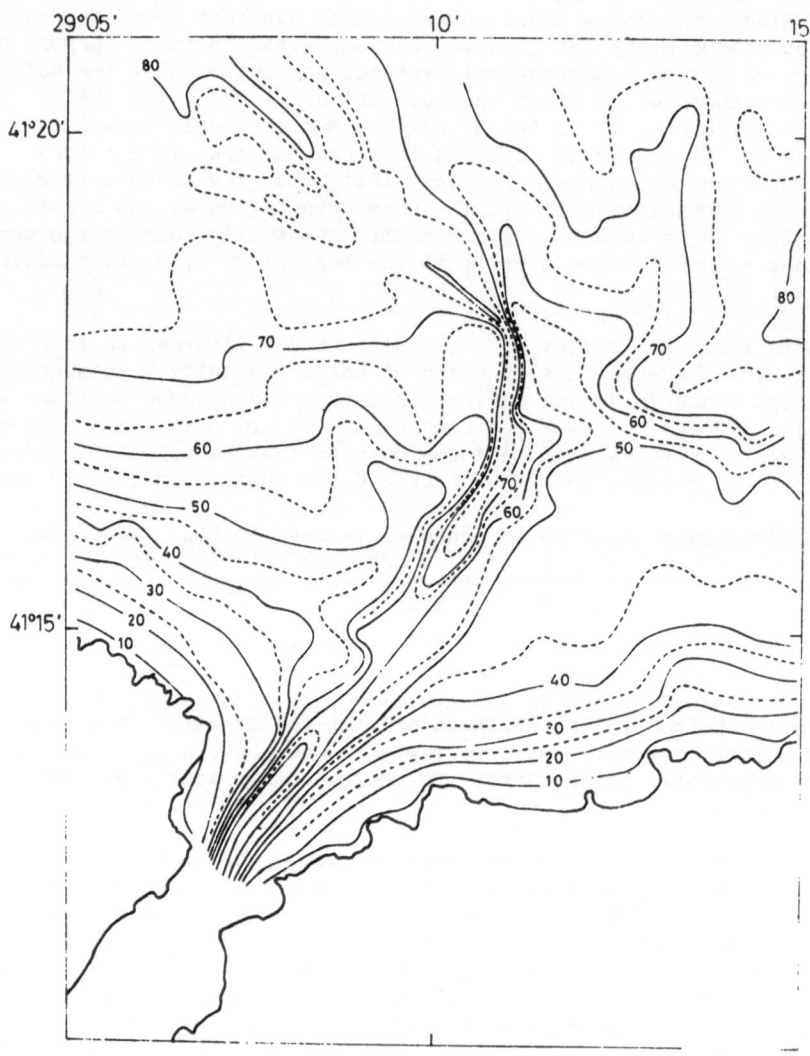

Figure 12. Bottom topography in the Bosphorus-Black sea junction region (depths in meters) (After Latif et al., 1989a).

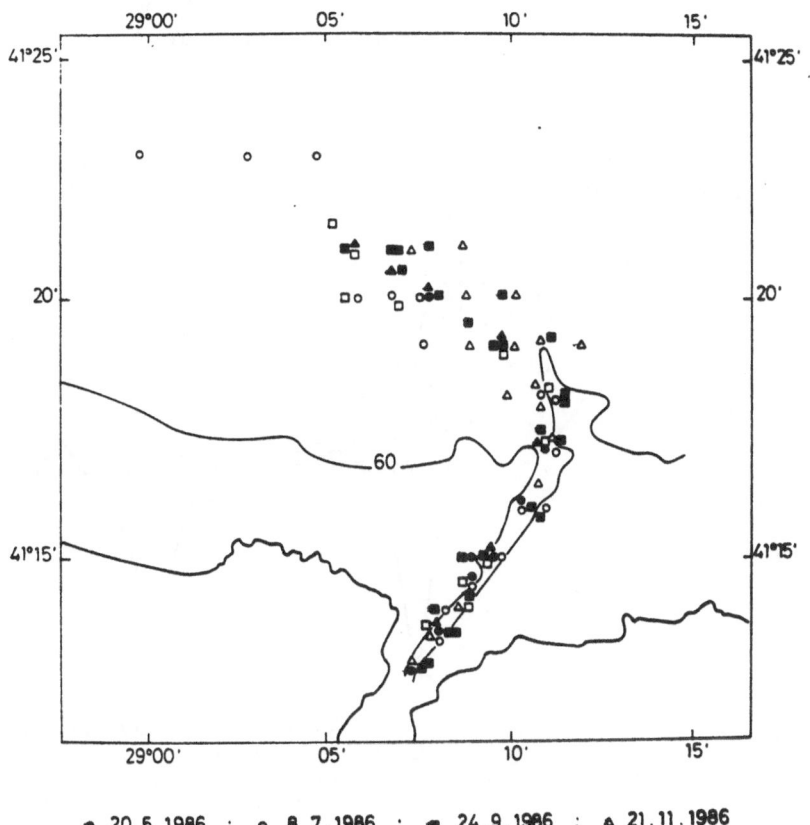

Figure 13. Distribution of bottom salinity values greater than 24 ppt from several cruises, superimposed on the contours defining the channels (After Latif et al., 1989a).

(Fig. 16) that, while in the channel, the Mediterranean water mixes only slightly and essentially vertically with the ambient water because of its confinement (Latif, *et al.*, 1989a, 1989b).   After leaving the channel, a more rapid dilution takes place.   The salinity decrease between the exit from the Strait and the end of the channel is 1.7 ppt, while the decrease in a comparable distance after leaving the channel is 3.2 ppt (from 35.3 to 32.1 ppt).   In the close proximity of the shelf break, the salinity of the effluent is reduced down to 22.6 ppt which does not differ much from the 22.4 ppt salinity of the Black Sea bottom waters.   Beyond the shelf, the Mediterranean flow is evidently

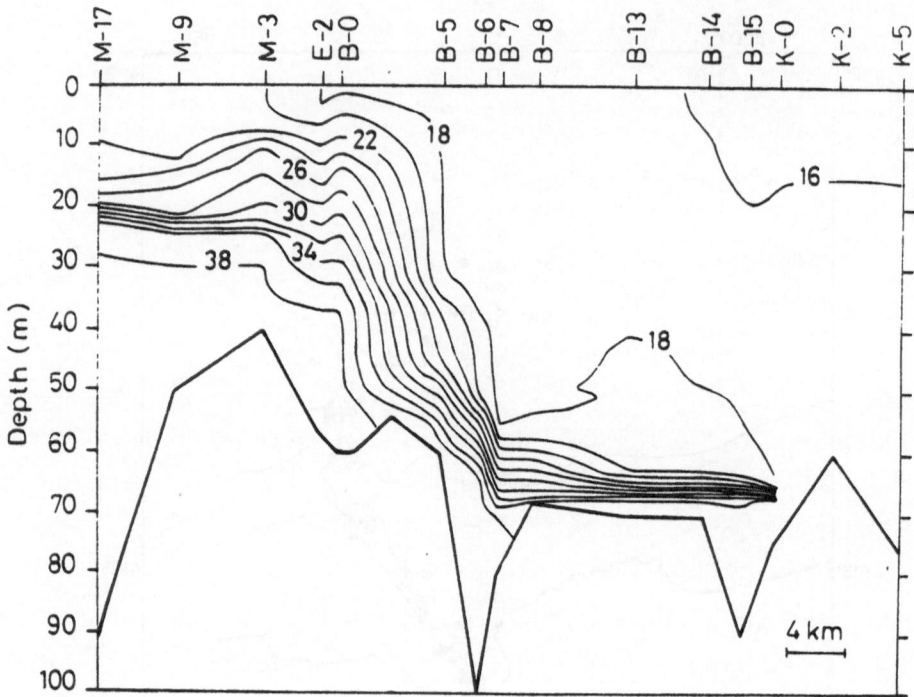

Figure 14. Salinity transect in the Bosphorus and its exit regions during a lower layer blocking, 13 March 1986 (After Latif, et al., 1989a)

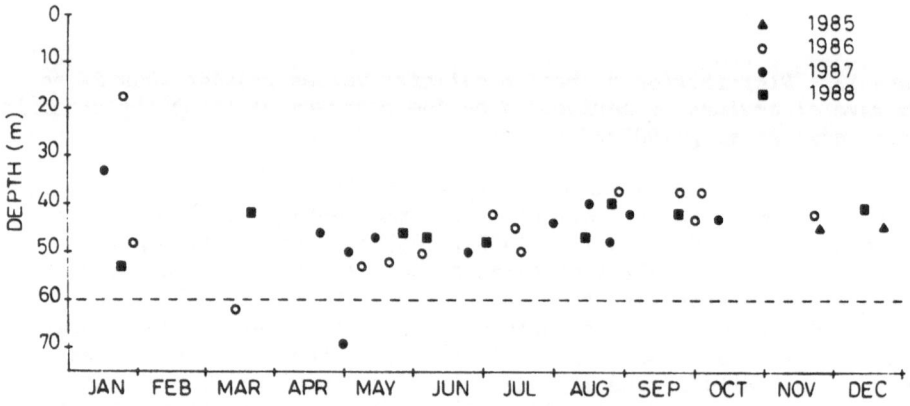

Figure 15. Depth of 20 ppt distribution in the northern exit of the Bosphorus. Depth of the sill (60m) is indicated by the dotted line. Total depth at this location is 75m (After Latif et al., 1989a).

incorporated into the prevailing eastward general circulation of the Black Sea (Tolmazin, 1985).

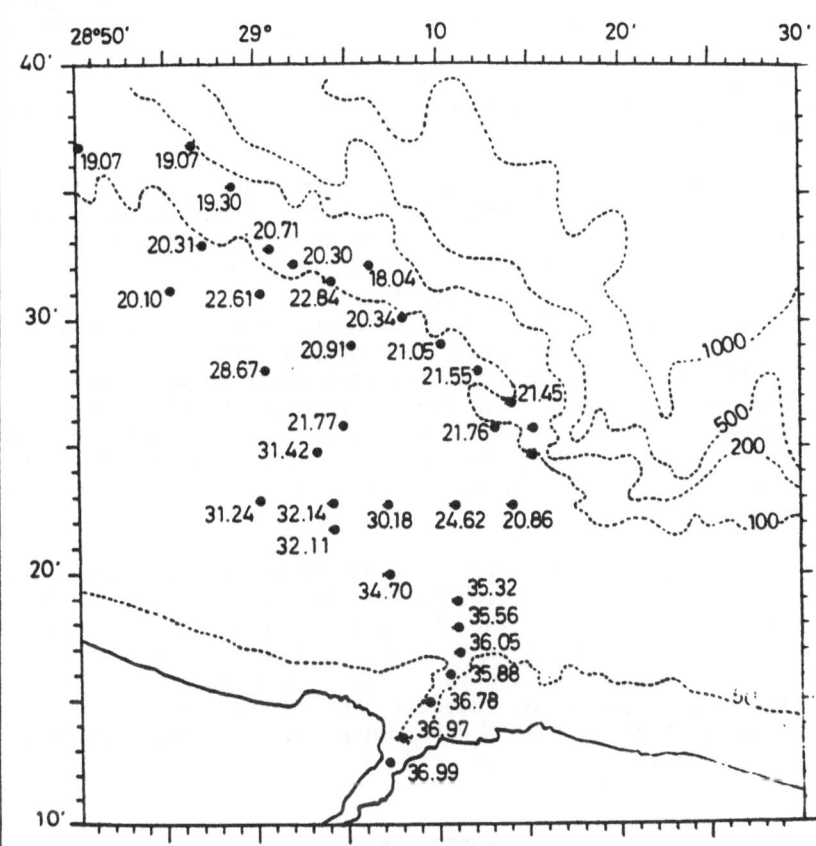

Figure 16. Distribution of bottom salinity values in the Black Sea shelf region during June 1988; the track of the Mediterranean effluent is indicated by high salinity values (After Latif, et al., 1989a).

9. SUMMARY AND CONCLUSIONS

The Turkish Strait System, joining two of the world's largest isolated seas with extremely different water mass properties, presents a wealth of oceanographic phenomena which have only recently begun to be understood.

The two layer flow regime in the straits displays seasonal variability as well as short-term fluctuations in response to meteorological forcing. The sea level difference is at its maximum in spring and early summer, and is minimum in winter, corresponding to the fresh water inflow changes in the Black Sea. Strong southerly winds cause a rapid rise in sea level in the southern part of the Bosphorus, while northerlies result in high levels at the other end. Blockage of either layer may occur under sufficiently strong and sustained winds; such episodes are, however, infrequent and short-lived. Specific cases for the blockage of the upper layer (e.g. during January 16, 1987) and of the lower layer (e.g. during March 13, 1986) have been presented in Latif et al. (1989a).

Internal hydraulic controls greatly influence the flow regime and intensity of mixing in the Turkish straits. The Bosphorus possesses maximal exchange because of the presence of the combination of the constriction and the northern sill and also the combination of the southern sill and the abrupt expansion to the Sea of Marmara. The internal hydraulic adjustments of the Bosphorus exchange flow are however time dependent and can generate a complex situation. In the Dardanelles, the upper layer attains critical state in the Nara contraction region, however, since the lower layer is not subject to an internal hydraulic control at its Marmara exit, a submaximal exchange exists. Intense vertical mixing in regions of supercritical flows and internal hydraulic jumps associated with the controls are inferred from property transects.

The Bosphorus upper layer enters the Sea of Marmara in a jet-like flow and forms an anticyclonic eddy on the Thracian side; the scale of the eddy varies dependent upon the strength of the outflow and the prevailing winds. At the northern end, the Mediterranean flow entering the Black Sea is transported onto the shelf region confined in a narrow channel with steep banks in the seabed for about 10km; at the termination of the channel, the Mediterranean water spreads out in a thin layer, becomes highly diluted, and is incorporated into the prevailing Black Sea circulation.

Estimates of the exchange between the Bosphorus and the Black Sea reported in the literature vary widely. In the present study, the time-averaged exchanges of the various components of the System have been estimated using steady state salt and mass conservation equations. Recent measurements were utilized for the average salinity values at the junctions of the system.

The present estimates of the Bosphorus flows differ from most previous ones. This is due to the utilization of different net fresh water input values for the Black Sea. The present estimates are, however, consistent with the separate and independent numerical computations that are based on the sea level differences at the two ends of the Bosphorus.

The computations show that the upper layer volume flux increases significantly between its exit at the Bosphorus - Marmara junction and its entrance into the Dardanelles, and again during its transit through the Strait. The increases in the flux, together with the increases in salinities, indicate entrainment of the subhalocline waters into the surface layer during its traverse through the Sea of Marmara and the Dardanelles. Indeed, the most striking aspect of the water budget of the system is that a major portion of the underflow entering from the Aegean Sea is returned back before reaching the Black Sea.

The upward mixing within the Sea of Marmara is due to the influence of winds down to the halocline, the internal hydraulic jump at the southern entrance of the Bosphorus as well as the surface jet emanating from the Bosphorus. These mechanisms lead to the entrainment of the saltier subhalocline waters into the surface layer. Similarly, the upward mixing in the Dardanelles is induced by an internal hydraulic jump in the Nara constriction region.

The Turkish Straits System, as seen from the material presented here, possesses a number of interesting processes whose further study should be scientifically rewarding. In particular, the distribution of the bottom waters on and beyond the Black Sea shelf, the dynamics and mixing in the Bosphorus-Marmara junction region, interaction between the Dardenelles and the Aegean Sea, and the possible formation of water mass as related to the outflow form topics for future research which, it is believed, would contribute much towards our understanding of the System as well as of interactions between the Mediterranean and the Black Seas.

ACKNOWLEDGEMENTS

This work was supported by the Department of Water and Sewer Works, Municipality of Istanbul as a part of a five year programme aimed at documenting the present state of health and the oceanography of the Turkish Straits. Partial support was also provided by the Scientific and Technical Research Council of Turkey.

APPENDIX

The differences between the present estimates of the Bosphorus flows and the previous ones are due to the differences in the Net Fresh Water Input (NFWI) values employed for the Black Sea. NFWI is equal to the volume fluxes of rainfall plus run-off minus the evaporation. The transports within the Turkish Straits are directly proportional to and crucially depend on the estimates of the NFWI to the Black Sea.

The estimates of the Black Sea rainfall, run-off and evaporative fluxes and NFWI as well as the Bosphorus flows from various sources are given in Table 1. It is seen in Table 1 that the various estimates of the run-off as well as the precipitation do not significantly differ from each other. With the exception of Pora and Oros (1974), however, the precipitation estimate of Özturgut (1971) significantly differs from the others. The higher precipitation value of Özturgut (1971) leads to nearly 300 km^3 of NFWI to the Black Sea and, within the accuracy of the estimates, this is sufficiently close to NFWI values by Bruevich (1963) and Pora and Oros (1974) but, again, significantly different from others.

Özturgut's (1971) estimate of the rainfall is based on a mean annual precipitation of 71.4 cm found by averaging the rainfall measured at the Black Sea coastal stations. Multiplication of this average precipitation by the surface area of 4.2×10^5 km^2 of the BLack Sea yields 300 km^3/yr. The result of a separate computation that utilizes the long-term rainfall data recorded at stations along the perimeter of the Black Sea (provided in *Weather in the Black Sea*, 1963 and *The Black Sea Pilot*, 1969) agrees well with Ozturgut's value for precipitation. Needless to say, the utilization of rainfall data based on measurements at coastal stations for estimating precipitation at sea leaves much to be desired. What is surprising, however, is that the utilization of the annual rainfall distribution over the surface of the Black Sea as given in the Morskoi Atlas (1950) leads to an average precipitation of 294 km^3/yr. This is in agreement with Ozturgut (1971).

In view of the wide range of estimates of NFWI to the Black Sea, we resort to the results obtained by different methods. Sümer and Bakioğlu (1981) and Oguz *et al.* (1989) have computed numerically the volume fluxes in the upper (:Q1) and the lower (:Q2) layers of the Bosphorus as a function of the difference (:Δh) at the two ends of the strait. Their results are reproduced in Figs. 17a,b. In studying Fig. 7 it should firstly be borne in mind that, based on sufficiently long measurements (Cecen et al., 1981), the sea level difference across the Bosphorus varies in time with a mean of 33 cm and standard deviation of 13 cm. In other words, the mean range of :Δh is [20-46] cm. Secondly, the mean NFWI to the Black Sea is positive so that the portion of the Fig. 17a that is of interest here is the region where Q1>Q2, that is to the right of the intersections of Q1 and Q2 curves.

Now, in Q1>Q2 region, the minimum volume flux for the upper layer is 400 km³/yr, corresponding to the intersection of Q1 and Q2 curves. This value is nearly equal to the Bosphorus out-flows reported in Table 1. But the corresponding lower layer flow is almost twice the inflows given in the same table. In fact, all the estimates, including the present one, in Table 1 imply a Q1/Q2 ratio approximately equal to two (since this ratio must be equal to the salinity ratio of the two layers). The Δh value where this ratio is satisfied is 31 cm, with Q1=570 km³/yr and Q2=285 km³/yr, implying a NFWI=285 km³/yr which is remarkably close to that found by using the estimates of Ozturgut.

Oğuz et. al. (1989) results (Fig. 17b) lead to similar values; Q1/Q2=2 is satisfied where NFWI=300 km³/yr with Δh=27cm.

This does not prove, however, that Özturgut's rainfall value is correct but rather that possible combination of errors incurred in evaporation and precipitation estimates can yield to underestimated values of NFWI. For example, consider the rainfall and run-off estimates from the Entsiklopedia Okean Atmosfera (1983) given in Table 1. If instead of using 354 km³/yr for the evaporation, a value of 254 km³, as carefully computed by Kochikov (1961) is employed, the resulting NFWI would have been 300 km³/yr!

We point out in passing that the differences in the estimates of the net fresh water input also affect fluxes across the halocline of the Black Sea. With the present estimates, the upward and the downward fluxes are 1133 and 821 km³/yr, respectively (Fig. 6). Using Merz's (in Möller, 1928) data, Fonselius (1974) reports an upward flux of 700 km³/yr and a downward flux of 500 km³/yr.

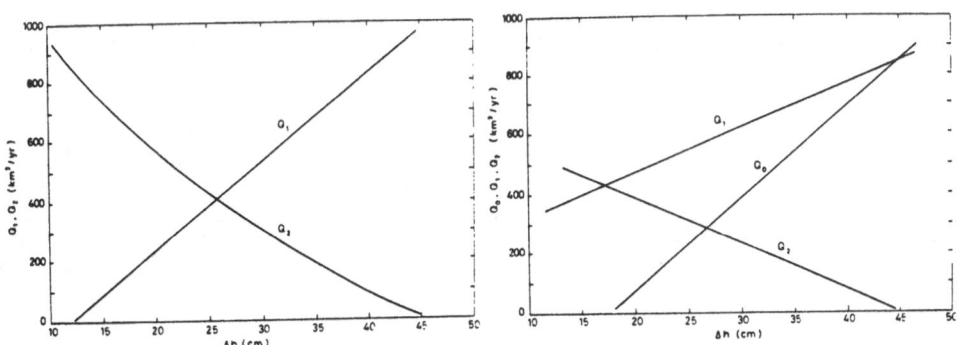

Figure 17a, b. Variations of upper and lower layer flows (Q₁, Q₂) as a function of the sea surface difference across the Bosphorus computed by (a) Sümer and Bakioğlu (1981) (on the left), (b) Oğuz et al. (1989) (on the right).

TABLE 1. ESTIMATES OF THE BLACK SEA BUDGET (km3/yr)

	RAIN FALL	RUN_ OFF	EVAPO_ RATION	NET FRESH WATER INPUT	BOSP. INFLOW Q2	INFLOW FROM AZOV	TOTAL INFLOW	BOSP. OUT_ FLOW Q1	OUTFLOW TO AZOV	TOTAL OUT_ FLOW
MERZ	231	328	354	205	193	_	398	398	_	398
ZENKEVICH (1947)	145	320	319	146	202	_	348	348	_	348
NEUMANN AND ROSEMAN (1954)	240	428	462	206	193	_	399	398		398
CASPERS (1957)	234	320	354	200	_	_	_	_	_	_
LEONOV (1960)	230	309	365	174	193	95	462	392	70	462
BRUKVICH (1960)	225	350	350	255	175	_	430	400	_	400
SOLYANKIN (1963)	119	346	332	133	176	53	362	340	32	372
OKEANOGRAFICESKAIA ENTSIKLOPEDIA (1966)	_	400	_	_	202	_	_	398	59	_
TIXERONT (1970)	181	400	392	189	211	_	400	400	_	400
OZTURGUT (1971)	300	352	353	299	249		548	548		548
SERPOIANU (1973)	120	336	340	116	123	53	292	260	32	292
PORA AND OROS(1974)	254	294	301	247	229	38	514	485	29	514
ENTSIKLOPEDIA OKEAN_ ATMOSFERA (1983)	234	320	354	200	188	_	388	388	_	388
BONDAR (1986)	119	364	332	151	202	50	403	371	32	403
PRESENT STUDY	300	352	353	300	312	_	612	612	_	612

REFERENCES

Armi, L. and D. M. Farmer, (1985): The internal hydraulics of the Straits of Gibraltar and associated sills and narrows. *Oceanol. Acta*, 8, 37-46.

Armi, L. and D. M. Farmer, (1987): A generalization of the concept of maximal exchange in a strait. *J. Geophy. Res.*, 92, 14679-14680.

Armi, L. and D. M. Farmer, (1988): The Flow of Mediterranean water through the Strait of Gibraltar. *Progress in Oceanography*, 21, 1-41.

Baştürk, Ö, C. Saydam, I. Salihoğlu, A. Yılmaz, (1986): Oceanography of the Turkish Straits, First Annual Report, Institute of Marine Sciences. Middle East Technical University, Vol. III, 86pp.

Baştürk, Ö, C. Saydam, A. Yılmaz, I. Salihoğlu, (1988): Oceanography of the Turkish Straits, Second Annual Report, Institute of Marine Sciences. Middle East Technical University, Vol. II, 130pp.

Beardsley, R.C., and J. Hart, (1978): A simple theoretical model for the flow of an estuary onto a continental shelf. *J. Geophys. Res.*, 83, 873-883.

The Black Sea Pilot, (1969): 11th Ed., Hydrographer of the Navy, Taunton, Somerset, Great Britain.

Bogdanova, A.K., (1965): Seasonal fluctuations in the inflow and distribution of the Mediterranean waters in the Black Sea, in Fomin, L.M. (Ed.) Basic Features of the Geological Structure, of the Hydrologic Regime and Biology of the Mediterranean Sea, pp.131-139, Academy of Sciences, USSR, Moscow. English transl., 1969 Institute of Modern Languages, Washington, D.C.

Bogdanova, A.K. and V.N. Stepanov, (1974): Hydrodynamic estimate of the blocking conditions of the lower Bosphorus current. *Oceanology*, 14, 37-40.

Bondar, C., (1986): Considerations on the water balance of the Black Sea. In: The proceedings of the international meeting on the chemical and physical oceanography of the Black Sea. Report on the chemistry of seawater. University of Sweden, Göteborg.

Bruevich, S. V., (1960): On water and salt balance of the Black Sea. In Proceedings of Institute of Oceanology. Nauka, Moskow, 42, pp.3-21 (in Russian).

Büyükay, M., (1989): The surface and internal oscillations in the Bosphorus, related to meteorological forces. M.Sc Thesis, Institute of Marine Sciences, Middle East Technical University, 169pp.

Büyüközden, A., H. Yüce and T. Bayraktar, (1985): Investigation of the Mediterranean Water Along the Bosphorus and the Black Sea, Doğa Bilim Dergisi, Seri B, 9, 312-324, (in Turkish).

Caspers, H. (1957): Black Sea and Sea of Azov, In: Treatise of Marine Ecology and Paleocology, Hedgpeth, J. (Ed.), Geol. Soc. Am. Mem. 67, 1, pp.803-890.

Çecen, K., M. Beyazit, M. Sümer, S. Güçlüer, M. Doğusal, and H. Yüce, (1981): Oceanographic and Hydraulic Investigation of the Bosphorus: Section I, Final report, submitted to the Irrigation Unit of the Turkish Scientific and Technical Research Council, Istanbul Technical University, Istanbul, 166pp, (in Turkish).

DAMOC, (1971): Master Plan and Feasibility Report for Water Supply and

58

Sewerage for the Istanbul Region, V.III., prepared by DAMOC
Consorsium, for WHO, Los Angeles.

Defant, A., (1961): Physical Oceanography, V.I, Pergamon Press, Oxford,
729pp.

De Filippi, G.L., L. Iovenitti and A. Akyarlı, (1986): Current analysis
in the Marmara-Bosphorus Junction, 1ᵐᵗ AIOM (Associazione di
Ingegneria Offshore e Marina) Congress, Venice.

Entsiklopedia Okean-Atmosfera, (1983): Gidrometeoizdat, Leningrad.

Farmer, D. M. and Armi, L., (1986): Maximal two-layer exchange over a
sill and through the combination of a sill and contraction with
barotropic flow. J. Fluid Mech., 164, 53-76.

Fonselius, S.H., (1974): Phosphorus in Black Sea, in Degens, E.T. and
D.A. Ross (Eds.), The Black Sea Geology, Chemistry and Biology,
pp.249- 278, Am. Assoc. Pet.Geol. Memoir 20, Tulsa, Oklahoma.

Gunnerson, C.G., and E. Ozturgut, (1974): The Bosphorus. in Degens,
E.T.and D.A. Ross (Eds.). The Black Sea-Geology,Chemistry and
Biology, pp.99-113, Am. Assoc. Pet. Geol. Memoir 20, Tulsa, Oklahoma.

Johns, B. and T. Oğuz, (1989): The modelling of the flow of water
through the Bosphorus. Dyn. Atm. and Oceans (in press).

Kochikov, V.N., (1961): Evaporation from the surface of the Black Sea.
Rapp. Pr.-Verb. Reun. CIESM, 16, 639-642.

Lacombe, H., J.C. Gascard, J. Gonella and J.P. Bethoux, (1981): Response
of the Mediterranean to the water and energy fluxes across its
surface on seasonal and interannual scales. Oceanol. Acta, 4,
247-255.

Latif, M. A., E. Özsoy, T. Oğuz, Ü. Ünluata, Özden Baştürk, C. Saydam,
A. Yılmaz, and I. Salihoglu, (1989a): Oceanographic Characteristics of
the Region Surrounding the Northern Entrance of the Bosphorus as
Related to the Planned Sewage Outfalls, First Annual Report, Institute
of Marine Sciences, Middle East Technical University, 111pp.

Latif, M. A., E. Özsoy, T. Oguz and Ü. Ünlüata, (1989b): Observation of
the Mediterranean inflow into the Black Sea. Submitted for
publication to Deep Sea Research.

Leonov, D., (1960): Regional Oceanography; Bering, Okhotsk, Japan,
Caspian, and Black Seas. In: Snezlrinskii, B.A. (Ed.),
Gidrometeorologicheskoe Izd., Leningrad, 765pp.

McGraw-Hill Encyclopedia of ocean and atmospheric science, (1980):
Editor in Chief: Sybil, P.P., McGraw-Hill Co., New York, 580pp.

Merz, A. (1918): Die strömungen des Bosporus und Dardanellen, Verh.
Deutsch. Geogr. Tages. 20

Möller, L., (1928): Alfred Merz' hydrographische untersuchungen in
Bosphorus and Dardanellen, Veroff. Inst. Meeresk., Berlin Univ., Neue
Folge A, V.18, 284pp.

Morskoi Atlas, (1950): V.II (Black Sea). U.S.S.R. Admiralty.

Neuman, J. and N. Rosenan, (1954): The Black Sea energy balance and
evaporation. Trans. Am. Geophys. Union, 35, 767-774.

Nof, D., (1978): On geostrophic adjustment in sea straits and wide
estuaries: Theory and laboratory experiments. Part II- Two-layer
system. J. Phys. Oceanogr., 8, 861-872.

Oğuz, T. and H. I. Sur, (1989): A two-layer model of water exchange
through the Dardanalles Strait. Oceanol. Acta, 12, 23-31.

Oğuz, T., E. Özsoy, M. A. Latif, Ü. Ünluata, (1989): Modelling of
  hydraulically controlled exchange flow in the Bosphorus Strait.
  Submitted for publication to *J. Phys. Oceanog*.
Okeanograficeskaia Entsiklopedia (1966): Gidrometeoizdat, Leningrad.
Ovchinnikov, I.M., (1974): On the water balance of the Mediterranean
  Sea. *Oceanology*, 14, 198-202.
Östlund, H.G., (1969): Expedition Odysseus 65: Tritium and Radio Carbon
  in the Mediterranean and Black Seas. Tech. Rep. Univ. Miami, Ins. Mar.
  Sci., 27pp. Unpublished manuscript.
Östlund, H. G., (1986): Renewal rates of the Black Sea deep water,
  Extended Abstract, presented at meeting on "The chemical and physical
  oceanography of the Black Sea", Göteborg.
Özsoy, E., T. Oğuz, M. A. Latif, and Ü. Ünlüata, (1986): Oceanography of
  the Turkish Straits, First Annual Report, Institute of Marine
  Sciences, Middle East Technical University, Vol.I. 269pp.
Özsoy, E., T. Oğuz, M. A. Latif, Ü. Ünlüata, H. I. Sur and Ş. Beşiktepe,
  (1988): Oceanography of the Turkish Straits, Second Annual Report,
  Institute of Marine Sciences, Middle East Technical University,
  Vol.I. 110pp.
Özturgut, E., (1971): Physical Oceanographic Study of the Bosphorus.
  Ph.D. thesis, Inst. Geography, University of Istanbul, (in Turkish).
Pektaş, H., (1953): Surface currents in the Bosphorus and the Sea of
  Marmara, Hydrobiology, Ser. A, v.1, no.4, pp.154-169. Publication of
  Hydrobiology Research Institute, Faculty of Sciences, University of
  Istanbul, (in Turkish).
Pora, A. E. and I. Oros, (1974): Limnologie si oceanologie (Limnology
  and Oceanology). Editura didactica si pedagogica, Bucuresti.
Preller, R., (1985): Numerical Model Study of Circulation in the Alboran
  Sea. Mesoscale Air-Sea Interaction Group, Technical Report, Florida
  State University, Tallahassee, 125pp.
Scholten, R. (1974): The role of the Bosphorus in Black Sea chemistry
  and sedimentation, In: the Black Sea. Its geology, chemistry and
  biology. Degens, E.T. and D.A. Ross, editors, A.A.P.G. Memoir, Tulsa,
  Oklahoma, pp.115-126.
Serpoianu, G., (1973): Le bilan hydrologique de la mer Noire. Cercetări
  marine, I.R.C.M. No.5-6, pp.145-153.
Solyankin, E.V., (1963): On the water balance of the Black Sea.
  Okeanologiya, 3, 986-993.
Sümer, B.M. and M. Bakioğlu, (1981): Sea-Strait Flow with Special
  Reference to Bosphorus. Tech. Rap., Faculty of Civil Engineering,
  Technical University of İstanbul, 25pp.
Sverdrup, H. U., M. W. Johnson and R. H. Fleming, (1946): The Oceans,
  Prentice-Hall C., New York, 1087pp.
Tixeront, F., (1970): Le bilan hydrologique de la Mar Noire et de la
  Mediterranee. *Cahiers Oceanog.*, 22, 227-237.
Tolmazin, D., (1981): Two dimensional circulation in straits. *Oceans'81*,
  820-823, Proceedings of IEEE Conference, September 1981, Boston.
Tolmazin, D., (1985): Changing coastal oceanography of the Black Sea, II:
  Mediterranean effluent. *Progress in Oceanography*, 15, 277-316.
Trewartha, G. T., (1968): An introduction to climate, McGraw-Hill Co.
  408pp.

Ullyott, P., and O. Ilgaz, (1943): Observations on the Bosphorus, I: A definition of standard conditions throughout the year. *Hydrobiology, Ser.B*, 8, 229–255, Publication of the Hydrobiological Research Institute Faculty of Science, University of İstanbul.

Ullyott, P., and O. Ilgaz, (1946): The hydrography of the Bosphorus: An introduction. *Geogr. Rev.*, 36, 44–66.

Ünlüata, Ü. and T. Oğuz, (1983): A review of the dynamical aspects of the Bosphorus. Review paper, presented at the NATO Workshop on the Atmospheric and Oceanic Circulation of the Mediterranean, La Spezia, Italy, Sept. 1983, 42pp.

Ünlüata, Ü., and E. Özsoy, (1986): Oceanography of the Turkish Straits, First Annual Report, Vol. II, Health of the Turkish Straits I: Oxygen Deficiency of the Sea of Marmara, Institute of Marine Sciences, Middle East Technical University, 88pp.

Yüce, H., (1986): İstanbul Boğazında su seviyesi değişimlerinin incelenmesi. Bülten, İstanbul Üniversitesi Deniz Bilimleri ve Cografya Enstitüsü, Cilt 2, Sayı 3, pp.67–78.

Wang, D.P., (1987): The Strait surface outflow. *J.Geophys. Res.*, 92, 10807–10825.

Weather in the Black Sea, (1963): Meteorological Office, H.M. Stationery Office, London.

Whitehead, J.A., and A.R. Miller, (1979): Laboratory simulation of the gyre in the Alboran Sea. *J. Geophys. Res.*, 90, 7045–7060.

Zenkevich, L.A., (1947): The fauna and the biological productivity of the sea. In: Morya SSSR, V.2, Moskow, Izd. Sovetskaya Nauka, pp.519–538.

A REVIEW OF THE PHYSICAL OCEANOGRAPHY OF FRAM STRAIT

KENNETH HUNKINS
Lamont-Doherty Geological Observatory
  of Columbia University
Palisades, New York 10964

ABSTRACT. Fram Strait is the broad and deep gap (width, 450 km; sill
depth, 2700 m) separating Greenland and Spitsbergen. There is an
exchange through it of cold, fresh Arctic waters and warm, saline
Atlantic waters. Narrow coastal boundary currents on either side
flowing in opposite directions are important elements of the exchange.
Mesoscale eddies are abundant in this region but their part in transport through the strait is uncertain. This appears to be primarily a
convective circulation driven by density differences between the
Arctic Ocean and Greenland Sea with wind forcing playing a minor part.
It is possible that atmospheric pressure gradients and tidal rectification also contribute to the exchange although their importance has
not yet been demonstrated. There has been considerable success in
describing the ice exchange with numerical models which incorporate a
restricted ocean model. Laboratory experiments provide an alternative
method which has proved useful in giving insight into the relative
roles of boundary currents and eddies in transport through wide straits.

INTRODUCTION

Within the broad gap separating Greenland and Spitsbergen cold waters
of low salinity from the Arctic Ocean come into contact with warmer
waters of higher salinity from the Atlantic Ocean. There is an exchange of water and ice through this opening which has important
effects on regional climate and possibly on global climate. The name
Fram Strait has been given to this gap in recognition of the epic and
highly successful expedition led by the Norwegian explorer and
oceanographer, Fridtjof Nansen. After drifting for three years locked
in arctic pack ice, the expedition's ship Fram emerged into open water
in 1896 and sailed southward through Fram Strait on its homeward
journey. Thus the ship's name is most appropriate for Fram Strait but
the use of the word strait might be questioned since this opening is
much wider than the narrow constrictions usually designated by the
term strait. The name Fram Strait is now too well established to

L. J. Pratt (ed.), The Physical Oceanography of Sea Straits, 61–93.
© 1990 Kluwer Academic Publishers.

consider changing it, but it should be kept in mind that at the narrowest section there is a distance of 450 km between Nordostrundingen, the northeast cape of Greenland, and West Spitsbergen. Fram Strait is deep as well as wide with a span of 100 km at the 2000 m depth level (Fig. 1).

Fig. 1 – Bathymetry of the polar regions. Fram Strait links the Greenland Sea with the Eurasian Basin of the Arctic Ocean (Carmack, 1986).

The existence of a deep connection between the Arctic Ocean and
Greenland Sea was first conclusively established by a Soviet expedition
in 1956 (Laktionov, 1959) and as a result of this and later U.S.
surveys the sill depth is now known to be about 2,600 m (Fig. 2). This
opening originated as a fracture zone which offsets the mid-ocean ridge
and bottom topography is accordingly rugged. The Molloy Deep which is
a circular depression 30 km in diameter with a maximum depth of 5,570 m,
greater than any known in either the Greenland Sea or Arctic Ocean,
is only one example of the complex relief found in the center of the
Strait which also includes seamounts and ridges as well as depressions.
The bottom of the Molloy Deep is located a distance of only 20 km from
a seamount with a crest at 1,481 m.

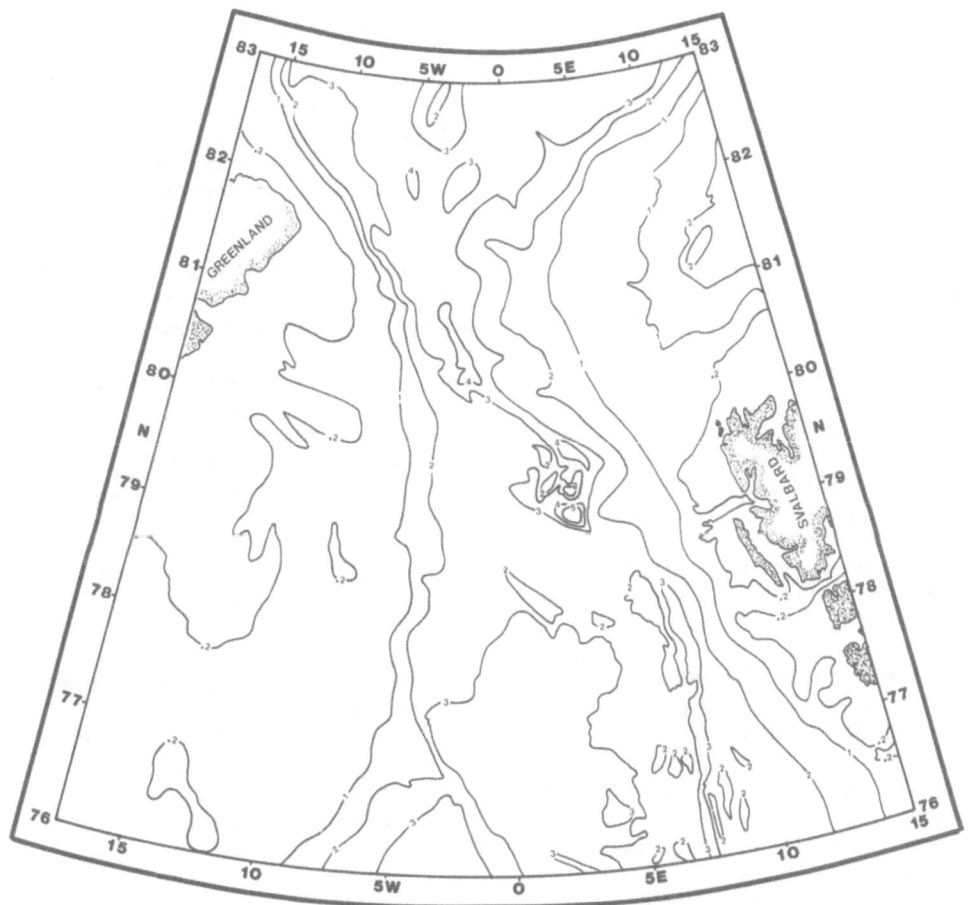

Fig. 2 - Bathymetry of Fram Strait (adapted from Perry and Fleming,
1986).

Ice Drift

Sea ice is exported from the Arctic Ocean through Fram Strait in a stream which travels southward along the east coast of Greenland. The other side of the strait along the coast of West Spitsbergen is largely free of ice. The extent of ice in Fram Strait is now fairly well known since it can be monitored in a number of ways: with visual observations from aircraft and ship, with satellite microwave and infrared imagery and with satellite-tracked buoys (Figs. 3 and 4). The arctic ice edge makes seasonal changes which are small in comparison with antarctic pack ice. Drifting ice not only serves as a valuable indicator of surface ocean behavior but also plays an important part in heat exchange, for although ice constitutes only a small fraction of the water exchange its large heat capacity gives it an importance in heat exchange beyond its small volume percentage. A recent estimate of mean annual ice discharge based on buoy drift tracks is $0.159 \times 10^6 \mathrm{m}^3/\mathrm{s}$ with a variation from a minimum of $0.09 \times 10^6 \mathrm{m}^3/\mathrm{s}$ in summer to a maximum of $0.19 \times 10^6 \mathrm{m}^3/\mathrm{s}$ in winter (Vinje and Finnekåsa, 1986). Since water temperatures are near the freezing point, little melting occurs in the East Greenland Current (EGC), but on the east side of the Strait southward moving sea ice encounters the warm West Spitsbergen Current (WSC) flowing northward and rapid melting occurs there. A balance between ice advection and melting maintains a relatively stationary ice edge west and north of Spitsbergen. Early whaling ships were able to take refuge from pack ice in this area, known as Whalers' Bay, which often remains open throughout the winter.

Motion of a single ice floe is controlled by air stress on the upper surface, water stress on the lower surface and Coriolis force. In a field of adjacent ice floes there is also an internal ice stress which develops in response to convergence or divergence of the pack. In the central Arctic Ocean winds and ocean currents are of roughly equal importance to ice drift (Thorndike and Colony, 1982). Mean wind stress has a southward component over the entire width of Fram Strait and tends to drive ice into the Greenland Sea. Advection by ocean currents acts in the same direction and is even more important, accounting for 80% of the observed southward drift according to Vinje and Finnekåsa (1986). These same authors estimate the average thickness of ice in the strait to be 4.0 m which is similar to ice thickness in the Eurasian Basin according to McLaren (1989). Ice drift in the Arctic Ocean north of Fram Strait is everywhere directed toward the opening with mean speeds of about 10 km/day down the center of the strait (Fig. 5). As Fram Strait is approached, the ice field both converges and accelerates. Since the ice velocity field in this region chiefly reflects surface currents, it is therefore a useful indicator of them.

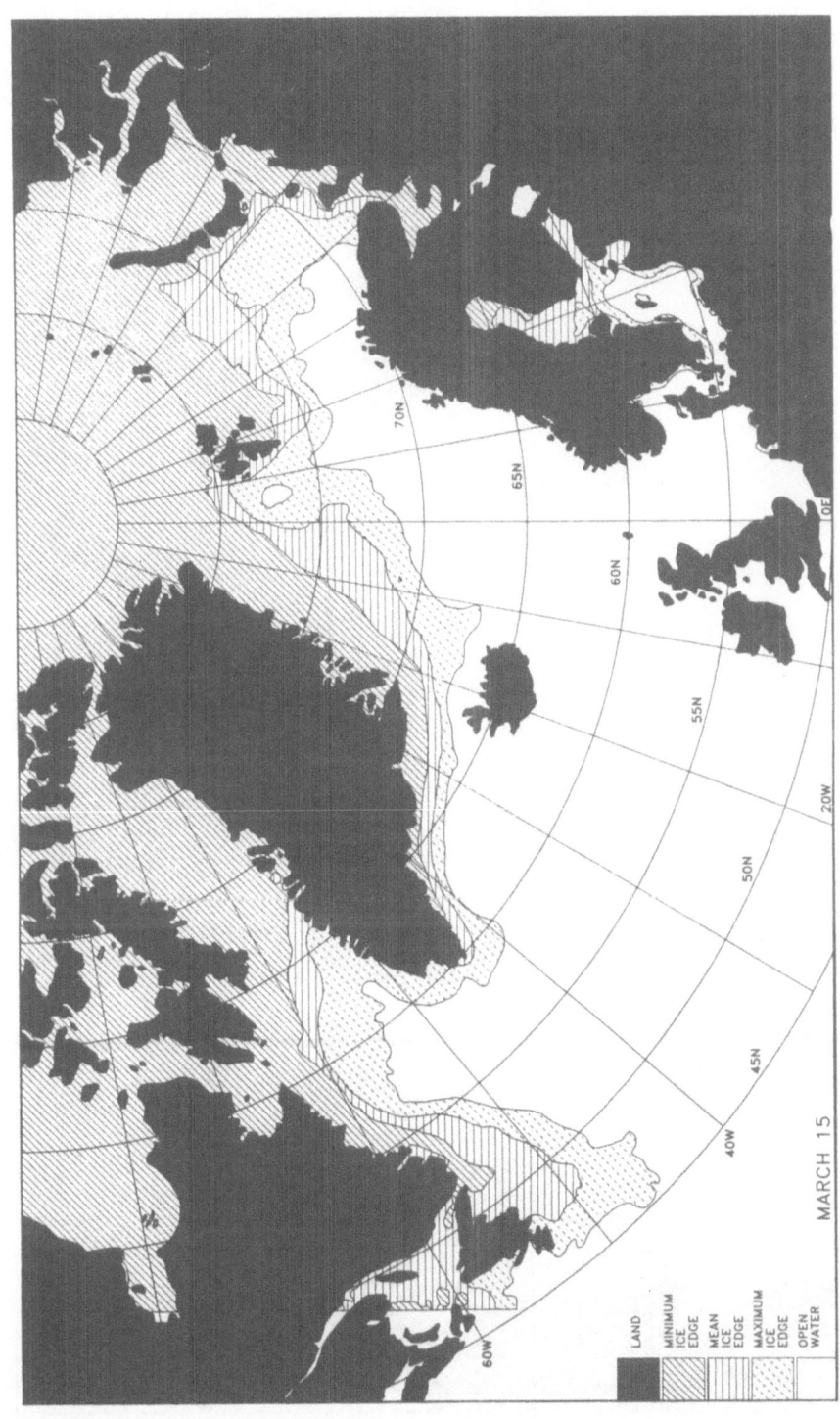

MAXIMUM — MEAN — MINIMUM ICE EDGES

Fig. 3 – Maximum, minimum and mean extent of sea ice on March 15 for the decade, 1972–1982 (NOC, 1986).

66

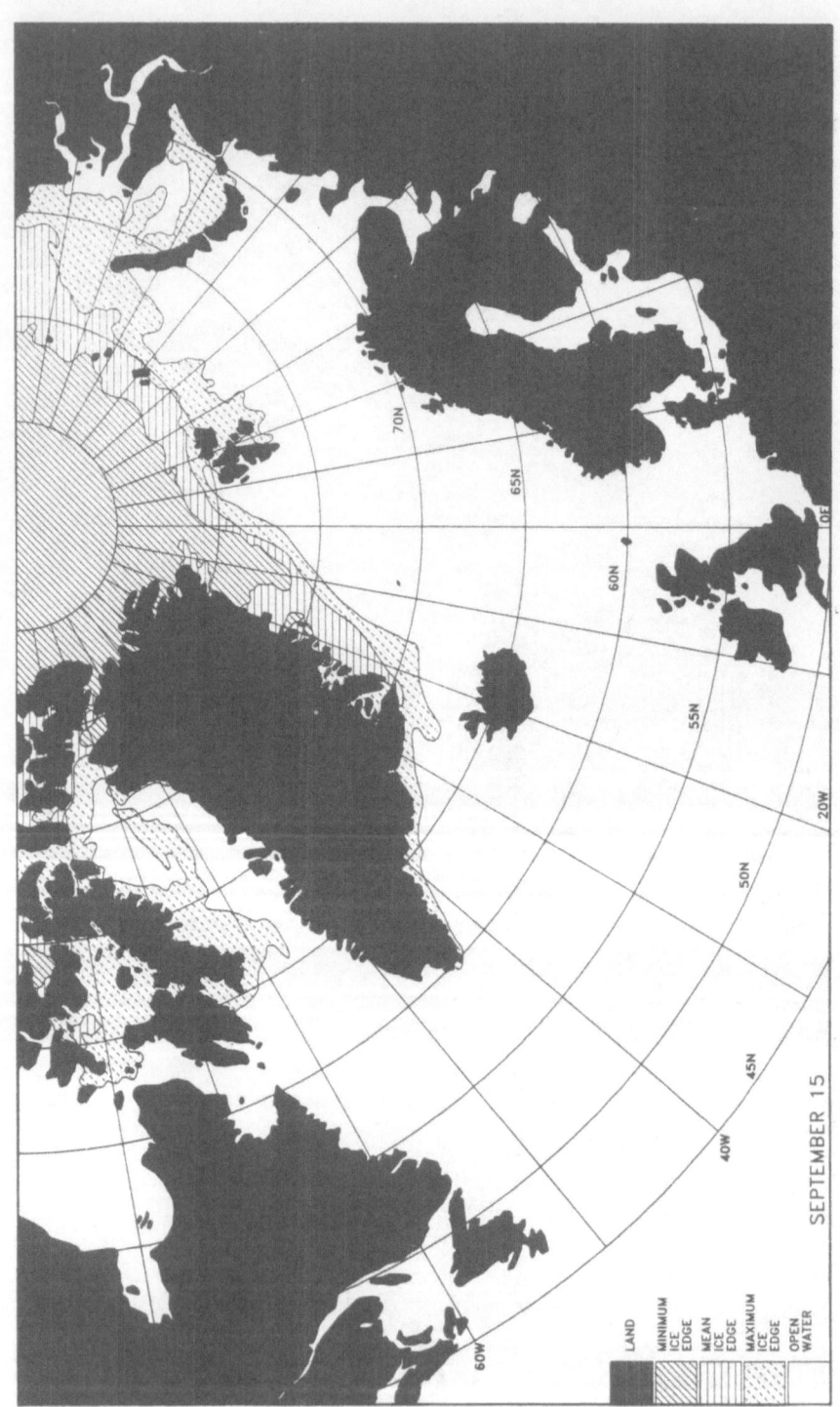

MAXIMUM — MEAN — MINIMUM ICE EDGES

SEPTEMBER 15

Fig. 4 – Maximum, minimum and mean extent of sea ice on Sept. 15 for the decade, 1972-1982 (NOC, 1986).

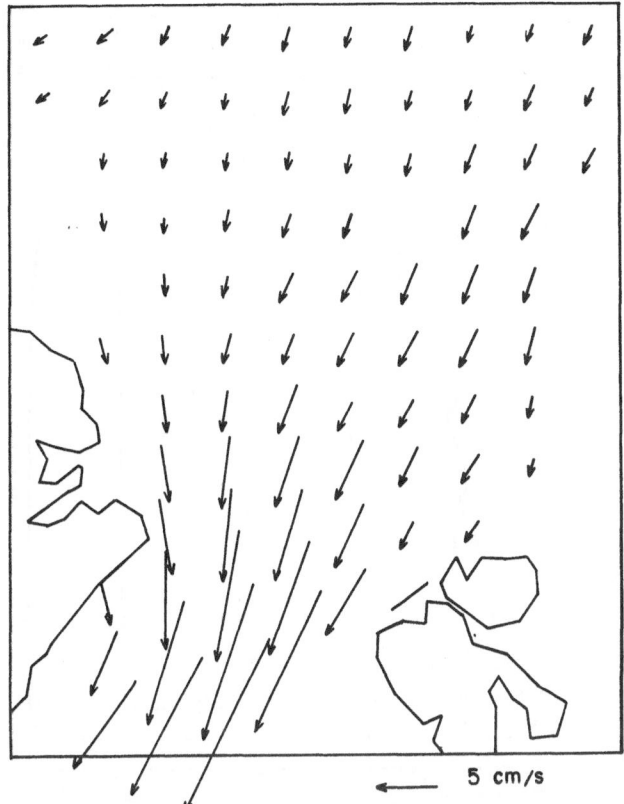

Fig. 5 - Optimal estimate of mean annual ice velocity (Moritz and
Colony, 1988).

Hydrography

Within the perennially ice-covered Arctic Ocean north of Fram Strait,
surface waters are low in salinity, typically 31-32 ppt, with
temperatures at the freezing point appropriate to the particular
salinity. The upper layer is well mixed in both salinity and
temperature to a depth of 20 to 60 m (Fig. 6). As sea ice grows downward during winter, the crystal matrix extrudes heavy brine which
convectively mixes the upper layer. When melting occurs during summer
the upper layer tends to stratify but a thin layer (<10 m) is kept
mixed by the mechanical stirring of drifting ice (Lemke and Manley,
1984). Below the surface layer, salinity increases rapidly to a depth
of 200 m and then more slowly below that depth. Temperature, however,
usually remains constant down to about 100 m. This cold upper pycnocline is apparently formed during ice formation on the continental
shelves during winter when heavy brine sinks and flows off the shelf
as a lateral injection into the water column (Aagaard et al., 1981).
Below 100 m temperatures also increase with depth in water which
clearly has its origin in the Atlantic Ocean. North of Fram Strait at
depths of 200-300 m there is a subsurface temperature maximum of 2°C.

68

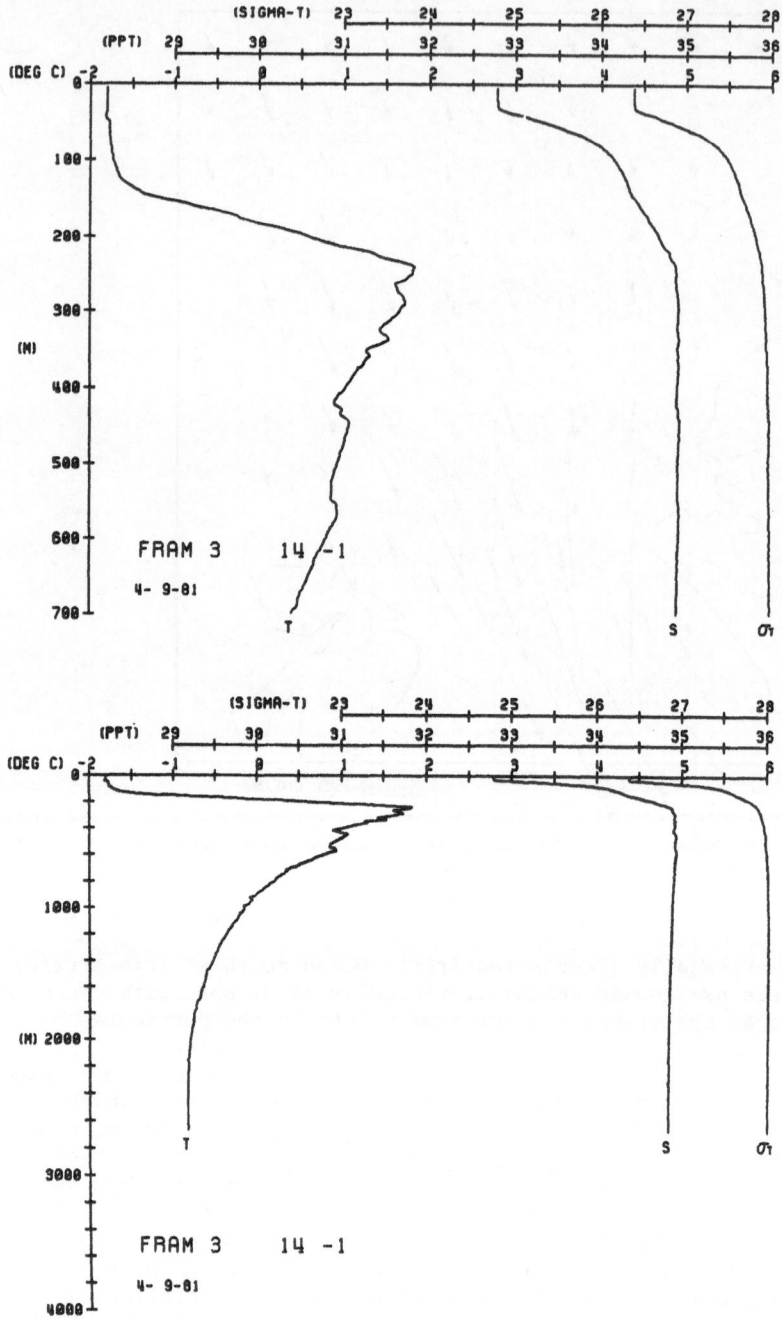

Fig. 6 – Temperature, salinity and sigma-t profiles from Fram III
ice camp. April 9, 1981, 83°23'N 9°28'E (Manley and Camp, 1985).

This characteristic maximum occurs throughout the Arctic Ocean but at lower temperature and with increasing depth as distance from Fram Strait increases. A warm core extends as a subsurface tongue along the shelf edge north of the USSR (Fig. 7). Beneath these Atlantic Waters, which are somewhat arbitrarily defined as waters with temperature greater than 0°C, lie Arctic Deep Waters. At depths between 600 m to 2000 m salinity remains nearly constant at 34.94 ppt as temperature decreases slightly with depth. Below 2000 m temperature increases at the adiabatic gradient. This deep water mass occupies more than one-half of the volume of the Arctic Ocean and has been considered to originate primarily as Atlantic waters entering from the south although there may be some mixing with shelf water of high salinity (Aagaard et al., 1985). In the central Arctic Ocean itself strong salinity stratification near the surface prevents the formation of deep water.

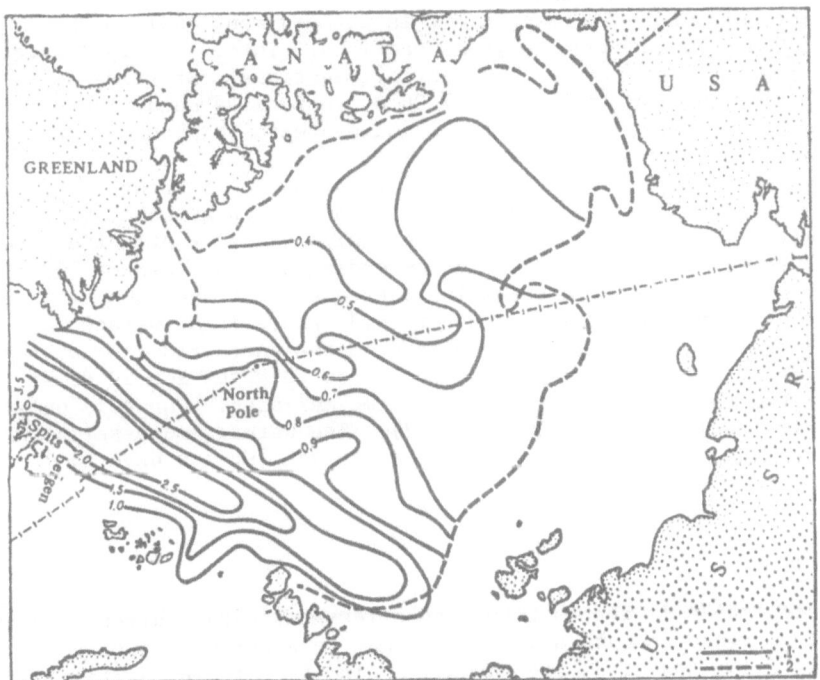

Fig. 7 - Distribution of maximum temperatures in subsurface Atlantic Water of the Arctic Ocean. Solid lines are isotherms in °C. Dashed lines are continental shelf boundaries. Dashed and barred line is limit of Soviet sovereignty. (Treshnikov et al., 1977).

In contrast with the Arctic Ocean, the central Greenland Sea has only a weak stratification. A doming of temperature, salinity and density surfaces in this sea characterizes the cyclonic Greenland gyre (Fig. 8). During intense winter cooling it is considered likely that surface waters near the center of the Greenland Sea are cooled sufficiently to sink and overturn the entire water column to produce deep water although the process has not been observed directly. Swift (1986) has reviewed in detail the characteristics of water masses in this region.

A hydrographic section extending across the Arctic Ocean and Greenland-Norwegian Sea from Alaska to Norway passes through Fram Strait and shows the transition between the Arctic Ocean and Greenland Sea water masses in the Strait (Figs. 9 and 10). The juxtaposition of highly stratified Arctic Ocean waters with waters of low static stability in the Greenland Sea is emphasized in this meridional section down the center of the strait.

Water structure in Fram Strait is three-dimensional and a different aspect is shown in zonal sections. Three principal water masses can be identified in cross-strait sections. Starting at the western side, in the profile at 77°30'N, there is a cold, low salinity layer extending to a depth of 100 m and eastward to about 5°W where it terminates in a surface front (Fig. 11). This is the outflowing polar water of the EGC bounded by the East Greenland Polar Front (EGPF). Immediately east of this front there is the Return Atlantic Current (RAC) with a subsurface core of warmer, high salinity Atlantic Water between depths of 50 and 600 m and extending over a width of 100 km. In the center of the strait the isolines are domed over a nearly homogeneous water column. On the eastern side there is a surface layer of Atlantic Water extending to 600 m marking the northward flow of the WSC.

Farther north in a zonal section at 80°20'N the EGC is still evident but the EGPF is east of the Prime Meridian. There is only a slight indication of the RAC (Fig. 12). The warmer, more saline waters composing the WSC are no longer at the surface but now have their maximum at 200-300 m.

Circulation

Since ice motion is due to both wind stress and ocean current advection, surface currents beneath drifting ice may be determined indirectly if both ice drift and winds are known. A wind factor representing the response of ice to wind is known from other studies in the Arctic Ocean (Thorndike and Colony, 1982). The geostrophic wind can be determined from atmospheric pressure data. Then the wind-driven ice drift is calculated from geostrophic wind data. Vectorial subtraction of the calculated wind drift from the observed ice drift then leaves the surface current as a residual. Surface currents determined by this method are shown in Fig. 13. There is a tendency for current velocities to increase southward through Fram Strait. Maximum velocities are found near the edge of the Greenland shelf.

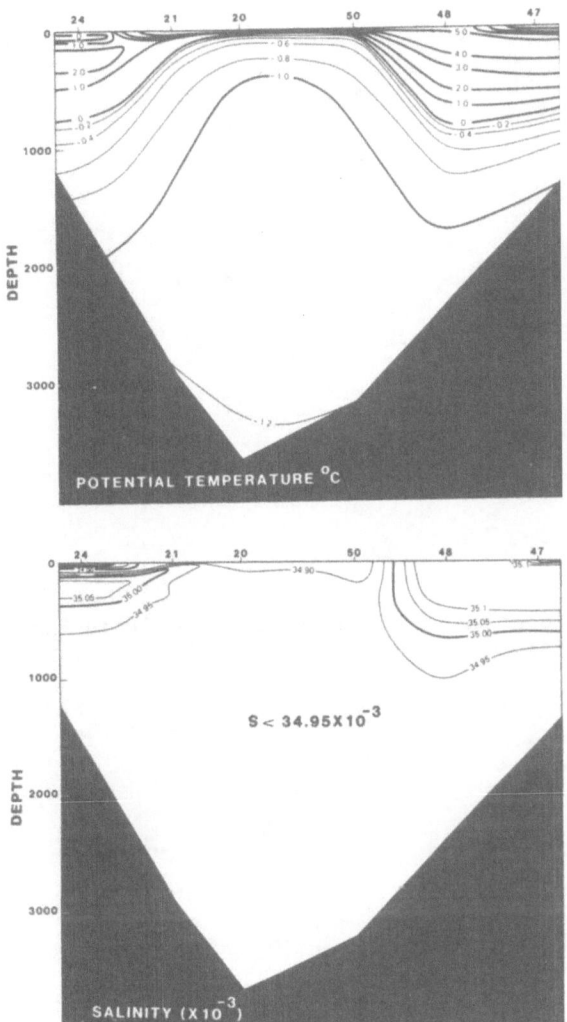

Fig. 8 - Potential temperature and salinity section across the
Greenland Sea (Carmack, 1986).

Fig. 9 - Map showing location of oceanographic profile in Fig. 10
(Aagaard et al., 1985).

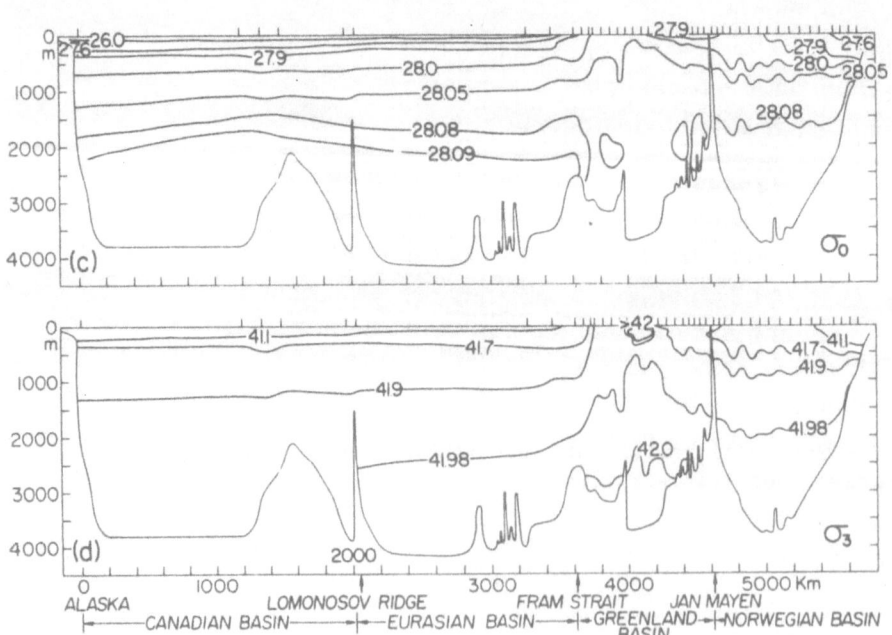

Fig. 10 - Density section along line shown in Fig. 9 (Aagaard et al.,
1985).

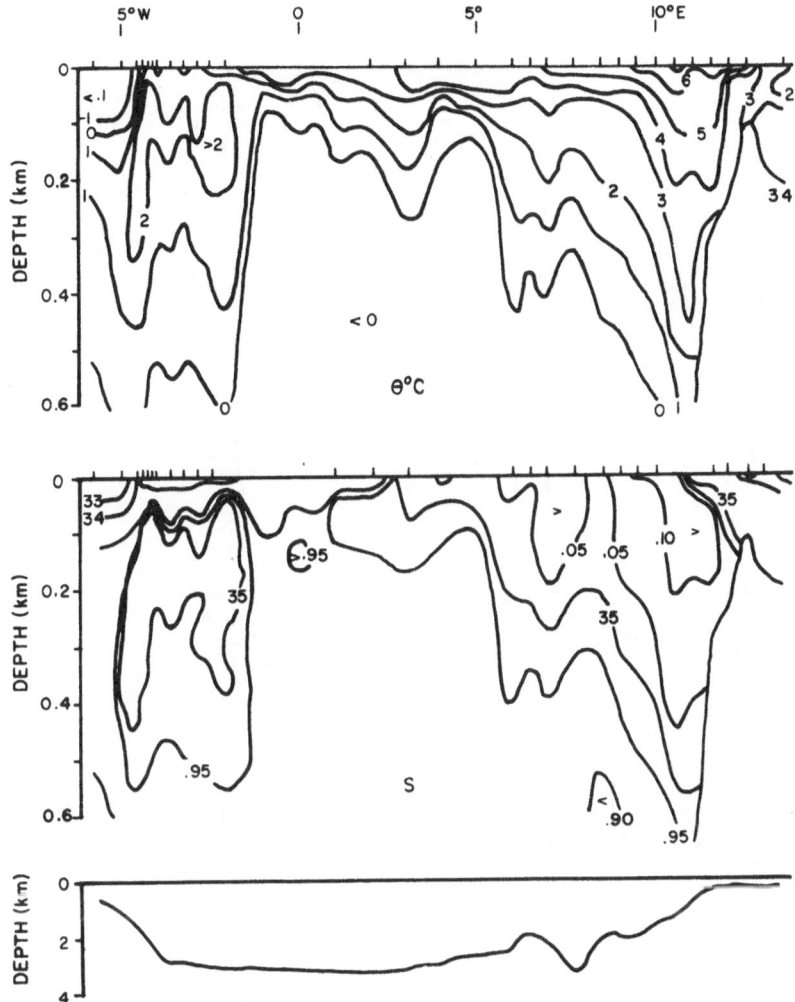

Fig. 11 – Vertical distribution of temperature and salinity in the upper 600 m along a zonal section at 77°30'N. Bottom topography is shown in the lower panel (Quadfasel et al., 1987).

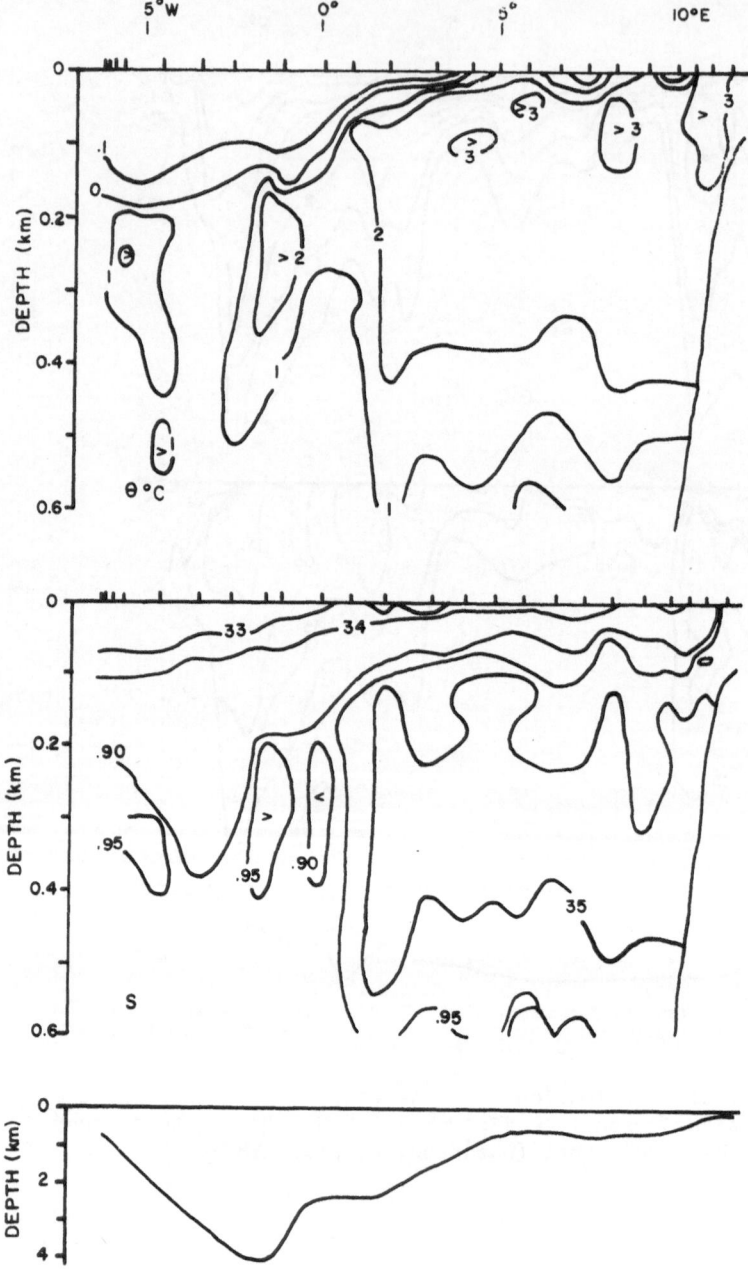

Fig. 12 – Vertical distribution of temperature and salinity in the upper 600 m along a zonal section at 80°20'N. Bottom topography is shown in the lower panel (Quadfasel et al., 1987).

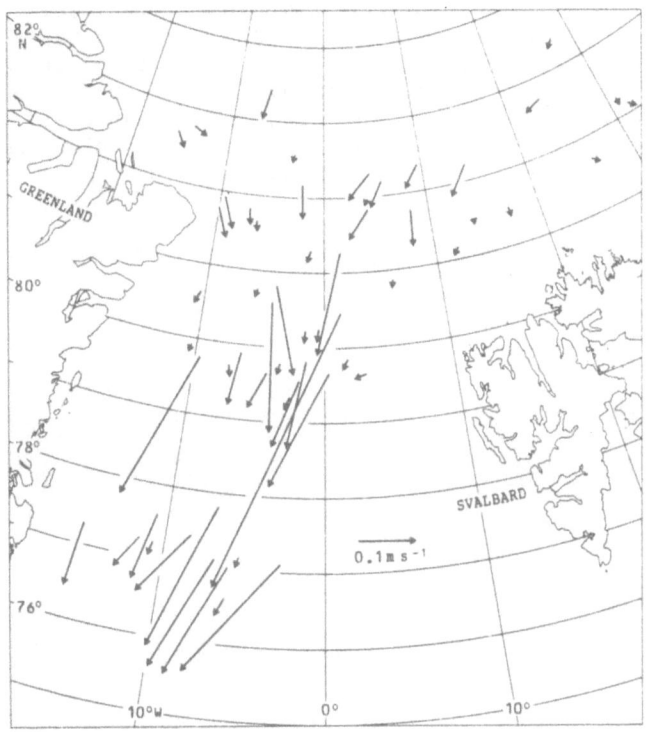

Fig. 13 - Average ocean surface currents determined by subtracting the wind-induced motion from observed ice motion (Vinje and Finnekåsa, 1986).

Other techniques must be used to measure currents in ice-free areas and below the surface. In regions of high baroclinicity, where deep currents may be assumed small such as in the Arctic Ocean and western Fram Strait, the geostrophic method is effective for determining average surface currents. Dynamic topography of the surface of the Arctic Ocean shows currents flowing across the Eurasian Basin and southward through most of Fram Strait (Fig. 14). Only along the Spitsbergen coast is there a narrow northward current.

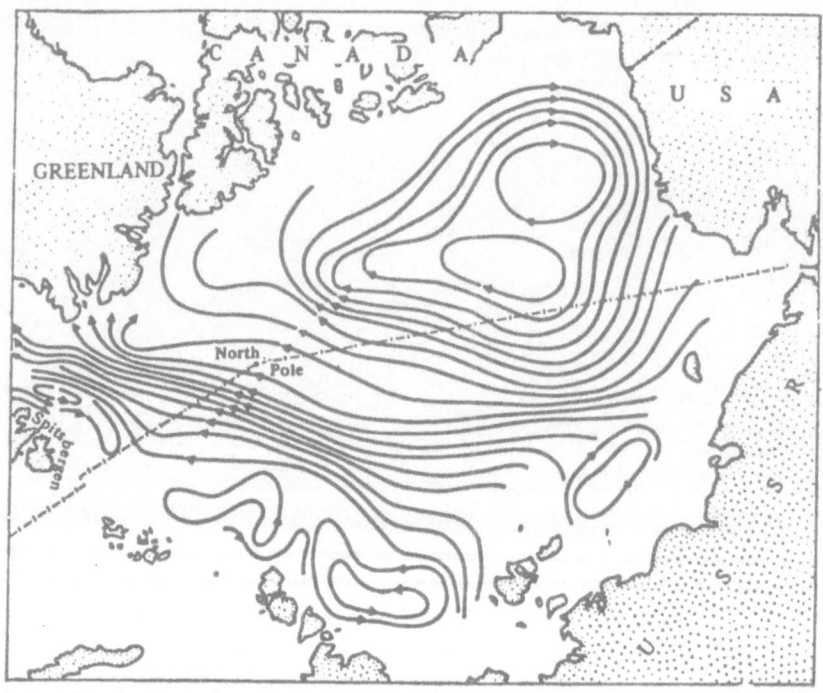

Fig. 14 - Schematic surface dynamic topography of the Arctic Basin. Dashed and barred line is limit of Soviet sovereignty. (Treshnikov et al., 1977).

South of Fram Strait in the Greenland Sea there is a large cyclonic gyre with a northern lobe extending into the strait (Fig. 15). This gyre must be primarily a wind-driven feature since calculations of Sverdrup transport reproduce the observed circulation fairly well (Aagaard, 1970).

Direct measurements of currents with instrumented moorings were begun in the WSC in 1971 by a group from the University of Washington (Aagaard et al., 1973) and resumed in 1976 with Norwegian cooperation (Aagaard, 1982). Recently moorings have also been maintained in the EGC (Foldvik et al., 1988; Manley et al., 1987a). Observations from moorings have an advantage over the geostrophic method in giving the total flow, including the barotropic as well as baroclinic component of current. They are generally most useful in assessing temporal variability since the practical limitations on mooring spacing and record length do not usually allow much confidence in transport calculations from them. A recent value of $3 \times 10^6 m^3$/s for southward transport above 700 m in the EGC at 79°N has been determined by Foldvik et al. (1988) based on results from an array of nine instruments maintained for one year. The transport was about equally divided between the barotropic and baroclinic modes. An estimate of $5.6 \times 10^6 m^3$/s was made by Hanzlick (1983) for transport in the WSC, who found considerable spatial variability with cross-stream scales of 10-20 km in this current.

A number of investigators have produced transport estimates for Fram Strait based on hydrographic data alone using various budget considerations. A recent and thorough study was made by Rudels (1987) who used two detailed hydrographic sections across the strait and constraints based on mass and salt balance requirements to find unique transport values including both barotropic and baroclinic components. His results for the "Ymer" section are shown in Table 1. These calculations placed certain restraints on deep water exchange and allowed for a residual net transport which is compensated by flow through other Arctic Ocean openings, including Bering Strait, the Barents Sea and Canadian Archipelago. These values are in general agreement with most of the earlier estimates. Better values in the future must await improved surveys and moorings with more extensive sampling in time and space.

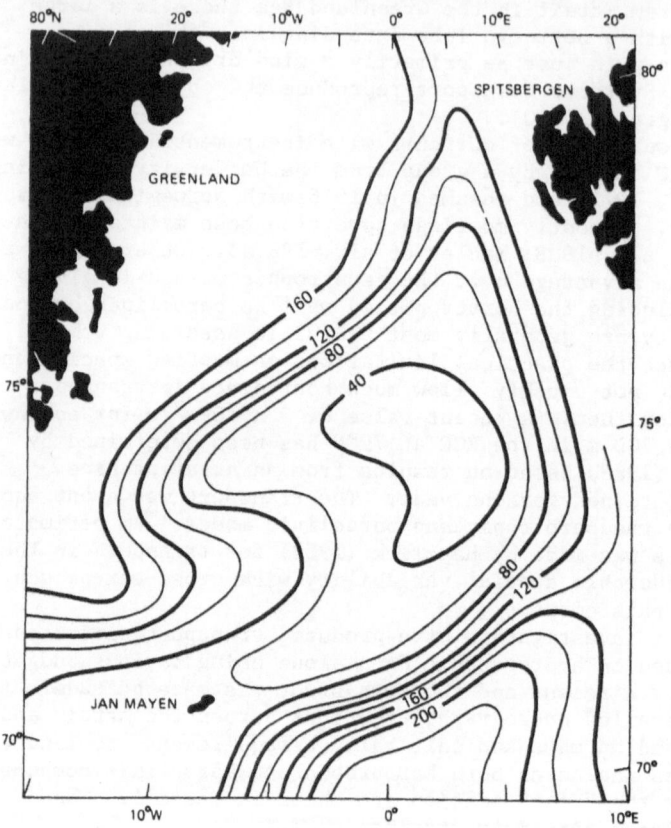

Fig. 15 - Dynamic topography (dyn. cm., 0/800 db) of the Greenland
Sea (Carmack, 1986).

TABLE I. Transport through Fram Strait (Adapted from Rudels, 1987).

West Spitsbergen Current	
Atlantic Water	1.9×10^9 kg/s
Deep Water	1.1
East Greenland Current	
Polar Surface Water	-0.9
Atlantic and Modified Atlantic Water	-1.7
Deep Water	-1.5
Ice	-0.08
Net Transport	-1.2

    Surface drifters at depths of 100 and 200 m were tracked by Gascard et al. (1988) during the MIZEX 84 project. One group of drifters launched near Spitsbergen drifted southwestward across Fram Strait until they reached the Greenland shelf edge and then followed bottom contours southward with the EGC. These data emphasize the presence of a cross-strait surface flow. A few drifters launched off the south coast of Spitsbergen travelled northward in the WSC. A buoy drogued at 30 m was tracked by satellite during Winter MIZEX in 1987 as it travelled north-ward in the open Greenland Sea, then turned west and south into the EGC (MIZEX '87 Group, 1989). Again, like the mooring results, surface drifter and Sofar float data emphasize the presence of eddies nearly everywhere in Fram Strait.
    Circulation in the strait is represented schematically in Fig. 16. The flow of ice and polar water generally follows the ice edge south-westward across the strait and then southward parallel to the Greenland shelf margin. Atlantic Water travels northward parallel to the Spitsbergen shelf edge until it divides into two main branches. One branch recirculates southward as the RAC. Another branch turns eastward around the north coast of Spitsbergen. This latter branch becomes the subsurface Atlantic boundary current shown in Fig. 7 and supplies the Atlantic Layer which spreads throughout the Arctic Ocean. Its high-salinity core can be traced along the shelf edge east of Spitsbergen (Anderson et al., 1989). Still a third branch, not shown in the diagram, following the contours of the Yermak Plateau north of Spitsbergen has been suggested by Perkin and Lewis (1984) on the basis of hydrographic profiles, but its existence has been shown to be doubt-ful on the basis of later more detailed hydrographic surveys (Bourke et al., 1988).

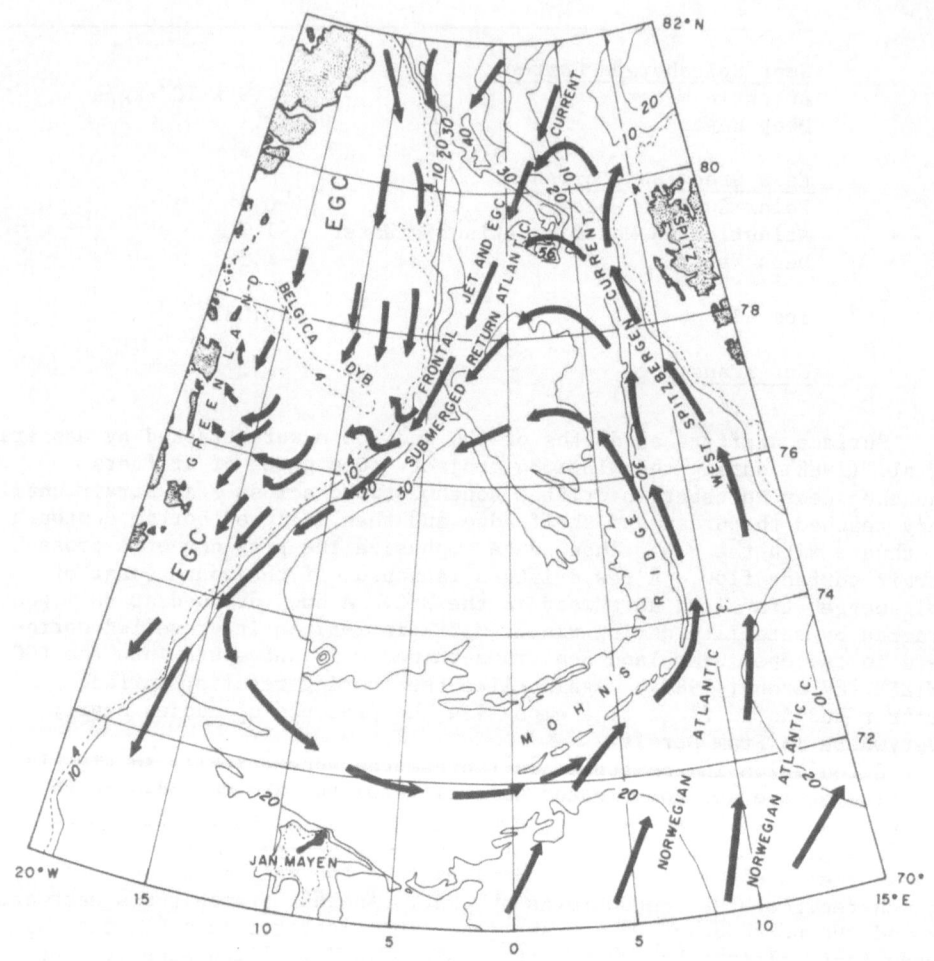

Fig. 16 - Schematic circulation in Fram Strait. Arrows represent mean
currents. Light lines are depth contours in hundreds of meters
(Paquette et al., 1985).

Mesoscale Variability

There is a wide range of mesoscale motion and hydrographic structure in
Fram Strait which is superimposed on the circulation and mean oceanographic fields. Although much of the eddy structure is transient in
time and propagating with the mean currents, the most detailed study
of mesoscale motion has probably been given to a feature which appears
to be nearly stationary in position. Evidence from Landsat imagery
shows that eddies exist all along the ice edge but one in particular
has been observed repeatedly in the same location centered near 79°40'N
1°00'E above the complex topography in the vicinity of the Molloy Deep.
This cyclonic eddy with a diameter of 60 km has been explained in
terms of topographic vortex stretching by Smith et al. (1984), but it
has also been suggested that it originates in the baroclinic
instability of the Polar Front (Wadhams and Squire, 1983). A detailed
study showed the baroclinicity extending to depths of at least 900 m
with a swirl transport of $10^6 m^3/s$ (Bourke et al., 1987). The most
recent survey during the CEAREX 1989 expedition failed however to detect
this eddy (Manley, pers. comm.) so it is evidently not a permanent
feature but perhaps only a frequently recurring one.
     Intensive oceanographic sampling from ships, aircraft, satellites
and ice camps during the Marginal Ice Zone experiment of 1983 and 1984
described the mesoscale spatial variability in this region and revealed
a total of 14 eddies occurring in the ice-free ocean, along the ice
edge and beneath the ice cover (Johannessen et al., 1987; Manley et al.,
1987b). Typical eddy diameter was 20 to 40 km and typical rotation
was cyclonic with maximum rotational speed often found below the surface
at velocities as great as 40 cm/s. Five of the eddies were advected
with the mean flow at translation speeds of 1-5 km/day. The others
remained in nearly the same locations, apparently trapped over bottom
topography. Several of these eddies were also tracked with surface
Argos buoys and with subsurface SOFAR floats (Gascard et al., 1988).
A new feature, a vortex pair, was seen with SAR imagery during the
Winter MIZEX of 1987. This mushroom-shaped current carried a jet of
ice into the open water east of the EGPF (MIZEX '87 Group, 1989).
     Year-long current records from the WSC showed that wind forcing
was an important contributor to variability there (Hanzlick, 1983).
Comparison with numerical models suggested that baroclinic, but not
barotropic, instability was an important contributor to variability
on time scales of 3 to 4½ days and length scales of 30 to 50 km. Other
moorings deployed for one month on either side of the EGPF (Manley et
al., 1987a) showed strong wind influence east of the front, but west
of the front in the EGC fluctuations were related to eddies generated
upstream and advected with the mean flow. A variety of mesoscale
structure with scales of days to weeks was seen to be advected by
the EGC by Foldvik et al. (1988) in a study of four year-long records
from moorings in the EGC. Analysis of this last data set led to the
conclusion that although there was abundant eddy motion, its
contribution to heat flux across the EGPF was small which in turn
suggests that the contribution of baroclinic instability to eddy

generation was small.

The processes by which this mesoscale motion is generated and the
source locale of the eddies are still a matter for discussion. It is
now generally accepted that there is a semipermanent eddy in the
vicinity of the Molloy Deep which is trapped in that location by the
complex bottom topography, but it is not clear whether the eddy was
formed there by vortex stretching or whether eddies formed elsewhere
were advected into that location and then trapped there. The only
serious contenders as sources for the propagating eddy field appear to
be either barotropic or baroclinic instability or a combination of the
two. Other possible sources such as differential Ekman pumping along
a meandering ice edge (Hakkinen, 1986) and internal ice dynamics
(Killworth and Paldor, 1985) have been suggested but have received
little support yet from field evidence.

A theoretical study of boundary current instability by Jones (1977)
defines a critical width for a two-layer system in which the upper
layer forms a wedge abutting against a vertical wall. For widths
greater than this critical value, the current is unstable. Wadhams
et al. (1979) calculated the critical width in the EGC to be 25 km
and considered the EGC to be much broader than this and hence unstable.
The same theory has been used to reach the opposite conclusion by
Gascard et al. (1988) since the MIZEX studies characterized the EGC
as a jetlike feature slightly less than 25 km wide and therefore stable
according to the same theory. A more comprehensive analysis must however take into account the stabilizing influence of the bottom slope.
Comparison of a theory for stability of a front intersecting the
continental slope (Flagg and Beardsley, 1978) with MIZEX observations
by Manley et al. (1987a) led to the conclusion that the EGPF is stable.
These results, along with the conclusion that baroclinic instability
is unimportant, reached by Foldvik et al. (1988) on the basis of low
eddy heat exchange, indicate that baroclinic instability within the
EGC is not likely to be the source of the eddies observed there.

Although the EGPF along the Greenland shelf edge is evidently
baroclinically stable, the more northern part of the front which
crosses the deep part of the Strait between 79° and 80°N may be unstable. A baroclinic instability theory for an exponentially-stratified
ocean with exponential shear was compared with a zonal section at $87\frac{1}{2}°$N
by Hunkins (1981) who found the cross-strait flow to be unstable with
an e-folding time of 18 days for the fastest growing half wavelength
of 30 km.

In the case of the WSC, Hanzlick (1983) determined through a
comparison of observations with a two-layer theory that baroclinic
instability was a possible source of variability on time scales of
3-4 days and wavelengths of 35-40 km although there was no pronounced
spectral indication to support this suggestion. On the other hand
barotropic instability was concluded to be of little importance in
the WSC.

A plausible hypothesis seems to be that most of the eddies in the
EGC have been generated by baroclinic instability in the WSC and cross-
strait front, then carried into the EGC by the mean flow where they are

passively advected southward.

Forces driving the large-scale circulation

In the Arctic Ocean there is a net production of freshwater as a result of river runoff and excess precipitation over evaporation while in temperate latitudes of the Atlantic Ocean there is a net production of salt since evaporation exceeds precipitation there. The density difference thus created leads to a meridional convective circulation. There is an opposing effect of temperature on density which is important south of Fram Strait but haline effects dominate within and north of the strait where temperatures are relatively low. Density-driven exchange was considered to be an important part of Fram Strait circulation by Nansen (1902) and has continued to be considered important by many investigators since.

Prevailing northerly winds over the strait drive ice and the upper water layers southward through the strait. A recent study of ice drift within the strait showed however, that ocean currents account for two to five times as much mean ice motion as do winds (Moritz and Colony, 1988). So local winds are only a minor factor in forcing ice and ocean currents in the ice-covered western side of the strait. On the eastern side where open water persists up to 80-82°N, the WSC flows northward against the opposing mean winds. Fram Strait circulation therefore must be forced by a mechanism other than local winds.

Although local wind effects are apparently of small importance within Fram Strait, they are an important forcing agent in the Norwegian-Greenland Sea to the south. These more distant winds were held by Greisman and Aagaard (1979) to be the cause of exchange in the strait even though local winds are ineffectual. The geostrophic circulation of the Norwegian-Greenland Sea closely resembles the Sverdrup transport derived from the curl of observed mean wind stress for a basin with a level floor (Aagaard, 1970). The gyre is presumably closed on the western side by that part of the EGC south of Fram Strait. It is speculated that some of the Sverdrup flow is ducted out of the directly wind-driven area by topography and steered into the Arctic Ocean as the WSC. This hypothesis of indirect wind forcing does not seem to have been tested yet although numerical models with realistic topography would be able to check aspects of the hypothesis.

The EGC at 79°N within the strait is observed to have a strong barotropic component in recordings made with moored current meters over both year-long and monthly periods (Foldvik et al., 1988; Manley et al. 1987a). This is considered evidence for considerable wind forcing since density-driven convective circulation is expected to be essentially baroclinic. The year-long records showed the barotropic component to be comparable with the baroclinic which is puzzling since the ice drift studies mentioned earlier show wind forcing to be only about one half as important as current advection.

Another possible driving force is the atmospheric pressure gradient which is in the correct sense to force a southward barotropic current.

Preliminary calculations using mean pressure data show the resulting current to be small (less than 1 cm/s) but adequate to account for the net transport of 1.2×10^9 kg/s calculated by Rudels (Table 1).

The generation of mean currents within Fram Strait by the nonlinear rectification of tidal currents within frictional boundary layers has not yet received serious consideration. The arctic seas are for the most part characterized by tides of small amplitude with a rise and fall of less than one foot along the coast. Recently however, diurnal tidal currents with peak velocities of 30 cm/s have been observed over the flanks of the Yermak Plateau north of Fram Strait (Hunkins, 1986). The tidal regime is predominantly semidiurnal in this region so these diurnal tidal currents were unexpected. Theoretical studies support an interpretation of resonant forcing by local topography of the weak deep-sea diurnal tides (Hunkins, 1986; Chapman, 1989). Tidal oscillations of this magnitude may be large enough to generate significant residual currents through nonlinear effects in boundary layers. Such residual currents have been demonstrated in estuaries and over such features as Georges Bank but their importance as a contributing mechanism to the exchange through Fram Strait still awaits demonstration.

Modeling

Numerical models of coupled air-sea-ice circulation have been developed with the objective of improving long-term forecasting of sea ice thickness and extent. Two of these models cover both the Arctic Ocean and Norwegian-Greenland Sea and thus include Fram Strait. They incorporate multi-level oceans and reproduce observed sea ice coverage with considerable realism. In the model of Hibler and Bryan (1987) the ocean is diagnostically constrained on a three-year time scale to remain close to observed temperature and salinity values. The model of Semtner (1987) is similar in many respects but the ocean is not so constrained, allowing long-term ocean changes to develop. His model is driven by atmospheric forcing based on observed monthly mean values and by prescribed inflow through the Faeroes-Shetland Channel. A 20 year integration produces an outflow through Fram Strait in the upper layers and an inflow in the deep layers. Both ice and currents in the upper level converge and accelerate as the opening is approached. The Atlantic Water inflow originates in both the Norwegian Coastal Current and the Greenland gyre. After passing through the strait, it travels along the Siberian shelf edge in conformity with present ideas about Atlantic Water intrusion into the Arctic Ocean. Thus combined wind- and thermohaline-forcing yields a two-layer exchange. Published results focus on the ice cover but more details on the strait currents could undoubtedly be derived from past or future runs of the model.

Another type of model developed by Stigebrandt (1981) is based on those used in estuarine research. Water, salt and heat exchange are balanced and an ice cover is included. Currents are assumed to be geostrophic and entrainment is modeled using the Kato-Phillips formula. Specified parameters are river runoff, inflow through Bering Strait and

ice export. The model was employed to predict ice cover under
various conditions including the decrease in freshwater input which
would occur if Siberian river water were to be diverted southward
for agriculture as has been proposed by Soviet planners.

Laboratory Models

The wide range of ocean phenomena in Fram Strait encourages the use of
analytical and laboratory models as well as numerical models to gain
understanding of specific physical processes. Little work in these
types of modeling has been specifically oriented toward this strait
although there is a wealth of studies on shelf edge dynamics,
instability, frontal dynamics and convection which has relevance to
the problems in this region. Rotating tank experiments are especially
useful for visualizing fluid behavior and clarifying special aspects
of exchange processes. A version of the EGC has been demonstrated in
laboratory experiments by Wadhams et al. (1979) in which dyed freshwater is initially contained in a cylinder at the center of a tank
filled with salt water. When the cylinder is removed freshwater spreads
radially outward over the salt water until it encounters a radial
barrier. A narrow boundary current then flows outward (southward)
along this meridional wall. The width of the boundary current is close
to the radius of deformation. Based on the resemblances between the
laboratory boundary current and the EGC they concluded that the EGC
is driven by density differences. This laboratory boundary current
closely resembles that in a steady frictionless analytical model in
which potential vorticity is uniform in the upper layer (Manley et al.,
(1987a). The interface and velocity have exponential profiles with
e-folding widths scaled by the radius of deformation. Using values
appropriate to the EGC, a volume transport of $1.1 \times 10^6 m^3/s$ is found.
The transport of this uniform potential vorticity model is in reasonable
agreement with the value of $0.9 \times 10^6 m^3/s$ for the EGC polar water
arrived at by Rudels (Table 1) on the basis of budget calculation.
However, it is much less than the $3 \times 10^6 m^3/s$ found from mooring data
by Foldvik et al. (1988).
     Experiments using a rotating tank divided into two basins separated
by a wide gap have been conducted by the author and Jack Whitehead to
model aspects of the exchange through Fram Strait. These were lock-
exchange experiments with a sliding barrier initially separating layers
of different density which had been dyed for visualization. After
removal of the barrier and passage of the initial surge there was a
period during which quasi-steady boundary currents develop. The first
experiments used a single shallow layer on one side of the barrier
overlying a deep layer common to both basins. Results demonstrated
the importance of boundary currents and produced an outflow which had
some of the characteristics of the EGC, confirming the earlier work
of Wadhams et al. (1979) but with a geometry more closely resembling
Fram Strait. Next two shallow layers of different densities were placed
on either side of the barrier, again over a common deep layer. The deep

layer restricts the dynamics of the experiment to an f-plane since scaling analysis shows that the beta-effect is not significant in Fram Strait (Quadfasel et al., 1987). Experiments were run in pairs with first the freshwater layer dyed and then the intermediate density layer dyed but with all other conditions identical for each pair. This allowed the flow pattern for each layer to be traced separately.

The appearance of the shallow layer in one of the experiments one minute after the barrier was removed is shown in Figs. 17 and 18. Boundary currents were observed to flow along the right-hand wall into the opposite basin on both sides of the gap in a symmetrical pattern mimicking the EGC and WSC. These boundary currents along the walls were close to the internal radius of deformation in width and relatively stable, traveling completely around the rim and returning along the opposite wall with little change.

At the moment the barrier is lifted the cross-strait front is straight but it soon develops lobes which transfer waters through the strait into opposite basins. These lobes eventually evolve into isolated vortices. The free front across the strait may be considered a segment of a circular front of the type studied by Griffiths and Linden (1982). They found continuous circular fronts to be unstable both experimentally and theoretically. Their experiments suggest that in our experiment we can expect instability of the cross-strait front with a wavelenth of 56 cm for maximum amplification. This value is close to the gap width of 40 cm and the instability of this free front consequently seems to be of the same type observed in a continuous circular front. The lobe in contact with the left-hand wall usually develops most rapidly probably due to the inertial effects explored numerically by Hermann et al. (1989).

In Fig. 17 the dyed upper freshwater intrudes into the other basin as a wall boundary current and as an anticyclonic vortex with a diameter on the order of the gap width. In the intermediate layer, which is dyed in Fig. 18, there is a smaller anticyclonic vortex developing along the left-hand wall with respect to that layer. Also a backward-breaking wave can be seen developing into a vortex pair. In these experiments there is an inverse relationship between instability of the cross-strait front and the radius of deformation with the smaller values associated with greater instability. This qualitative observation is in accord with several stability analyses for somewhat similar situations although no analysis yet exists which can be directly compared with our results.

Problems for future study

There remain a number of questions on the mechanisms which generate and control the exchange through Fram Strait. One problem concerns the relative importance of wind- and buoyancy-forcing. Another concerns the part that the mesoscale variability, which is so widely observed in this region, plays in the transfer of salt and heat between the Arctic and Atlantic Oceans. Some of the exchange occurs in the relatively coherent boundary currents and some by advected eddies but

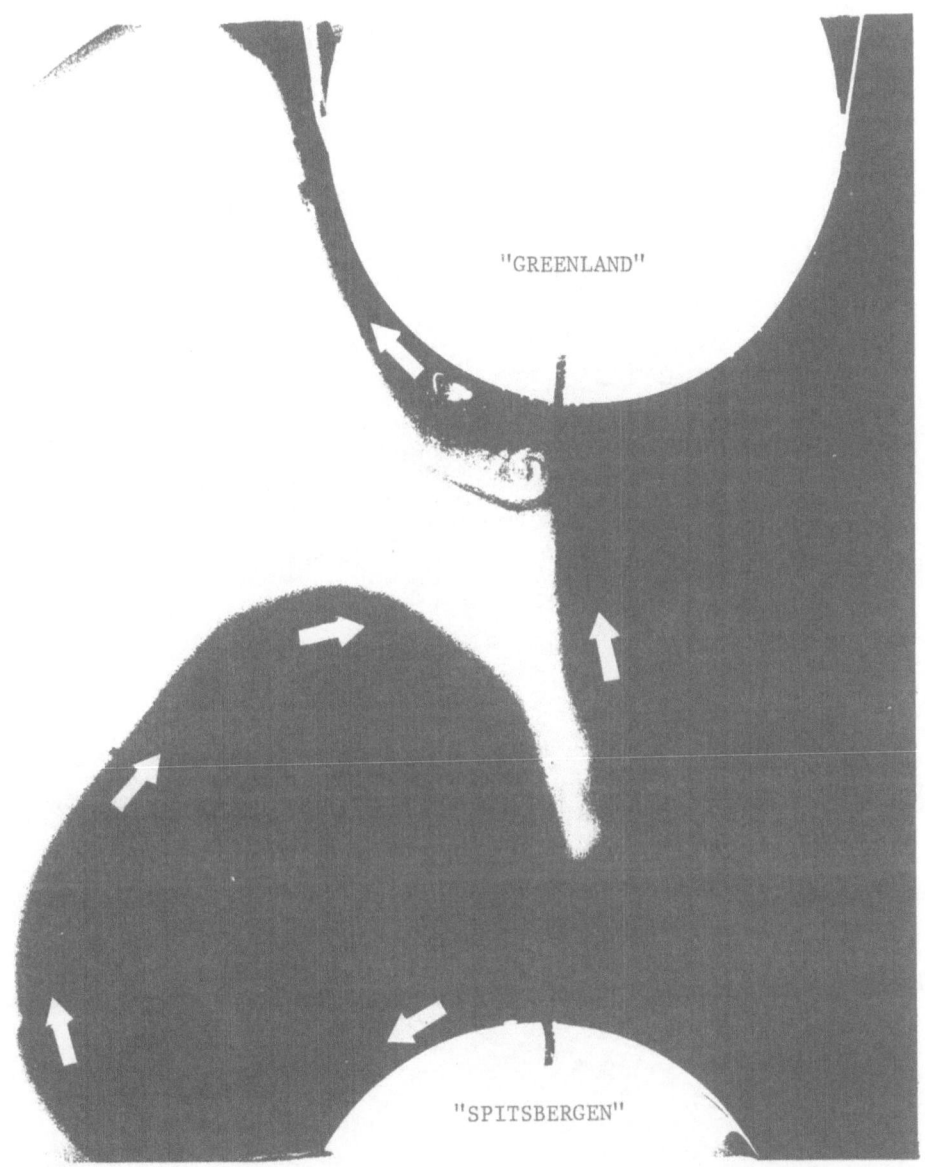

Fig. 17 – Lock exchange through a broad gap, h = 5.2 cm, H = 25.2 cm, f = 0.82 Hz, $\Delta\rho_1/\rho$ = 0.004, $\Delta\rho_2/\rho$ = 0.0022, R_c = 5.5 cm. Dyed freshwater layer on right. Clear intermediate layer on left. Note "EGC" wall boundary current in upper left and unstable lobe in lower left.

88

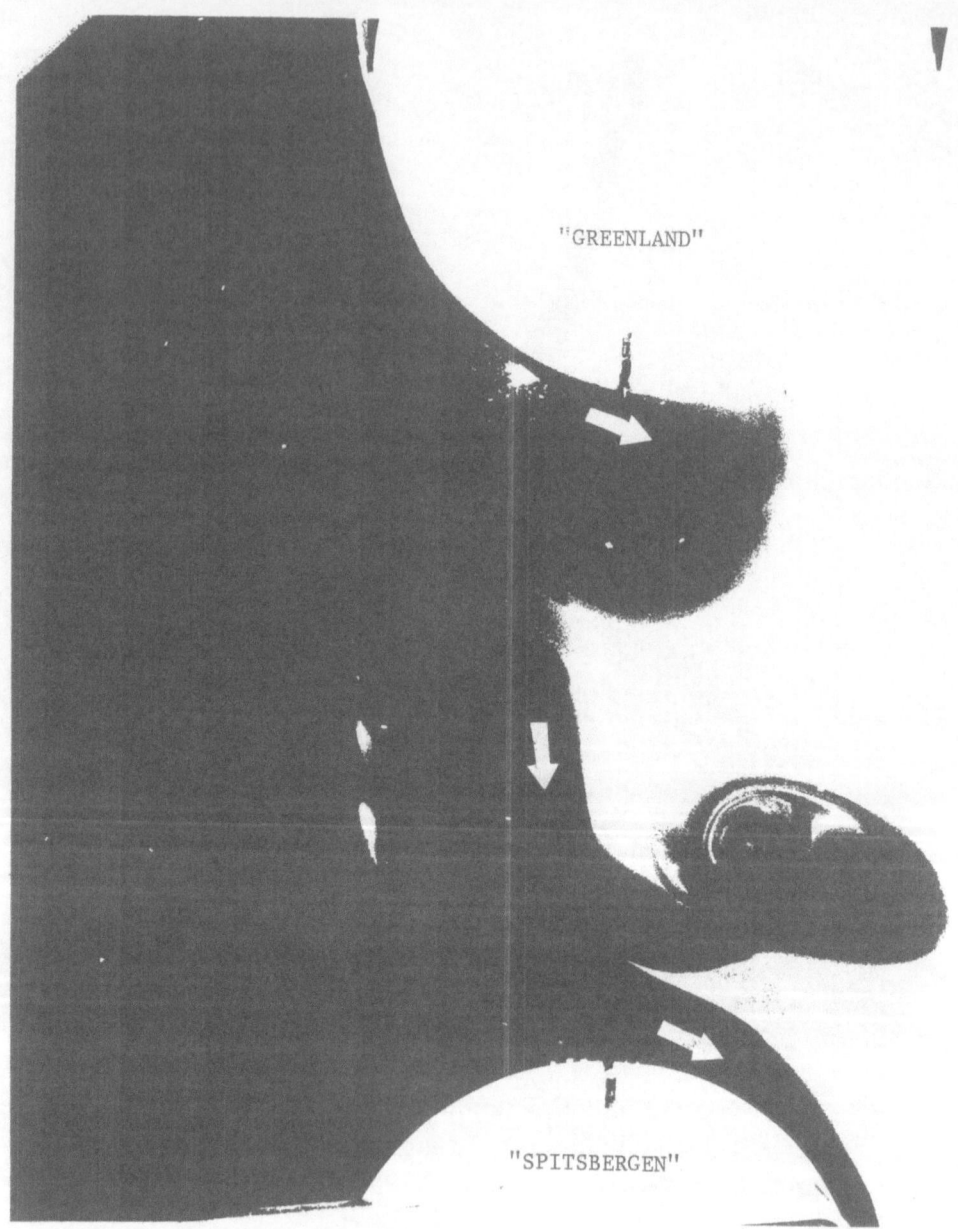

Fig. 18 - Dyed intermediate layer on left. Clear freshwater layer on
right. Note "Atlantic" subsurface boundary current along wall in
lower right and unstable vortices developing on cross-strait front.

we have little idea which method is dominant. The generation of mesoscale eddies has only been researched slightly in this region and
further work on their origin will likely help answer the previous
question on their importance in exchange. Then there are specific
geographic features of the circulation in Fram Strait which invite
inquiry. What controls the recirculation of Atlantic Water? Is it
only the remote northern extremity of the wind-driven Greenland gyre
or is it a result of particular topographic features within the strait?
Also the shelves seem important in guiding the EGC and WSC. Do they
play an essential part or would the exchange be little changed if
there were only wall boundaries.

The presence of sea ice may be a critical factor in maintaining
the mean location of the EGPF and the observed circulation. On the
eastern side of Fram Strait, warm water flowing northward meets the
pack ice driven southward by the prevailing winds. The freshwater
produced by the melting is apparently driven westward as an Ekman flow
which then becomes entrained in the EGC (Untersteiner, 1988). The
simplest concept embracing this idea is that all of the freshwater
from rivers is frozen and crosses the Arctic Ocean without mixing to
be finally released again as freshwater north of Spitsbergen. In
terms of estuary dynamics this concept may be thought of as an
"undermixed" model in contrast to the "overmixed" model which has been
applied to the Strait of Gibraltar (Bryden and Stommel, 1984).

Summary

While Fram Strait is much wider than those constrictions usually
designated by the term strait, it does share in common with narrower
straits many features such as coastal boundary currents, instabilities
and recirculation. The exchange through Fram Strait is primarily
by narrow boundary currents: the EGC carries cold fresher polar waters
southward while the WSC carries warmer, more saline Atlantic waters
northward. There is considerable mesoscale variability in this
strait including both transient and semipermanent eddies but the
importance of mesoscale features to the exchange is still not known.
It appears that most of the eddies in the EGC are generated in the
cross-strait front which is baroclinically unstable and then advected
southward. Density differences between the fresher Arctic Ocean and
more saline Atlantic Ocean is considered to be the primary driving
force although wind forcing also plays a supporting part. Differences
in atmospheric pressure between the Arctic Ocean and Greenland Sea
may be the driving force for the net southward transport through
Fram Strait. It is also possible that some of the exchange may be
caused by tidal rectification since unusually large diurnal tidal
currents have been observed over the Yermak Plateau just north of the
strait. No quantitative studies have been made but it is possible
that the residual circulation resulting from nonlinear rectification
of these diurnal currents contributes to the mean exchange.

Sea ice is a unique aspect of Fram Strait especially on the
western side where the EGC carries ice far southward along the coast

of Greenland. On the eastern side, the warm WSC melts sea ice, maintaining the ice margin at the latitude of Spitsbergen. Melting sea ice contributes to the freshwater supply and is undoubtedly of importance in maintaining the front across the strait.

Numerical models have been developed which include the ice cover as well as fairly realistic hydrography and topography. The results reported have been chiefly concerned with behavior of the ice pack. Laboratory experiments have been run which model the buoyancy exchange through an idealized broad gap. These experiments reproduce some of the observed features such as boundary currents resembling the EGC and WSC as well as instability and the production of vortices.

Acknowledgements

The author is grateful to the Office of Naval Research for their support of the research which made this review possible. I am indebted to colleagues too numerous to mention whose conversations with me have helped to form the contents of this paper. It is a pleasure to recognize the typing and drafting of Mei Be Hunkins. This is contribution no. 4556 of Lamont-Doherty Geological Observatory of Columbia University.

References

Aagaard, K. (1970) 'Wind-driven transports in the Greenland and
Norwegian Seas', Deep-Sea Res. 17, 281-291.
Aagaard, K. (1982) Chap. 3 'Inflow from the Atlantic Ocean to
the Polar Basin', in L. Rey (ed.), The Arctic Ocean, John Wiley,
N.Y., pp. 69-81.
Aagaard, K., Darnall, C. and Greisman, P. (1973) 'Year-long
current measurements in the Greenland-Spitsbergen passage',
Deep-Sea Res. 20, 743-6.
Aagaard, K., Coachman, L. and Carmack, E. (1981) 'On the
halocline of the Arctic Ocean', Deep-Sea Res. 28A, 529-545.
Aagaard, K., Swift, J. and Carmack, E. (1985) 'Thermohaline
circulation in the Arctic Mediterranean Seas', J. Geophys. Res.
90(C5), 4833-4846.
Anderson, L., Jones, E., Koltermann, K., Schlosser, P., Swift, J.
and Wallace, D. (1989) 'The first oceanographic section across
the Nansen Basin in the Arctic Ocean', Deep-Sea Res. 36, 475-482.
Bourke, R., Tunnicliffe, M., Newton, J., Paquette, R. and
Manley, T. (1987) 'Eddy near Molloy Deep revisited', J. Geophys.
Res. 92(C7), 6773-6776.
Bourke, R., Wiegel, A.M. and Paquette, R. (1988) 'The westward
turning of the West Spitsbergen Current', J. Geophys. Res.
93(C11), 14,065-14,077.
Bryden, H. and Stommel, H. (1984) 'Limiting processes that
determine basic features of the circulation in the Mediterranean
Sea', Oceanologica Acta 7(3), 289-296.
Carmack, E. (1986) Chap. 10 'Circulation and mixing in ice-covered
waters', in Untersteiner, N. (ed.), The Geophysics of Sea Ice,
Plenum Publ. Corp., New York, 641-712.
Chapman, D. (1989) 'Enhanced subinertial diurnal tides over
isolated topographic features', Deep-Sea Res. 36, 815-824.
Flagg, C. and Beardsley, R. (1978) 'On the stability of the shelf
water/slope water front south of New England', J. Geophys. Res.
83(C9), 4623-4631.
Foldvik, A., Aagaard, K. and Tørressen, T. (1988) 'On the velocity
field of the East Greenland Current', Deep-Sea Res. 35(8),
1335-1354.
Gascard, J.-C., Kergomard, C., Jeannin, P.-F. and Fily, M. (1988)
'Diagnostic study of the Fram Strait Marginal Ice Zone during
Summer from 1983 and 1984 Marginal Ice Zone Experiment
Lagrangian observations', J. Geophys. Res. 93(C4), 3613-3641.
Greisman, P. and Aagaard, K. (1979) 'Seasonal variability of the
West Spitsbergen Current', Institute of Oceanographic Sciences,
Wormley, U.K., unpublished document, Ocean Modeling 19, 3-5.
Griffiths, R. and Linden, P. (1982) 'Laboratory experiments on
fronts', Geophys. Astrophys. Fluid Dynamics 19, 159-187.
Hakkinen, S. (1986) 'Ice banding as a response of the coupled
ice-ocean system to temporally varying winds', J. Geophys.
Res. 91, 5047-5053.

Hanzlick, D.J. (1983) 'The West Spitsbergen Current: transport, forcing and variability', Ph.D. Dissertation, Univ. of Wash., 127 pp.

Hermann, A., Rhines, P. and Johnson, E. (1989) 'Non-linear Rossby waves in a channel: beyond Kelvin waves', J. Fluid Mech. 205, 469–502.

Hibler III, W. and Bryan, K. (1987) 'A diagnostic ice-ocean model', J. Phys. Oceanogr. 17, 987–1015.

Hunkins, K. (1981) 'Arctic ocean eddies and baroclinic instability', Technical Report CU-2-81, Lamont-Doherty Geol. Obs., 39 pp.

Hunkins, K. (1986) 'Anomalous diurnal tidal currents on the Yermak Plateau', J. Mar. Res. 44, 51–69.

Johannessesn, J.A., Johannessen, O.M., Svendsen, E., Shuchman, R., Manley, T., Campbell, W.J., Josberger, E.G., Sandven, S., Gascard, J.C., Olaussen, T., Davidson, K. and Van Leer, J. (1987) 'Mesoscale eddies in the Fram Strait Marginal Ice Zone during the 1983 and 1984 Marginal Ice Zone experiments', J. Geophys. Res. 92(C7), 6754–6772.

Jones, S. (1977) 'Instabilities and wave interactions in a rotating two-layer fluid', Ph.D. Thesis, Univ. of Cambridge, 295 pp.

Killworth, P. and Paldor, N. (1985) 'A model of sea-ice front instabilities', J. Geophys. Res. 90, 883–888.

Laktionov, A.F. (1959) 'Bottom topography of the Greenland Sea in the region of Nansen's Sill', Priroda 101, 95–97. (Eng. Transl. by Hope, E.R., DRB Canada).

Lemke, P. and Manley, T. (1984) 'The seasonal variation of the mixed layer and the pycnocline under polar sea ice', J. Geophys. Res. 89(C4), 6494–6504.

Manley, T. and Camp, D. (1985) 'Physical oceanography report: camp-based and helicopter-based STD data from the drifting ice station Fram III', Tech. Rpt., L-DGO-85-8, 335 pp.

Manley, T., Hunkins, K. and Muench, R. (1987a) 'Current regimes across the East Greenland Polar Front at 78°40'N during summer 1984', J. Geophys. Res. 92(C7), 6741–6753.

Manley, T., Villanueva, J., Gascard, J., Jeannin, P., Hunkins, K. and Van Leer, J. (1987b) 'Mesoscale oceanographic processes beneath the ice of Fram Strait', Science 236, 432–434.

McLaren, A. (1988) 'The under-ice thickness distribution of the Arctic Basin as recorded in 1958 and 1970', J. Geophys. Res. 94(C4), 4971–4983.

MIZEX '87 Group (1989) 'MIZEX East 1987', EOS 70(17), 545.

Moritz, R.E. and Colony, R. (1988) 'Statistics of sea ice motion, Fram Strait to North Pole', Proc. Seventh Int'l Conf. on Offshore Mechanics and Arctic Engineering, Vol. IV, Amer. Soc. Mech. Eng., New York, 75–82.

Nansen, F. (1902) The Norwegian North Polar Expedition, 1893–1896, Sci. Results vol. III, Longman, Green and Co., London (Reprinted by Greenwood Press, 1969).

NOC (1986) Sea ice climatic atlas: Vol. II, Arctic East NAVAIR
    50-1c-541, Naval Oceanography Command, NSTL, MS, 39529-5000.
Paquette, R., Bourke, R., Newton, J. and Perdue, W. (1985) 'The
    East Greenland Polar Front in autumn', J. Geophys. Res. 90(C3),
    4866-4882.
Perkin, R.G. and Lewis, E.L. (1984) 'Mixing in the West Spitsbergen
    Current', J. Phys. Oceanogr. 14, 1315-1325.
Perry, R.K. and Fleming, H.S. (1986) 'Bathymetry of the Arctic
    Ocean', Naval Research Lab.-Acoustics Div., Map and Chart
    Series MC-56, The Geol. Soc. of Amer., Inc., P.O. Box 9140,
    Boulder, CO, 80301.
Quadfasel, D., Gascard, J.-C. and Koltermann, K. P. (1987) 'Large-
    scale oceanography in Fram Strait during the 1984 Marginal Ice
    Zone Experiment', J. Geophys. Res. 92(C7), 6719-6728.
Rudels, B. (1987) 'On the mass balance of the Polar Ocean, with
    special emphasis on the Fram Strait', Norsk Polarinstitutt,
    Skrifter 188, 53 pp.
Semtner, A. (1987) 'A numerical study of sea ice and ocean
    circulation in the Arctic', J. Phys. Oceanogr. 17, 1077-1099.
Smith, D., Morison, J., Johannessen, J. and Untersteiner, N. (1984)
    'Topographic generation of an eddy at the edge of the East
    Greenland Current', J. Geophys. Res. 89(C5), 8205-8208.
Stigebrandt, A. (1981) 'A model for the thickness and salinity of
    the upper layer in the Arctic Ocean and the relationship
    between the ice thickness and some external parameters',
    J. Phys. Oceanogr. 11, 1407-1422.
Swift, J.H. (1986) Chap. 5 'The Arctic Waters', in Hurdle, B. (ed.),
    The Nordic Seas, Springer-Verlag, New York, pp. 129-153.
Thorndike, A. S. and Colony, R. (1982) 'Sea ice motion in response
    to geostrophic winds', J. Geophys. Res. 87(C8), 5845-5852.
Treshnikov, A.F., Nikiforov, Ye.G. and Blinov, N.J. (1977)
    'Results of oceanological investigations by the "North Pole"
    drifting stations', Polar Geography 1(1), 22-40.
Untersteiner, N. (1988) 'On the ice and heat balance in Fram
    Strait', J. Geophys. Res. 93(C1), 527-531.
Vinje, T. and Finneskåsa, O. (1986) 'The ice transport through
    Fram Strait', Skrifter Nr. 186, Norsk Polarinstitutt, Oslo,
    39 pp.
Wadhams, P., Gill, A. and Linden, P. (1979) 'Transects by
    submarine of the East Greenland Polar Front', Deep-Sea Res.
    26A, 1311-1327.
Wadhams, P. and Squire, V.A. (1983) 'An ice-water vortex at the
    edge of the East Greenland Current', J. Geophys. Res. 88(C5),
    2770-2780.

TIDAL CURRENTS AND TRANSIENT PHENOMENA IN THE STRAIT OF MESSINA: A REVIEW

F. BIGNAMI
Dipartimento di Fisica
Università degli Studi di Roma "La Sapienza"
P.le A. Moro, 2
00185 Rome
Italy

E. SALUSTI
INFN, Dipartimento di Fisica
Università degli Studi di Roma "La Sapienza"
P.le A. Moro, 2
00185 Rome
Italy

ABSTRACT. The strait of Messina, separating Sicily from the Italian peninsula, has been studied since Homer's times. Scientists have always been fascinated by its strong currents generated by the opposition of tidal phase between the two contiguous Tyrrhenian and Ionian basins. Tidal barotropic and baroclinic currents, up to 2 m/s, transient barotropic eddies and nonlinear wave trains as detected from remote and in situ observations and in the light of numerical models, are here analyzed and discussed.

1. Introduction

The strait of Messina separates the Italian peninsula from the island of Sicily. The mean depth is about 120 m and the sill depth is 80 m. The smallest cross section (0.3 Km^2) is precisely over its sill, between Punta Pezzo and Ganzirri where the strait is $\approx$ 3 Km wide. The bottom slope running north into the Tyrrhenian Sea is much steeper than that of the southern side of the sill (Ionian Sea, fig. 1). Both slopes are regular.

The strait of Messina has been known since ancient times as an area of strong currents and vortices. The first documented hypothesis to explain the violent currents in the strait of Messina was made by Homer (around 800 B. C.) who thought that they were caused by two monsters, Scylla and Charybdis (Homer, Odyssey, 12th song, lines 80 - 114). Later, Aristotle (384 - 322 B. C.) tried to explain the oceanographic phenomena in this strait in terms of hollows in the sea floor and of an interaction between two opposing wind - generated currents (Aristotle, Problemata Physica, chap. 23).

More than 2000 years elapsed before modern scientific oceanographic measurements were carried out in this area. The French vice - consul in Messina, Ribaud, in 1824 A. D. gave a fairly detailed description of the currents in the strait, but no modern oceanographic

95

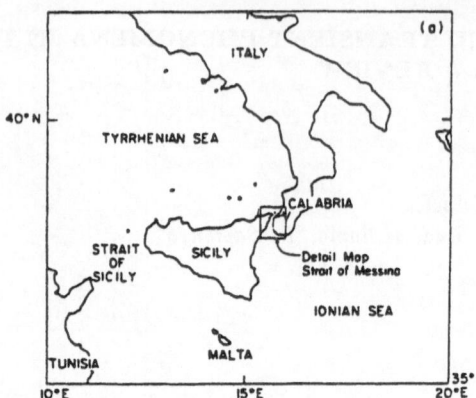

Figure 1a. Geographical location of the strait of Messina (after Hopkins et al., 1984).

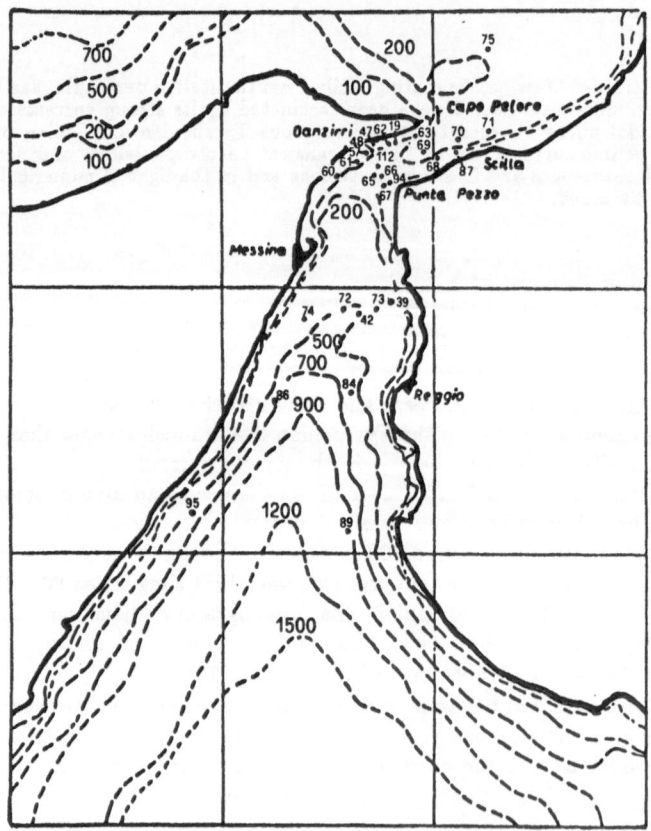

Figure 1b. Geographical description of the strait of Messina (after Vercelli, 1925) with the locations of Vercelli's current meter stations.

measurements were carried out before 1922 and 1923. During two cruises onboard the research vessel Marsigli, F. Vercelli made the first extensive oceanographical survey of this area. His data are still considered to be the most detailed and systematic set available on the strait of Messina.

The analysis and interpretation of Vercelli's data has been performed by Vercelli himself (1925), Defant (1940, 1961) and more recently by Brandolini et al. (1980) and by Hopkins et al. (1984): the current distribution within the strait of Messina is largely a result of semidiurnal tides. At first glance this is surprising, since the amplitude of the tide is known to be very small in the Mediterranean Sea, i.e. of the order of 10 cm. But in the strait of Messina a large gradient of tidal amplitude is encountered because the tides in the basin north of the strait ($\approx$ 16 cm) and south of the strait ($\approx$ 10 cm) are out of phase by almost 5 hours in $\approx$ 10 km (see also fig. 4).

Recently, Hopkins et al. (1984) pointed out that a complex barotropic - baroclinic phenomenon occurs inside the strait of Messina. In the neighboring basins, two different water layers (Atlantic water over Levantine Intermediate water, respectively AW and LIW in the following) can be found. The T - S diagram of fig. 2 shows the differences between these two water types as found on the two sides of the strait. Usually, the interface between these two water layers is at a depth of $\approx$ 150 m (Vercelli and Picotti, 1925). But near and above the sill of the strait, using time - averages of the existing data, Vercelli (1925) observed: (i) an upper layer of water moving southward ($\approx$ 10 cm/s as time - averaged

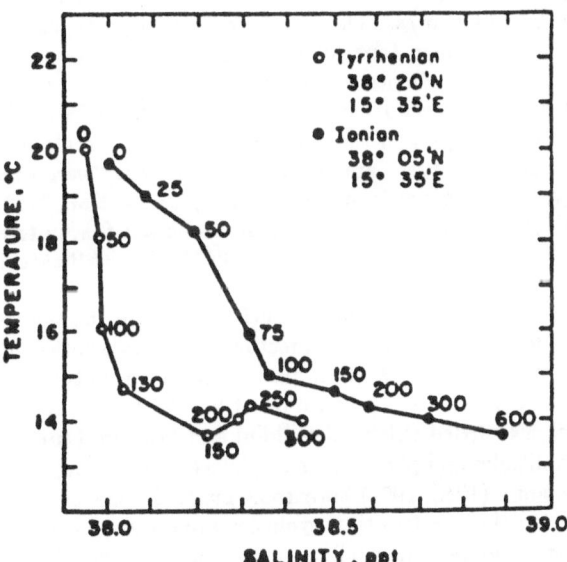

Figure 2. The vertical T - S structure from two hydrographic stations taken by Vercelli (after Hopkins et al., 1984).

velocity), (ii) a time - averaged interface at $\approx$ 30 m depth i. e. an uplift of $\approx$ 120 m from the adjoining basins, (iii) a lower layer of denser water moving northward ($\approx$ 12 cm/s time - averaged velocity).

This complex situation, closely matching Stigebrandt's (1977) treatment of barotropic forcing in baroclinic flow constrictions, can lead to strong currents. The general hydrographic situation has been discussed by Colacino et al. (1980). The tidal barotropic aspect of water motion was analysed by Defant (1961; fig. 4). More recently, Del Ricco (1982) applied a baroclinic two - dimensional vertical numerical model to Vercelli's March data set in order to determine the time - evolution of the interface between these two water layers. Hopkins et al. (1984) used Vercelli's original data to construct a fairly realistic description of the time evolution at the interface (see fig. 6). It is interesting to note that these theoretical computations showed the presence of internal waves of ≈ 100 m amplitude immediately south of the sill (a region where Vercelli was unable to make measurements because of nautical traffic). Del Ricco used a March hydrographic data set and found that the interface could also make jumps of ≈ 400 m depth in the same zone. The steady interface position over the sill was computed by Salusti et al. (1988; fig. 10).

As far as small scale features are concerned, Di Sarra, et al. (1987) described field observations of the eddies and turbulence related to the semidiurnal time - evolution of this interface. Their data are in good agreement with Del Ricco's (1982) and Hopkins et al.'s (1984) theoretical reconstructions. In particular, sonar imagery revealed the presence of small scale barotropic eddies of mixed water extending from the bottom to the surface in the sill area. Di Sarra et al. (1987) have called them "Bignami columns" after one of the authors of this paper, who first observed them. They are ≈ 400 m wide and the temperature inside them is ≈ 0.7 °C lower than the environment; their swirl velocity has been estimated to be ≈ 0.6 m/s. These features are of particular interest as a marine equivalent of eddies observed in tank experiments of steady flows over obstacles (Wei et al., 1975; Baines; 1977, 1979; Baines and Hoinka, 1985).

The presence of internal waves inside the strait is confirmed by experimental evidence outside the strait. Indeed, part of the energy of the large - scale internal movements is radiated away: packets of large - amplitude internal waves were observed north of the strait, near Cape Vaticano (Alpers and Salusti, 1983; Griffa et al., 1986; Sapia and Salusti, 1987; Nicolo' et al., 1989). The existence of these waves was first inferred from a Synthetic Aperture Radar (SAR) image obtained from the SEASAT - SAR satellite at an altitude of 800 Km on Sept. 15, 1978.

This satellite image also revealed another surface effect of the tide: bores or "tagli", as they are defined by local fishermen. These features are seen as areas of surface roughness, i.e. of low reflectivity, and appear as dark narrow strips across the breadth of the strait.

The present review begins with a description of Vercelli's data, together with the results of the spectral analysis carried out by Castaldini and Franzini (1979), giving estimates of the M_2 and M_4 amplitudes and phases from stations occupied longer than one day (section 2). In section 3 Defant's (1940, 1961) barotropic model for the tidal currents is discussed. The studies concerning the interface time evolution are reviewed in section 4. Then internal waves are described, as viewed by satellite and as observed with "in situ" measurements, both north and south of the strait (section 5). In particular, the internal non - linear waves are discussed in the light of the Korteweg - de Vries equation (Osborne and Burch, 1980). We conclude with a last section (section 6) dedicated to the open problems concerning the dynamics of the strait and an Appendix dedicated to detailed description of recent measurements concerning internal waves in the strait.

2. Vercelli's measurements

During two long cruises (August 11 - November 8, 1922 and April 2 - June 20, 1923), Vercelli (1925) studied the horizontal water motion and made some hydrographic samplings in more than 100 anchor stations (fig. 1b, table 1), of which 12 lasted for more than one day and one for 15 days (station 1) . The horizontal velocities were measured hourly at depths of 5, 10, 30, 50, 100 and 200 m. Vercelli used Ekman current meters (surface currents), Boccardo current meters (intermediate depths) and Mertz current meters for the lower depths. The degree of fit achieved by Vercelli's analysis is indicated by the comparison given in fig. 3 in which his tidal curve is superimposed on the more recent observations (Massi et al., 1979). Unfortunately his results have been overlooked, probably because his papers were never translated from Italian.

Vercelli Station Number	Bottom Depth	Time and Length.° hours	Tidal Component as Per-centage of Observed Energy	Depth of Observation	M_2 Amplitude. cm/s	M_2 Phase. hours	M_4 Amplitude. cm/s	M_4 Phase hours
73	373	May 15. 1923 (11)	— — — — —	5 10 30 100 175	87.5 78.1 55.2 26.7 20.2	−1.3 −1.3 −0.6 0.7 −5.3		
74	450	May 16. 1923 (10)	— — — — —	5 10 30 100 175	55.7 44.1 36.0 13.8 25.9	−0.3 −1.0 −0.5 1.2 0.1		
75	240	May 18. 1923 (10)	— — — — —	5 10 30 100 175	12.8 16.7 15.5 26.7 20.2	2.3 2.2 1.0 0.7 −5.3		
84	600	May 22. 1923 (10)	— — —	5 10 30	25.5 41.0 23.9	−1.3 −1.4 −1.7		
86	640	May 23. 1923 (8)	— — — .. —	5 10 30 100 175	27.0 22.0 18.0 12.0 9.0	0.9 1.0 2.2 −1.4 4.2		
87	86	May 31. 1923 (30)	70% 70% 60% 70%	5 10 30 50	91.3 85.0 89.1 77.5	1.0 1.1 0.9 1.2	3.5 4.1 8.3 6.0	−2.3 0.7 1.5 1.0
94	75	June 7.8. 1923 (36)	70% 50% 50% 60%	5 10 30 60	61.4 58.4 96.3 71.1	−5.7 −5.9 −5.8 5.8	11.1 5.8 7.5 2.3	−2.6 −1.4 2.7 1.7

Table 1. M_2 and M_4 tidal velocity characteristics observed in Vercelli's anchor stations lasting more than one day. Horizontal velocities were measured at hourly intervals at depths of 5, 10, 30, 100 and 175 m. These tidal harmonics were computed by Vercelli using the Darwin method (Table 1 continued on the following page).

By applying the Darwin method to time series of horizontal along strait velocities, Vercelli computed tidal harmonics for the M_2, M_4, S_2, K_1 and O_1 constituents. Then, by

Vercelli Station Number	Bottom Depth	Time and Length,* hours	Tidal Component as Percentage of Observed Energy	Depth of Observation	M_2 Amplitude, cm/s	M_2 Phase, hours	M_4 Amplitude, cm/s	M_4 Phase, hours
1	106	Aug 16-30, 1922 (336)	90%	5	131.0	4.5	11.0	1.5
			85%	10	126.0	4.5	11.0	1.4
			90%	20	122.0	4.5	12.9	1.3
			90%	30	116.0	4.4	10.0	1.0
			90%	50	115.0	4.4	9.0	0.5
			90%	90	107.0	4.5	8.0	0.0
57	88	April 2-4, 1923	50%	5	140.0	0.4	16.0	2.7
			30%	10	137.1	0.3	15.6	2.6
			40%	30	138.3	0.2	10.0	-2.9
			40%	75	127.9	0.2	19.8	-2.6
60	65	April 12-13, 1923 (24)	80%	5	51.6	2.4	3.3	2.1
			75%	10	41.7	2.6	4.8	1.8
			70%	25	38.7	2.6	1.5	2.1
			75%	50	36.4	2.6	3.8	1.2
61	80	April 16-17, 1923 (29)	30%	5	113.8	1.1	5.7	2.3
			30%	10	104.6	1.0	3.7	-2.4
			30%	20	94.6	1.1	11.8	-2.9
			40%	50	94.2	1.1	5.8	-1.6
66	110	April 28-30, 1923 (54)	20%	5	190.7	2.7	6.3	-0.1
			30%	10	191.3	1.9	1.9	0.5
			30%	30	192.5	2.9	2.9	-2.3
			60%	90	181.9	2.6	6.2	2.8
67	70	May 2-4, 1923 (44)	40%	5	232.3	0.3	19.7	-3.0
			50%	10	231.4	0.4	21.7	-1.9
			40%	25	228.5	0.4	21.2	-2.0
				50	221.8	0.2	10.7	-2.3
68	85	May 6-7, 1923 (30)	70%	5	28.0	-3.1	7.2	-2.7
			50%	10	39.5	-3.5	3.7	-2.5
			65%	30	47.7	-3.4	5.8	2.5
			60%	60	49.5	-3.5	3.3	-2.1
69	222	May 8, 1923 (10)	—	5	17.6	5.8		
			—	10	20.9	-5.5		
			—	30	22.4	-5.4		
			—	100	47.2	4.7		
			—	170	55.6	4.7		
70	300	May 11, 1923 (10)	—	5	30.3	3.9		
			—	10	32.8	4.1		
			—	30	22.5	3.2		
			—	100	47.2	2.3		
			—	200	61.9	2.0		
71	306	May 12, 1923 (11)	—	5	17.1	4.1		
			—	10	19.1	4.1		
			—	30	16.0	4.2		
			—	100	14.3	3.7		
			—	200	26.4	3.2		
72	520	May 14, 1923 (7)	—	5	12.4	-0.4		
			—	10	10.9	-0.6		
			—	30	45.0	0.0		
			—	100	12.0	0.6		
			—	200	20.0	4.5		

Table 1. (cont'd)

extrapolating the data, he was able to compute the K_2, P_1 and N_2 constituents (Table 2). Castaldini and Franzini (1979), obtained similar results using modern computational methods. The reduction in variance of Vercelli's data is about 85 %. Using a Richard sea level recorder, Vercelli also measured the sea level at Milazzo, Lipari, Tropea, Faro, Reggio and Villa San Giovanni. At each station at least one 30 - day tidal series was taken from May 1922 to October 1923 (Table 3). About the velocity field, Castaldini and Franzini (1979) show that over the sill, the M_4 amplitude is nearly 10% of the M_2 component and decreases rapidly away from the sill (Table 1, also to be discussed in the following).

Mosetti (1988) has described and discussed recent current meter measurements around the sill.

3. Defant's barotropic tidal model, "tagli" and "macchie d'olio"

The tides inside the strait of Messina were studied by Defant (1940, 1961) using a barotropic model. He computed tidal elevations and horizontal currents by stepwise integration of the one - dimensional, inviscid, linearized barotropic equations of motion. South and north of the sill, the boundary conditions are that the M_2 tide behaves like that of the Ionian and Tyrrhenian Seas (Table 3). His results showed that tidal heights decreased from 10 cm in the Ionian Sea to zero at the sill and then rose quickly to the Tyrrhenian value of ≈ 16 cm (fig. 4). These abrupt changes provide the large sea level slopes necessary to drive the tidal speeds of ≈ 2 m/s, as calculated by Defant and observed by Vercelli. Defant's tidal phase was 5.9 hours, while the observed value was 4.9 hours. This discrepancy is probably due to his omission of turbulent dissipation which is obviously strong through the strait of Messina. Using the observed phases, Defant was able to reproduce a good flow sequence

Depths	A_o	M_2 A	Φ	M_4 A	Φ	S_2 A	Φ	Depths	K_1 A	Φ	P_1 A	Φ	O_1 A	Φ	N_2 A	Φ
5	-8,4	131	131	11	86	35	148	5	34	74	10	74	14	58	20	125
10	-3,4	126	130	11	79	34	150	10	29	74	10	74	11	73	19	124
20	-2,9	124	130	13	76	30	150	20	29	73	10	73	11	91	19	124
30	0,5	116	129	10	60	35	139	30	30	64	10	64	13	97	17	123
50	8,0	115	129	9	32	36	132	50	29	59	10	59	15	100	17	123
90	13	107	130	8	0	31	137	90	31	66	10	66	13	100	16	124

Table 2. Tidal velocity harmonics calculated by Vercelli by means of the Darwin method and extrapolating techniques. Amplitudes (A) are expressed in cm/s, phases (Φ) are in degrees.

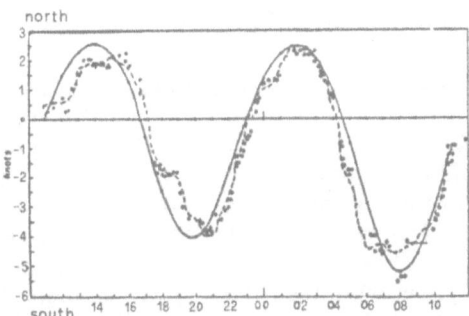

Figure 3. Comparison between Massi et al.'s (1979) experimental values (dotted line), Vercelli's (1925) prediction (solid line) and Massi et al.'s (1979) prediction (dashed line) (after Massi et al. 1979).

Stations	Amplitudes, cm							Phases, deg						
	M_2	M_4	S_2	K_2	K_1	P_1	O_1	M_2	M_4	S_2	K_2	K_1	P_1	O_1
Milazzo (M)	12 1	—	4.7	1.3	3.2	1 1	0.9	262	—	287	287	222	222	126
Lipari (L)	12 1	—	4 9	1.3	2.7	0.9	0 9	258	—	286	286	212	212	162
Tropea (T)	14 6	—	5 2	1.4	4.1	1.4	1 0	274	—	296	296	219	219	160
Faro (F)	5 5	—	3.1	0.8	2.3	0 8	1.0	269	—	314	314	232	232	251
Villa San Giovanni (V)	3.3	1.9	1.3	0.3	1.2	0 4	0.2	116	72	104	104	48	48	—
Reggio (R)	6.2	—	3 0	0.8	1.6	0 6	0.8	95	—	100	100	57	57	50

Table 3. Sea level recordings taken by Vercelli at Milazzo, Lipari, Tropea, Faro, Reggio Calabria and Villa San Giovanni. In each station at least one 30 - day sea level time series was taken. Measurements were made between May 1922 and October 1923.

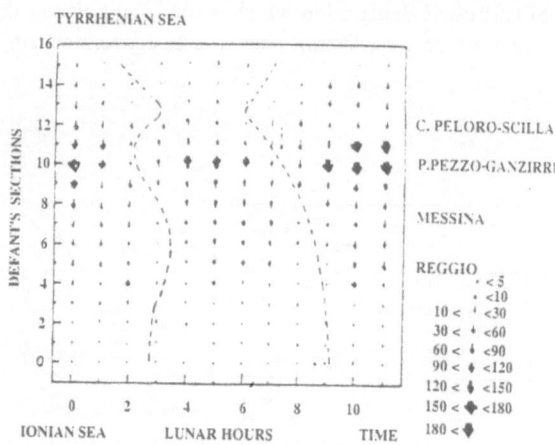

Figure 4. The distribution of tidal height, its amplitude and phase, and the tidal current speed as computed by Defant (1940) through the strait of Messina (after Hopkins et al., 1984).

(fig. 5). Between 3 and 9 hours after the upper or lower lunar meridional transit, the current on the sill flows northwards ("rema montante"), and the flow is reversed between 9 and 3 hours ("rema scendente"). In the vicinity of the sections, Punta Pezzo to Ganzirri and Capo Peloro to Scilla, where the strait bends somewhat, convergences occur from 2.5 to 4 hours as the flow shifts from north to south and again on reversal from 6.5 to 8 hours. An interesting consequence of these convergence zones is that they act as generation sites for strong bores, locally called "tagli" ("cuts"; Vercelli, 1925; Mazzarelli, 1938). These appear as zones of abrupt increase of sea surface elevation and roughness, which can be damped or amplified by opposing winds as well as spring tides. This surface manifestation is greater for the "rema scendente" phase, when the lighter Tyrrhenian water flows southwards over the heavier Ionian water. Current reversals also give rise to turbulent eddies (Defant, 1940; 1961). According to Defant, the larger eddies have a vertical axis and develop at three locations: off Capo Peloro, Scilla (where the two monsters Scylla and Charybdis were thought to reside!) and near the Messina harbor entrance (fig. 1). The water at the center of the smaller eddies has a smooth, oily appearance so they are locally referred to as "macchia d'olio" ("oil patch").

4. Interface time evolution: theory and sonar observations

Defant treats the Messina tidal currents as barotropic. In fact, the stations over the sill show little baroclinicity, being hydrographically homogeneous and practically depth - independent in speed. However, sea surface temperature displays large variations in time, since the sill is alternately occupied by different water masses: the interface separating them fluctuates to such a degree that it alternately touches the sea bottom and the air - sea surface. Hence Messina is somewhat analogous to the case of the Dröbak Sill in the Oslofjord (Norway) discussed by Stigebrandt (1977). Also in the strait of Gibraltar (Lacombe and Richez, 1983) the interface over the sill has ≈ 140 m vertical excursions, but without breaking the surface or touching the sea bottom.

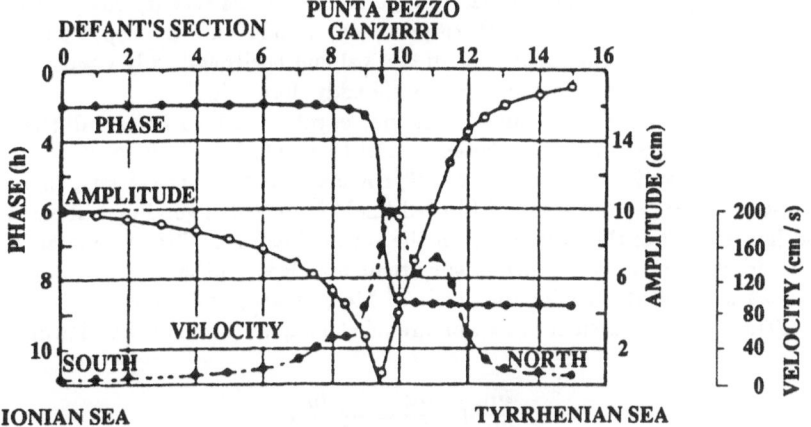

Figure 5. The distribution of surface tidal speeds through the strait of Messina over a semidiurnal tidal cycle (after Defant, 1961).

We now give a brief description of three models of the time - dependent and steady trends of the interface across the sill.

A simplified viscous baroclinic numerical model for the motion in a strait is used by Del Ricco (1982): starting from the equations of motion, continuity and salt balance, he first integrates the various parameters in the cross - strait direction y. The continuity equation is also depth integrated. In formulae this gives the continuity equation:

$$\frac{\partial}{\partial x} \int_{-\eta}^{d} Bu dz + B \frac{\partial \eta}{\partial t} \big|_{z=-\eta} = 0, \tag{1}$$

the salt balance equation

$$\frac{\partial}{\partial t}(Bs) + \frac{\partial}{\partial x}(Bus) + \frac{\partial}{\partial z}(Bws) - \frac{\partial}{\partial x}(BK_x\frac{\partial s}{\partial x}) - \frac{\partial}{\partial z}(BK_z\frac{\partial s}{\partial z}) = 0 \qquad (2)$$

and the momentum equation

$$\frac{\partial}{\partial t}(Bu) + \frac{\partial}{\partial x}(Bu^2) + \frac{\partial}{\partial z}(Buw) =$$

$$= -gB\frac{\bar{\rho}}{\rho}\frac{\partial \eta}{\partial x} - g\frac{B}{\rho}(z+\eta)\frac{\partial \rho}{\partial x} + \frac{\partial}{\partial x}\left(BN_x\frac{\partial u}{\partial x}\right) + \frac{\partial}{\partial z}\left(BN_z\frac{\partial u}{\partial z}\right) \qquad (3)$$

where: the depth z is taken positive downwards; η, s and p are the elevation of the water surface, the salinity and pressure, respectively; u and w are the velocity components in the axial and vertical directions (x and z), respectively; d and B are the depth and width of the strait; N_z and N_x are the coefficients of vertical and horizontal eddy viscosity; K_z and K_x are the coefficients of vertical and horizontal eddy diffusivity.

This model, adapted to the strait of Messina, is solved with a finite - difference grid method. The boundary conditions chosen by Del Ricco (1982) are the observed salinities and tidal surface elevations at the Ionic and Tyrrhenian mouths of the strait. The resulting interface positions are shown in fig. 6 (dotted line). They are in good agreement with the observed intersection of the interface with the sea surface in the "rema montante" phase and its lowering below the sill depth in the "rema scendente" phase.

Hopkins et al. (1984) consider the strait of Messina as a two - layer system with an interface $H(x,t)$. The equations of motion are much simpler (Stommel and Farmer, 1952):

$$\frac{\partial u}{\partial t} + u\frac{\partial u}{\partial x} = -g\frac{\partial \eta}{\partial x} \qquad (4)$$

$$\frac{\partial U}{\partial t} + U\frac{\partial U}{\partial x} = g\frac{\Delta \rho}{\rho_1}\frac{\partial H}{\partial x} - \frac{\rho_1}{\rho_2}g\frac{\partial \eta}{\partial x} \qquad (5)$$

where u and ρ_1 are the axial speed and density in the upper layer and U and ρ_2 are the corresponding quantities for the lower layer; $\Delta \rho = \rho_2 - \rho_1$ and η is the surface elevation. Friction and the Coriolis force are neglected in this approximation.

The expression for the interface $H(x,t)$ is heuristically obtained by introducing Vercelli's M_2 tidal velocities in equations (4) and (5), after neglecting the quadratic terms. The time - evolution of the interface is shown in fig. 6 (dashed line). Good agreement with observations is achieved also in this model although some differences can be seen with respect to the calculations by Del Ricco (1982). The evolution of the interface was experimentally studied during the JANE 84 cruise (October 25 - 30, 1984) on board the Italian R/V Bannock (Di Sarra et al., 1987). A KODEN Fish Finder sonar (28 and 200 kHz) detected the interface (Farmer and Freeland, 1983) while the ship was traveling at $\approx$ 8 knots along the centerline

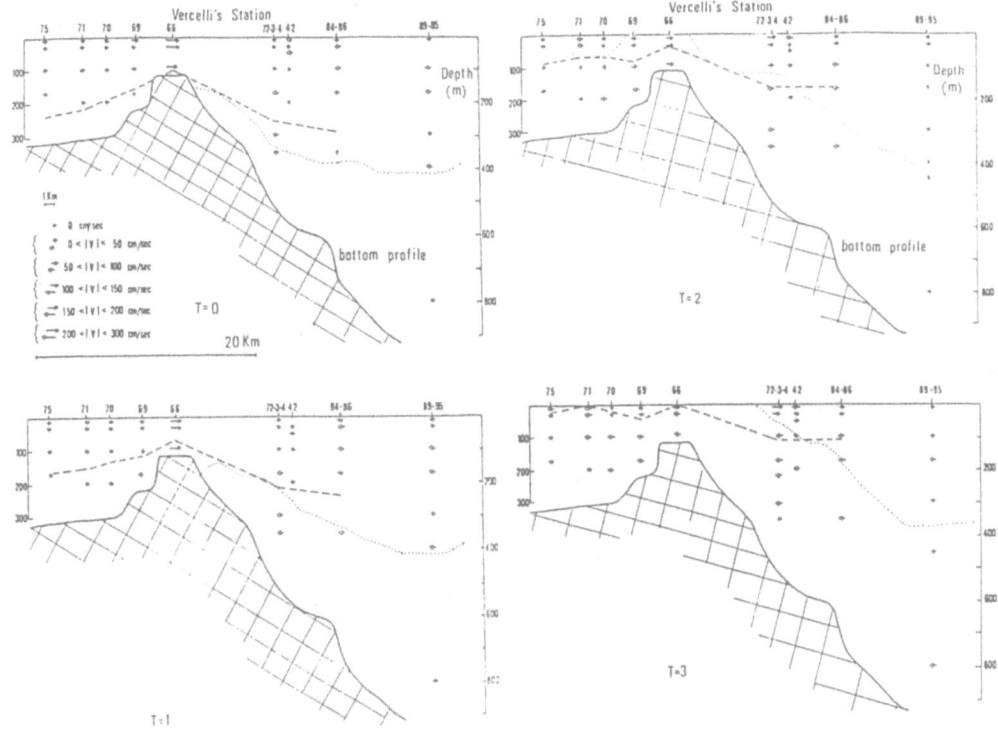

Figure 6. Hourly interface time evolution. Comparison between the calculations by Del Ricco (1982, dotted line) and by Hopkins et al. (1984, dashed line). Arrows indicate velocities observed by Vercelli (1925) (after Salusti et al., 1988).

of the strait. Figs. 7a - d are the drawings made from the KODEN Fish Finder images, the observed interface and the corresponding interfaces obtained by theoretical models (Del Ricco, 1982; Hopkins et al., 1984). A photograph of the KODEN image is shown in fig. 8. It is satisfying to find general agreement between these observations and the results of Hopkins et al. (1984) and Del Ricco (1982). However, some interesting small - scale differences were observed, such as barotropic columnar disturbances, i. e. Bignami columns (probably vortices due to flow - bottom interaction). On October 29, 1984 about 8 h 40 min GMT (tidal current of $\approx$ 1 m/s) one of these columns was observed on the

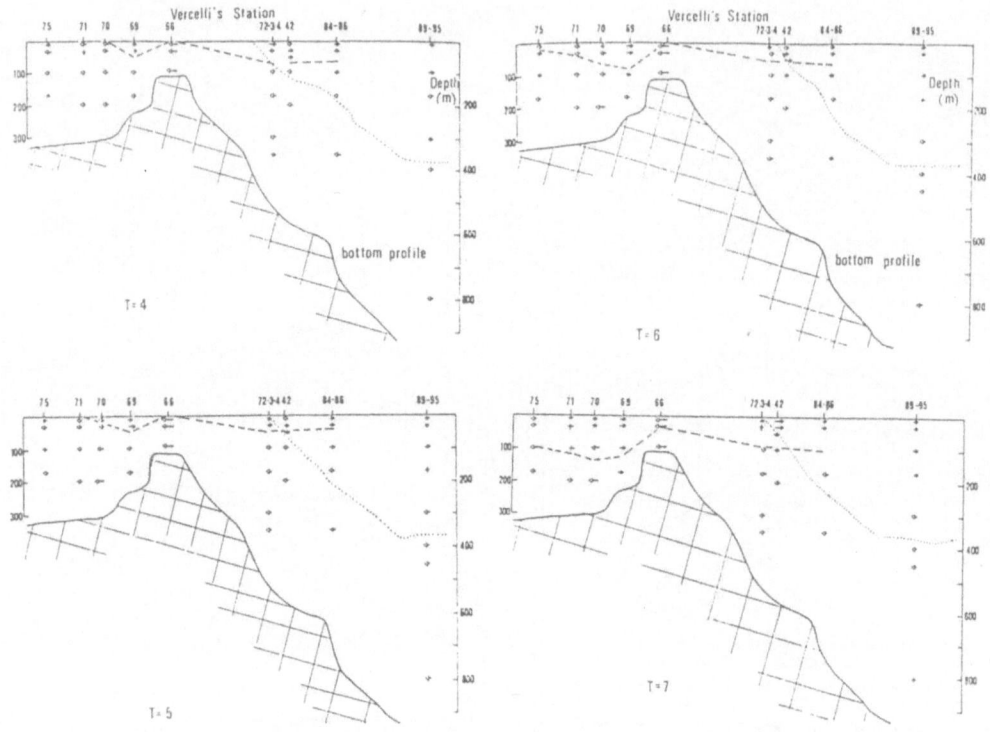

Figure 6. (cont'd)

KODEN Fish Finder: a vigorous $\approx$ 400 m wide eddy extending from the sea surface to the bottom at $\approx$ 250 m. It appeared like a regular white column on the KODEN video (fig. 8). The ship was appropriately located and a moored temperature time series (13 minutes) measurement was performed by stopping the CTD at a depth of 20 m. The graph relative to this measurement (fig. 9) allows the time interval in which the column encountered the moored CTD to be identified clearly. The thermal jump was 0.7 °C; the resulting swirl velocity was 0.6 m/s.

Di Sarra et al. (1987) discuss these eddies in the light of existing theories. For steady flows in a stratified fluid passing over an obstacle (Long, 1955), the role of the baroclinic

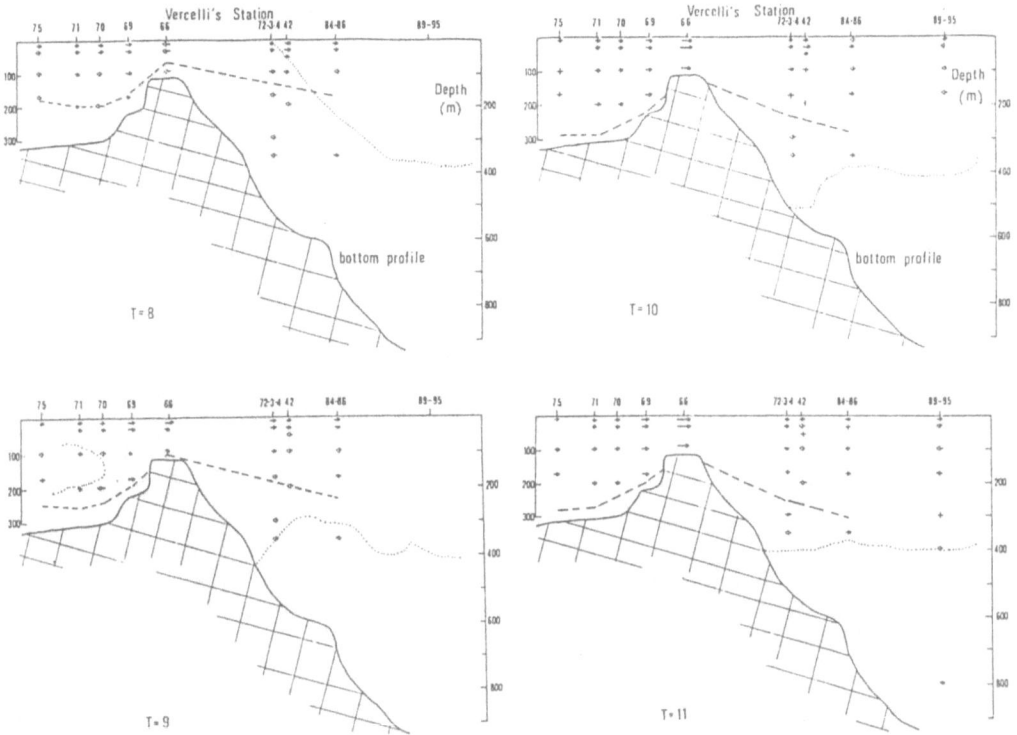

Figure 6. (cont'd)

Froude number, defined as

$$Fr = \frac{U}{C_0} \approx \frac{U}{N_d},$$ (6)

is of primary importance (U is the horizontal velocity, C_0 is the phase velocity of internal waves, N is the Brunt - Väisälä frequency and d is the bottom depth). At Messina, U can be as high as 2 - 3 m/s, while $C_0 \approx 0.5 - 1.0$ m/s. Di Sarra et al. (1987) found that near the sill the flow is often supercritical, but they also saw that if one assumes that the flow is barotropic (in the region of the sill) and flux is conserved (over the whole water column),

108

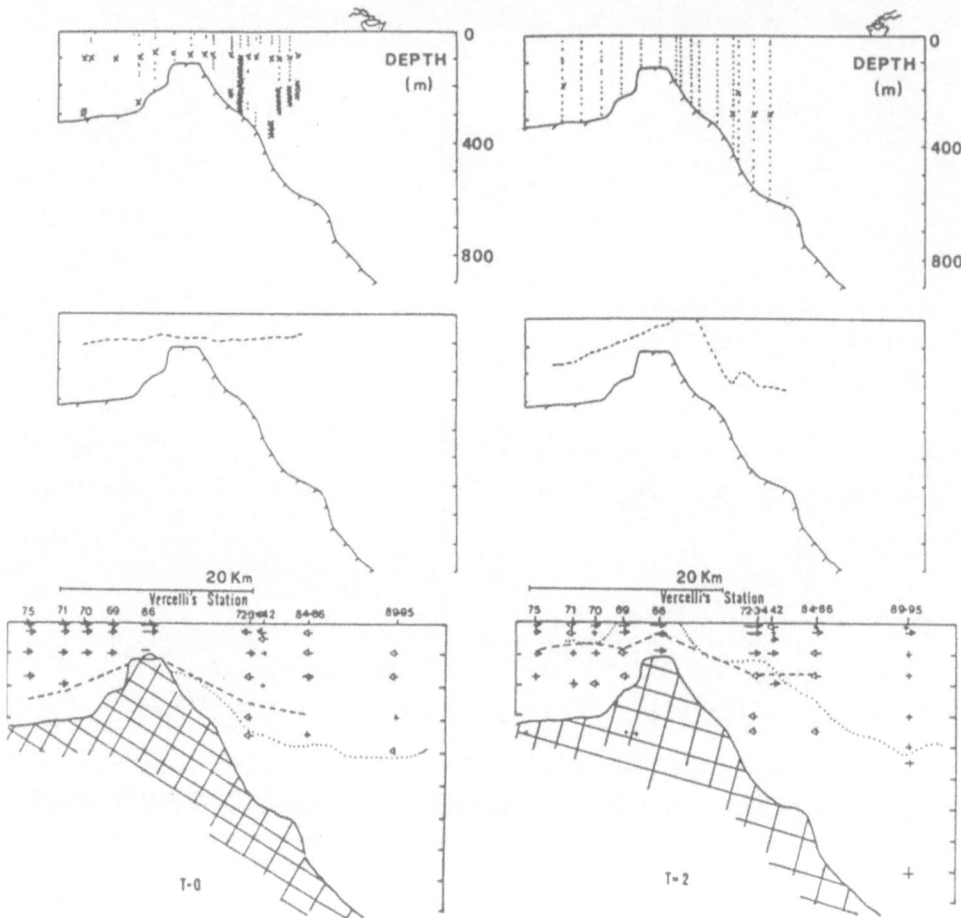

Figure 7 (a - d). Drawings of the KODEN Fish Finder images (top) displaying the interface (isolated crosses) and the columnar disturbances (vertical series of crosses) observed over the sill of the strait of Messina. Reconstruction of the interface from the Fish Finder images (middle) and comparison with the tidal situation (see fig. 6; after Di Sarra et al., 1987).

the flow becomes subcritical at distances of ≈ 8 km from the sill. Columnar disturbances were observed over the sill only for subcritical flows, as theoretically predicted by McIntyre (1972) and confirmed by tank experiments (Baines, 1977, 1979; Baines and Hoinka, 1985; Wei et al., 1975). In fig. 7a the northward flow is seen to be supercritical over the sill (d ≈ 70 m), but subcritical where these disturbances are present (d ≈ 250 m). Figs. 7b and 7c show columnar disturbances which propagate upstream for a northward weak flow and downstream disturbances in a southward flow, respectively. During the still water phase (fig. 7d) no columnar disturbances appear.

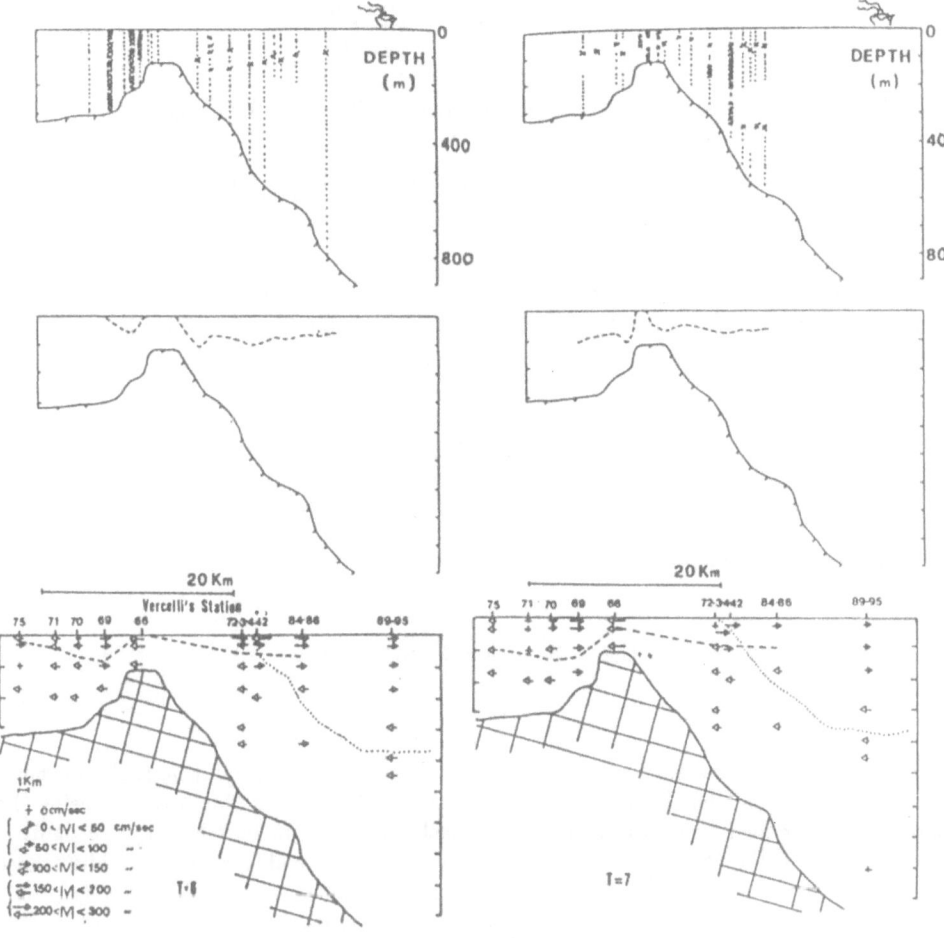

Figure 7. (cont'd)

These observations cannot be considered as definitive, since they were obtained under conditions of heavy nautical traffic of ferries, commercial ships, etc. On the other hand, a comforting agreement with deductions obtained from the steady flow theories has been obtained. Di Sarra et al. (1987) discuss fully this point.

Finally, in order to analyze the time - averaged depth of the interface between the two layers, Salusti et al. (1988) also determined the time - averaged trend of the interface, starting from equations (4) and (5). Taking a time average of these two equations over a period of time which was long in comparison with the semidiurnal tidal period, they obtained:

$$\bar{u}\frac{d\bar{u}}{dx} = -g\frac{d\bar{\eta}}{dx} \tag{7}$$

$$\overline{U}\frac{d\overline{U}}{dx} = -\frac{\rho_1}{\rho_2}g\frac{d\overline{\eta}}{dx} + g\frac{\Delta\rho}{\rho_2}\frac{d\overline{H}}{dx} \tag{8}$$

where the overbar denotes time averaging. Combining (7) and (8), this gives:

$$\frac{d}{dx}(\overline{U}^2 - \overline{u}^2) = 2g\left(1 - \frac{\rho_1}{\rho_2}\right)\left(\frac{d\overline{\eta}}{dx} + \frac{d\overline{H}}{dx}\right) \approx 2g\frac{\Delta\rho}{\rho_2}\frac{d\overline{H}}{dx} \tag{9}$$

since in the strait of Messina $dH/dx \approx 10^3(d\eta/dx)$. This gives the time - averaged along - strait rise of the interface. The time - dependent geostrophic tilt (cross - strait) angle τ of the interface is found to be:

$$\tau = f\frac{u(t) - U(t)}{g\dfrac{\Delta\rho}{\rho_1}}. \tag{10}$$

The time - average of equation (10) gives

$$\overline{\tau} = f\frac{\overline{u} - \overline{U}}{g\dfrac{\Delta\rho}{\rho_1}} \approx 3 \cdot 10^{-4} \approx 0.$$

Obviously, lateral boundary layers are not considered in deriving equation (10) and therefore the most near - shore stations (where lateral boundary layers play a role, causing the onset of the so called "correnti bastarde" i.e. backwater) cannot be taken into account.

It has to be stressed that this relation holds for all reasonably straight straits whose widths are much smaller than the Rossby internal deformation radius. The steady profile of the interface obtained by introducing Vercelli's velocity data (Table 2) in equation (10) is shown in fig. 10. The profile has most of the main expected features such as the uplift in the vicinity of the sill.

5. Internal waves and tidal bores: observations

The first hint of the presence of internal waves in the strait of Messina was given by satellite imagery (Alpers and Salusti, 1983). Fig. 11 shows a Synthetic Aperture Radar image of the area. Fig. 12 is a schematic map of the imaged scene in which the most important features are marked. The most pronounced oceanic features are the three rings in the Tyrrhenian Sea (Golfo di Gioia), ≈ 30 km north of the strait of Messina. They are most clearly detectable in the dark region, a sea area with low radar backscattering. The spacing of the rings varies between 700 and 1800 m; it is broadest at the front (north) and decreases towards the rear (south). The spacing of the rings is narrower near the coast (east) than

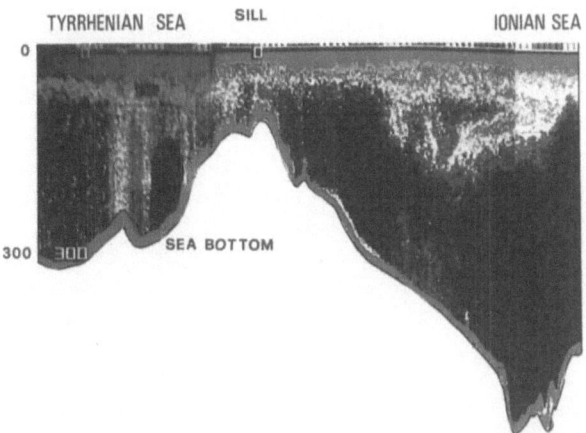

Figure 8. Photograph of a KODEN Fish Finder image taken on Oct. 26, 1984, 17 h 00 min GMT (Jane 84 cruise) over the sill of the strait of Messina. The white areas are the vertical columnar disturbances and the interface.

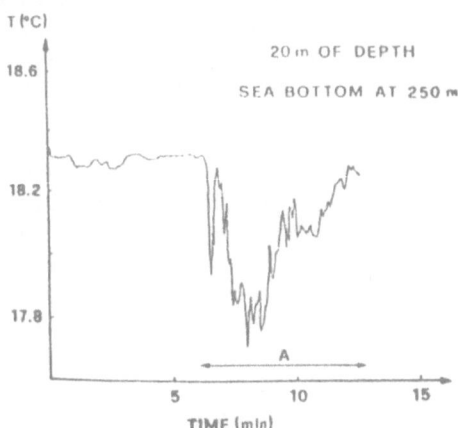

Figure 9. Temperature time series taken with a moored CTD at a depth of 20 m (38 ° 15' 58" N, 15° 40' 00" E, Oct. 29, 1984,08 h 40 min GMT, JANE 84 cruise). For the indicated time interval (A) the CTD was inside the columnar disturbance (after Di Sarra et al., 1987).

farther out to sea (west). Alpers and Salusti (1983) argue that these rings are surface evidence of internal wave trains generated in the strait.

Another feature discernable on both images consists of bright bands connecting Punta Pezzo in Calabria to the Sicilian coast with other less bright bands. They are surface manifestations of "tagli" which are also visible to the naked eye from a boat. They mark

112

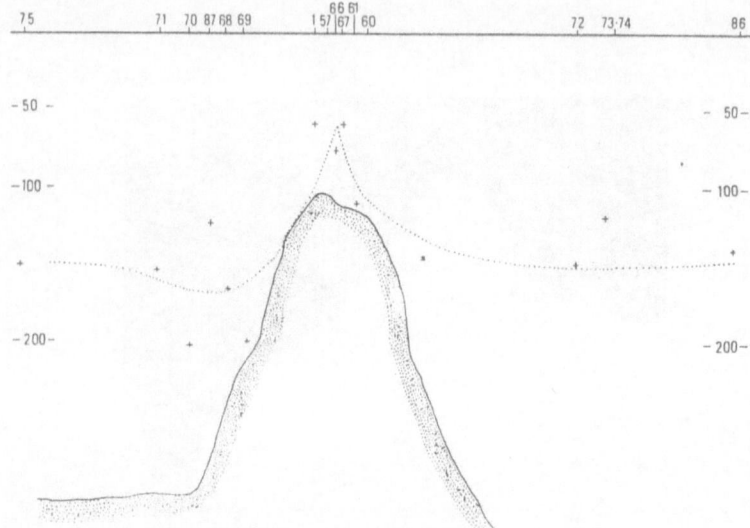

Figure 10. Interface steady profile (after Salusti et al. 1988).

boundaries of water masses of different hydrodynamic, hydrographical and biological properties: the local fishermen regulate their departure time for fishing expeditions according to their passage. The position of the "tagli" on the SEASAT image is in good agreement with what Vercelli (1925) reported on April 28, 1923, 19 h 00 min local time.

With reference to the three circular patterns on the SEASAT image, Alpers and Salusti (1983) hypothesize that they are the surface signature of a large - amplitude internal wave train, i. e. a packet of internal solitary waves generated inside the strait of Messina. They also assume that these internal waves were generated when tidal flow inversion occurs, that is, when thermocline displacement takes place (Gardner et al., 1967; Osborne and Burch, 1980). Reversal of the tidal flow direction occurred at 23 h 30 min GMT on Sept. 14, i. e. 8 hours and 47 minutes before the SEASAT - SAR image was taken. Since the distance of the leading edge of the internal soliton train is 33 km from the sill, propagation velocity $c = 1.0$ m/s is obtained. Alpers and Salusti (1983) compared this value with theoretical estimates from the Korteweg and de Vries (1895) theory:

$$c = c_0 \left(1 + \frac{1}{2}\frac{h_2 - h_1}{h_1 h_2}\eta_0\right) \tag{11}$$

where

$$c_0 \approx g \left(\frac{\Delta\rho}{\rho_2}\frac{h_1 h_2}{H_1 + h_2}\right)^{\frac{1}{2}} \approx 0.66 m/s \tag{12}$$

is the phase speed of a small amplitude internal wave, h_1 and h_2 are the thicknesses of the

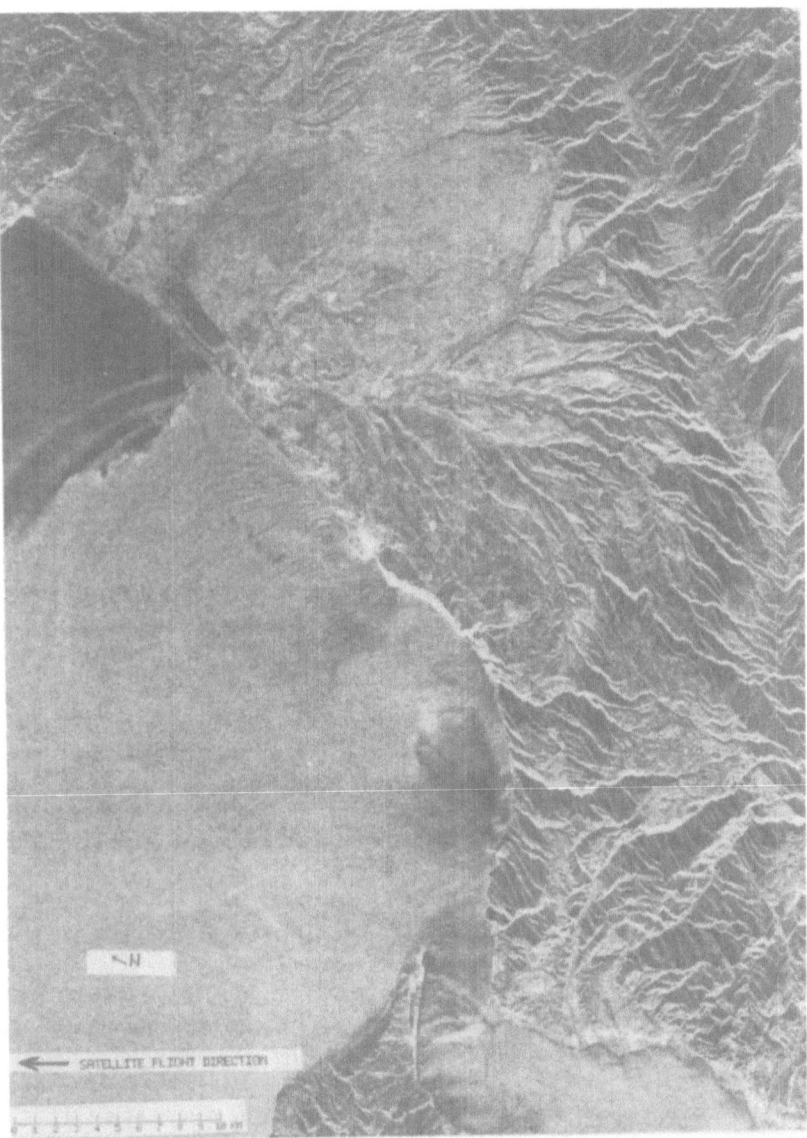

Figure 11. SEASAT - SAR image of the area of the strait of Messina (40 km x 40 km) with the frame center located at 38° 20' 46" N and 15° 49' 56" E, orbit 1149, Sept. 15, 1978, 08 h 17 min GMT. According to the tide table of the strait, still water occurred at 23 h 30 min GMT on Sept. 14; maximum current (between 5 and 5.5 kts) from the Tyrrhenian Sea into the Ionian occurring at 08 h 07 min GMT the next day. The point at the bottom center is Capo Peloro (see fig. 1b). The concentric rings are the signature of internal waves propagating into the Gulf of Gioia. Tidal bores can also be seen inside the strait (after Alpers and Salusti, 1983).

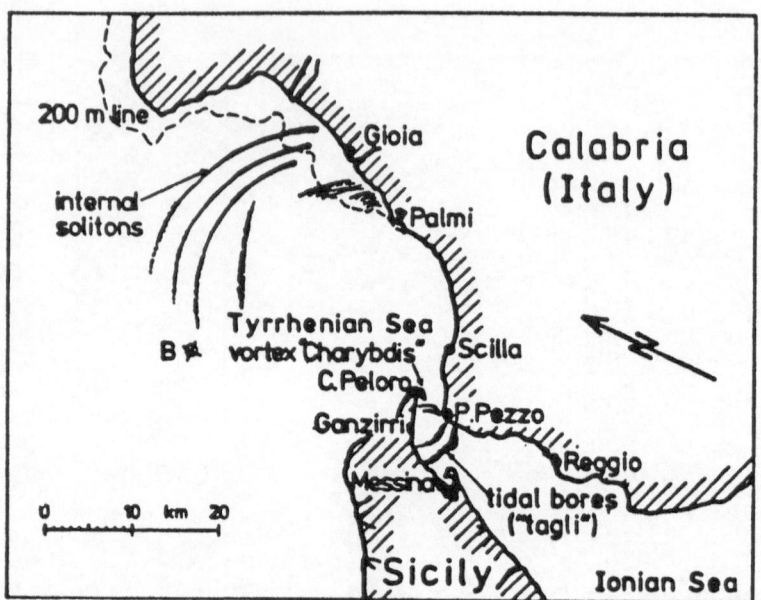

Figure 12. Schematic map of the SEASAT - SAR image (fig. 11). Point B is the position of R/V Bannock, where on Nov. 24 - 25, 1980, in situ measurements were carried out (after Alpers and Salusti, 1983).

two superposed layers respectively of densities ρ_1 and ρ_2, estimated from historical data. The value of c is smaller than the value inferred by them from the SEASAT image (1.0 m/s), although this is not a definitive inconsistency in view of the uncertainties in all of the estimates involved. The three rings' focus is not located exactly at the Tyrrhenian mouth of the strait of Messina but approximately 10 km east of this point, probably because of a current flowing in the eastward direction.

Unfortunately no "in situ" measurements were carried out during this SEASAT overflight, but under the stimulus of this SEASAT image some cruises were organized later with the aim of detecting tidal internal wave trains in the strait of Messina.

In situ measurements referring to two cruises which followed these SEASAT - SAR observations, are now discussed (Alpers and Salusti, 1983; Sapia and Salusti, 1987). During the JUDITH 80 cruise on board the Italian CNR vessel R/V Bannock, measurements were taken at station B in the Gulf of Gioia (fig. 12). The measurements consisted of temperature and salinity time series at a depth of 43 m recorded by means of a CTD (fig. 13). The wave train comprises eight waves and has a duration of approximately 2 hours. A second wave train was observed approximately 11 h 30 min later at a depth of 46 m at a slightly different position because of the ship's drift. The tidal current reversed from a southward to a northward direction on November 24, 1980 at 15 h 38 min local time, and on November 25 at 3 h 1 min. The time interval between these two current reversals is 11 h 22 min. This is in good agreement with the 11 h 30 min time lag between the arrivals of the two wave trains at station B and thus supports the hypothesis that the internal waves

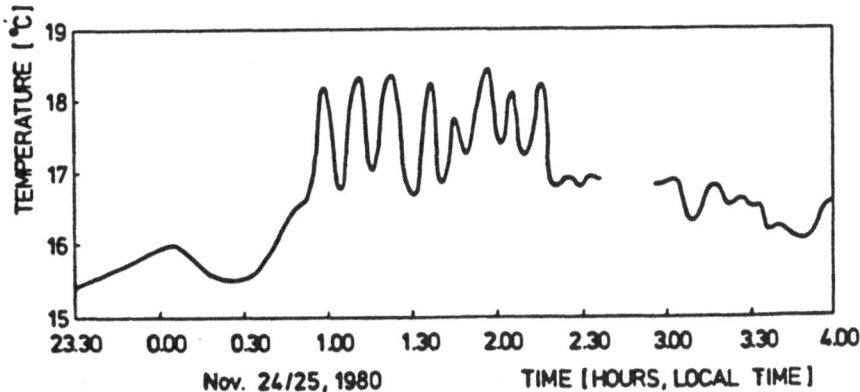

Figure 13. Temperature profile measured on Nov. 24 - 25, 1980, at point B (see fig. 12) at a depth of 43 m (after Alpers and Salusti, 1983).

are generated by the tide at the sill. The velocity for the wave trains, estimated using equation (11), is $c \approx 0.82$ m/s. This value agrees quite well with the propagation velocity of 0.80 m/s computed above. The interpretation of the circular striation pattern on the SEASAT image in terms of internal waves was moreover checked by comparing the image wavelengths with the in situ measured wave periods multiplied by the propagation speed c. For $c = 0.8$ m/s and for the measured range of periods (8 min - 34 min) one obtains wave lengths ranging from 384 m to 1732 m, which is consistent with the satellite image estimations.

Two years after these first observations, during the PRIME cruise (May 26 - June 6 1982), again on board R/V Bannock, two calibrated CTDs (a Neil Brown and a towable Guildline) and two thermistor chains (75 m long, time resolution 0.3 min) were employed in the area shown in fig. 14. Details about these measurements are given in the Appendix since they are published elsewhere.

It is sufficient here to stress that south of the strait moored observations detected rather long - lasting signals (10 - 30 minutes' duration with peak amplitudes of 1 - 2 °C, figs. 15 and 16), but north of the strait both moored and towed observations showed also "quicker" large amplitude signals (5 minutes' duration, 4 °C peak amplitude, figs. 17 and 18). These measurements were taken in a particularly calm sea, therefore without surface or thermocline perturbations.

There are two more indications of these waves south of the strait. Di Sarra et al. (1987) first observed a 100 m large vertical displacement of the interface in the zone immediately south of the strait. Also Sapia and Salusti (1987) and Nicolo' et al. (1989) detected large amplitude isolated signals south of the sill and as far south as Syracuse. In more detail, south of the strait (38° 00' N, 15° 34' E, October 22 - 24, 1987; figs. 19 and 20) Nicolo' et al. (1989) observed three clear packets of internal waves (fig. 20) with a "classical" shape: a series of large -amplitude oscillations followed by a dispersive tail (Osborne and Burch, 1980). The generation point of the internal solitary waves moving southwards from the strait is not clearly identifiable, as has been pointed out by Nicolo' et al., (1989).

It has to be added that two quasi - stationary currents are generated by this tidal phe- nomenon: a surface current flowing southwards over the sharp Sicilian shelf for more than

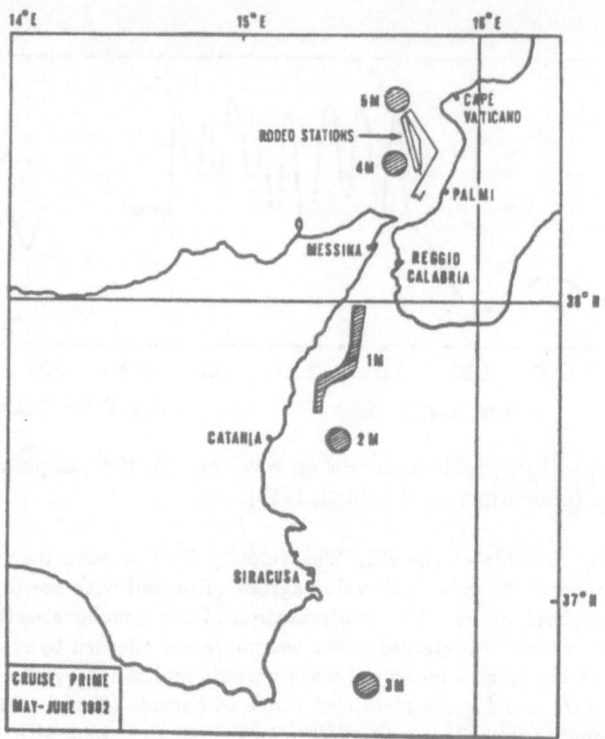

Figure 14. Geographical position of the stations of the PRIME 82 cruise (after Sapia and Salusti, 1987).

100 km (Böhm, Magazzu', Wald and Zoccolotti, 1987) and a deeper current flowing north-wards (Marullo and Santoleri, 1986). To conclude, on June 5 and 6, 1983, from temperature records at 30 m depth, south of Capo Vaticano, a strong surface thermal front (5 °C/8 km) was noted around 38° 30' N, 15° 43' E (Marullo and Santoleri, 1986), in agreement with the SEASAT image.

6. Open problems

Although the strait of Messina has been studied for many centuries, some interesting major questions remain unsolved. While Hopkins et al. (1984) gave an idea of the along - strait time evolution of the interface between surface Atlantic and Levantine waters, the cross - strait variations have not been fully studied.

Also of importance is the role of the horizontal and vertical vortices: Abbate et al.(1982) looked for them in the Vercelli data set but this attempt merely marked the beginning of the present research.

Coming down to a smaller scale, i.e. that of internal solitary waves, a first question concerns the equation governing their time evolution: KdV, Benjamin - Ono (Sapia and

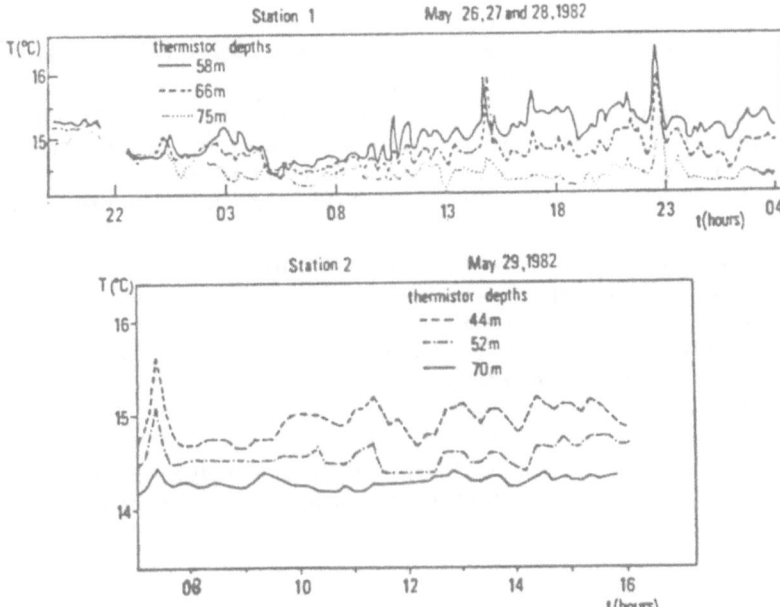

Figure 15. Thermistor chain data at stations 1 and 2 on the southern side of the strait of Messina (after Sapia and Salusti, 1987).

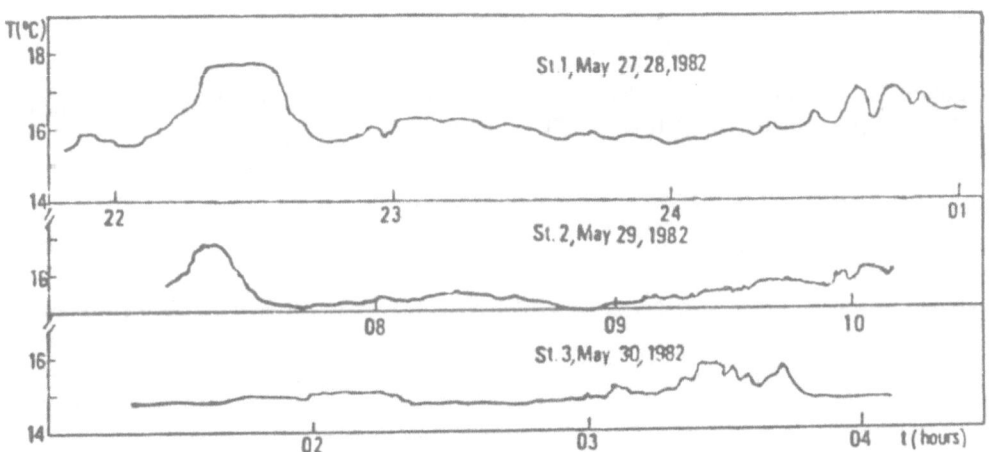

Figure 16. Temperature records of the Guildline moored at the fixed depth of 30 m in stations 1, 2 and 3 (after Sapia and Salusti, 1987).

Salusti, 1987), "finite depth" or what? One might also infer that the internal waves detected from the SEASAT image (fig. 11) could be affected by coastal effects since the rings fade in the seaward direction. Sapia and Salusti (1987) discuss this difficult point. One may also wonder why the internal wave packets observed by Alpers and Salusti (1983), Griffa et al. (1986) and Sapia and Salusti (1987), have such different shapes. The difference between the shape of the internal solitary waves south of the strait, as observed by Sapia and Salusti (1987) and by Nicolo' et al. (1989), and that of the waves observed north of the strait is a further open question.

Also, north of the strait, what is the nature of the large zone of low radar backscattering in the SEASAT - SAR image discussed by Alpers and Salusti (1983)? Guardiani et al. (1988), following an idea of Marullo and Santoleri (1986), hypothesized that this patch, observable South of Capo Vaticano, is due to a front generated by the breaking of internal solitary waves over the sloping bottom of this cape.

Finally, the turbulent phenomenology of the strait has still to be fully investigated since this approach is important both for the dynamics and the mixing and is practically unknown: knowledge of the bottom boundary layers, vertical eddy mixing rates and frictional effects is only partial and does not allow a systematic analysis for the moment.

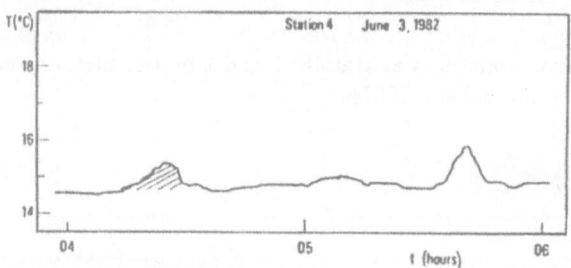

Figure 17. Temperature record of the Guildline moored at the fixed depth of 25 m in station 4 (after Sapia and Salusti, 1987).

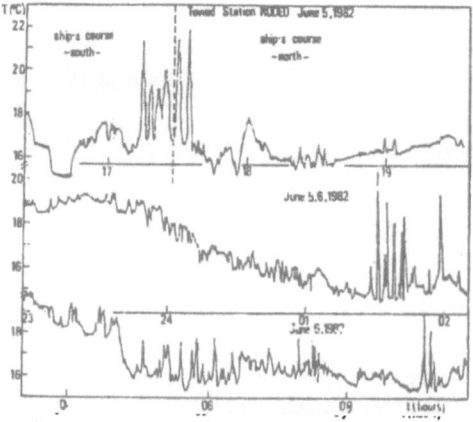

Figure 18. Temperature jumps observed on June 5 and 6, 1982 (station Rodeo), by the towed Guildline at a depth of 17 m; the ship's velocity was 4.0 ± 0.1 knots, moving southwards.

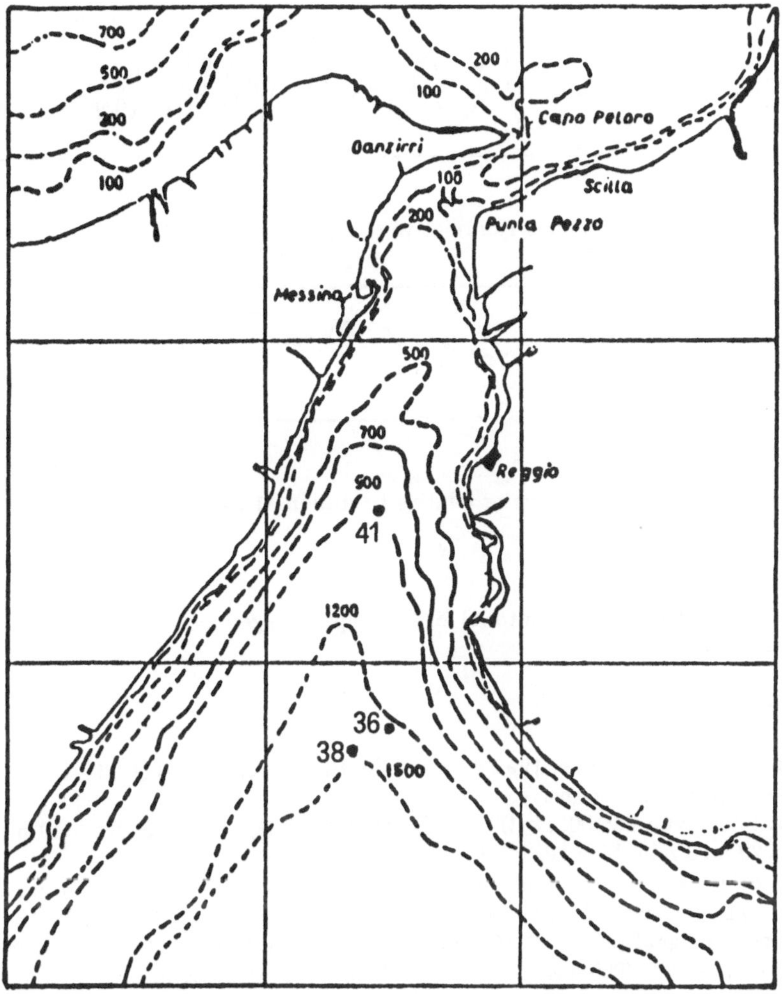

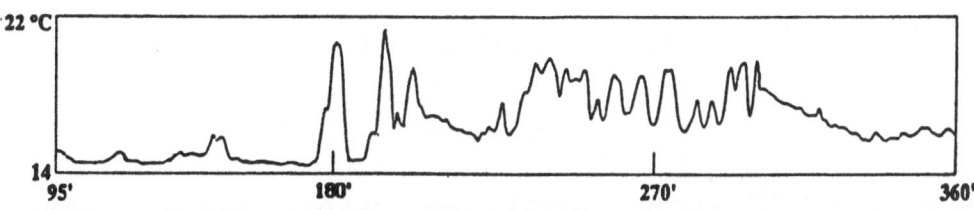

Figure 19. Working zone and stations' location of the CACTUS 87 cruise (October 13 - 30, 1987) for the measurements in the strait of Messina.

Figure 20 a. Temperature time series relative to station 36 (fig. 19).

120

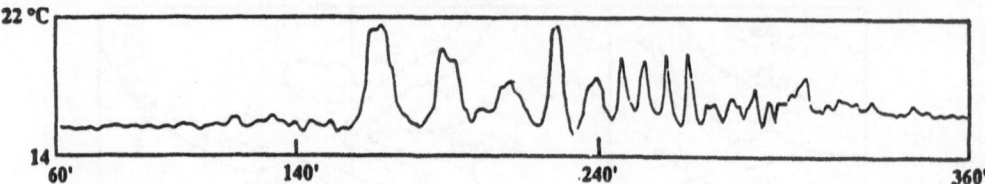

Figure 20 b. Temperature time series relative to station 38 (fig. 19).

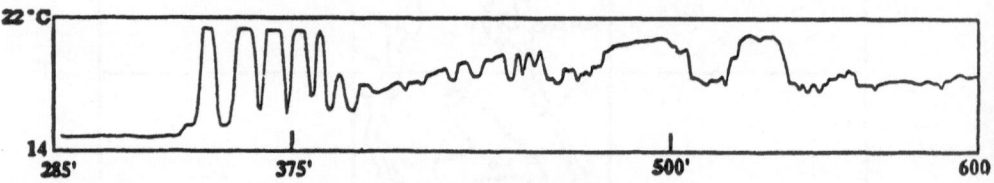

Figure 20 c. Temperature time series relative to station 41 (fig. 19).

ACKNOWLEDGEMENTS

We wish to thank Dr. Rosalia Santoleri for her helpful suggestions and the two referees for their fruitful criticisms.

APPENDIX

We now present in more detail the results of the PRIME cruise (May 26 - June 6, 1982) concerning internal wave trains.
a) Station 1 (between 38° 00' N, 15° 30' E and 37° 37' N, 15° 19' E, from 19 h 00 min, May 26 to 22 h 00 min, May 28, local time).
The thermocline was detected at a depth of $\simeq 30m$ by preliminary CTD casts. The Guildline CTD was located at the thermocline and both this CTD and the thermistor chain detected a signal ($\Delta T = 1.3°C$, figs. 15 and 16) lasting for 20 min, at about 23 h local time on May 27. This corresponds to a lowering of the interface $a_1 \simeq 17m$. During these measurements a north wind ($\simeq 15m/s$) caused the ship to drift southwards at the relatively constant velocity of $c_{drift} \simeq 0.4m/s$, rather close to the theoretical estimate of the "linear" velocity of the wave $c_0 \simeq 0.5m/s$ (equation 12). The estimated velocity (equation 13) is $c \simeq 0.63m/s$. The evaluated wavelength gives

$$\lambda_1 = (c - c_{drift})(20min) = 280m$$

and the period was

$$\tau_1 = \lambda_1/c_1 = 7min.$$

In Messina the tide was weak.
b) Station 2 (37° 32' N, 15° 20' E, from 3 h 00 min to 16 h 00 min local time, May 29). The procedure used and the data obtained were similar to those of station 1, the thermocline depth was $h_2 \simeq 30m$; the ship's southward drift was $\simeq 0.39m/s$. The instruments detected

a thermal jump of $\Delta T = 1.5°C$, corresponding to a lowering of the interface of $a_2 \simeq 11m$, which lasted 11 min, between 7 h 00 min and 8 h 00 min (figs. 15 and 16). An estimate of the linear wave velocity is $c_0 \simeq 0.62m/s$. The corresponding wavelength was $\lambda_2 \simeq 150m$ and $\tau_2 \simeq 4min$. In Messina the tide was weak.

c) Station 3 (36° 44' N, 15° 32' E from 21 h 00 min, May 29 to 8 h 00 min, May 30).

The thermocline was at $h_3 \simeq 20m$. The ship's southward drift in this southernmost station reached $0.6m/s$. A strong dispersed signal was observed, lasting $\simeq 30min$, between 3 h and 4 h on May 30, corresponding to $a_3 \simeq 10m$ (fig. 16). The estimated velocity was $c_0 \simeq 0.5m/s$ for the linear theory and $c_3 \simeq 0.65m/s$ for the nonlinear theory. The resulting wavelength is $\lambda_3 \simeq 90m$, and $\tau_3 \simeq 2.5min$. These should be considered as only rough estimates since the wave's velocity was much too close to the ship's drift. The tidal current in Messina was weak.

d) Station 4 (38° 27' N, 15° 40' E, between 17 h 30 min, June 2 and 23 h 00 min, June 3). This was the first station north of the sill of the strait of Messina for this cruise and was of particular interest since no wind or surface waves were present and the interface was particularly flat. A strong thermocline was present at $h_4 \simeq 25m$. At 5 h 30 min on June 3, the temperature at 25 m jumped $\Delta T \simeq 1.1°C$ (fig. 17), corresponding to an interface lowering of $a_4 \simeq 10m$; it lasted 6 - 7 min. The corresponding linear velocity was $c_0 \simeq 0.6m/s$ and the nonlinear velocity was estimated as $c_4 \simeq 0.72m/s$. Similarly, $\lambda_4 \simeq 280m$, and $\tau_4 \simeq 7min$. In the strait the tidal current was at its lowest value relative to its strong periodicity. The other signal at about 04 h 30 min hours was not considered because it was recorded while the ship was in motion.

e) Station 5 (38° 38' N, 15° 41' E, between 1 h 00 min and 7 h 00 min, June 4).

Wind, surface and internal waves were absent and a strong thermocline was present. Again, the tidal current in Messina was weak.

f) "Rodeo" Station.

The measurements were performed here by towing the Guildline CTD after mounting additional ailerons on the instrument so that it could glide at a depth of $\simeq 17m$ with a ship's velocity of $\simeq 2.0 \pm 0.1m/s$, between 15 h 00 min, June 5 and 10 h 00 min, June 6. The towing transects north of the strait (fig. 15) were repeated 4 times.

The first set of signals was observed at 17 h 00 min on June 5 (fig. 18). A corresponding surface phenomenon, namely different optical properties of the air - sea surface, propagating with apparently constant velocity, was visible to the ship's crew. This surface signal was somewhat dispersed (several segments $\simeq 1km$ wide and $\simeq 100m$ long) and was apparently moving northwards at a velocity of 0.5 - 1.0 m/s. Subsequently, during a southward displacement of the ship, a strong set of internal waves was observed (1 h 30 min, June 6, as shown in fig. 18). It comprised of 4 - 5 internal waves, with decreasing amplitudes, followed by a more dispersed signal. After 1 hour the measurement was repeated, at a distance of 3 km, during a northward transect; only the last 3 - 4 waves were observable. Since the ship and the waves were both moving northwards it was possible to make more precise observations. The largest wave had $\Delta T = 5.5°C$, corresponding to an interface lowering of $a_R = 25m$. The other waves had amplitudes of 20, 15 and 13.5 m; they lasted $\simeq 2min$ each and the whole signal lasted 1/2 hour. The corresponding length was $\simeq 140m$. Later, between 9 h 00 min and 10 h 00 min on June 6, three internal waves were observed in the same position, in order of decreasing amplitude, the largest measuring $a_R = 15m$ (fig. 18). In Messina the tidal currents were at their highest values with respect to the 14 day modulation.

During the CACTUS 87 cruise (October 1987) Nicolo' et al. (1989) observed three packets of internal waves south of the strait (fig. 19), with the following characteristics.

a) Station 36 (37° 58' N, 15° 35' E, October 22, 1987 between 9 h 00 min and 16 h 55 min GMT).

The thermocline was detected, by means of a preliminary CTD cast, at a depth of $\simeq 37m$. Here a temperature time series was recorded with the same CTD and three signals were seen to occur in a 30 min time interval (fig. 20 a). Temperature jumps were of $4.2^\circ C$, $3.6^\circ C$ and $2.3^\circ C$, corresponding to lowering of the interface (38 m, 18 m and 13 m respectively). A long tail approximately $2^\circ C$ above the average can also be seen. The estimated linear velocity is $c_0 \simeq 0.5m/s$ (equation 12), while nonlinear velocity estimates for each of the three signals give $c_1 \simeq 0.8m/s$, $c_2 \simeq 0.7m/s$ and $c_3 \simeq 0.6m/s$ (equation 11). Wave lengths were computed for each event, also taking into account the ship's drift, which in this case was northwards at a speed of $c_d = 0.2m/s$. One has:

$$c = \lambda/(\tau + \tau_g)$$

where τ is the duration of each signal and $\tau_g = c_d\tau/c$ is the time correction factor due to the ship's drift.

b) Station 38 (37° 59' N, 15° 33' E, October 22 - 23, 1987, between 22 h 54 min and 5 h 54 min GMT).

The thermocline was located at $\simeq 45m$. Nine signals and a tail were observed during 170 min the highest of which is a $6^\circ C$ thermal jump, corresponding to a lowering of the interface of $\simeq 40m$ (fig. 20 b). Non linear speed $c = 0.6m/s$. The lowest amplitude ($\simeq 2.8C$ corresponding to a lowering of $12.7m$) refers to the last peak before the tail; here $c = 0.5m/s$. The linear velocity for these waves is $c_0 = 0.4m/s$. The ship's northward drift velocity was $c_d = 0.1m/s$.

c) Station 41 (38° 04' N, 15° 34' E, October 24, 1987, between 1 h 00 min and 4 h 00 min GMT).

The thermocline was detected at 45 m. The data show 5 signals and a large tail (fig. 20 c) which cover a time interval of 50 min. All but the last peak have a top cutoff; this is probably due to the fact that the interface was lowered past the CTD. The characteristics of the longest wave were a thermal jump of $4.7^\circ C$, an interface lowering of 38 m and $c \simeq 0.9m/s$. Those of the smallest one were: $2^\circ C$, $8.0m$ and $0.7m/s$, respectively. The linear velocity of the waves was $c_0 = 0.6m/s$ and the ship drifted southwards at a speed of about $0.1m/s$.

REFERENCES

Abbate, M., Dalu, G. A. and Salusti E. (1982) 'Energy containing eddies in the strait of Messina', Il Nuovo Cimento, 5C (5), 571 - 585.

Alpers, W. and Salusti, E. (1983) 'Scylla and Charybdis observed from space', J. Geophys. Res., 88, 1800 - 1808.

Baines, P. G. (1977) 'Upstream influence and Long's model in stratified flow', J. Fluid Mech., 82, 147 - 220.

Baines, P. G. (1979) 'Observations of stratified flow over two dimensional obstacles in fluid of finite depth', Tellus, 31, 351 - 371.

Baines, P. G. and Hoinka, K. (1985) 'Stratified flow over two dimensional topography in fluid of infinite depth, a laboratory simulation', J. Atmos. Sci., 42 (15), 1614 - 1630.

Bhm, E., Magazzú, G., Wald, L. and Zoccolotti, M. L. (1987) 'Coastal currents on the Sicilian shelf south of Messina', Oceanologica Acta, 10, 2, 137 - 142.

Brandolini, M., Franzini, L. and Salusti, E. (1980) 'On the tides in the strait of Messina', Il Nuovo Cimento, 3C (6), 671 - 695.

Castaldini, M. and Franzini, L. (1979) 'On the currents on the Messina's strait: a modern treatment of historical Vercelli's data', Il Nuovo Cimento, 2C, 569 - 584.

Colacino, M., Garzoli, S. and Salusti, E. (1980) 'Currents and counter - currents in the western Mediterranean straits', Il Nuovo Cimento, 4C, 123 - 144.

Defant, A. (1940) 'Scilla e Cariddi e le correnti di marea nello Stretto di Messina', Geofisica Pura Applicata, 2, 93.

Defant, A. (1961) 'Physical oceanography', Pergamon Press, New York.

Del Ricco, R. (1982) 'Numerical model of the internal circulation of a strait under the influence of the tides, and its application to the Messina strait', Il Nuovo Cimento, 5C (1), 21 - 45.

Di Sarra, A., Pace, A. and Salusti, E. (1987) 'Long internal waves and columnar disturbances in the strait of Messina', J. Geophys. Res., 92, 6495 - 6500.

Farmer, D. M. and Freeland, H. J. (1983) 'The physical oceanography of fjords', Prog. Oceanogr., 12, 147 - 159.

Gardner, C. S., Greene, J. M., Kruskal, M. D. and Miura, R. M. (1967) 'Method of solving the Korteweg - de Vries equation', Phys. Rev. Lett., 19, 1095 - 1097.

Griffa, A., Marullo, S., Santoleri, R., Viola, A. and Paschini, E. (1986) 'Preliminary observations of large amplitude tidal internal waves near the strait of Messina', Cont Shelf Res., 6 (5), 677 - 687.

Guardiani, G., Pace, A. and Salusti, E. (1988) 'Preliminary observations of turbulence due to the collapse of internal solitary waves in the Gulf of Gioia, North of the strait of Messina', Bollettino di Oceanologia Teorica ed Applicata, 6 (1), 3 - 14.

Hopkins, T. S., Salusti, E. and Settimi, D. (1984) 'Tidal forcing of the water mass interface in the strait of Messina', J. Geophys. Res., 89, 2013 - 2024.

Korteweg, D. J., Vries, G. de (1895) 'On the change of long waves advancing in a rectangular canal and a new type of long stationary waves' Phil. Mag., 5, 422.

Lacombe, H., Richez, C. (1983) 'Regime of the strait of Gibraltar', in J. C. J. Nihoul (ed), Hydrodynamics of semi - enclosed seas, Elsevier, New York, pp. 13 - 73.

Long, R. R. (1955) 'Some aspects of the flow of stratified fluids, III, continuous density gradients', Tellus, 7, 341 - 357.

Marullo, S. and Santoleri, R. (1986) 'Fronts and internal currents at the northern mouth of the strait of Messina', Il Nuovo Cimento, 9C, 701 - 714.

Massi, M., Salusti, E. and Stocchino, C. (1979) 'On the currents in the strait of Messina', Il Nuovo Cimento, 2C, 543 - 548.

Mazzarelli, G. (1938) 'Vortici, tagli e altri fenomeni delle correnti nello Stretto di Messina', Atti Reale Accademia Peloritana, Messina, vol. XL.

McIntyre, M. F. (1972) 'On Long's hypothesis of no upstream influence in uniformly stratified or rotating fluid', J. Fluid Mech., 52, 209 - 243.

Mosetti, F. (1988) 'Some news on the currents in the strait of Messina', Bollettino di Oceanologia Teorica ed Applicata, 6 (3), 119 - 179.

Nicolo', L., Salusti, E. (1989) 'Satellite and "in situ" observations of tidal large amplitude internal waves south of the strait of Messina, Mediterranean sea', work in progress.

Osborne, A. R. and Burch, T. L. (1980) 'Internal solitons in the Andaman Sea', Science, 208, 451 - 460.

Salusti, E., San Emeterio, J., Zambianchi, E. (1988) 'Steady interface rising due to tidal effects in the strait of Messina', Bollettino di Oceanologia Teorica ed Applicata, 6 (1), 43 - 56.

Sapia, A. and Salusti, E. (1987), 'Observation of nonlinear internal solitary wave trains at the northern and southern mouths of the strait of Messina', Deep Sea Res., 34 (7), 1081 - 1092.

Stigebrandt, A. (1977) 'On the effect of barotropic current fluctuations on the two - layer transport capacity of a constriction', Journal of Physical Oceanography, 7, 118 - 122.

Stommel, H. and Farmer, H. G. (1952) 'On the nature of estuarine circulation. Part I.', Woods Hole Oceanographic Institution, ref. no. 52 - 88.

Vercelli, F. (1925) 'Il regime delle correnti e delle maree nello stretto di Messina', Commissione Internazionale del Mediterraneo, Venice, Italy.

Vercelli, F. and Picotti, M. (1925) 'Il regime fisico - chimico delle acque nello stretto di Messina', Commissione Internazionale del Mediterraneo, Venice, Italy.

Wei, S. N., Kao, T. W. and Pao, H. P. (1975) 'Experimental study of upstream influence in the two - dimensional flow of a stratified flow over an obstacle', Geophys. Fluid Dyn., 6, 315 - 336.

MEASUREMENTS AND MODELLING IN THE GREAT BELT:
A UNIQUE OPPORTUNITY FOR MODEL VERIFICATION

David M. Farmer   Jacob Steen Møller
Institute of Ocean Sciences   Danish Hydraulic Institute
9860 West Saanich Road   Agern Allé 5
Sydney, B.C.   2970 Hørsholm
Canada V8L 4B2   Denmark

ABSTRACT. Environmental studies initiated in conjunction with construction of a bridge across the Great Belt will offer some remarkable opportunities for scientific investigations. In this paper we provide some background to the studies and outline the oceanographic aspects of the environmental plan and its implementation. The combination of a two-dimensional two-layer numerical model of the strait and a plan for comprehensive field measurements provides a unique chance for verifying the model, and thus deepening our understanding both of the modelling approach and of the fluid dynamics of the strait. In anticipation of the major field program some limited data were collected in a pilot study. Preliminary comparisons are encouraging.

1. Introduction

The Great Belt (Storebælt in Danish) is a channel approximately 18 km wide that divides Denmark into Jutland and Funen to the west and Zealand to the east. It is the largest of the three channels linking the Baltic to the Kattegat and North Sea (Figure 1.1). A major construction project (The Great Belt Link) is now underway to build a rail and road link across the strait and thus replace the busy cross-Belt ferry routes. The strategic location of the link and environmental concern about potential interference in the exchange of water between the Baltic and Kattegat have led to a comprehensive environmental plan.

A primary purpose of the present paper is to draw to the attention of the scientific community the scope of this project and the unique opportunity it presents for the study of stratified flows in straits. We outline the oceanographic aspects of the environmental plan and its implementation, including numerical modelling, oceanographic measurements in the strait and a comparison of model predictions with preliminary field observations. As background to the subsequent discussion a brief outline of the construction plan and of the overall hydrographic setting is presented in §2. A key concept in this plan is the 'zero solution'

125

L. J. Pratt (ed.), The Physical Oceanography of Sea Straits, 125–152.
© 1990 *Kluwer Academic Publishers.*

discussed by Ottesen-Hansen & Møller (1989). The basic approach to be taken with respect to compensating for blocking effects of the construction works is summarised in §3.

Both in concept and execution this study departs from traditional environmental programs associated with industrial developments. The concept of fully compensating any adjustments to the exchange flow due to engineering works is novel and places special demands on the understanding of stratified flow in the strait and on the required field measurement program. Scale estimates of the influence of the construction works on the mean flow show that even in the absence of compensation dredging the overall effect of the construction would be small: it would therefore be unrealistic to rely only on direct measurements of the flow,

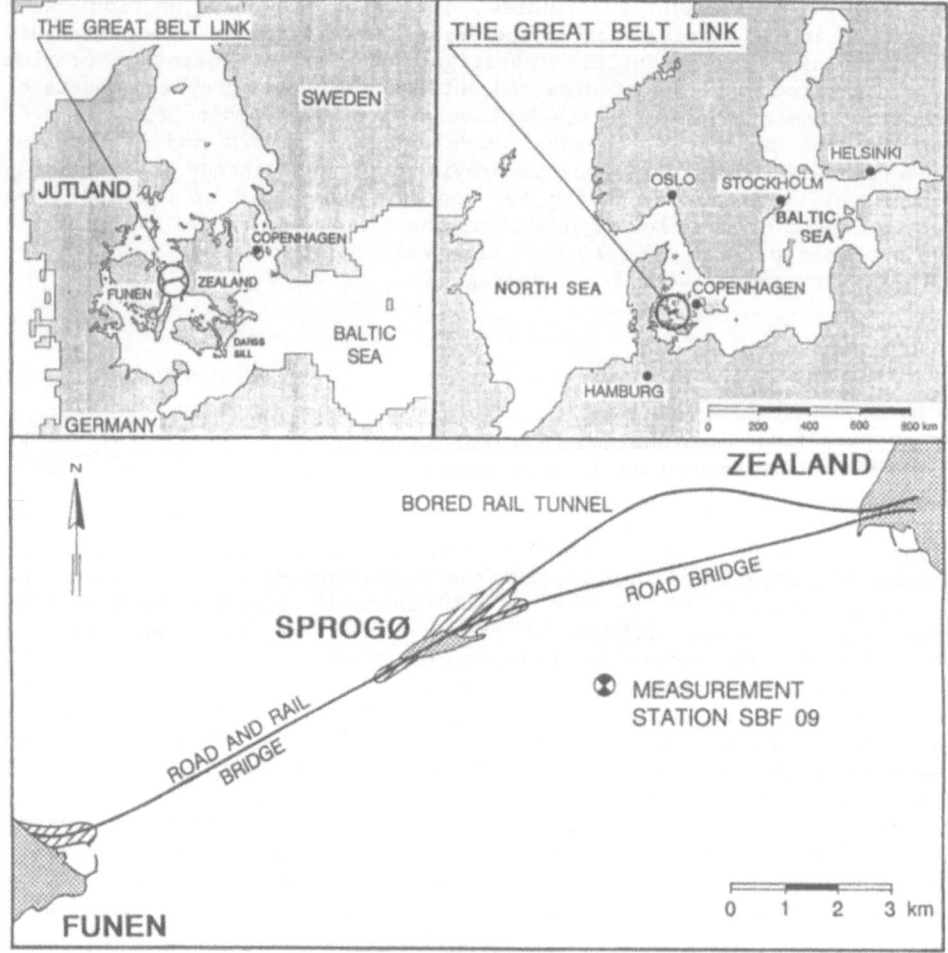

Figure 1.1. Chart showing the Great Belt Link.

either to identify such changes or to guide the required excavations, but of course the local effects in the immediate vicinity of construction will be quite large. For this reason our plan has been to use a numerical model and then to thoroughly test the model with measurements obtained in an intensive observational program. With the confidence gained from such tests the model may then be used to calculate the appropriate compensation adjustments.

It should be emphasised that no attempt is being made to model any long term changes in the Baltic that might arise from the blocking. The goal is to compensate the potential blocking with dredging calculated as accurately as possible with existing modelling technology. Thus, to within the accuracy of our calculations, there will be no change imposed on the water exchange, and therefore no change is predicted for the Baltic.

The proposed observational program which is discussed in §4, makes use of novel measurement approaches and is probably one of the most comprehensive that has ever been undertaken in a major strait. At the time of writing (June 1989) this program is just beginning; it will continue throughout the construction period (1989-1995).

The modelling effort represents a substantial undertaking. Together with analytical and laboratory investigations it is discussed in §5. Two-dimensional two-layer models have of course been used previously. The primary difficulty in applying such models to a topographically and hydrodynamically complex environment such as the Great Belt, is lack of knowledge of the boundary conditions. As discussed in §3, the application of the model to the 'zero solution' neatly circumvents the requirement for knowledge of precise and detailed boundary conditions, because a correct compensation for blocking of the flow by the bridge implies, by definition, that there will be no far-field influence; thus any reasonable boundary conditions outside of the immediate region of construction and compensation excavations may be used to test whether the compensation dredging is appropriate. But the validity of the model itself still needs to be tested and it is this need that has in part motivated the design of a comprehensive field measurement program.

A preliminary observational program was carried out in November 1987. The purpose here was to gain experience with different measurement approaches and to acquire an initial set of data with which to compare the model. In this pilot study the extent of data available is much less than that which will be obtained in the full measurement program. The data are nevertheless sufficient to carry out some preliminary tests of the model and these tests are also discussed in §5.

From an oceanographic point of view, a comparison of model output and observations constitutes an ambitious challenge to our ability to describe and understand the details of stratified flow in the Great Belt. A comprehensive analysis of the data over different seasons, and thus for various degrees of stratification, should lead to new insights on factors influencing exchange through this major strait.

2. The Great Belt Link Project and Its Hydrographic Setting

The project will consist of a combined road and rail bridge across the western Channel. At the island of Sprogø (see Figure 2.1) the traffic is divided into rail and road. The road link between Sjælland and Sprogø will consist of an elevated bridge while the rail link will consist of a bored tunnel.

The Great Belt is a broad and shallow strait separating the relatively saline North Sea from the fresher Baltic. It is the largest of the Danish straits and is estimated to account for 70% of the exchange between the Baltic and the open ocean, Jacobsen 1980. The exchange of water is of course of fundamental importance to the maintenance of water quality throughout the Baltic (Fonselius 1969 and 1988, Welander 1974, Pedersen 1978 and 1982, Pedersen and Møller 1981, Stigebrandt and Wulff 1987); it is this consideration that has led to the design of the 'zero solution' concept and has motivated the extensive hydrodynamic modelling and measurement program discussed below.

The hydrography of the Danish straits has been subject to study since the turn of the century (Knudsen 1899, Jacobsen 1910, 1913 and 1925). In this early period the basic features of the flow were revealed. Prior to the the comprehensive Danish Belt Project (1980a, 1980b,1981) several sporadic field investigations and theoretical works exist and attempts to model the exchange were made. (Würtki, (1954), Svansson (1972), Bertelsen and Warren (1977), Matthäus et al. (1983), Stigebrandt

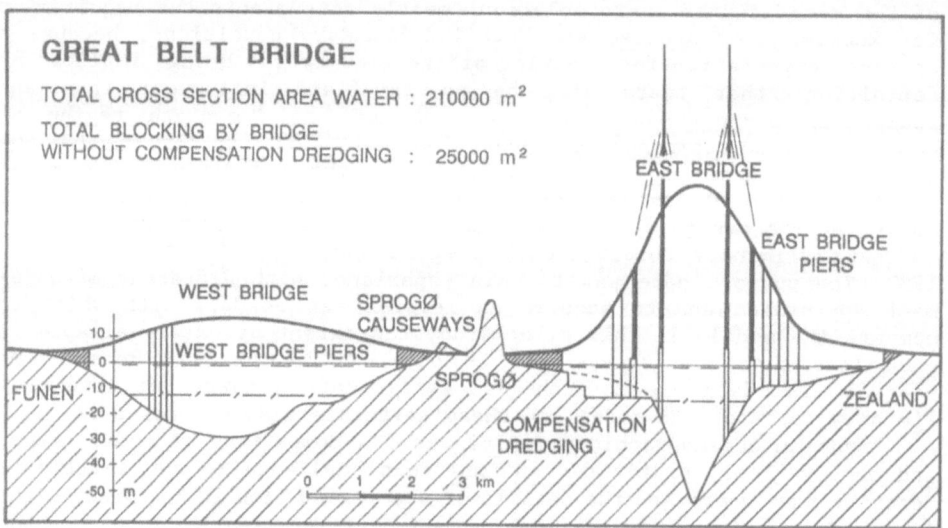

Figure. 2.1. Overall layout of the Great Belt Link. In addition to the bridge piers, substantial causeways in addition to an artificial island just north of Sprogø contribute to a blocking of the flow. Compensation dredging is planned in the eastern channel so to as offset this effect.

1983)). But with the Belt project a large and long term monitoring program was launched. Based on this wide material a brief summary of the hydrography of the Great Belt is given here.

For much of the year the strait is stratified. There is a background density driven flow in which the saline water flows southward toward the Baltic along the sea-floor, beneath a northward flowing fresher layer. This residual baroclinic flow is however strongly forced by a barotropic component which is driven by larger scale weather systems moving across Scandinavia so that both layers typically move together, but at different speeds, for a few days in one direction and then in the other. A modest tidal signal is superimposed on this meteorologically forced flow.

The Great Belt forms part of a large estuary comprising the Baltic Sea, Danish Straits and southern portion of the Kattegat. The outer boundary condition for the resulting estuarine circulation occurs in the form of fronts separating the fresher Baltic outflow from saline water moving in from the North Sea. The fronts are just the surface outcroppings of the interface dividing the two layers; they move back and forth through the Straits under the influence of the barotropic forcing (see Figure 2.2) and may be sites of significant mixing.

CURRENT CONDITION, GREAT BELT

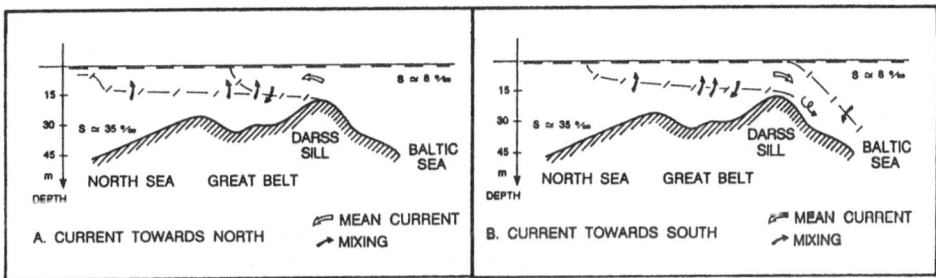

Figure 2.2.   Sketch of stratification in the Great Belt and neighbouring waters during outflow and inflow events.

Current measurements in the Strait illustrate the strongly correlated movements of upper and lower layers. However when the barotropic component is weak the baroclinic circulation can assert itself as a bi-directional flow with the fresh layer moving north above the southward flowing deeper layer. Figure 2.3 shows a scatter diagram in which current measurements at 7m in the surface layer are plotted against those at 30m in the deeper layer. While much of the data is consistent with a positive correlation, a significant fraction of the points lie in the second quadrant implying that for part of the time there is bidirectional flow corresponding to the estuarine circulation.

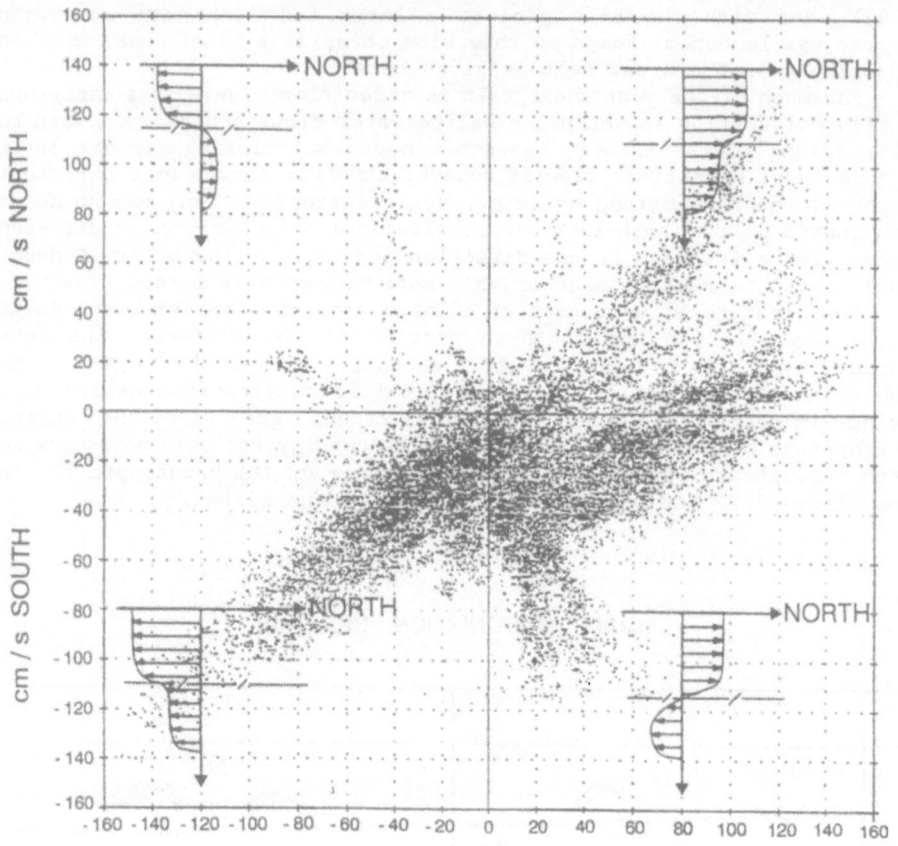

Figure 2.3.   This scatter plot is derived from currents measured in the
               eastern channel at 7m depth in the upper layer and 30m in
               the lower layer. The currents are resolved into North-South
               components, with data from the deeper instrument scaled on
               the vertical axis and the upper layer instrument on the
               horizontal axis. Data are from the period 7 May - 2 June
               and 4 October - 1 November, 1977. Data from The Belt
               Project (1980a).

        The density gradient is dominated by the salinity distribution.
Maximum stratification occurs in June and July, with minimum values in
the winter months (Figure 2.4).   By combining available current
measurements with the known variations in stratification, estimates of
the internal Froude number can be made (Figure 2.5). It appears that for
most of the year a subcritical two-layer flow is the rule. For about 15%
of the time this two-layer flow is hydraulically controlled. The control
however is not of the bidirectional sort, but arises from strong

barotropic forcing on the two layers as they move through constricted areas of the Strait. For less than 1% of the time the flow is well mixed; this condition holds for extreme events, such as occasional inflows arising from winter storms.

A scale analysis of the momentum equation for each layer has been used to estimate the relative balance of forces. (DHI/LIC 1988). The sea surface slope typically provides the dominant driving force for the upper layer, except during transitional periods when wind stress and the horizontal density gradient become important. For the lower layer, the barotropic pressure gradient is also typically dominant, with interfacial stress contributing in the same direction as the surface layer current. This explains the fact that both layers tend to move in the same direction except during transitional periods. When the barotropic component is weak, the interfacial slope can dominate, driving the salty lower layer southward.

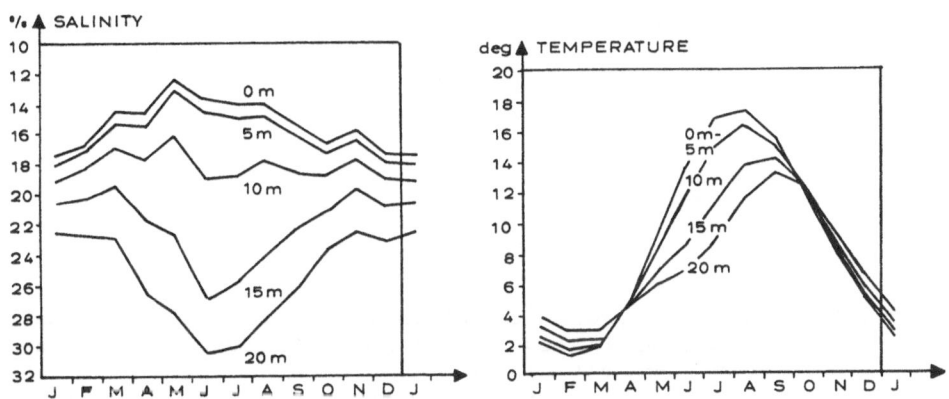

Figure 2.4.   Seasonal variation in salinity and temperature. Monthly mean values are shown for different depths at the Halsskov Rev Lightship East of Sprogø for the period 1931-60. Strong mixing between the layers occurs during stormy periods in the autumn and winter. During summer there is less wind, reduced mixing and consequently stronger stratification. Data from The Belt Project 1980a.

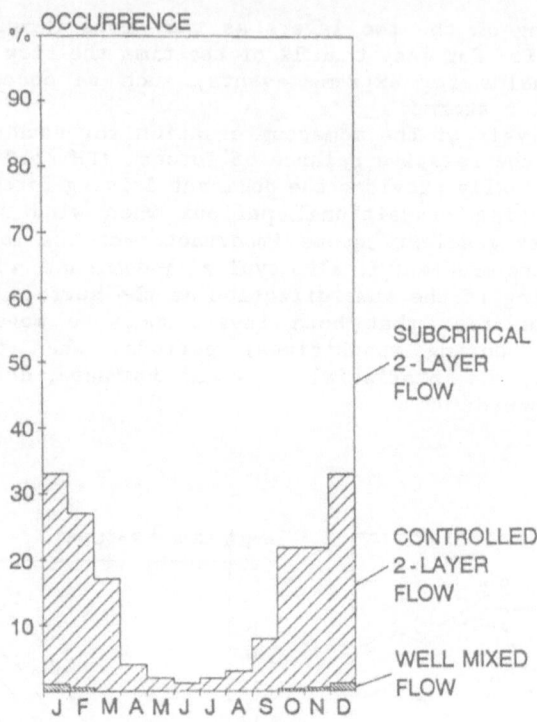

Figure 2.5.   Estimated probability distribution for different hydraulic
              conditions in the Great Belt, including subcritical 2-layer
              flow, controlled 2-layer flow and well mixed flow. It must
              be emphasized that this estimated distribution may be
              substantially revised as data from the comprehensive
              environmental monitoring program become available.

3. Application of the Zero Solution Concept

The bridge is being constructed under the constraints, imposed by
political agreement, that:

1. The discharge through the Belt remain unchanged by the crossing;

2. The salt balance for the Baltic remain unchanged by the crossing;

This political agreement requires a scientific interpretation. As
discussed by Ottesen-Hansen and Møller (1989) and DHI/LIC (1988), the
interpretation is based on the assumption that the stratification can be
approximated by a 2-layer flow. With this approximation the requirement
is that the deviation in discharge for each layer, and that the mixing

between the two layers, remain unchanged. Note that it is not sufficient that just the discharge should remain unchanged. An alteration in mixing will alter the salinity of each layer, and indeed it has recently been estimated (Jürgensen, 1988 and 1990, and Jacobsen 1988) that mixing by ship traffic has resulted in a noticeable increase in the salinity of the Baltic.

The above interpretation depends upon an arbitrary choice for the 2-layer representation. Observations indicate that both a near surface zone and the deeper part of the water column are usually well mixed. On occasion there may be two pycnoclines, although usually only a single, well defined pycnocline occurs (Figure 3.1). The two-layer representation is found by choosing the depth at which the salinity is half-way between the surface and bottom values. This approximation breaks down as the stratification gets very weak in the winter, but in these conditions the flow is more appropriately considered to be single layer.

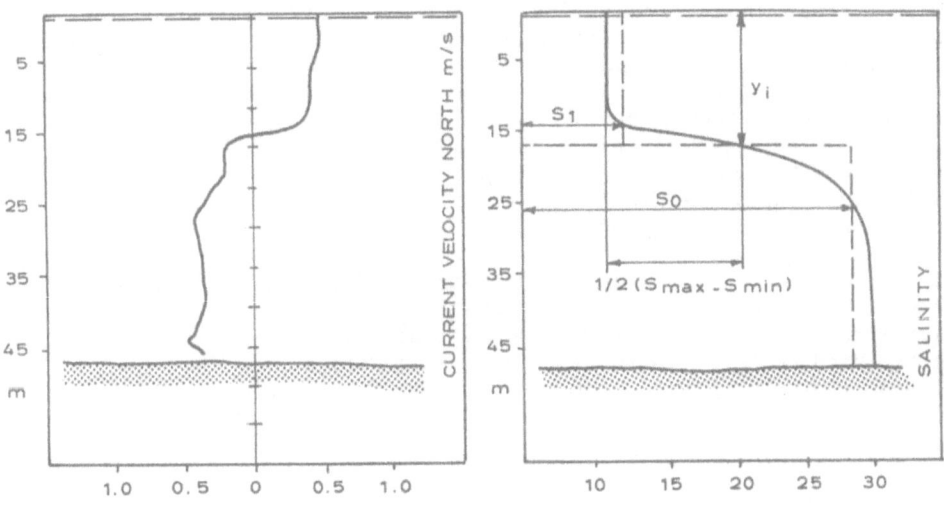

Figure 3.1.   Example of current and salinity profiles. 30 October 1987. North East of Sprogø. Densities and velocities in each layer are determined as mean values within the layers defined by y_i (interface depth), where y_i is determined as the depth in which the salinity equals $S(y_i) = 0,5 (S_{max} + S_{min})$.

The intention is to avoid far-field hydrographic influence. Only in this way is it possible to avoid essentially unpredictable environmental effects (biological and hydrodynamic). Some near-field influences are of course inevitable. But the concept of a zero-solution discussed by Ottesen-Hansen and Møller (1989) involves compensating for the flow resistance introduced by ramps, bridge piers, etc. so that, to within the limits of our present ability to determine it, the two-layer exchange remains unaffected and thus the far field consequences are eliminated.

From the point of view of the present discussions, an important aspect of the application of this solution is that a prototype flow regime for a design period of 10 days is used against which to optimise an excavation plan. The model flow regime includes a typical meteorologically forced northward flow followed by a southward flow. The optimisation procedure consists of adjusting the modelled compensation excavations incrementally, running the model for the same boundary conditions and finding the deviation from the prototype flow.

The model calculations are not limited to a single 10-day prototype flow, and in fact tests have been carried out for a wide range of flow conditions. For example tests have been carried out for substantially different values of interface depth, barotropic pressure gradient and wind-stress. The 10-day period was chosen as being representative of stratified conditions including both subcritical and controlled regimes. Its relatively short duration makes it practicable to run a very large number of simulations, which is necessary for optimising the dredging calculation. Although the well-mixed flow is not represented in the design period separate calculations show that the dredging required for satisfactory compensation of a 2-layer flow is fully applicable for the well-mixed case. The final zero solution will be thoroughly checked for much longer period simulations.

An advantage of the chosen scheme is that since a 'zero solution' is sought, that is, a solution with minimal deviation from the prototype flow, the precise boundary conditions are not important. They only become important if the deviation from a 'zero solution' is significant. On the other hand it is critical to the success of the scheme that the model be a good representation of the physics. The extensive measurement program to be carried out in the strait will provide an opportunity for model verification.

4. Environmental Monitoring Program

A comprehensive monitoring program has been initiated in the Great Belt, with the following purposes:

(1)   To establish additional reference data on flow in the strait prior to the bridge construction.

(2)   To provide a comprehensive data set for verifying and if necessary, tuning the numerical model.

(3)   To provide adequate data for running forecast models required in the day-to-day planning of sensitive aspects of the construction.

(4)   Verification and control of the effects of compensation dredging.

(5)   Verification of effects in the nearby waters of Langelandssund and nearfield impact studies.

The program includes both moored instrumentation and measurements from a vessel. Figure 4.1 shows the general layout. The overall design of the program is closely linked with the two-layer numerical model discussed below. In particular, measurements of the required boundary conditions (sea-surface height and interface height) are obtained at the northern and southern boundaries of that part of the strait included in the model. A total of 9 water level gauges is included. These water level measurements, along with observations from other selected sites, are communicated directly by radio, cable or public telephone link to a central data station in Sprogø. Real time measurement in this way is essential to provide up-to-date input for environmental forecasting.

Remote current profiling using acoustic Doppler instruments will take place both in the eastern and western channels (stations 7 and 9) with real time data link by cable. A novel technique will be used to measure the interface level at the northern and southern boundaries. In addition to temperature-conductivity chains (with sensors at 7, 10, 12, 14, 16 and 19m) an echo-sounder will be used. Prior tests from a ship have shown that a well defined maximum in the acoustic backscatter occurs at the interface. The acoustic technique is still under development but good results are achieved when the interface is sharp and not too close to the surface or bottom (within 10% of depth). Both internally recording (stations 3 and 12) and real time measurements (stations 2, 7, 9, 11 and 13) of the interface depth will be obtained.

Additional instrumentation will be used to measure meteorological parameters (at Sprogø), waves (stations 5 and 7) and dissolved oxygen and turbidity at 7 and 9m depth (stations 2, 7, 9, and 11). The vessel M/S PIP will be equipped with an acoustic Doppler current profiler, an electro-magnetic current profiler, CTD with dissolved oxygen and turbidity sensors, acoustic interface sensor and a conventional echo-sounder. The vessel uses a Syledis positioning system. Additional measurements of currents, temperature and salinity, oxygen and interface level will be obtained in the sound West of Langelandssund.

While the recording instruments and real time observations will continue throughout the construction phase, there will also be certain periods of intensive ship-based measurements during 1989, 1990 and 1991. During the measurement periods, comprehensive observations of the flow in the vicinity of construction and dredged areas will allow the calculations of compensation dredging to be updated.

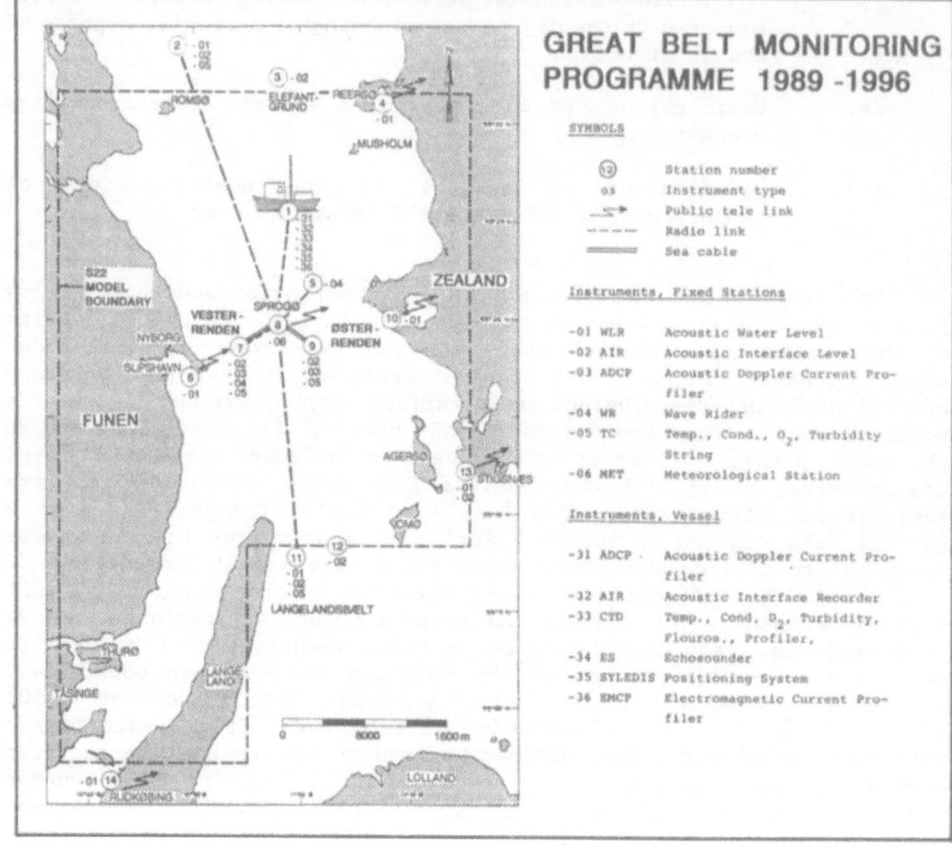

Figure 4.1.   Sketch showing locations of different instrumentation
included in the Master Plan for Environmental Monitoring.
Much of the data will be relayed to a central station at
Sprogø by cable, telephone and radio, allowing real-time
analysis and environmental forecasting.

5.   Modelling and Model Verification

Scale analyses and laboratory modelling

The primary tool for calculating the influence of the bridge on the
flow, and for determining the necessary compensation dredging, is a two-
layer, two-dimensional numerical model called System 22. Before this

model was set up, a large number of analytical investigations as well as some laboratory studies were carried out (DHI/LIC 1988). The purpose of these was to identify the overall magnitude of various effects and to ensure that the subsequent modelling would provide a satisfactory representation of the fluid dynamics of the Great Belt.

The scale analysis was carried out for a representative bridge design (the final design was not fixed until later), along with representative flow parameters established during field studies. The analysis showed that resistance to flow through the Great Belt is dominated by a combination of friction and expansion loss, the latter occurring just downstream of the contraction formed by the narrow channels on either side of Sprogø. The bridge construction will have the effect of increasing the flow resistance in the western channel by about 4% while reducing it, after completion of compensation dredging, by about 5% in the eastern channel. The net effect will be a small transfer in discharge from the west to the east channel.

In order to ensure that the zero solution is achieved it is necessary to keep the mixing unchanged. Mixing between two layers in a subcritical flow is appropriately formulated in terms of a flux Richardson number. Following Bo Pederson (1980), the flux Richardson number depends upon the production of turbulent kinetic energy. For given flux Richardson number, R_f the resultant entrainment across the interface can be found. The turbulent kinetic energy production can be calculated by integration, for example over the sea-floor, interface and sea surface. The resultant mixing can then be found for a given value of the flux Richardson number. When correctly formulated in terms of the turbulent kinetic energy production, estimates of R_f both in laboratory and field studies suggest a value of about 0.05. However it was not clear that this value would also be appropriate for the case of bridge piers in the two-layer flow. Consequently laboratory studies were carried out; these demonstrated that even for this case a value of 0.05 was appropriate. This demonstrates that the mixing is maintained unchanged by transferring the energy loss of the flow (turbulence production) from 'natural' processes like bottom friction and expansion losses to 'artificial' processes like pier resistance, which precisely is the goal of the compensation dredging. Nevertheless, scale calculations of mixing due to the bridge showed that it is small relative to other factors. For example the turbulent kinetic energy production due to the bridge piers in the upper layer is about 5% of that due to typical wind effects on the area within 10 km North and South of the bridge.

A result of the scale analysis of two layer flow in the strait was the finding that for a significant fraction of the time (Figure 2.5) the flow was hydraulically controlled. The analysis showed however that under these circumstances the flow typically passes through a gradually adjusting internal hydraulic jump and only resembles a fully turbulent jump in a few cases if at all. The hydraulic analysis further showed that it is normally only flow in the eastern channel that is controlled, with flow in the western channel remaining subcritical but frictionally limited so as to match the downstream conditions.

An uncertainty remains with respect to mixing in the controlled flow. The flux Richardson number formulation could not be expected to

138

apply in the rapidly varying flow associated with an internal hydraulic
jump and laboratory studies were conducted to investigate this case also.
(DHI/LIC 1988).

Numerical Model

Before the recent implementation of a two-layer numerical model in
the strait, various theoretical models had been developed for the Great
Belt and neighbouring seas (Figure 5.1). These include a local numerical
model (DHI/VKI, 1985) in which only the surface layer was described using
the DHI System 21 barotropic model. Model calculations were also
performed for the Kattegat (LIC, 1985). A model was also developed to
describe the mean current and mixing for long term equilibrium (~30
years). In this model, referred to as the Østersø model (LIC, 1984), the
mean transport of salt to and from the Baltic, east of the Darss sill,
was calculated.

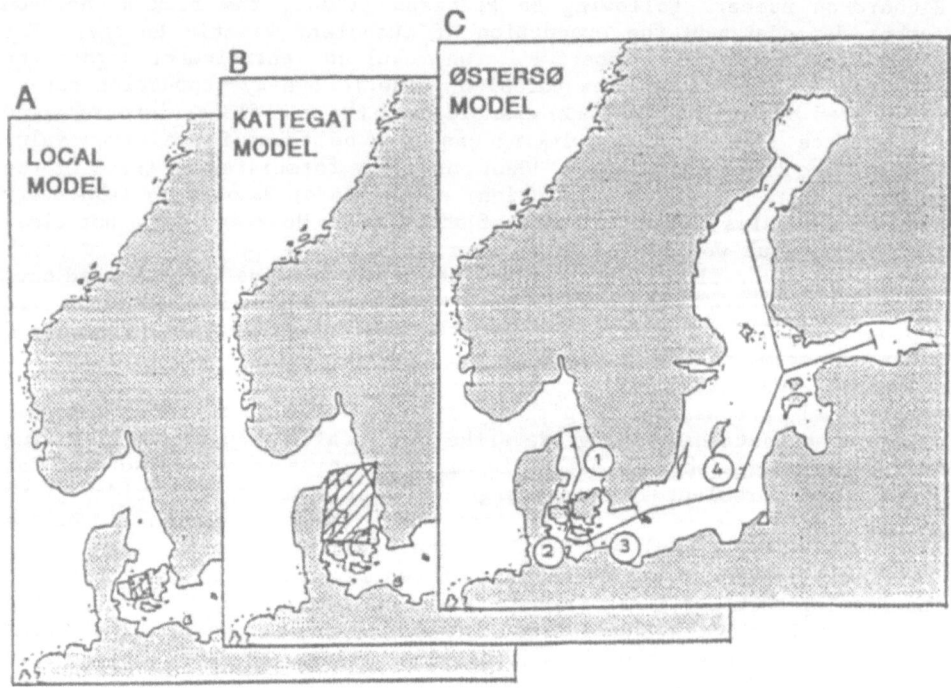

Figure 5.1.   Areas included in previous model studies related to the
               Great Belt.

The primary numerical model used in this study is a fully 2-dimensional 2-layer mathematical model developed at DHI and referred to as System 22. (For full details of this model the reader is referred to Abbott (1979) and DHI (1987)). The system generates a model of a given area by entry of appropriate bathymetric data. It computes the surface, interface levels, and flows in both layers in a rectangular grid in response to specified forcing functions, boundary and initial conditions. The basic equations are the continuity, x-momentum and y-momentum equations for each layer. These include the non-linear convective and cross-momentum terms. The model includes effects of rotation, wind shear stress, bed shear stress, interfacial shear stress, mixing, turbulent momentum dispersion, barometric pressure gradients and horizontal density gradients and changes within each layer. Production of turbulent kinetic energy is computed from the modelled flows as an overlay to S22. The equations are solved by implicit finite difference techniques with variables defined on a space-staggered rectangular grid.

In the present implementation the grid size was chosen to be 500m (a 250m grid is chosen for final calculations). Since the bridge piers are of order 30m, their impact on the flow was modelled by calculating the current induced drag force due to each pier and representing it by an equivalent shear stress contribution compatible with the S22 model momentum formulation.

The model calculations were carried out over an area that includes the central portion of the Great Belt where the construction is taking place. In order to provide more realistic boundary conditions the two-layer S22 model is located within a larger area that is described using the single layer model (S21). Sea-surface levels at the S22 model boundaries are transferred from the single-layer to the two-layer model. The areas included in each model are shown in Figure 5.2.

A number of model verification tests were conducted using both laboratory results and analytical solutions. An extensive series of tests were carried out to verify that the model performed adequately in the presence of flow separation downstream of a contraction. In particular, flow separation is expected in the neighbourhood of headlands and islands. Provided that the modelled area includes all of the separated flow the model performs well. These comparisons provide confidence in the ability of the model to reproduce well defined flow regimes such as can be modelled analytically or in the laboratory. Nevertheless the model's primary use requires that it successfully describe the much more complex environment of the Great Belt. The only realistic way of verifying its performance in this environment is by direct comparison with appropriate field data.

There is one important limitation to the S22 model: the numerical technique is unstable in the presence of supercritical flow. This limitation may be overcome with an appropriate reformulation of the basic equations and appropriate procedures for achieving this have been proposed. However, implementation and testing of these procedures have not at this time been carried out and an alternative measure has been adopted. A dissipative interface is used to dampen the instability in supercritical flow. Although there is some theoretical basis for this

140

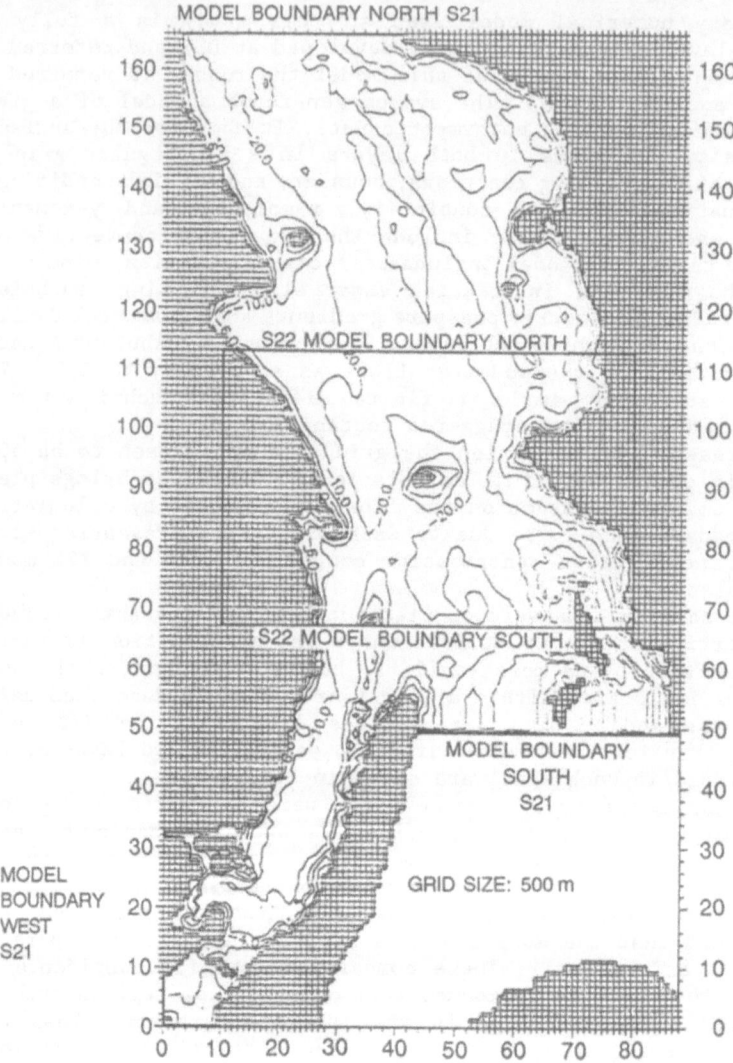

Figure 5.2   Areas included in the present model studies. The larger
area is modelled by the single layer S21 sytem using
available water level and meteorological data. This model
is then used to derive water level boundary conditions for
the 2-layer model S22, which encompasses the smaller area
of primary interest. The grid element spacing is presently
set at 500m, although finer resolution may be used if
required.

approach (Abbott 1979), tests with comparisons with analytical solutions were carried out. The tests showed that the model behaves well, provided that the jump is weak. This is nearly always the case in the Great Belt, so the dissipative solution is considered satisfactory in lieu of an appropriate adaptation of the numerical scheme.

An example of using the dissipative interface for flow over a sill is shown in Figure 5.3. The boundary condition is a steadily increasing discharge Q_0 and Q_1 and a steadily decreasing interface level at the down stream end of the flume.

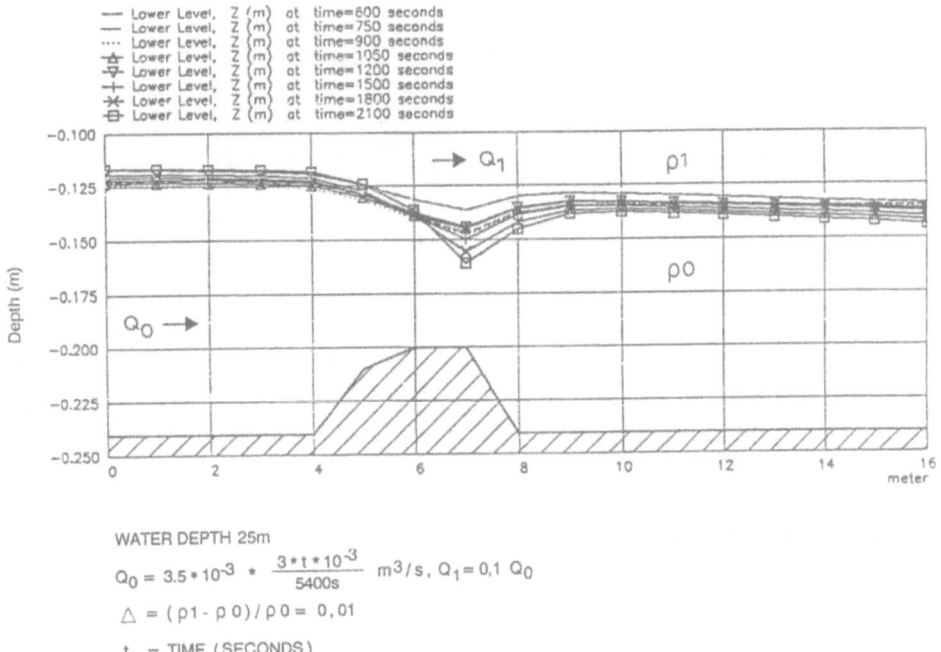

Figure 5.3.   Numerical simulation of flow over a sill.

In Figure 5.4 is shown a time series of the interface depth upstream and at the sill. After an initial wave (t < 600 sec.) the interface stabilises. On the figure is shown the quasi-steady analytical solution for the critical depth over the sill and the upstream interface level.

It is seen that the numerical solution breaks down at approximately t = 2300 sec. But it is interesting to note that for moderate flow (850 s < t < 2300 s) the flow is critical and the upstream interface depth is calculated correctly. This will be the typical situation in Storebælt where only weak jumps have been observed.

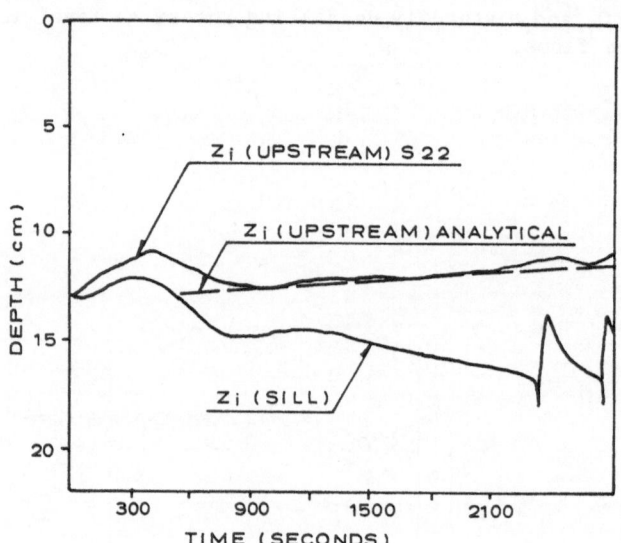

Figure 5.4.   Numerical simulation of flow over a sill. Time series of interface level upstream of sill and at sill. Analytical solution is quasi-steady.

In §5 an example of critical flow observed in Storebælt is shown together with the simulated flow.

Model Verification with Field Data

A preliminary field study was carried out in November 1987. It had originally been planned that the initial period (November 1-4) be used to tune the model, but comparisons indicated reasonable agreement between model and data, so model calibration was omitted and the full data set used for verification by direct uncalibrated comparison with measurements. Thus, the parameters originally chosen, for example for the shear stream formulation, were retained unchanged. Comprehensive boundary data, such as will be collected under the new environmental monitoring program, were not available. Nevertheless there was sufficient information to allow an initial comparison of data and model output. Figure 5.5 shows the general plan; current meters were installed at Stations A, B, C and D and there were a total of 6 water level recorders.

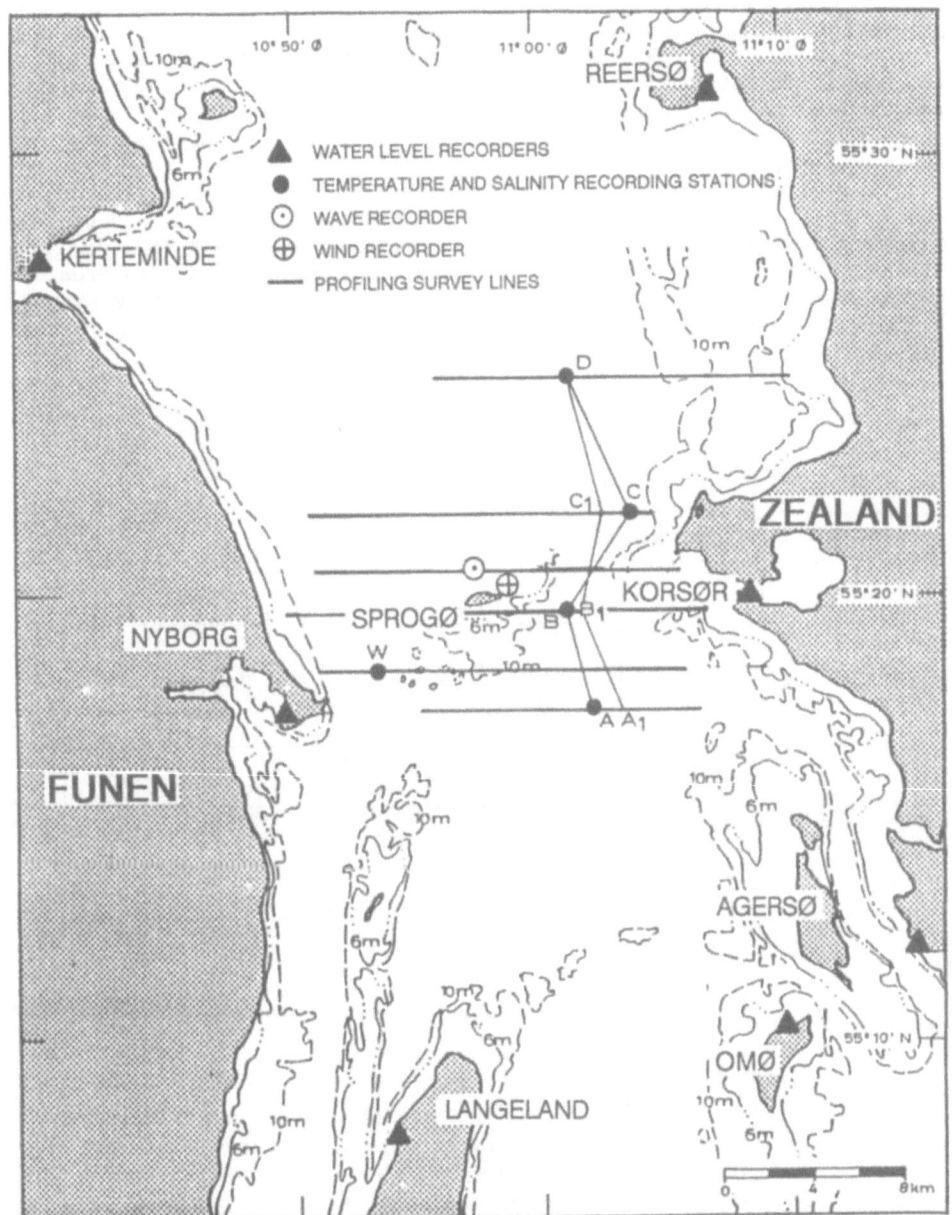

Figure 5.5.   Location of instrumentation used in the November 1987 pilot
study. Current meter records were obtained at Stations A,
B, C, D and W.

144

Care is required in the definition of interface depth and layer properties, given the continuous nature of the observed profiles, see Figure 3.1. Even for the quite limited period of available observations, the amount of data and model calculations is enormous and only a small representative sample can be shown here. We have attempted to develop means of presenting the data in such a way that the comparison between model and observations is most readily discerned.

Sea-level fluctuations were calculated with the barotropic model S21. Its effectiveness may be judged from Figure 5.6 where the height <u>difference</u> across the mid-point of the Great Belt (i.e. Korsør-Sliphavn) is shown during a short period of strong flow towards the north. Typical variability in sea level difference is 15-20 cm; the difference between modelled and observed sea-level difference is of order 1-2 cm.

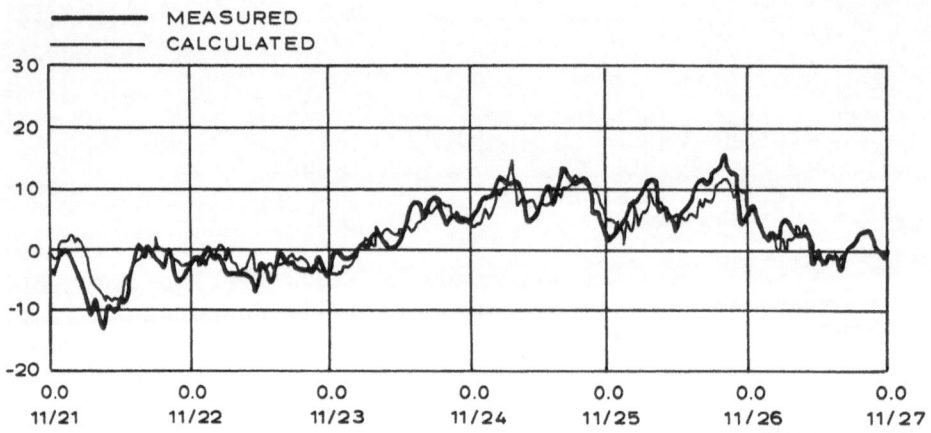

Figure 5.6.   Water level difference across the Great belt. Measured and calculated. Note that comparing measured and calculated water level difference is a more sensitive measure of model performance than comparison of water levels.

From the observations of meteorological and hydrographic parameters, 21-27 Nov. 1987, the two-layer model S22 was used to calculate the driving forces. For the chosen period, the north-south sea-surface slope produces the dominant term in the momentum balance (Figure 5.7).

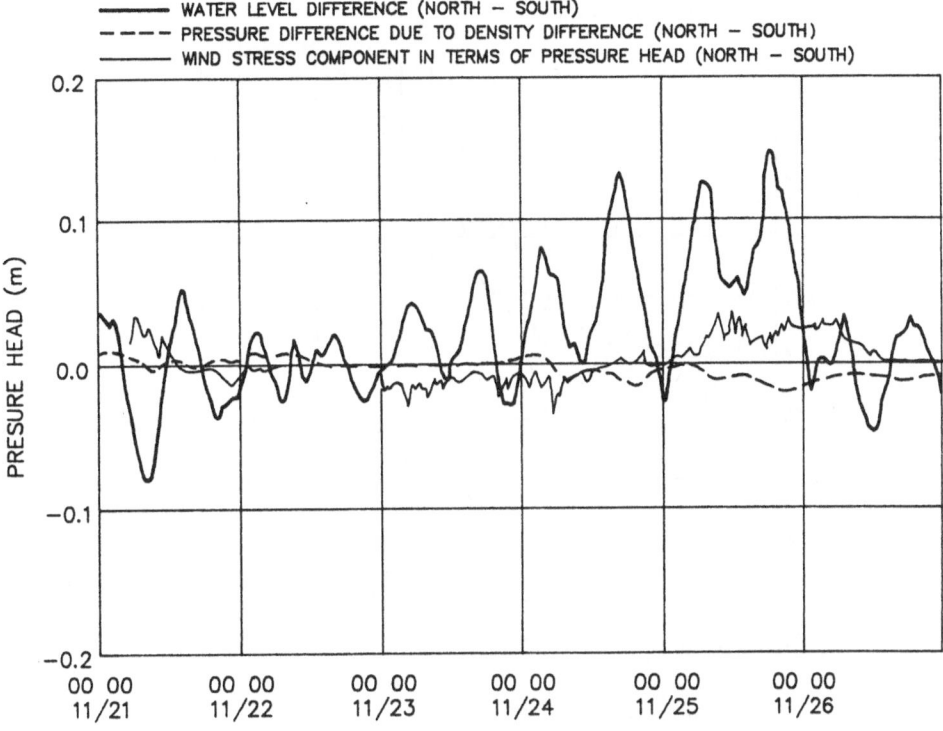

Figure 5.7. Relative magnitude of terms driving the two layer flow,
calculated using observations obtained during the 6 day
period 21-27 November, 1987. The sea-surface slope is
generally the dominant term.

Observed and calculated layer velocities are shown in Figure 5.8 for
both upper and lower layers. Except for a brief extreme overestimate on
November 24, the upper layer is modelled fairly well. In the lower layer
there is a tendency to underestimate the flow speed. The modelled depth
to the interface at Station B is compared with depths obtained with the
CTD in Figure 5.9; agreement is good. However it should be noted that
attempts to derive the interface depth by interpolation of conductivity
and temperature at the four discrete depths of the moored current meters
are of limited accuracy.

146

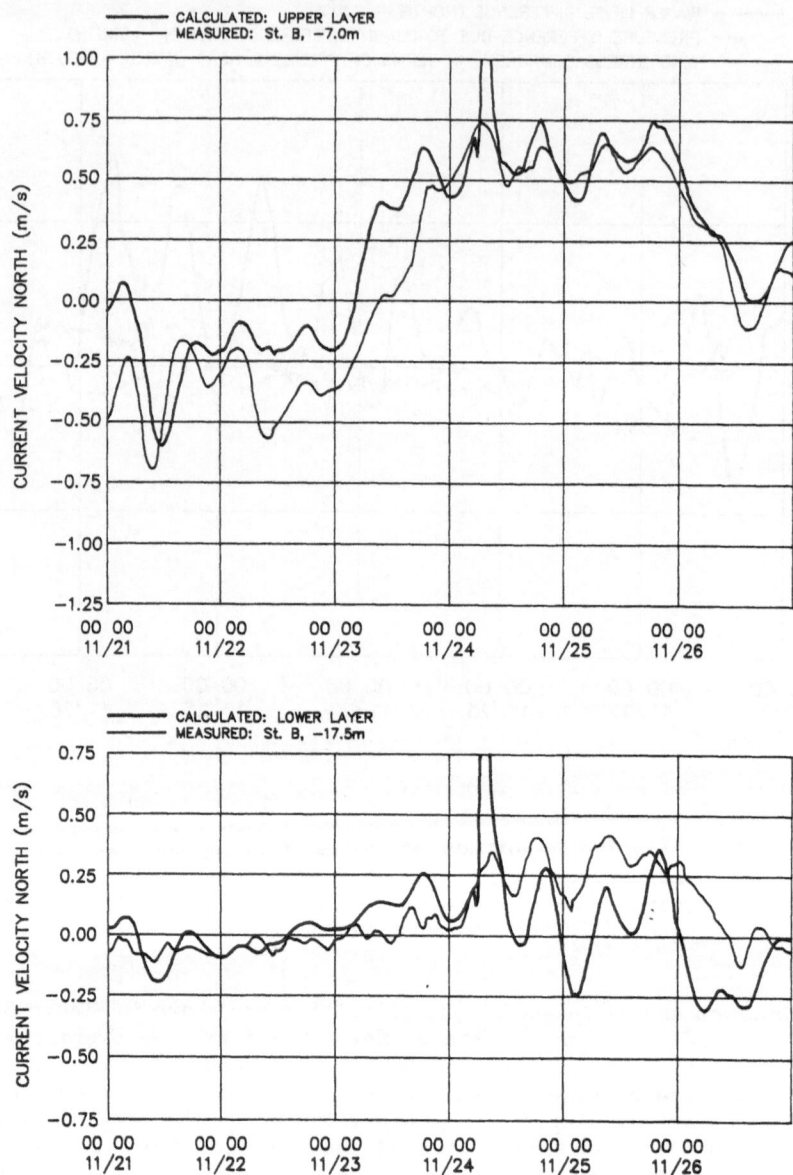

Figure 5.8.   Calculated (solid line) and observed (broken lines) currents at Station B for the upper and lower layer respectively during the period 21-26 November, 1987. Apart from a brief extreme discrepancy on 24 November, the calculated flow speeds are in reasonable accord with the observations, although somewhat biasing the current towards the south in the lower layer.

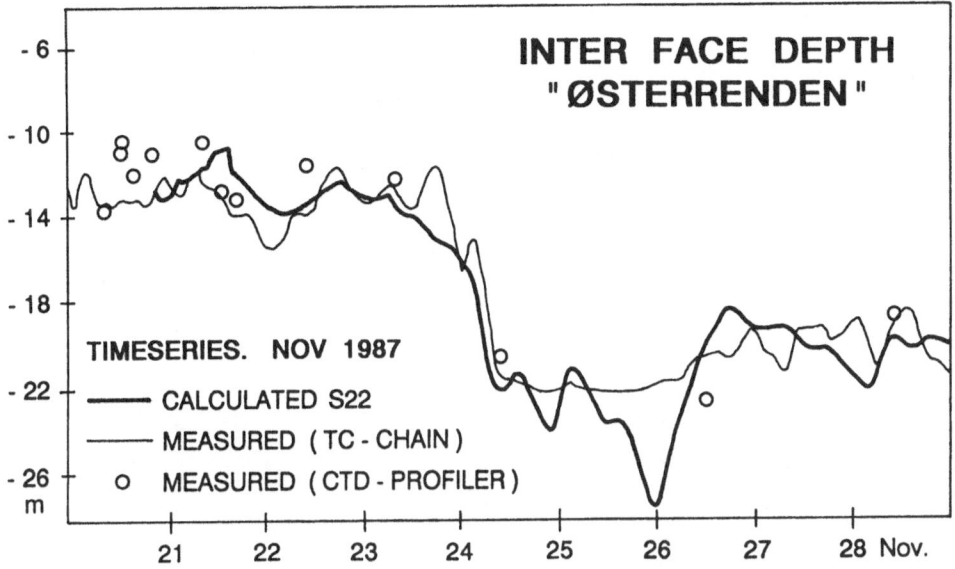

Figure 5.9.   Measured and calculated interface depth for Station B. Nov.
1987.

Measurements of the interface depth and current were also obtained from the M/S PIP as transects and sections across and along portions of the strait. The quality of the echo-sounding estimates of interface depth depends upon the strength and consistency of the target and the procedure for processing the signal. Normally a fairly clear indication is obtained but occasionally the results are somewhat scattered. Figure 5.10 shows a transect on 6 November when there was supercritical flow towards the south. The interface estimates are somewhat scattered in this example, although not inconsistent with the CTD interface depths, shown as '*' in the figure. Further developments of the acoustic technique have increased the resolution. Model interface depth is also shown as a dashed line connecting small open circles. It is interesting to note that the modelled and observed interface depth appear to be in good agreement for this case of hydraulic control.

Finally in Figure 5.10 an example of a spatial plot of current vectors for the upper layer is compared with corresponding vectors obtained from the Doppler profiler on M/S PIP as it traverses the channel on 6 November. The upper layer model velocities appear to be in close agreement with the observations.

148

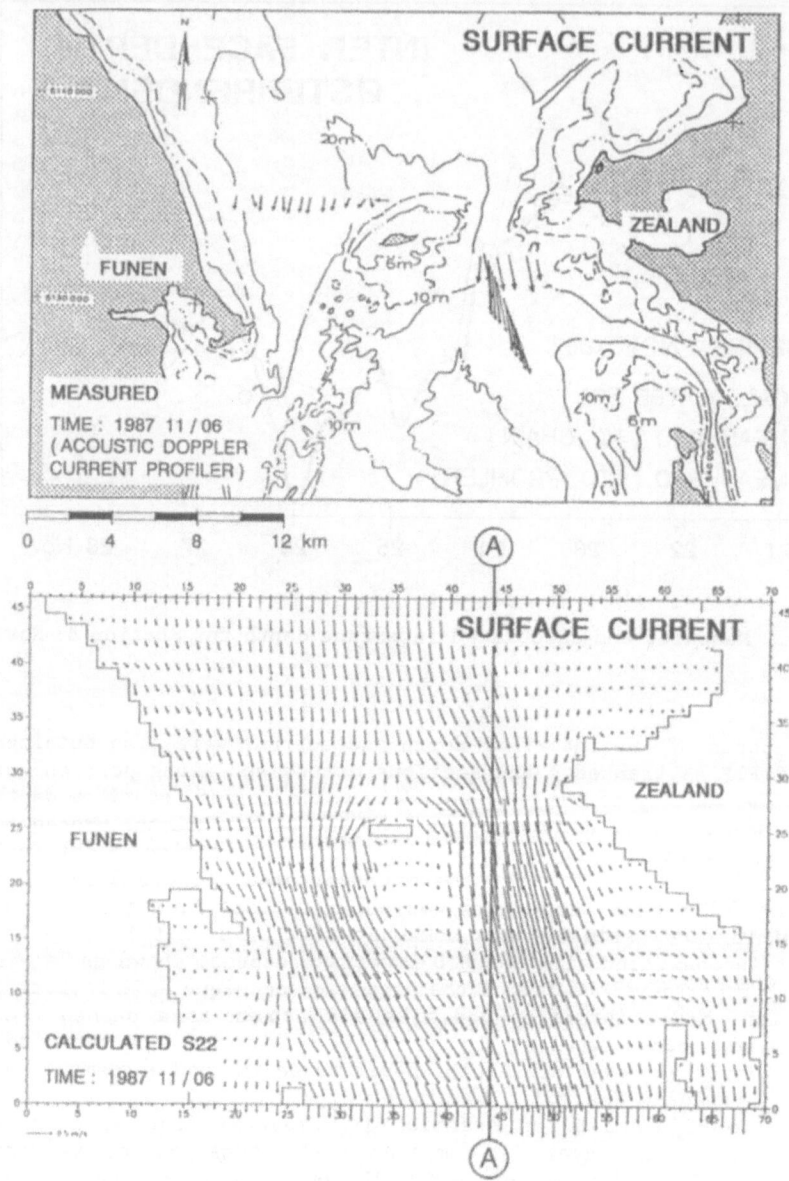

Figure 5.10. (to be continued).

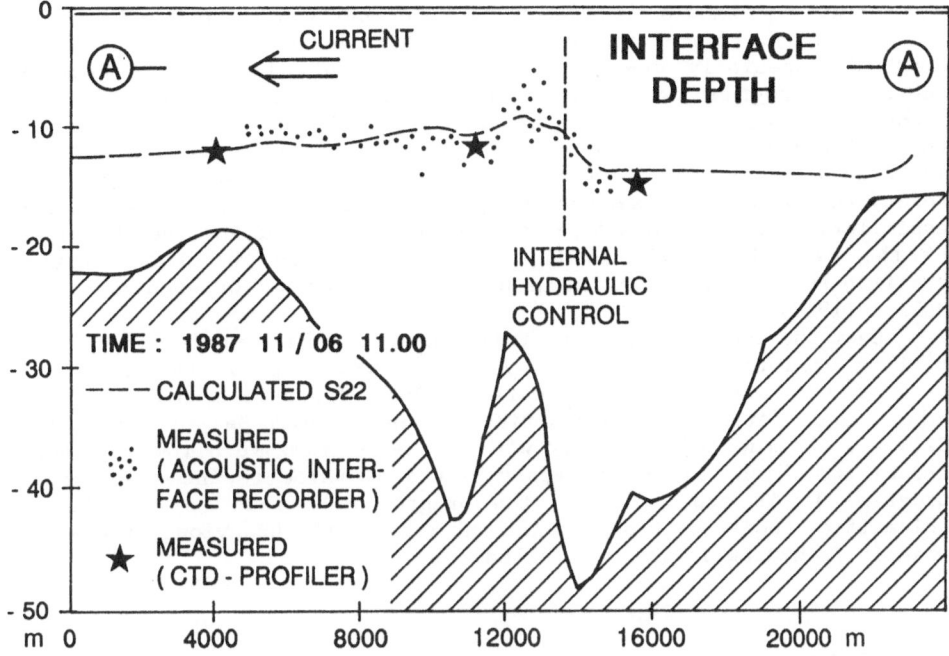

Figure 5.10.   Examples of measured and modelled flow parameters. A full documentation of model-field data comparisons is given in DHI/LIC 1988. Further reports will describe results of the full model verification program 1989-91.

6.   Concluding Remarks

The initial comparisons of model predictions and observations are in reasonable accord, and are probably as good as can be expected given the limited boundary data available. Together with the laboratory model and analytical verifications, this gives us reasonable confidence that the S22 model is a suitable tool for calculating the required compensation dredging. We emphasise that the zero solution calculation is insensitive to the choice of boundary conditions, but the accuracy of the model used to calculate the zero solution can only be verified by field observation, and such verification depends upon accurate measurements of boundary conditions.

The new monitoring plan will provide a much more comprehensive data base with which to examine model performance.   Indeed the quality and extent of the monitoring program provide a truly unique opportunity for testing two-dimensional stratified models.   The Great Belt Link Company has declared an open policy with respect to the observations, so the resulting data set will provide a bench-mark for future studies and model comparisons of stratified flow in straits. Moreover the duration of the program (1989-1996) will make the data invaluable for studies of the exchange between the Baltic and the Kattegat.

However it should be noted that the comparison between measured and computed data given above is of a qualitive nature. This is a serious drawback especially when it comes to use of the present complete monitoring program. Two questions arise, the first being scientific, the second of a technical nature; namely, how do we quantitatively compare the two types of data and express the error bounds of modelled and measured data, and how do we evaluate the uncertainty of the method with regard to reaching the zero solution ? The first question involves data assimilation procedures not yet developed. The second involves knowledge of the sensitivity of the Baltic to changes in the Great Belt. A preliminary approach is given in Jürgensen (1988) and Ottesen Hansen and Møller (1989).

Over the coming years, the full potential of the monitoring program will be used to test the two-layer model S22. The limitations of the present model must be better understood and, if possible, overcome. Also a barotropic forecast model covering the entire North Sea and Baltic (DHI S21) will be run interactively with the monitoring to allow a forecast capability which will be of great practical value to the construction program. It is expected that the program will lead to a number of scientific studies. Of particular interest is the behaviour of the flow when it is hydraulically controlled.

Acknowledgement

The environmental studies presented here were sponsored by Storebæltforbindelsen under the administration of Mr. Christian Tolstrup.

References

Abbott, M.B. (1979). "Computational Hydraulics. Elements of the theory of Free Surface Flows". Pitman Adv. Publish. Program. London.

Belt Project, The (1980a). "Physical Measurements in the open Danish Waters 1974-77". NAEP, Copenhagen.

Belt Project, The (1980b). "Water Exchange of the Baltic Measurements and Methods", by Torben S. Jacobsen. NAEP. Copenhagen.

Belt Project, The (1981). "Evaluation of the Physical, Chemical and biological Measurements", by G. Nielsen, T.S. Jacobsen, E. Gargas, E. Buch. NAEP, Copenhagen.

Bertelsen, J.A. and I.R. Warren (1977). "Two layer modelling of the Danish Belts: Collection and processing of data and calibration of the models". 17. Congr. Int. Ass. for Hydr. Res., Baden-Baden, pp. 379-386.

DHI/VKI. (1985). "Lokale Hydrauliske og Biologiske Effekter af Alternative Faste Forbindelser over Storebælt" (in Danish), Ministry of Public Works, Denmark.

DHI. (1987) "System 22 – Short Description", Danish Hydraulic Institute, Hørsholm, Denmark.

DHI/LIC. (1988). "Fixed link across Storebælt 'Hydraulic Effects - Verification and Documentation'", Report for A/S Storebælts-forbindelsen, by Danish Hydraulic Institute and LIC-Engineering, Copenhagen.

Fonselius, S. (1969). "Hydrography of the Baltic Deep Basins III", Fishery Board of Sweden, Series Hydrography, Report No. 23.

Fonselius, S. (1988). "Baltic salinity observations 1964-1987", Presented at 16th Conference of the Baltic Oceanographers, 1988, Kiel, 9 pages.

Jacobsen, J.P. (1910). "Gezeitenstroeme und Resultierende Stroeme im Grossen Belt". Medd. komm. Havundersøgelser, Ser. Hydr. Copenhagen, 1:14, 19 pages. (In German).

Jacobsen, J.P. (1913). "Stroemmessungen in der Tiefe in Dänischen Gewässern in den Jahren 1909, 1910 und 1911". Medd. Komm. Havundersøgelser, Ser. Hydr. Copenhagen, 2:3, 43 pages. (In German).

Jacobsen, J.P. (1925). "Die Wassereinsetzung durch den Öresund, den Grossen und den Kleinen Belt". Medd. Komm. Havundersøgelser. Ser. Hydr. Copenhagen, 2:9, 71 pages. (In German).

Jacobsen, T.S. (1980). "Recent results on the Sea Water Exchange of the Baltic". Journal of Hydrological Sciences. Vol. 7. No. 1-2, 1980.

Jacobsen, T.S. (1988). "Ship traffic - an important source of mixing in the Great Belt?", Presented at 16th Conference of the Baltic Oceanographers, 1988, Kiel, 21 pages.

Jürgensen, C. (1988). "Vertical mixing due to ship traffic and a consequence study for the Baltic", Presented at 16th Conference of the Baltic Oceanographers, 1988, Kiel, 15 pages.

Jürgensen, C. (1990). "Entrainment introduced by piers, dams and ships in a stratified channel flow". Series Paper No. 48. Inst. Hydrodyn. and Hydraulic. Engrg. Techn. Univ. Denmark.

Knudsen, M. (1899). "De hydrografiske forhold i de danske farvande indenfor Skagen i 1894-98". Kommis. videnskabelige undersøgelser i de danske farvande, bd. 2, hefte 2, s. 19-79. (In Danish).

LIC. (1984). "Effect on hydrography of the Baltic by construction of a Great Belt Crossing", Ministry of Transport, Denmark.

LIC. (1985). "On the effect of a Great Belt Link on the Hydrography of the Kattegat", (in Danish), Ministry of Public Works, Denmark.

Matthäus, W., H.V. Lass, E. Francke und R. Schwalbe (1983). "Zur Veränderlichkeit des Volumen - und Salztransports über die Darsser Schwelle". Gerlands Beitr. Geophysik, Leipzig 92, 5. S. 407-420. (In German).

Ottesen-Hansen, N.E. and J. S. Møller (1989). "Zero Blocking Solution for the Great Belt Link", in this volume.

Pedersen, Fl. Bo. (1978). "On the influence of a bridge across the Great Belt on the Hydrography of the Baltic Sea", Proceedings from 11th Conference of Baltic Oceanographers, Rostock, Vol. 1, pp. 366-377.

Pedersen, Fl. Bo. (1980). "A monograph on turbulent entrainment and friction in two-layer stratified flow", Tech. Univ. of Denmark.

Pedersen, Fl. Bo and J. S. Møller (1981). "Diversion of River Neva. How will it influence the Baltic Sea, the Belts and Cattegat", Nordic Hydrology, 12, 1981: 1-20.

Pedersen, Fl. Bo. (1982). "The sensitivity of the Baltic Sea to natural and man-made impact", Hydrodynamics of semi-enclosed seas. Procs. 13. International Liége colloquium on ocean hydrodynamics. Liége 1981. Ed. by J.C.J. Nihoul. Elsevier 1982. Elsevier oceanography series, 24, pp. 385-397.

Stigebrandt, A. (1983). "A Model for the Exchange of Water and Salt Between the Baltic and the Skagerrak". Jour. Phys. Ocean., Vol. 13. 1983.

Stigebrandt, A. and F. Wulff (1987). "A Model for the dynamics of nutrients and oxygen in the Baltic proper". J. Mar. Res. 45. pp. 729-759.

Svansson, A. (1972). "Canal models of sea level and salinity variations in the Baltic and adjacent waters". Fishery Board of Sweden. Ser. Hydr. Lund, 26, 71 pages.

Welander, P. (1974). "Two layer exchange in an estuary basin with special reference to the Baltic Sea". J. Phys. Ocean, 4. pp. 542-556.

Würtki, K. (1954). "Der grosse Salzeinbruch in die Ostsee im November und December 1951". Kieler Meeresforsch., 10:1, pp. 19-25.

ZERO BLOCKING SOLUTION FOR THE GREAT BELT LINK

N.-E. Ottesen Hansen
LICengineering A/S
Ehlersvej 24
DK-2900 Hellerup
Denmark

Jacob Steen Møller
Danish Hydraulic Institute
Agern Allé 5
DK-2970 Hørsholm
Denmark

ABSTRACT. This paper presents a case of strait and estuary management where the building of a main causeway and a bridge and tunnel system will be made in such a way that environmental effects are minimized.

The particular case is the Great Belt Link in Denmark spanning the Great Belt, the main artery for water exchange to the Baltic Sea. Due to the sensitive environment of the Baltic the government of Denmark has decreed that the water exchange to the Baltic shall remain the same after the construction of the link. This means that the added resistance caused by the narrowing of the flow cross sections due to the construction of causeways and from bridge piers shall be compensated in such a way that the water exchange to the Baltic remains unchanged.

It is determined that the compensation will be made by increasing the flow cross section of the link by dredging. Areas and volumes of dredging have been determined by numerical calculations. It is demonstrated that although the flow in the Great Belt is a very complicated two layer flow it is possible to define a dredging pattern which compensates for all conditions, whether subcritical or supercritical.

Background

For many years there have been plans in the Kingdom of Denmark to connect the islands with bridges, tunnels or causeways. Some of the projects may, if carried out, affect the free flow through the Danish Straits between the Baltic and the North Sea, Fig. 1.

In general not much attention is given to the impact of bridges or causeways on the surrounding flow. However, the peculiar hydrography of the Danish Straits, with a distinct layered flow, results in a flow with a very small hydraulic resistance. Therefore the total resistance may be sensitive to small changes in geometry. For instance, the hydraulic head difference between the Kattegat and the Baltic is normally only 0.2-0.6m

153

L. J. Pratt (ed.), The Physical Oceanography of Sea Straits, 153–169.
© 1990 Kluwer Academic Publishers.

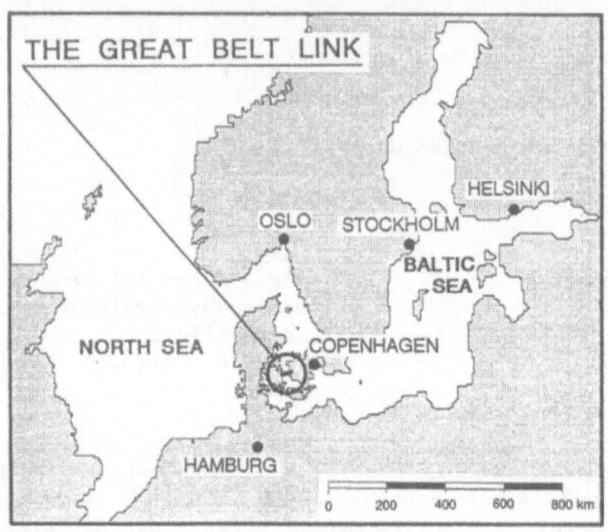

Figure 1.   General map of the area.

over a length of 300 km with a flow rate of 100.000 m^3/s. Further, bridges built in the Straits will have to be designed for heavy ship impact and for ice loads which make the bridge abutments large, leading to increased blocking.

Most of the hydraulic loss in the Great Belt takes place through 4-5 narrows where the current becomes strong. The fixed link across the Great Belt will span the most constricted narrows of the Great Belt - the Halsskov-Knudshoved narrows, see Fig. 2. Since a large part of the salt influx to the Baltic (60-70%) passes through this narrow, concern has been expressed that the construction of causeways, tunnels and bridges may exert a large flow resistance. It is likely that the consequent reduction in the water exchange would significantly alter the hydrography and environment of the Baltic. To avoid such a change it has been specified in Paragraph 5 of the Danish law for contracting the link that *"The work is to be carried out ... in such a way that the water flow through the Great Belt shall remain unchanged ... for the sake of the environment of the Baltic Sea"*. This somewhat unprecise design criterion has been interpreted by DHI/LIC (1987) to be understood as:

"- The discharge (m^3/s) through the Belt must not be changed by the crossing.

- The salt balance (kg salt/s) for the Baltic must not be changed by the crossing."

This so called "Zero Blocking Solution" will ensure that no environmental impact on the Baltic, be it hydrographical or biological, will originate in the construction of the link.

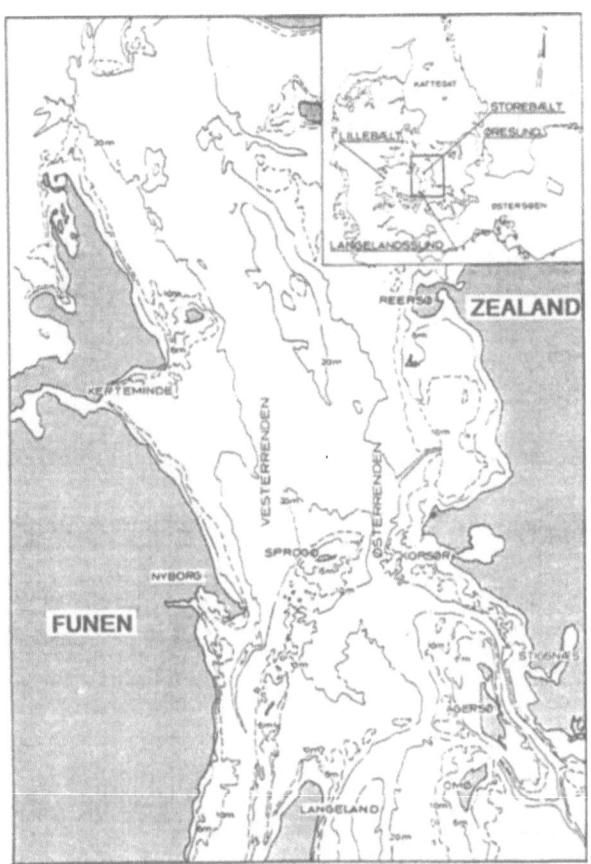

Figure 2   General map of the Link area

It is cheaper to build causeways than heavy duty structures like suspension bridges or immersed tunnels. Therefore there is an economical demand for letting part of the Great Belt link consist of causeways. This contradicts environmental interest which dictates as small a blocking of the flow as possible. A suitable compromise has been reached where causeways are only built in areas of water depth less than 6 metres and the blockage of these causeways and bridge piers has been compensated for by dredging.

Several proposals for compensation dredgings have been made. In the end it turned out that it would be most favourable to place the dredging in the Østerrende on the reef east of Sprogø Island, Figs. 3 and 4. This location was chosen because of high hydraulic efficiency of dredging and because materials from the area can be used for construction of causeways, etc. This implies that the constrictions of the flow in the Vesterrende will not be compensated in the Vesterrende (Western channel) itself but in the Østerrende (Eastern channel), though in such a way that the total flow through both the Østerrende and the Vesterrende remains unchanged.

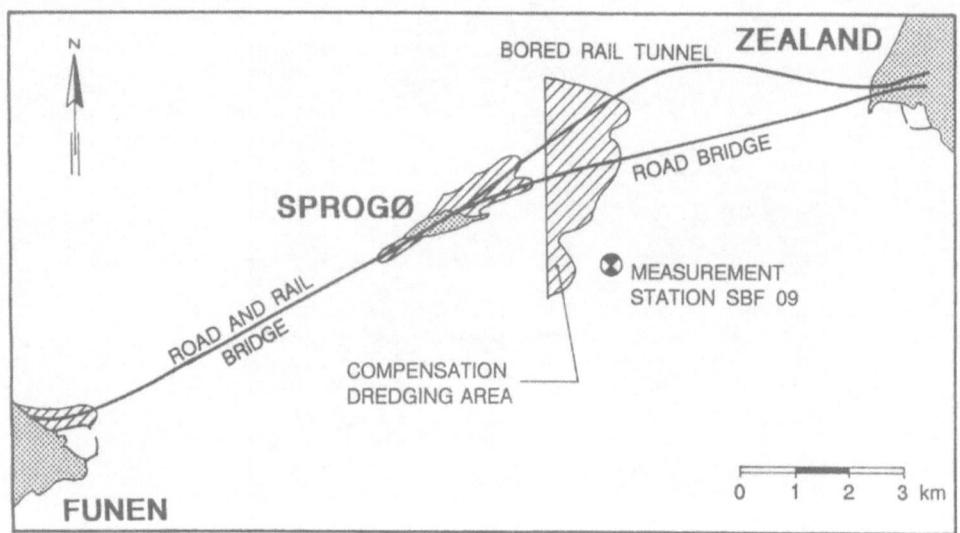

Figure 3   Typical project for the fixed link

Further, it is necessary to specify that the salt flux through the cross section and therefore the mixing between the upper brackish layer and the lower saline layer remains the same after the completion of the fixed link.

In order to document the prescribed effect of the compensation dredging, a series of investigations and some additional research have been completed:

- Hydraulic laboratory tests to determine the mixing between layers in stratified flow due to bridge piers, flow contractions and immersed tunnel elements.
- Hydraulic laboratory tests with subcritical and supercritical stratified flows in order to determine the types of flow which could be expected.
- Field measurements in the Halsskov-Knudshoved narrows to verify the findings from the laboratory tests and the preliminary hydraulic calculations.
- Set-up of a detailed two-layer mathematical model which can determine flow changes due to causeways and due to resistance from

bridge piers and immersed tunnel elements and which can further
determine the compensating effect of dredging.
-   Verification of the mathematical model by simulating a period in
    which field measurements took place and compare the calculated
    results with the measured results.
The results and conclusion of the above investigations are described in
the present paper except the verification of the model and the field
investigations which are described by Farmer and Møller, 1990.

HYDROGRAPHY

To obtain a non-blocking link across the Great Belt it is necessary
to have an intimate knowledge of the hydrography of the Baltic, Danish
Straits, and specifically of the Great Belt. With reference to Farmer and
Møller (1990) we briefly state the following:
Due to surplus of fresh water input to the Baltic Sea and due to the
narrow connection to the North Sea, the Baltic Sea is brackish and
stratified. The transport by mixing of oxygen into the bottom water is
restricted by the stratification of the Baltic. This has led to oxygen
depletion in large areas. The stratification of the Baltic and the oxygen
supply to the lower layers are maintained by the flow conditions in the
Danish Straits. The current in the Great Belt is governed by 1) The
meteorological forcing, 2) The tidal forcing, 3) The fresh water surplus
of the Baltic.
The fresh water surplus forces a mean current from the Baltic of 1.4
x 10^4 m^3/s. But the shifting weather conditions dominated by low pressure
travelling from the North Atlantic towards the east force the water to
oscillate in and out of the Baltic through the Danish straits with
irregular intervals. The amplitude of these oscillations reaches 2 x 10^5
m^3/s thus totally dominating the instantaneous current condition in the
Great Belt. The oscillating flow maintains the stratification of Great
Belt waters: By northbound current the Baltic brackish water will flow
into the Great Belt. The water will experience a very strong mixing. As
the brackish water is lighter than the saltier water coming from the
north, it will tend to lie on top of the water mass in the Belt. By
southbound current the surface layer will be forced southwards through
the Great Belt while mixing with the saltier bottom water. When passing
the Darss-sill the surface layer, which has got a high salt content on
account of the mixing, will plunge under the brackish Baltic water. At
the same time some of the bottom water is pulled over the sill. The
plunging salt water forms a dense bottom current and feeds the Baltic
lower layers with salt and oxygen.
The mixing between the layers in the Great Belt is caused by
turbulence generated by wind and current. The current in the upper and
lower layers of the Great Belt can flow in the same direction and
opposite to each other depending on the relative strengths of the driving
forces.

PRINCIPLES FOR ZERO BLOCKING SOLUTION

As described in Farmer and Møller (1990) the general behaviour of the flows between Kattegat and the Baltic are known in principle, but the details of the exchange over the sills at Darss and Drogden to the Baltic are still poorly understood. To prevent changes in the water exchange of the Baltic it is vital to ensure that the conditions at the Darss will remain unchanged. In order for these conditions to remain unchanged it is necessary to specify that the conditions at the boundaries of the Great Belt remain unchanged. In fact, changes are only allowed in a rather narrow area around the fixed link. In practice an area plus/minus 15 km north and south of the link is defined as the area in which changes are allowed, whereas areas outside this area shall remain unaffected. This area is called the near field in the following.

The problem is therefore reduced to calculate the effect of the link and the compensation dredging in this small area in such a way that flows across the boundaries to the near field remain unchanged. This criterion can be further refined with reference to the normal hydrodynamic laws by expressing that the total specific flow resistance within the near field shall remain the same.

However, in order to obtain a zero solution it is also necessary to specify that the mixing between layers remains the same. From hydrodynamic laboratory tests, field observations and theoretical analysis it has been found that the mixing between the layers in stratified flow is nearly proportional to the produced turbulent energy (Bo Pedersen, 1986). Bo Pedersen states that the Bulk Flux Richardson Number, R_f, defined as the ratio between the gain in potential as well as turbulent kinetic energy due to entrainment and the energy available for the turbulence production is a constant for subcritical, two layer flow. When this theory of constant efficiency of the mixing process is applied to the Great Belt, the problem of keeping the mixing unchanged is reduced to the problem of keeping the production of turbulent energy constant. We realise that the theory claiming a constant efficiency factor, R_f, for the mixing is debatable, however, several laboratory experiments and observations made in conjunction with the Great Belt link project support the simple assumption of a constant R_f. (See the paragraph below on mixing).

In the case of the flow through the Great Belt the production of turbulent energy is proprotional to the hydraulic energy loss. An important implication of the theory is that the mixing efficiency R_f is independent of the cause of the energy loss, be it bottom or interfacial shear, resistance from piers or expansion of flow lines and separation downstream of headlands and causeways. Only when the flow becomes supercritical is the efficiency increased.

The above findings are important because they state that so long as the hydraulic loss remains the same in areas with stratification then the mixing will remain the same.

The problem of determining compensation dredging is therefore reduced to determining the resistance from the link and then compensating for this resistance by dredging. Further, since the mixing is not important for the dynamics (balance of forces) it can be neglected in the equations of motion. Therefore the hydrodynamic calculations can be made to a sufficient degree of accuracy by a two-layered hydrodynamic model without including mixing for the near field.

ANALYSIS

The analysis centers on *resistance* due to expansion and friction loss and on *mixing* under subcritical and controlled conditions. Both shall remain the same in the near field over the typical time scales for in- and outflows to the Baltic. The analysis recognizes that changing salinities, interface levels, tilt of interfaces shall be taken into account and that the zero blocking solution shall be valid irrespective of all these different changes. All elements except controlled flow mixing are included in the numerical two-layer model used for the design of the compensation dredging. The model is the S22 model developed by Danish Hydraulic Institute ((Abbott, 1979), Farmer and Møller (1990)). An important advantage of S22 is that it takes the flow expansion in narrows correctly into account. This is important since it is one of the major contributions to the hydraulic resistances.

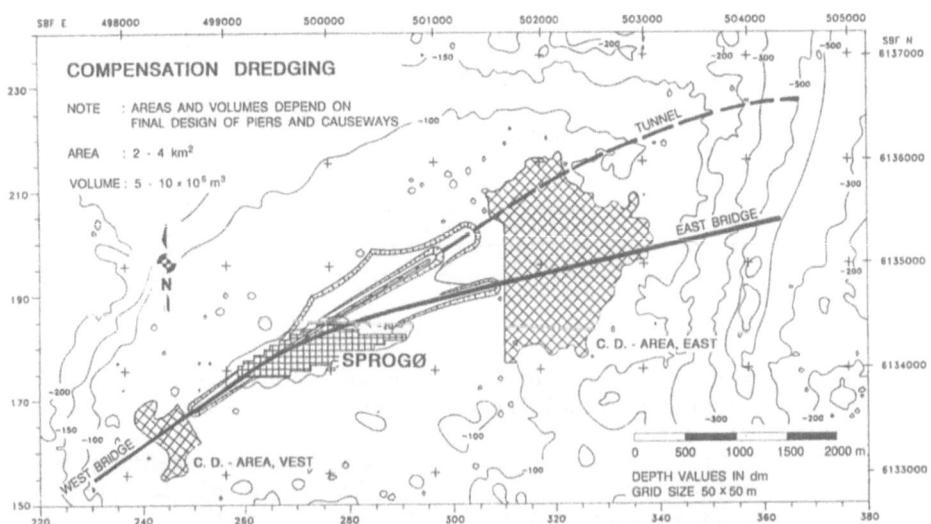

Figure 4.   Typical compensation dredging. The dredging is included in the numerical model by corresponding change in topography. DHI/LIC (1988).

FLOW RESISTANCE

The flow energy loss can most easily be dealt with directly by means of the numerical model. In the case of a link the resistances due to bridge abutments, bridge piers and protruding tunnel elements are simply determined by plugging their resistance characteristics into the mathematical model.

The problem now arises of how to determine the size and locations of the compensation dredging. This design work is carried out according to the following principle. Given the boundary conditions for the flow through the Great Belt the surface current is calculated with and without the link. By calculating the deviation in surface water flow, we have defined a measure of deviation from the zero solution. This measure will depend on the geometry of the link; large piers and long causeways will increase the deviation. The deviation is defined in Figure 5.

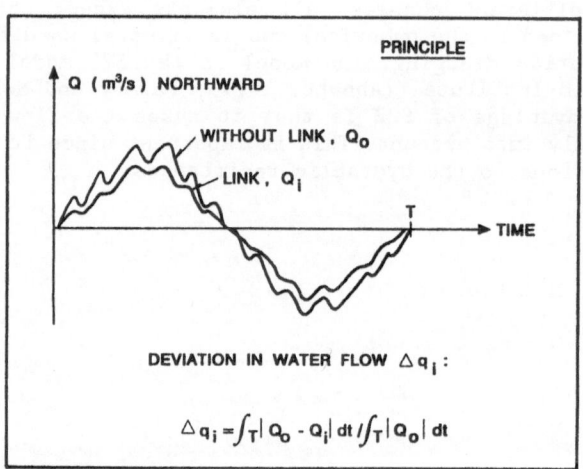

Figure 5.   Principle of determining discharge deviation.

Now for a given design of the link we can introduce a dredging scheme and calculate the resulting deviation caused by the combined link and dredging. If, for instance, the area to be dredged is kept constant the only parameter determining the efficiency of the dredging is the dredging elevation. By repeating the calculation of deviation for different dredging elevations the deviation is minimized.

Zero solution is reached for minimum deviation, see Fig. 6. The principle raises two theoretical concerns: 1) Not only the surface water flow, but also the bottom water flow and the mixing deviation must be minimized, 2) how are the boundary conditions for the flow defined?

The first problem is solved by calculating both the minimum deviation for discharge (both layers) and the minimum deviation for mixing (both layers). In the ideal case they will occur for the same dredging scheme. In practice they differ about ± 0,3 m in dredging level. This is within the practicable limits of dredging accuracy and a 'mean' level is chosen.

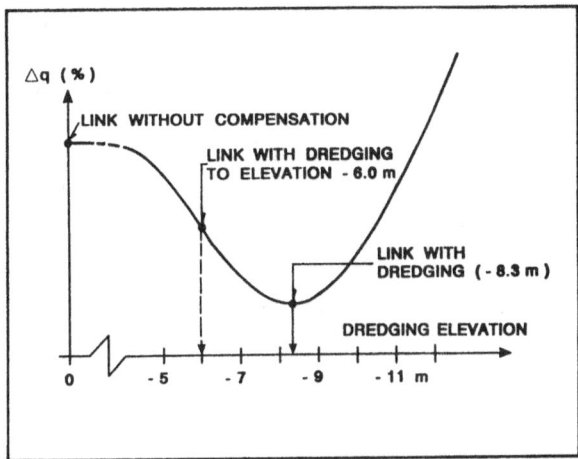

Figure 6.   Principle of defining zero solution dredging.

The second problem is solved by introducing a design period, T, where the flow field parameters vary in a statistically typical way. A set of boundary data selected from measured time series of water level, interface and wind is used as boundary conditions for the numerical model. The statistics of the calculated resultant exchange flows are compared with long term measured time series of exchange flows and a subset of T = 10 days showing statistics similar to the observed statistics is selected as boundary conditions for the design period.

The boundary conditions are not changed during the repetitive calculations necessary to define the minimum deviation. It can be argued that boundary conditions should be influenced by the resistance imposed by the construction. This, however, has no impact on the determination of the minimum deviation and hence the final zero solution dredging. To illustrate this we consider a simple conceptual model for channel flow.

In the model in Figure 7 the head loss ΔH_{AB} is the boundary condition for the model and the equation $\Delta H_{AB} = KQ^2$ is the model. ΔH_{AB} is determined by external forcing like wind fields and fresh water runoff in the basin and tidal motions in the sea. If K is increased due to construction works in the channel then ΔH_{AB} will be changed. By compensating the flow resistance K is decreased. At the point of zero solution ΔH_{AB} is unchanged. It is seen that by keeping ΔH_{AB} independent of K during the minimizing calculations the model will specify the correct zero solution.

An example of discharge calculation for the chosen design period without the link is shown in Fig. 8. In Fig. 9 is shown the change in discharges by inserting the link shown in Fig. 3. The effect of dredging is shown in Fig. 10, where the flow deviations are depicted as functions of the dredging depth for the dredging shown in Fig. 4.

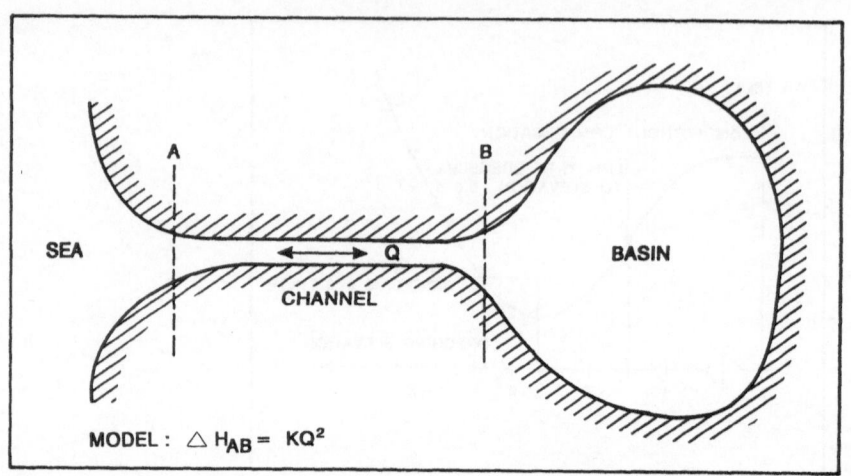

Figure 7.   Conceptual model ΔH_{AB} is the head loss between A and B. K is resistance factor, Q is discharge.

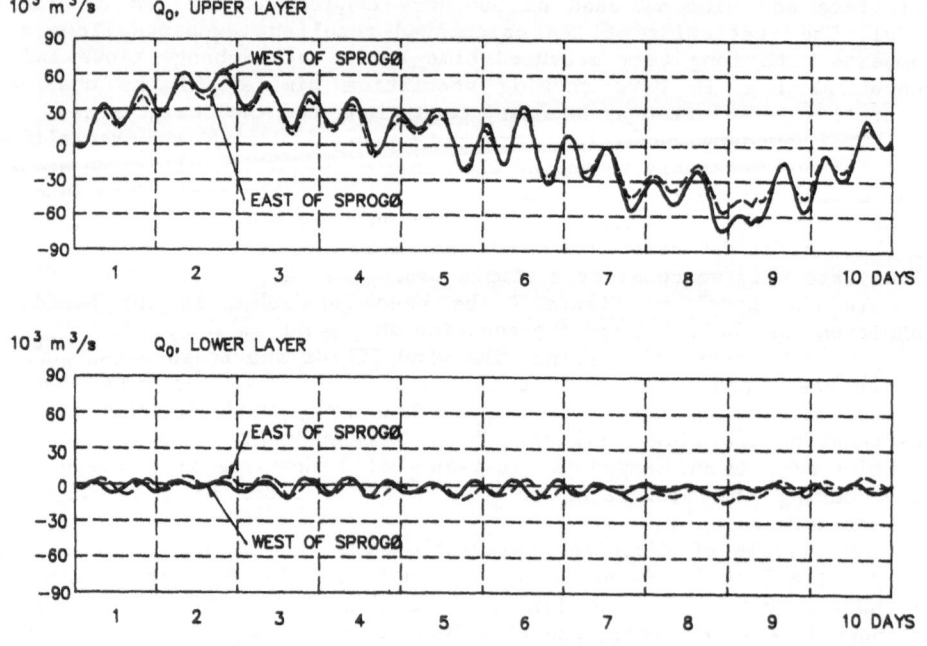

Figure 8.   Undisturbed calculated discharges, Q_o, for 10 day design period. Positive discharge is flow from the Baltic, Negative discharge is flow towards the Baltic. DHI/LIC (1988).

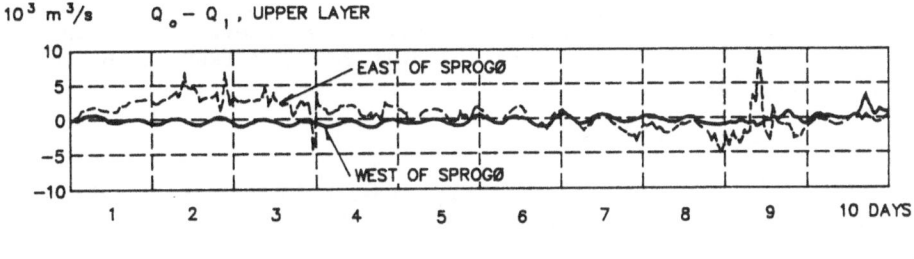

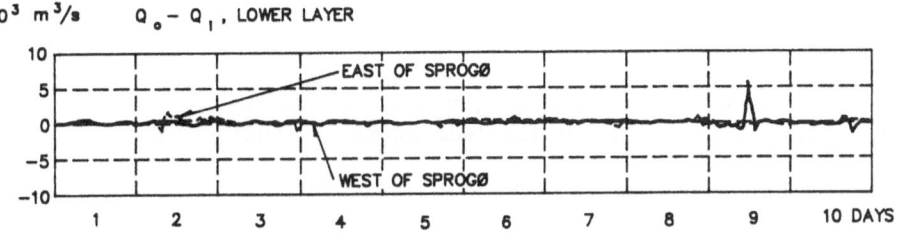

Figure 9. Difference between undisturbed discharge, Q_o, and discharge with link, but without compensation dredging, Q_i. DHI/LIC (1988).

MIXING, SUBCRITICAL FLOW

In the Great Belt where both layers are flowing the mixing between the layers is caused by entrainment from the lower layer to the upper layer and from the upper layer to the lower layer. This entrainment is calculated following the theory outlined by Bo Pedersen (1986). Bo Pedersen introduces the Bulk Flux Richardson Number, R_f, which for the two layer flow through the Great Belt can be reduced to:

$$R_f = 0,5\Delta\rho_o ghQ_e/PROD \qquad (1)$$

R_f is the Bulk Flux Richardson Number, which can be looked upon as an efficiency factor for the mixing process, $\Delta = (\rho_o - \rho_1)/\rho_o$, ρ_o is the density of the bottom layer, ρ_1 is the density of the surface layer, g is acceleration of gravity, $\frac{1}{2}h$ is the average lifting height of the entrained water, PROD is the production of the turbulent kinetic energy, and Q_e is the entrainment discharge.

The Bulk Flux Richardson Number, R_f, is evaluated by Bo Pedersen, 1986,

$$R_f = 0.045 \qquad (2)$$

When the above theory is applied to each layer in the Great Belt the following relation is obtained between the energy loss and the entrainment integrated over the total Belt Sea. The PROD can be related to the energy loss:

$$PROD = CQ\Delta H = CKQ^3 \qquad (3)$$

in which C is a constant. Combining Eqs 1, 2 and 3 the following expression for the entrainment is found.

$$Q_e = R_f CKQ^3/0.5\Delta\rho_o gh \qquad (4)$$

To keep the mixing (entrainment) unchanged requires not only that K is kept unchanged, but also that R_f must not be influenced by the link. Wind mixing is neglected as this mixing is not influenced by the link.

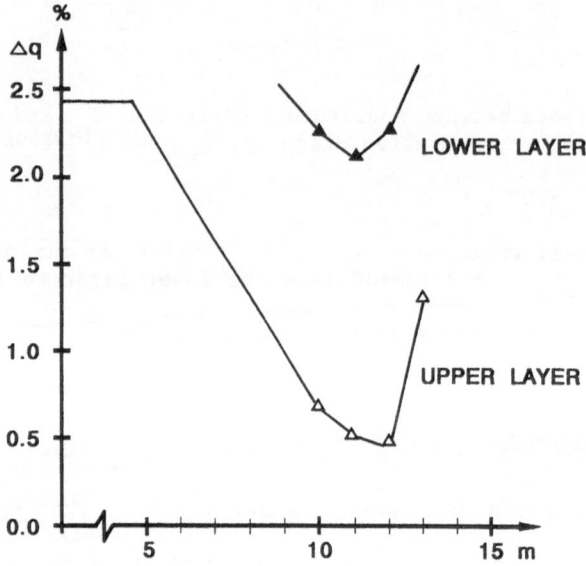

Figure 10.   Minimizing discharge deviation for a selected link, see Fig. 4. The abcissa is dredging depth as indicated in Fig. 4 and the ordinate is the flow deviation. DHI/LIC (1988).

A priori one may expect R_f to increase due to the turbulence generated by the bridge piers placed on the sea bottom. Experimental tests, however, showed that the value for R_f is not significantly changed by the piers (DHI/LIC, 1987). Therefore it can be concluded that when the zero blocking solution is reached for the flow resistance then a zero solution is also reached with respect to the mixing between layers.

The above arguments are only correct if the interfacial area in the near field remains more or less the same. The compensation dredging could alter this relationship if it is extended too much. Generally, the compensation dredgings are placed somewhat above the interface, but local strong currents combined with a strong tilting of the interface may cause the pycnocline to rise above the dredging area. In order to keep track of this another criterion is added to the resistance criterion, namely that the production of turbulent energy over the interface should remain the same before and after the link. The total turbulent energy produced in the near field and the interface area, A is given by

$$PROD = \int_T \int_A prod \, dAdt \qquad (5)$$

in which prod is the local production of turbulent energy. Eq. 5 can be divided into production for the upper layer and for the lower layer:

Upper layer

$$PROD_1 = PROD_w + PROD_i + PROD_p + PROD_e \qquad (6)$$

Lower layer

$$PROD_o = PROD_i + PROD_b + PROD_p + PROD_e \qquad (7)$$

in which

$PROD_w$: Wind generated turbulence term

$PROD_i$: Interface friction term

$PROD_b$: Bottom friction tern

$PROD_p$: Pier term

$PROD_e$: Flow expansion term

Each of the terms $PROD_1$ and $PROD_o$ are evaluated separately by a subroutine of the numerical model, see Fig. 11. Once this is done deviation of the PROD term, Δp, is treated like deviation of discharge, Δq, in order to determine the zero solution dredging with regard to mixing.

MIXING, SUPERCRITICAL FLOW

The findings of Bo Pedersen (1986) show that the simple approach of a constant efficiency factor, R_f, for the mixing is not fulfilled for supercritical flow. When the flow becomes supercritical the momentum balance is used to assess the mixing, Wilkinson (1970).

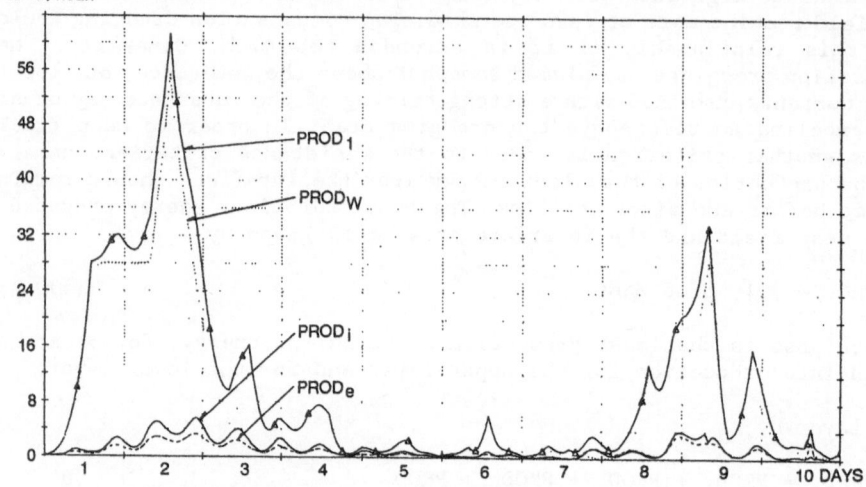

Figure 11. Example of calculated production of turbulent kinetic energy, PROD, for the design period. Integrated values for Great Belt ±10 km of the link. DHI/LIC (1989).

Expressing the momentum balance for the control volume shown in Fig. 12

$$I_o + P_o = F + I_D + P_D \tag{8}$$

where I and P are flow momentum and pressure integrated across the entire cross-section, index o is the southern boundary and D is the northern boundary of the near field, (Fig. 12). F is the combined resistance to the flow from the crossing and natural topography.

The individual terms in Eq. 8 are now examined. When the link is built a resistance is introduced due to the bridge-piers and the embankment which is compensated by dredging to keep F unchanged (to reach zero solution F will depend on the discharge and interface depth and be essentially independent of the redistribution of the mixing). I and P shall be unchanged for the zero solution. The supercritical flow situation is therefore characterized by having only one parameter which determines the mixing. This parameter is the total force F. When F is kept unchanged then the mixing is unchanged.

Applying this principle the flow through the link section means that the total force on Sprogø and the peninsulas of Knudshoved and Halsskov shall remain the same before and after the construction of the bridge.

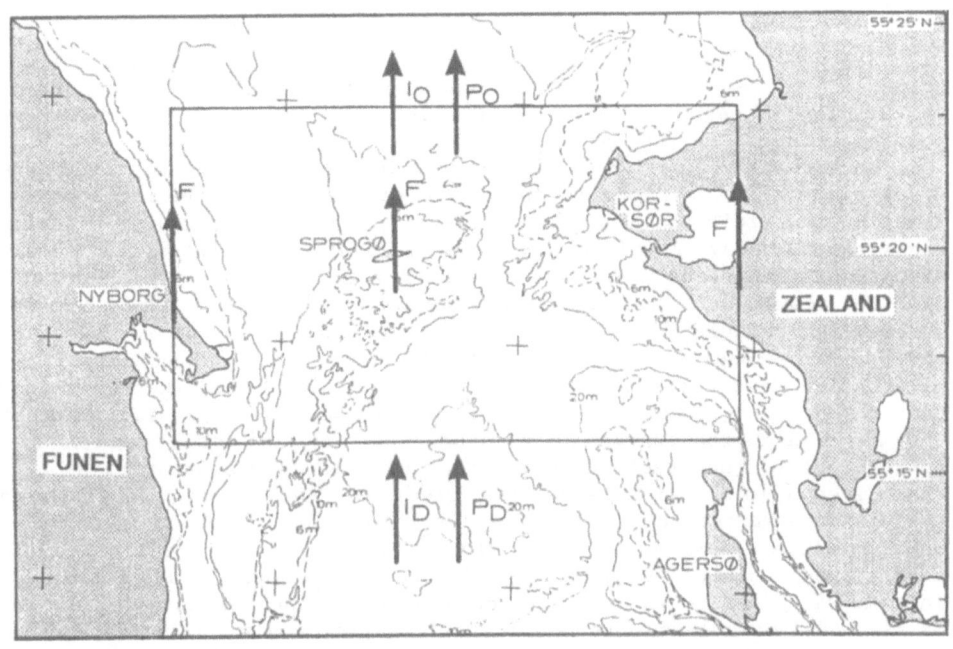

Figure 12.   Control volume for Halsskov-Knudshoved narrows.

In practice this means that the reaction force on the northern and
southern border shall remain the same before and after the construction
of the crossing. This is automatically true when the resistance criterion
is satisfied and therefore the zero solution criterion concerning
supercritical flow is satisfied as well.

It is noted that the above method assumes that the foreseen
redistribution in flow from Vesterrenden to Østerrenden does not
dramatically change the shape of the stratification profile downstream
of the flow control where the two flows merge. This assumption is
basically a part of the two-layer approach and believed to be of minor
importance with regard to the zero effect solution.

It is noted that the zero effect solution is aimed to ensure no
changes in the integrated mixing over the near field. Also, when the flow
is critical, the major part of the mixing will take place in the
subcritical part of the flow thus decreasing the sensitivity of the zero
solution to an incorrect representation of supercritical mixing. This
point is further supported by the fact that the field observations
suggest that the hydraulic jump in the Belt will in nearly all cases be
of a weak gradually adjusting type (Farmer and Møller, 1990).

168

SENSITIVITY EVALUATION

Since both the design as described above and the construction of the compensation dredging naturally includes some uncertainty it is important to establish a measure of this uncertainty and to assess the resultant possible changes in the Baltic.

Uncertainty of design is assessed by investigating the effect on the dredging level of changes in the model calibration parameters (bottom friction coefficients, interfacial friction coefficients, wind friction coefficient, eddy viscosity, boundary conditions, interface level, density difference between layers, design period).

The result of a standard design period calculation is shown in Fig. 10. Similar calculations are performed with changes in model parameters and boundary conditions. In total 10 different sets of parameters and alternative design periods were investigated. The investigation showed that the dredging depths varied within 0.5 m, which is very satisfactory and shows the robustness of the concept. It is noted that 0.5 m corresponds to the practicable accuracy of the dredging operation.

Given the uncertainty of the dredging level it is possible to evaluate the maximum uncertainty of the zero effect solution in terms of maximum deviation from the zero effect criteria for the Baltic Sea. To assess the possible effect on the Baltic a steady-state model is set up. This model (LIC, 1984) is based on Bo Pedersen and Møller (1981). The Baltic model calculates changes in Baltic salinities and layer depths due to changed resistance in the Great Belt. If the zero solution is reached exactly then there is no change in resistance and hence no change in the Baltic. But when the uncertainty of the dredging and hence the resistance is introduced we can calculate the response in the Baltic. The calculation shows that the potential change in surface layer salinity of the Baltic is compensated 90-110% by the compensation dredging: 90% if maximum uncertainty causes undercompensation, 110% if uncertainty causes overcompensation.

The sensitivity evaluation shows that the method of determining the zero solution is efficient.

REFERENCES[1]

Abbott, M.B. (1979). "Computational Hydraulics Elements of the Theory of Free Surface Flows". Pitman Adv. Publish. Program. London.

Bo Pedersen, F. (1986). "Environmental Hydraulics: Stratified Flows". Springer Verlag.

Bo Pedersen, F. and J.S. Møller (1981). "Diversion of the River Neva. How will it influence the Baltic Sea, the Belts and Cattegat". Nordic Hydrology, 12, p. 1-20.

DHI/LIC (1987). "Great Belt Crossing, Hydraulic Effects, Mixing Experiments". Danish Hydraulic Institute and LICengineering for AS Storebæltsforbindelsen.

DHI/LIC (1988). "Great Belt Crossing, Hydraulic Effects. Compensation Dredging. 1.6 Design Calculations". Danish Hydraulic Institute for AS Storebæltsforbindelsen. (A separate volume for each project alternative).

DHI/LIC (1989). "Great Belt Link, Hydraulic Effects. Documentation and Verification". Danish Hydraulic Institute for AS Storebæltsforbindelsen.

Farmer, D. and J.S. Møller (1990). "Measurements and Modelling in the Great Belt" in "The Physical Oceanography of Sea Straits". L. Pratt (ed.), NATO ARW series.

LIC (1984). "Effect on Hydrography of the Baltic by Construction of a Great Belt Crossing". LICengineering Ltd. for Ministry of Transport, Denmark.

Wilkinson, D.L. (1970). "Studies in Density Stratified Flows". Water Research Labortories, Univ. New South Wales, Australia. Report No. 118, 167 pp.

[1]It is recognised that the internal reports referenced here are not widely available; this appears to be an inevitable consequence of a program that is part of a major engineering project. However, the open policy now adopted by the Great Belt Link Company means that the information contained in them is accessible and for this reason it has been considered useful to include the bibliographic references at this stage.

THE FLOW THROUGH VITIAZ STRAIT AND ST. GEORGE'S CHANNEL, PAPUA NEW GUINEA

Eric Lindstrom
CSIRO Division of Oceanography
G.P.O. Box 1538
Hobart, Tasmania, 7001
Australia

Jeffrey Butt
CSIRO Division of Oceanography
G.P.O. Box 1538
Hobart, Tasmania, 7001
Australia

Roger Lukas
Joint Institute for Marine and Atmospheric Research
University of Hawaii, 1000 Pope Road
Honolulu, Hawaii, 96822

Stuart Godfrey
CSIRO Division of Oceanography
G.P.O. Box 1538
Hobart, Tasmania, 7001
Australia

ABSTRACT

Hydrographic and current meter measurements collected in 1985 and 1986 in Vitiaz Strait and St. George's Channel, Papua New Guinea are described. The subsurface, equatorward flowing current in Vitiaz Strait is remarkably strong (>1 m/s) and persistent throughout major seasonal wind shifts. Mass transport estimates through Vitiaz Strait of 8-14 Sverdrup are discussed. St. George's Channel mass transport is smaller (perhaps half) of that in Vitiaz Strait.

1. Introduction

In the region of Papua New Guinea the equatorial waters and the subtropical circulation of the southern hemisphere interact in a region of complex island arcs. There are numerous boundaries and constrictions in the circulation. Three sea straits are the only connection between the subtropical seas and equatorial ocean west of the Solomon Islands/Vanuatu island arcs (Fig. 1).

L. J. Pratt (ed.), The Physical Oceanography of Sea Straits, 171–189.
© 1990 Kluwer Academic Publishers.

172

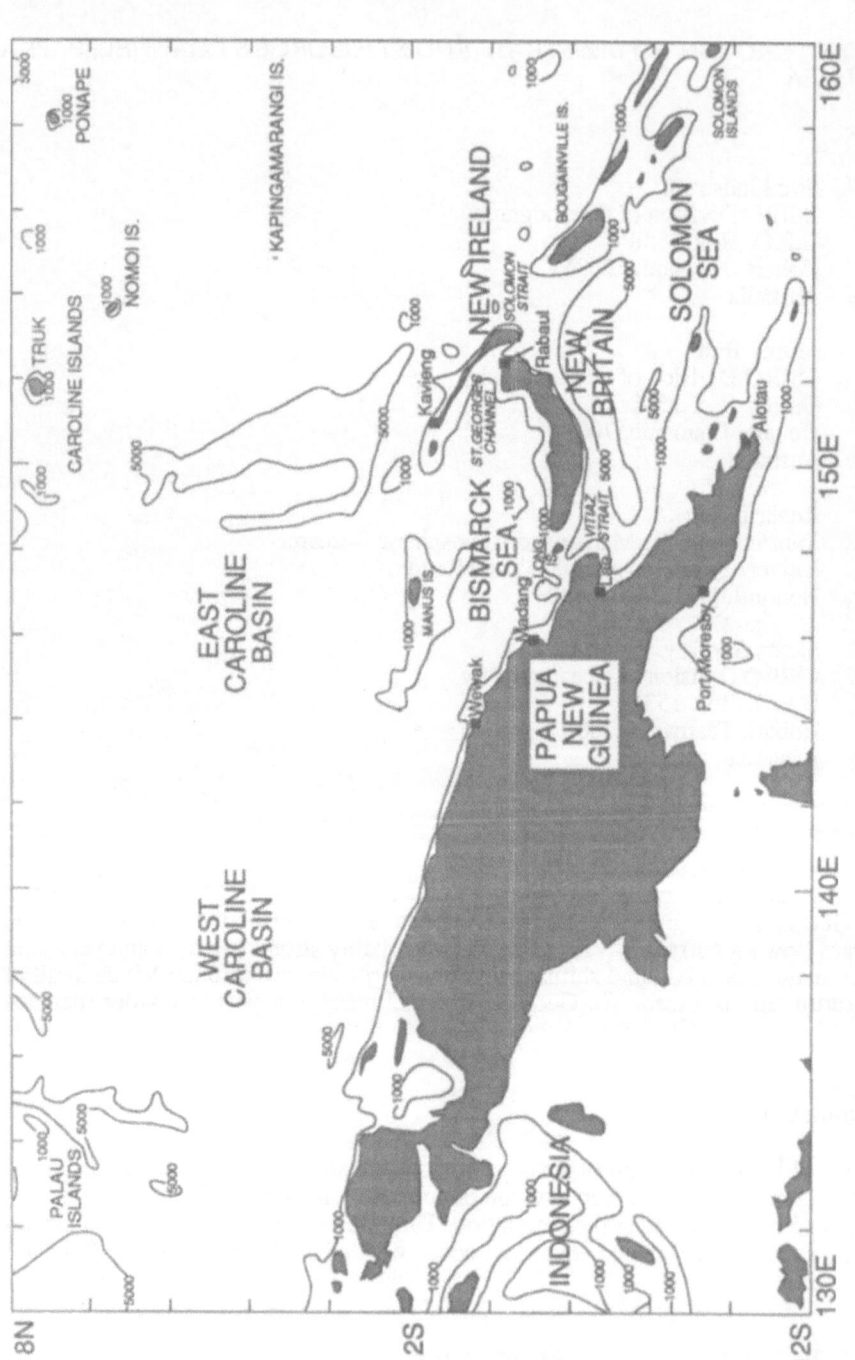

Figure 1.   Overall view of western equatorial Pacific showing the main geographic and ocean bottom topographic features as depicted by the 1000 and 5000 m isobaths.

Vitiaz Strait lies at the western end of New Britain and is the widest direct connection between the Solomon Sea to the south and the Bismarck Sea to the north. The major portion of the strait lying between the Papua New Guinea coast and Umboi Island is approximately 50 km wide and has an effective sill depth of around 1100 m. Between Umboi Island and the western end of New Britain lies a much shallower minor passage (Dampier Strait). Dampier Strait is not navigable and assumed to play a minimal role in the Solomon/Bismarck Sea interaction.

St. George's Channel is the other major connection between the Solomon and Bismarck Seas. It lies between the eastern side of New Britain and the southern end of New Ireland. This channel is approximately 35 km wide and is split into two channels at its northern end by the Duke of York Island group. The sill lying to the east of these islands is approximately 1000 m deep while that of the western channel is approximately 250 m.

1.1 HISTORICAL ANALYSES OF SOLOMON/BISMARCK SEA INTERACTION

As far as we are aware, there have been no previous direct measurements of the subsurface currents in Vitiaz Strait and St. George's Channel. However, various analyses of ocean surface currents have been made which indicated the seasonal variation of flow (Fig. 2; from Hisard et al., 1969) in these areas (Schott, 1939; Hisard et al., 1969; Colin et al., 1973; Ministry of the Defence of the USSR, 1974). These analyses suggest that surface flow in Vitiaz Strait reverses in response to the seasonal monsoon forcing; peak surface flow is towards the northwest (from the Solomon to Bismarck Seas) during July; maximum surface flow toward the southeast occurs during the peak of the northwest monsoon in January.

Water property maps of Tsuchiya (1968) and Rougerie and Donguy (1975) suggested that the subsurface flow through the straits was predominantly equatorward, feeding the equatorial region with thermocline waters characteristic of the southern hemisphere. Tsuchiya et al. (1989), using more recent data, confirmed this hypothesis. The reader is referred to Tsuchiya (1968) and Tsuchiya et al. (1989) for a full discussion of the upper ocean water property fields associated with the flow through Vitiaz Strait and St. George's Channel.

2. WEPOCS Measurement Program

One goal of the Western Equatorial Pacific Ocean Circulation Study (WEPOCS) was to gain further insight into the general circulation of the low latitude western Pacific. Two expeditions (WEPOCS 1 and 2) were conducted in the region (see Fig. 3) in June-July 1985 and January-February 1986 by researchers from Australia and the United States. The cruises were timed to coincide with the peaks of the Southeast Trades and the Northwest Monsoon. An overview of the WEPOCS measurement program and its primary results is given by Lindstrom et al. (1987).

As part of the extensive measurement program, particular attention was paid to measurement of flow through Vitiaz Strait and St.George's Channel. The purpose of these measurements was to test the hypothesis of Tsuchiya (1968) that the local source of southern hemisphere waters of the Equatorial Undercurrent was through Vitiaz Strait and St. George's Channel. As mentioned above, the confirmation of the source waters hypothesis is discussed by Tsuchiya et al. (1989).

The main observational components in the straits were Conductivity-Temperature-Depth (CTD) stations (including discrete sampling for O_2, NO_3, PO_4, SiO_2), numerous Acoustic Doppler Current Profiler (ADCP) transects from research vessels, single current meter moorings in each strait, and an array of pressure gauges in Vitiaz Strait.

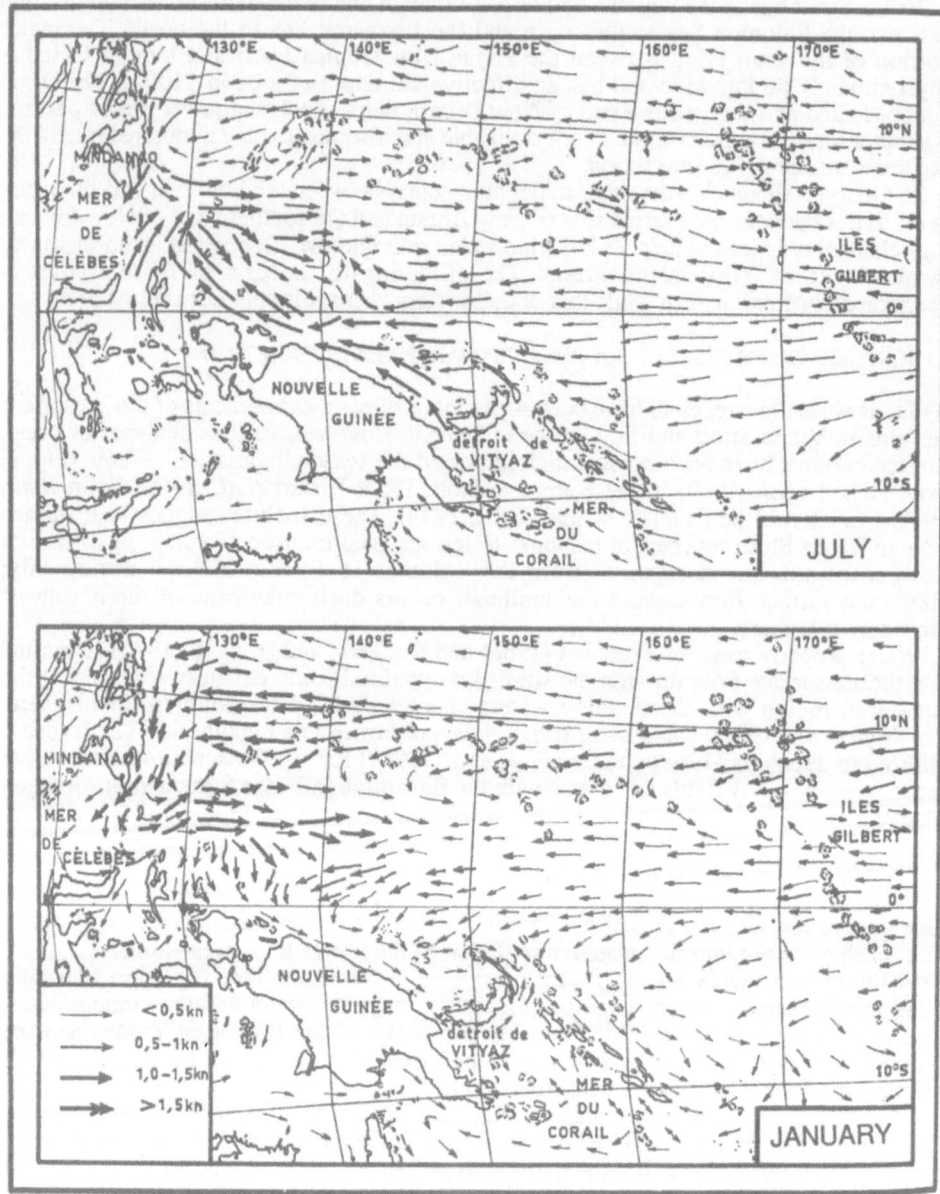

Figure 2. Surface currents north of Papua New Guinea in opposite monsoon seasons (from Morskoi Atlas as used by Hisard *et al*. 1969).

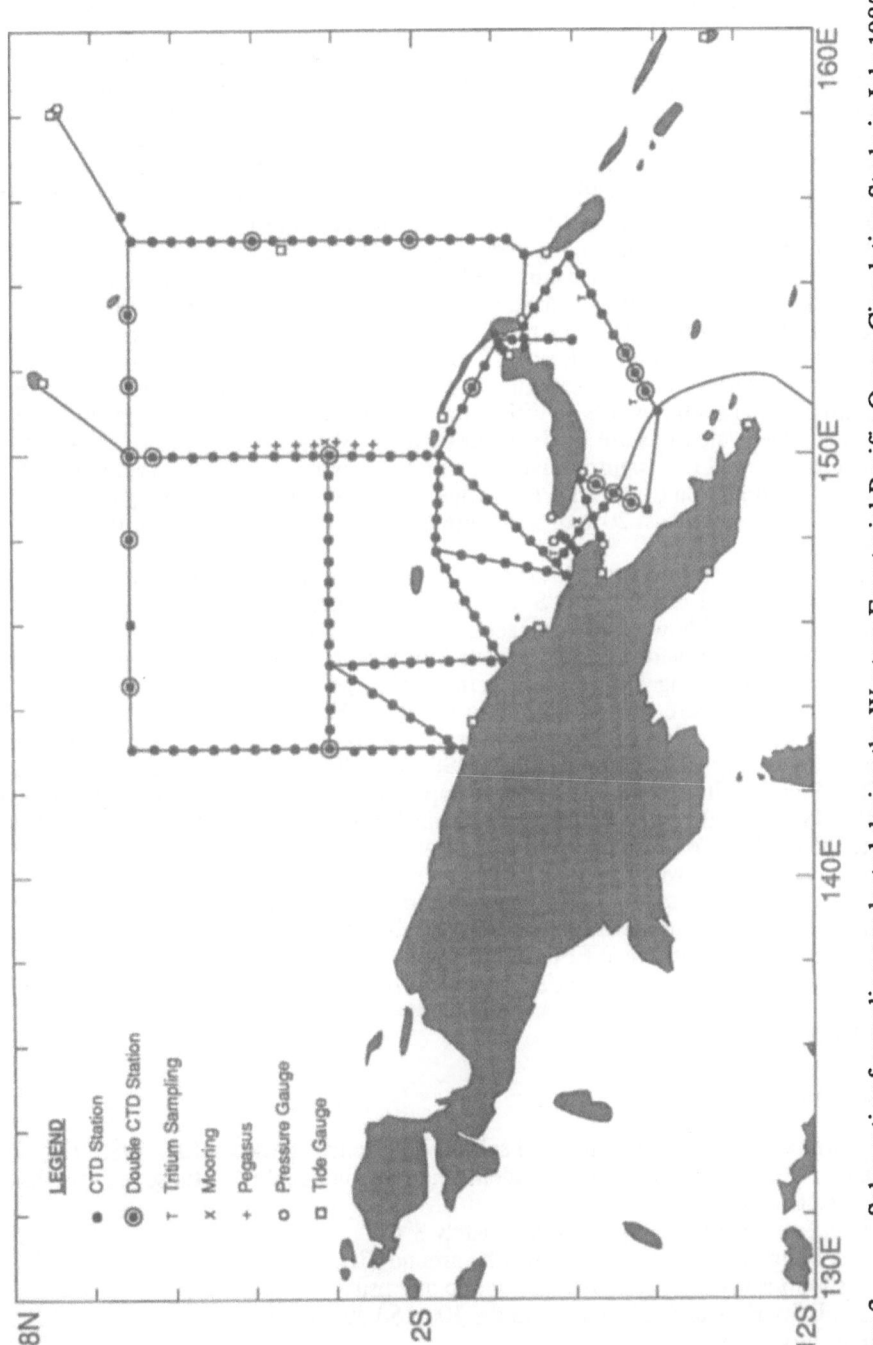

Figure 3.   Schematic of sampling conducted during the Western Equatorial Pacific Ocean Circulation Study in July 1985 (WEPOCS 1) and January 1986 (WEPOCS 2).

Figs. 4a, b indicate where the different types of sampling occurred in Vitiaz Strait and St. George's Channel.

CTD sections were completed across and along both straits during both cruises. Sections of temperature and salinity for both cruises of RV Franklin taken in Vitiaz Strait and St. George's Channel coincide with the primary ADCP sections discussed below. These sections are shown in Figs. 5-8. Station spacing was approximately 8 km, with casts being taken to within 50 m of the bottom. Twelve water samples were collected for each cast and were analyzed for salinity, dissolved oxygen, NO_3, PO_4 and SiO_2. Geostrophic velocities calculated from the CTD data are not shown here.

The Australian RV Franklin equipped with a 150 kHz RD Instruments Acoustic Doppler Current Profiler (capable of measuring velocity to 400 m depth) was used during multiple traverses of both Vitiaz Strait and St. George's Channel during WEPOCS 1 and 2. The ADCP provides current velocity relative to the ship. Both GPS and transit satellite navigation systems are used to determine the velocity of the ship by evenly spreading ship drift over the period between satellite fixes. Travelling at approximately 7 kts the RV Franklin covered each traverse of Vitiaz Strait or St. George's Channel in around 3 hours. Absolute water velocities are likely accurate to +/-10 cm/s. The ensembles of Doppler data were averaged over approximately 6 minutes, giving around 30 average profiles per traverse. Examination of the ADCP profiles showed that reliable velocity estimates were obtained provided at least 20% of transmitted echoes were received. Vertical binning of 8m was used.

ADCP traverses of Vitiaz Strait were carried out along the line from 147° 30.2'E, 6° 0.5'S to 147° 51.6'E, 5° 44.3'S where the channel width is 49.6 km wide. The shallower eastern side was not approachable, the result being that approximately 77% of the channel was sampled (Fig. 9). During July 1985 two traverses were made, while ten were carried out during January 1986. In St. George's Channel during both WEPOCS cruises, traverses were carried out along the line 152° 10.4'E, 4° 8.2'S to 152° 34.6'E, 3° 58.7'S, a distance of 47.8 km. On average the traverses covered approximately 90% of the channel (Fig. 10). During January 1986 eight 'full' traverses were carried out.

The Vitiaz Strait mooring was deployed approximately mid-channel (at 5°59.6'S, 147° 45.4'E, 882 m depth). The St. George's channel mooring was deployed approximately 2 km to the northeast of Duke of York Island (at 4° 7.6'S, 152° 29.4'E, 903 m depth). Each subsurface mooring was equipped with three Aanderaa current meters which recorded half-hourly averages of depth, temperature, current speed and direction. The deployment period was approximately 6 months (Vitiaz Strait 24/7/85 to 8/2/86, St. George's Channel 29/7/85 to 5/2/86) between WEPOCS cruises. Only the Vitiaz Strait data are discussed in this paper. Two of the St. George's Channel current meter records were incomplete.

3. Results

3.1. CTD SECTIONS

3.1.1. *Vitiaz Strait.* Temperature and salinity differences between July 1985 (Fig. 5) and January 1986 (Fig. 6) are clearly evident, the most noticeable of these are (for January compared to July):
• surface layer warming of approximately 3°C
• cooling throughout 100-500 dbar by around 1°C
• significantly lower surface salinity, up to 1 psu
• salinity rises of about 0.6 psu in the 100-150 dbar region.
Also, in July 1985:

177

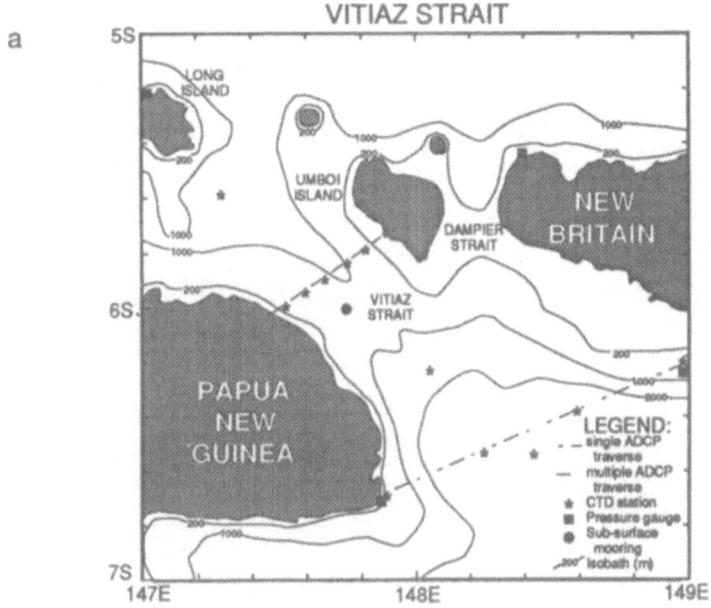

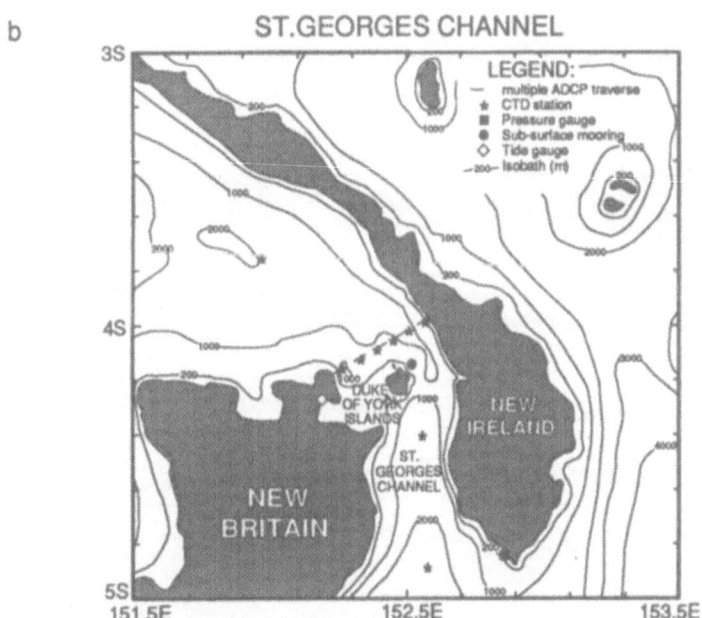

Figure 4.   Plan view of a) Vitiaz Strait and b) St. George's Channel showing
the primary transects for CTD and ADCP sampling during
WEPOCS.

178

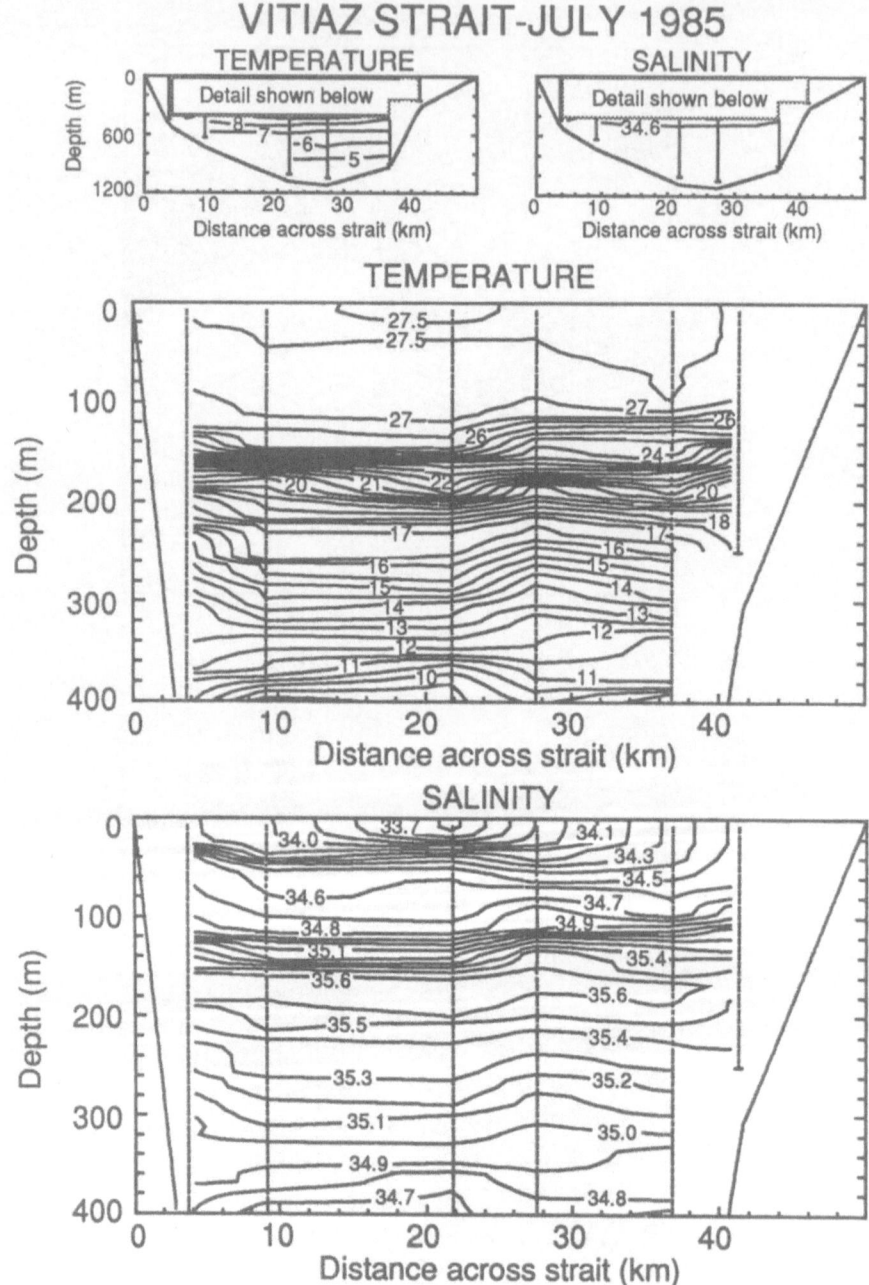

Figure 5.   CTD data from Vitiaz Strait from July 1985 showing cross-strait sections of temperature (°C) and salinity (psu).

179

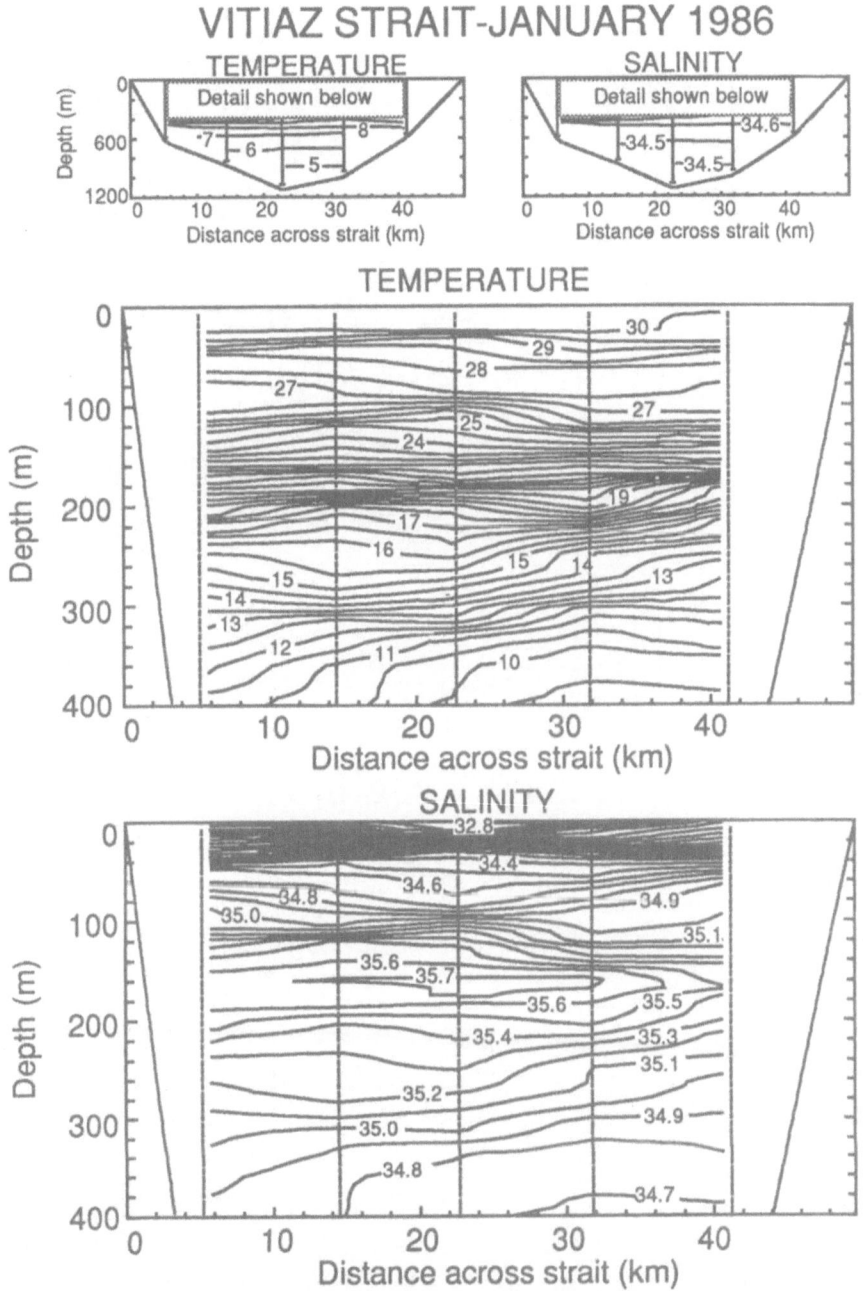

Figure 6.   CTD data from Vitiaz Strait from January 1986 showing cross-strait
               sections of temperature (°C) and salinity (psu).

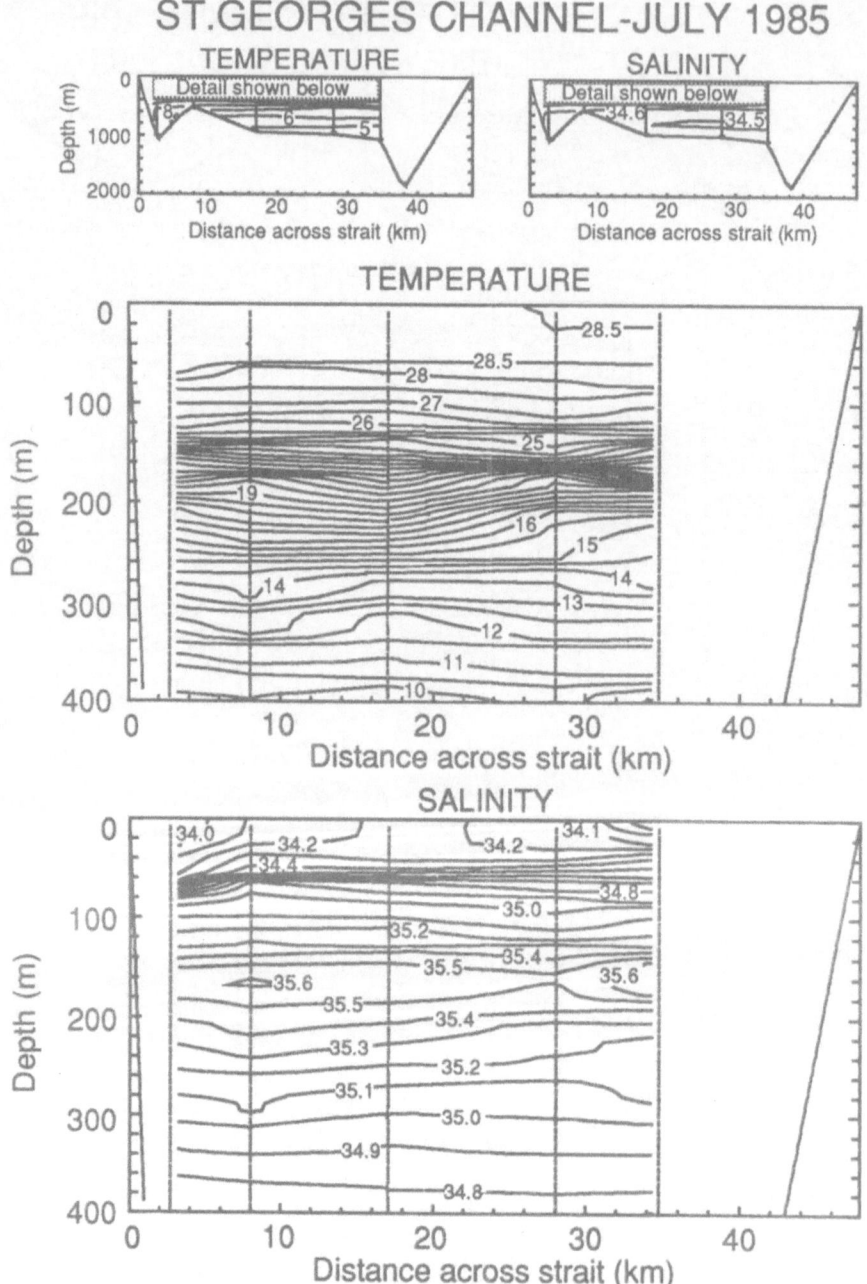

Figure 7.   CTD data from St. George's Channel from July 1985 showing cross-strait sections of temperature (°C) and salinity (psu).

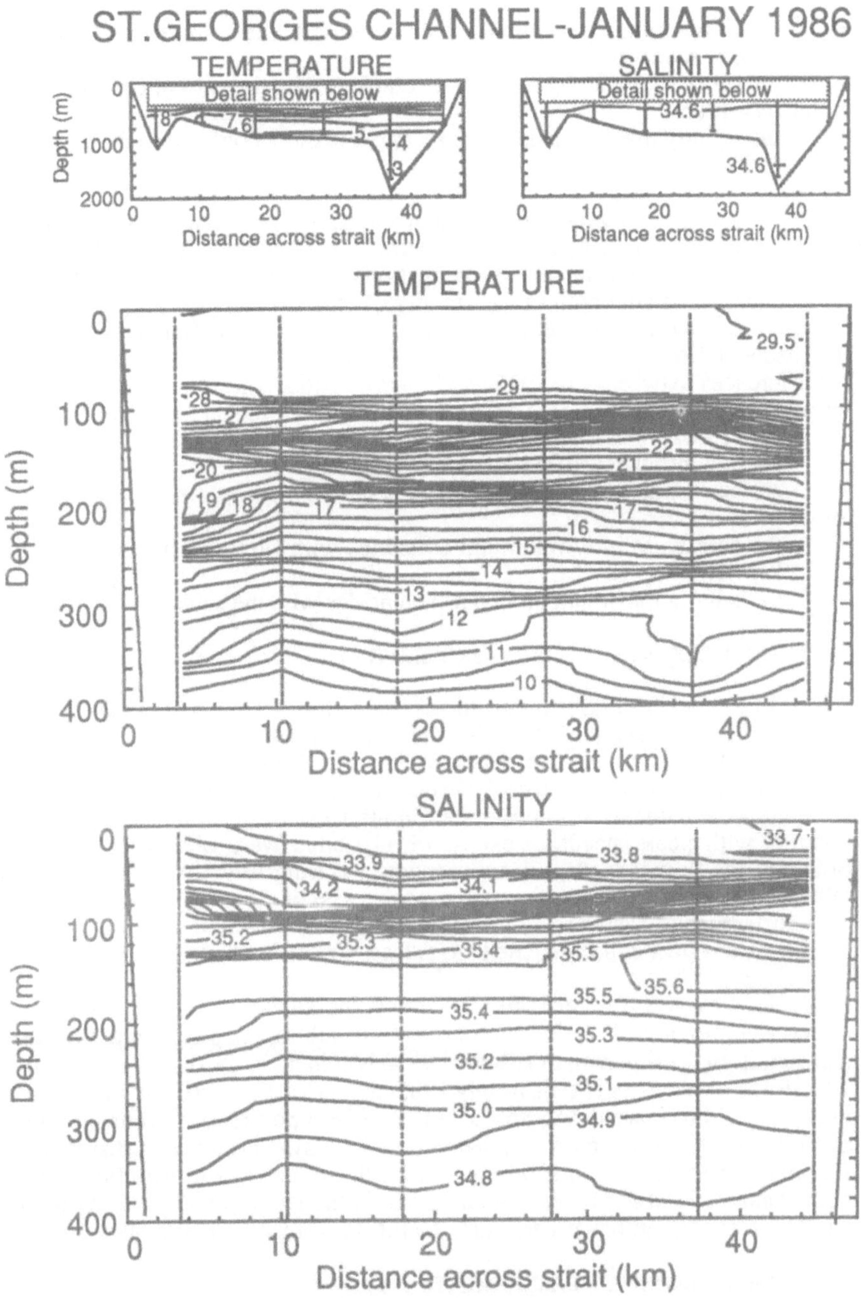

ST.GEORGES CHANNEL-JANUARY 1986

Figure 8.   CTD data from St. George's Channel from January 1986 showing cross-strait sections of temperature (°C) and salinity (psu).

- a surface salinity minimum coincides with the mid-channel Bismarck Sea to Solomon Sea flow
- the subsurface salinity maximum coincides with the current maximum (Fig. 9).

In January 1986:
- the salinity maximum near 200 m depth is more confined to the western side of the channel.

3.1.2 *St. George's Channel*. A notable feature in these temperature and salinity sections (Figs. 7 and 8) is the near vertical isotherms and isohalines west of the seamount along the western side of the channel. Examination of bathymetric data indicates this feature to be a crater with a sill depth in the vicinity of 590 m. Physical properties of the resident water correspond to those of approximately 590 dbar in the Solomon Sea or 510 dbar in the Bismarck Sea.

Differences between July 1985 and January 1986 are not as marked as for Vitiaz Strait, however many similar features are present, namely for January compared to July.
- surface warming is approximately 1°C
- cooling throughout 100-400 dbar by around 1°C
- surface salinity falls of approximately 0.4-0.5 psu.

Also in July 1985:
- the surface salinity minimum coincides with flow toward the Solomon Sea (see Fig. 10)
- the salinity maximum of 35.6 psu coincides with the subsurface current maxima.

3.2. ACOUSTIC DOPPLER CURRENT PROFILING

3.2.1. *Vitiaz Strait*. Fig. 9 shows the average along-channel flows (defined as positive along 304°T). The sections show that the flow is predominantly from the Solomon Sea to the Bismarck Sea. A broad current core with embedded maxima lies in the vicinity of 140 to 240 m depth. During July 1985 the core resides mid-channel, whereas during January 1986 the flow tends to be concentrated in the western portion of the channel. Velocity maxima with peak speeds in excess of 110 cm/s were observed. Bismarck Sea to Solomon Sea flow is evident in both sections. During July 1985 this flow takes the form of a mid-channel surface current with a peak speed of close to 40 cm/s. In January 1986 a near-surface southeastward flow resides in the eastern half of the strait and cores of southeastward flow appear to extend to at least 400 m depth along the eastern boundary.

3.2.2. *St. George's Channel*. Fig. 10 shows the average along-channel flows (defined as positive along 317°T) for each cruise. In all sections a strong current maximum (flowing from the Solomon to Bismarck Sea) is evident in the deeper eastern passage at the 100-200 m level. Peak speeds in July 1985 are in excess of 110 cm/s with average speeds in January 1986 of around 85 cm/s. In the western portion of the channel (west of Duke of York Island) another current maximum exists, however, it is less sharply defined in the July data. May 1988 data shows a strong subsurface flow toward the Bismarck Sea in both channels. Between these two maxima, in the lee of the Duke of York Island group the flow is significantly weaker, and in fact it is reversed during January 1986.

3.3. MOORED VELOCITY MEASUREMENTS

3.3.1. *Vitiaz Strait*. The approximate position of the mooring with respect to the ADCP measurements is shown in Fig. 11; the shallowest meter lies beyond the depth to which reliable ADCP data was obtained. The time series for the along-channel speed (Fig. 12)

183

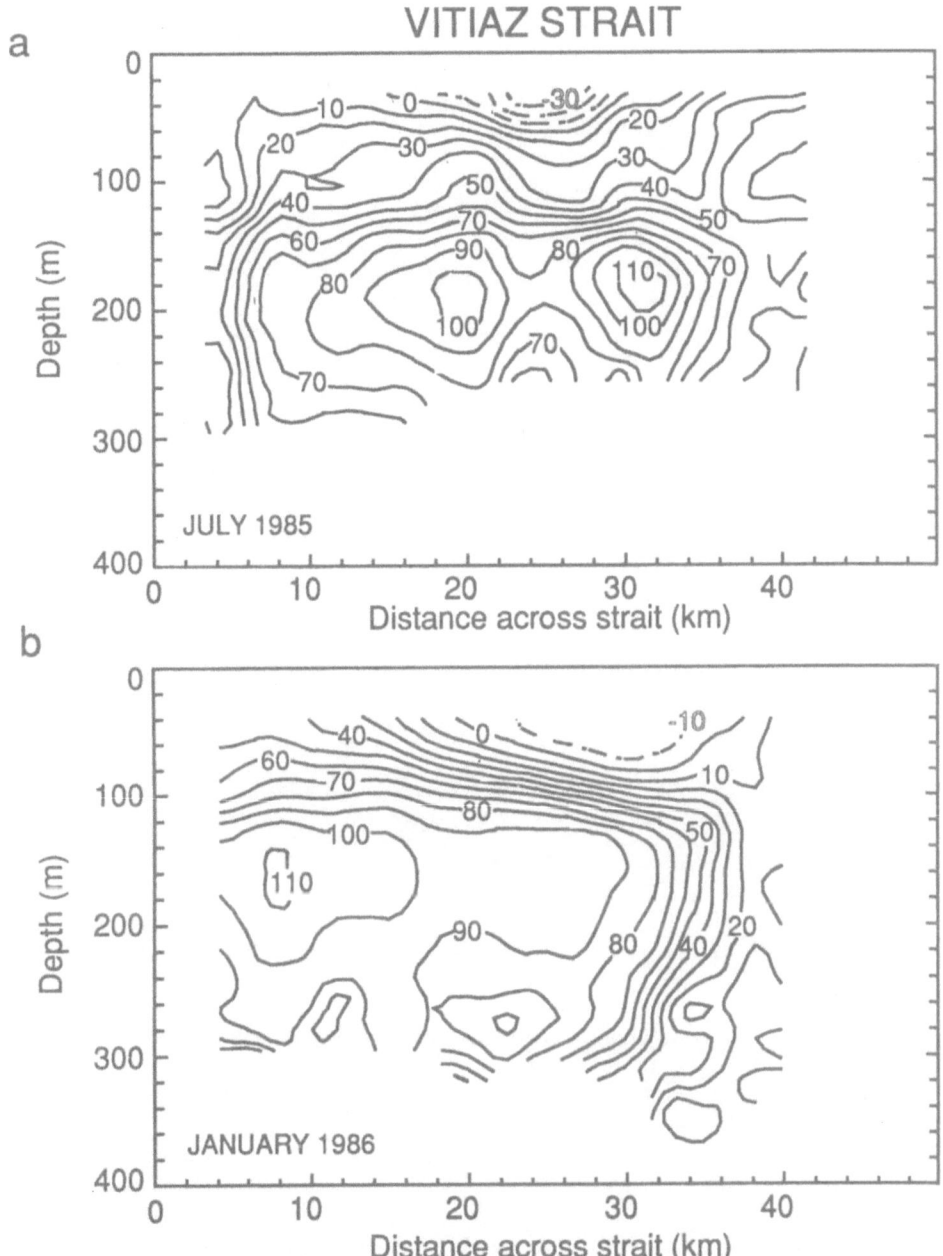

Figure 9. Along-channel speed (cm/sec) in Vitiaz Strait obtained from shipboard acoustic Doppler current profiler measurements in a) July 1985 and b) January 1986.

184

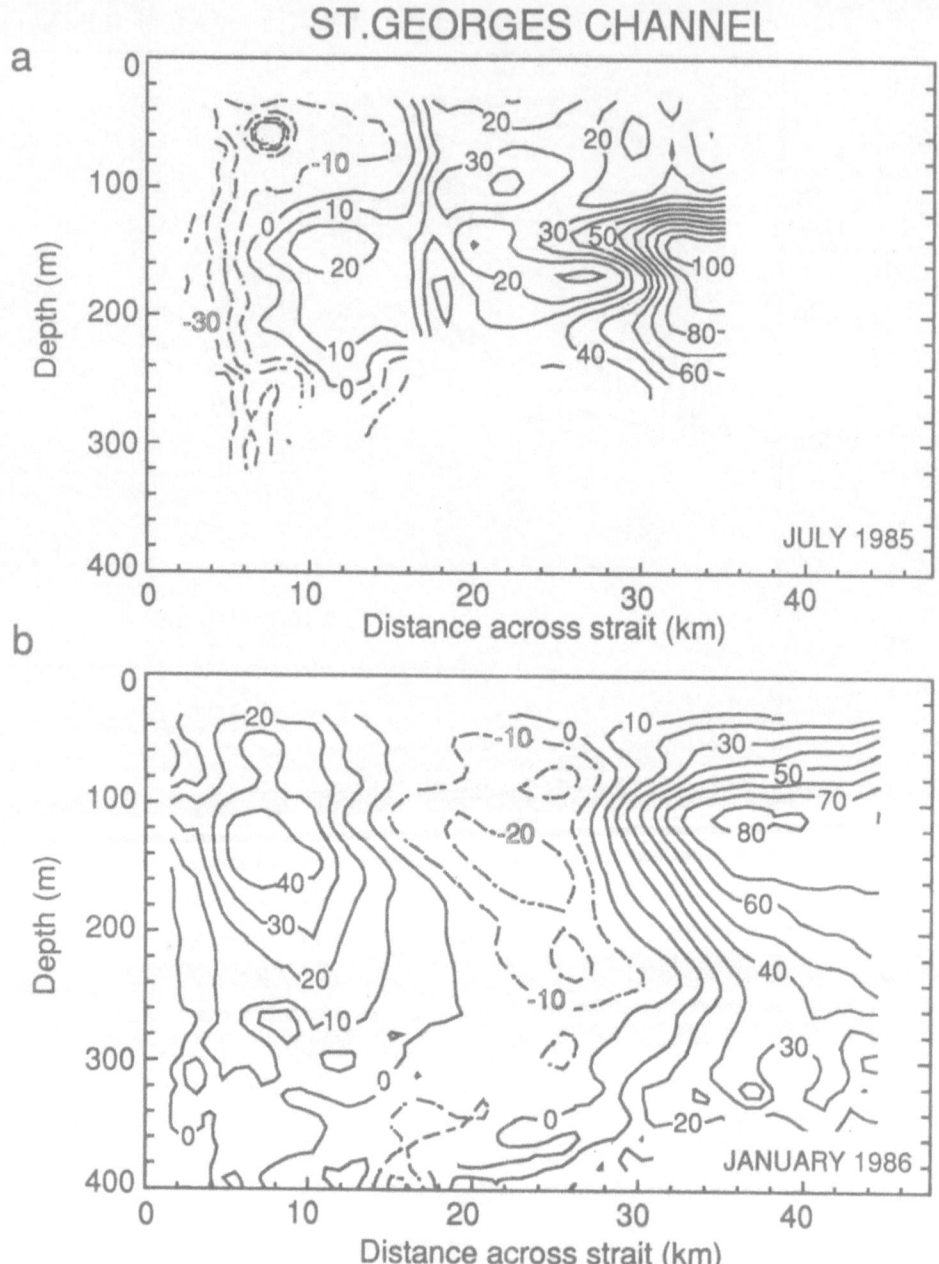

Figure 10.   Along-channel speed (cm/sec) in St. George's Channel obtained
from shipboard acoustic Doppler current profiler measurement in a)
July 1985 and b) January 1986.

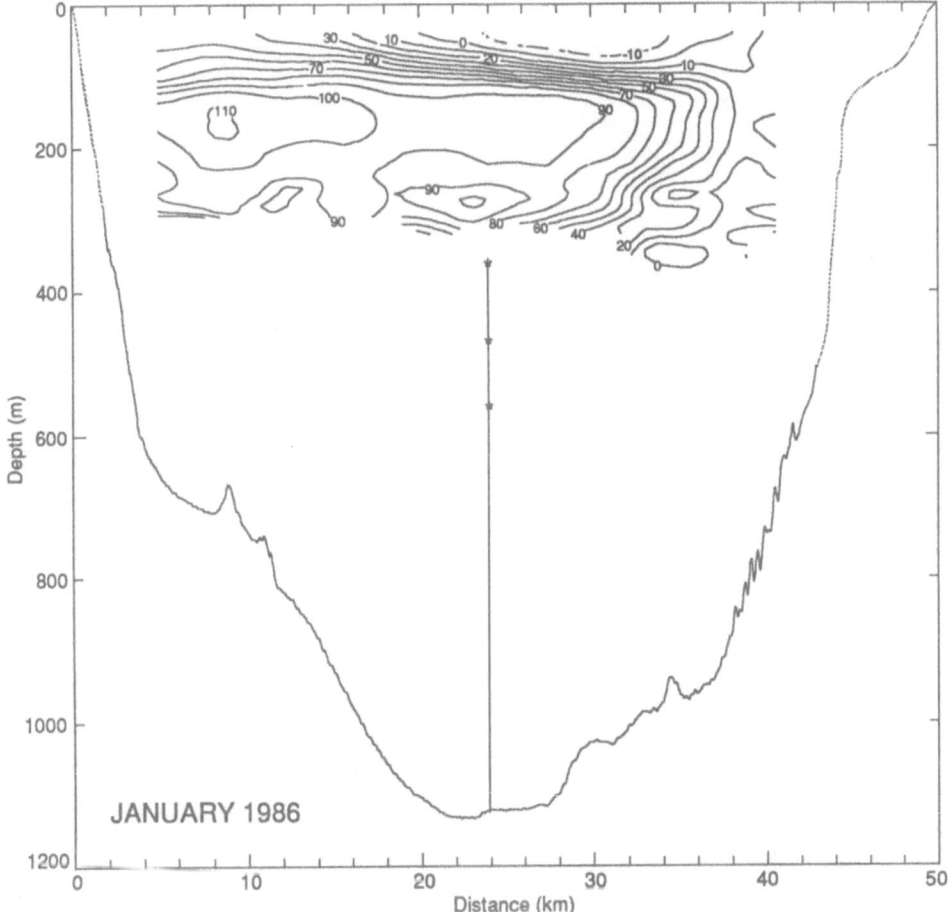

Figure 11.   Average along-channel velocity (cm/s) in Vitiaz Strait during
               January 1986 (as in Figure 9b). This figure shows the relationship
               of the ADCP data to the cross-channel topography of the strait and
               the depth of the moored measurements at the mid-channel current
               meter mooring.

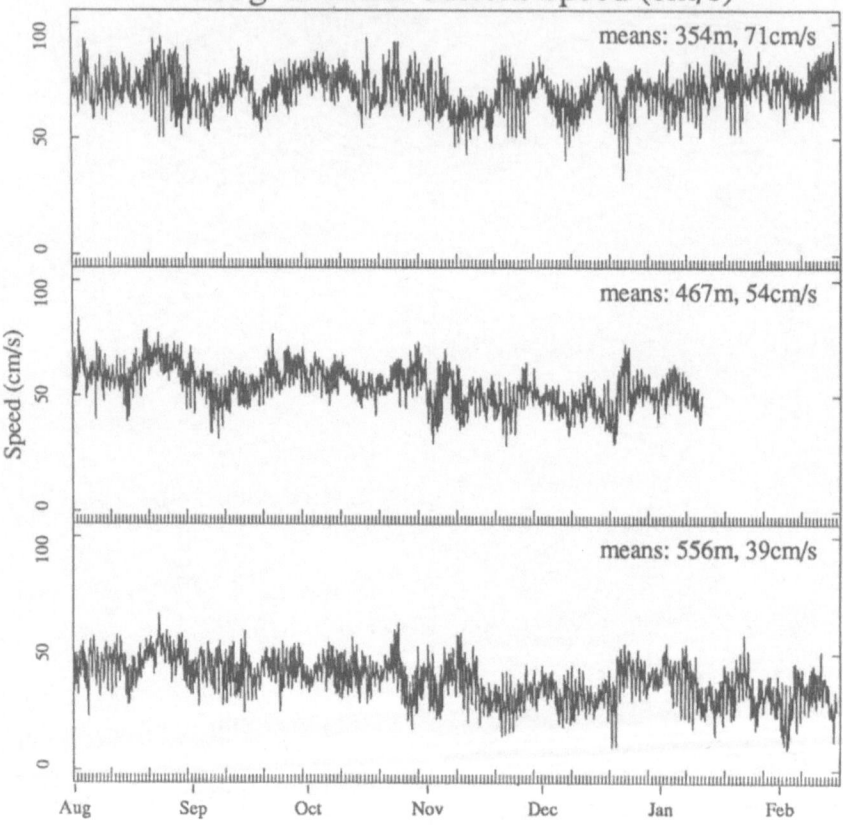

Figure 12.   Along-channel velocity measurements from a single mooring in Vitiaz Strait from July 1985 until February 1986.

indicate strong, steady flow from the Solomon Sea to the Bismarck Sea. Because of the current strength, the mooring was "blown over." The average depth of the current meters was 354 m, 467 m, and 556 m compared to their design depths of 125 m, 225 m, and 350 m, respectively. Average currents sampled by the current meters were 71 cm/s, 54 cm/s, and 39 cm/s, respectively.

4.   Discussion

A summary of transport estimates in Sverdrups ($1 \text{ Sv} = 10^6 \text{ m}^3 \text{ s}^{-1}$) from Vitiaz Strait and St. George's Channel for the 20-300 m depth range are shown in Fig. 13. Estimates are derived for the depth range of available overlapping data (20-300 m) using the ADCP transects, geostrophic transports from CTD sections, and from geostrophic sections derived from XBT data (not shown) using the mean T-S relationship. The steadiness of the six month Vitiaz Strait velocity time series from the single mooring, which extends between the two monsoon peaks, suggests that the ADCP transects at either end of the record may be representative of the typical subsurface flow conditions in the strait. Furthermore, the ADCP measurements, which span most of Vitiaz Strait, are the best for making an estimate of mass transport through the strait in the upper 300 m. The estimates indicate a range of transport values for both of the Straits, from 4-11 Sv in Vitiaz Strait and 1-11 Sv in St . George's Channel. However the largest estimates and the greatest variation are derived from the CTD measurements. These values may be fallacious because of non-geostrophic effects induced by patches of extremely low salinity water found in the straits (due to rainfall or runoff). The XBT transects do not show the same effect because of the use of the mean temperature salinity relationship and are in closer agreement with the ADCP results. Discounting the CTD derived transports, the range of estimates for transports in the upper 20-300 m for Vitiaz Strait is reduced to 4-7 Sv and St. George's Channel to 1-4 Sv.

A direct estimate of transport through Vitiaz Strait below 300 m is more difficult to determine because the single mooring is not representative of the average cross-strait structure. A detailed assessment is not provided here. However, the strength of the current observed at mid-depths over six months suggests a total transport of at least 8-14 Sv. S. Godfrey (private communication) has assessed the western boundary current transports at 6°S based on global Sverdrup balance calculations and found a transport of 23 Sv (including any equatorward boundary current off New Ireland). This is in surprisingly good agreement with the total Vitiaz and St. George's (20-300 m only) transport of 9-18 Sv conservatively estimated from our results. A better estimate of Vitiaz Strait transport could be obtained through a more extensive moored measurement program.

5.   Conclusions

The evidence presented here suggests that the subsurface flow through Vitiaz Strait from the Solomon Sea to the Bismarck Sea is extremely strong and persistent. We conservatively estimate the transport through the Strait at 8-14 Sv based on reliable ADCP measurements and limited current meter data. This work, combined with that of Tsuchiya et al. (1989), suggests that the Vitiaz Strait mass transport is a significant fraction of the equatorward flowing western boundary current in the low latitude western South Pacific and that it acts as a year-round source for the southern hemisphere thermocline waters observed in the Equatorial Undercurrent.

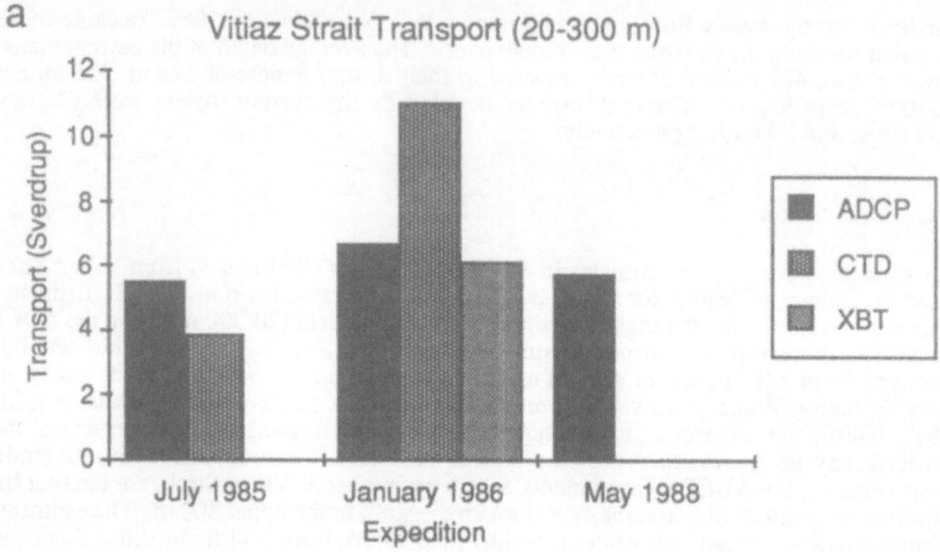

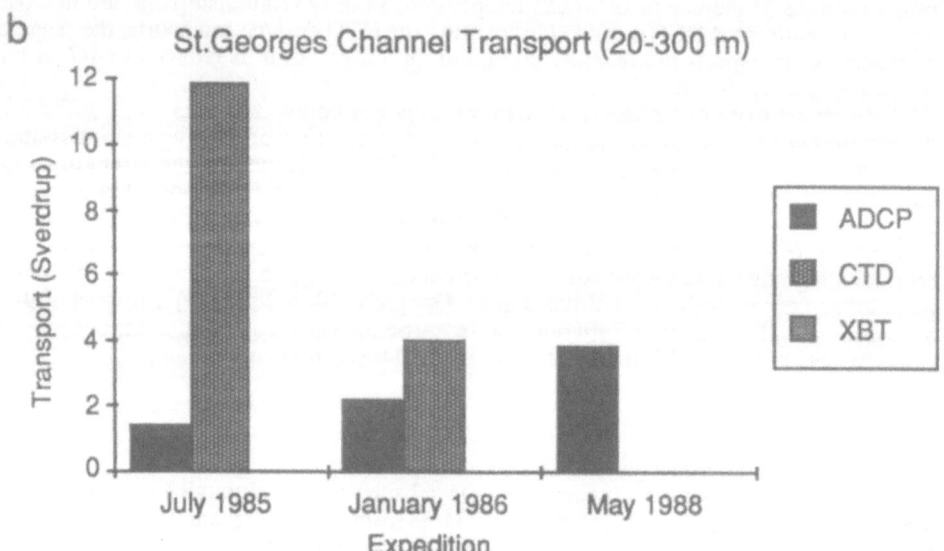

Figure 13.   Summary of mass transport measurements for the 20-300 m level in
a) Vitiaz Strait and b) St. George's Channel.

189

References

Colin, C., J.R. Donguy, C.Henin, C.Oudot, and B.Wauthy (1973) Upper waters north of New Guinea in 1971. *The Kuroshio III, Proceedings of the Third CSK Symposium, Bangkok, Thailand*, National Research Council of Thailand, 132-149.

Hisard, P., Y. Magnier, and B. Wauthy (1969) Comparison of the hydrographic structure of equatorial waters north of New Guinea and at 170°E. *J. Mar. Res.*, 27, 191-205.

Lindstrom, E.J., R. Lukas, R. Fine, E. Firing, S. Godfrey, G. Meyers, and M. Tsuchiya (1987) The Western Equatorial Pacific Ocean Circulation Study. *Nature*, 330, 533-537.

Ministry of the Defence of the USSR (1974) World Ocean Atlas: Pacific Ocean. Pergamon Press, Oxford.

Rougerie, F. and J.-R. Donguy (1975) La Mer du Corail en regime d'alize de sud-est. *Cahiers ORSTOM, Serie Oceanographie*, 13, 49-67.

Schott, G. (1939) Die Aquatorialen Stromungen des westlichen Stillen Ozeans. *Annalen der Hydrographie und Maritem Meteorologie*, 67, 247-257.

Tsuchiya, M. (1968) Upper waters of the intertropical Pacific Ocean. *Johns Hopkins Oceanographic Studies*, 4, 50 pp.

Tsuchiya, M., R. Lukas, R. Fine, E. Firing, and E. Lindstrom (1989) Source Waters of the Pacific Equatorial Undercurrent. *Progress in Oceanography* (accepted).

A REVIEW OF RECENT PHYSICAL INVESTIGATIONS ON THE STRAITS AROUND THE JAPANESE ISLANDS

TAKASHIGE SUGIMOTO
Ocean Research Institute, University of Tokyo
1-15-1, Minamidai, Nakano-ku, Tokyo, 164 Japan

ABSTRACT. Studies on the oceanographic structures and their short term, seasonal and year-to-year variations in the straits around the Japanese Islands are overviewed. Focus is upon the effects of through-flows such as the Kuroshio and the Tsushima–Tsugaru–Soya Warm Current. The vertical and horizontal structure of the currents in and around the straits is discussed, as are the physical mechanisms controlling volume transport. Attention is drawn to key research issues.

1. Introduction — Straits and Ocean Currents Around the Japanese Islands

The Japanese Islands are surrounded by the Pacific Ocean and the three characteristic marginal seas shown in Fig. 1. The latter are the relatively shallow East China Sea, the deep Japan Sea (which contains a homogeneous cold water mass deeper than about 200 m), and the Okhotsk Sea. These four are connected to each other by shallow straits. The straits around the Japan Sea include the Tsushima, Tsugaru and Soya, through which the Tsushima Current system passes. The others are the strait northeast of Taiwan, the Tokara Strait, and the gaps in the Izu Ridge. These gaps affect the path of the Kuroshio and the water characteristics south of Japan. The Okhotsk Sea and the Pacific Ocean are connected by the straits between the Kuril Islands. Part of the East Kamchatka Current enters into the Okhotsk Sea through the gaps between the northeastern Kuril Islands and forms a cyclonic gyre. The mixed, low salinity water flows out from the gaps between the southwestern Kuril Islands, forming the coastal and surface Oyashio water.

2. Hydrographic Structures and Their Variability in the Tsushima, Tsugaru and Soya Straits

2.1. THE TSUSHIMA STRAIT (KOREAN STRAIT)

The water of the Tsushima Warm Current originates over the East China Sea shelf. Figure 2 shows the distribution of mean volume transport of tidal residual currents in the Tsushima Strait in August, obtained using historical current meter data (Miita, 1976; Mi-

191

L. J. Pratt (ed.), The Physical Oceanography of Sea Straits, 191–209.
© 1990 *Kluwer Academic Publishers.*

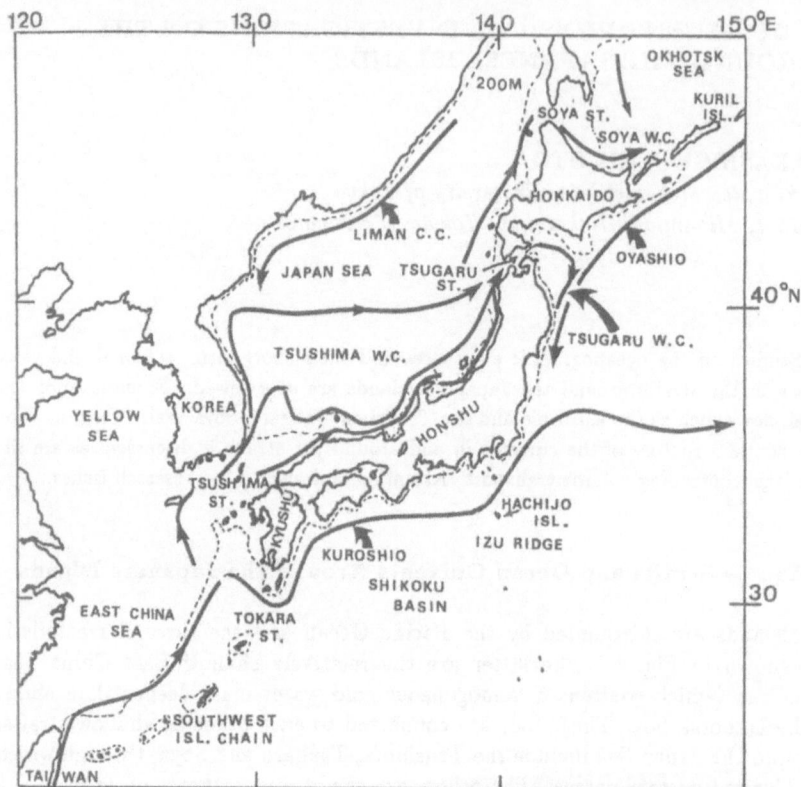

Fig. 1 Bathymetry, straits and ocean currents with their names in the adjacent seas around the Japanese Islands.

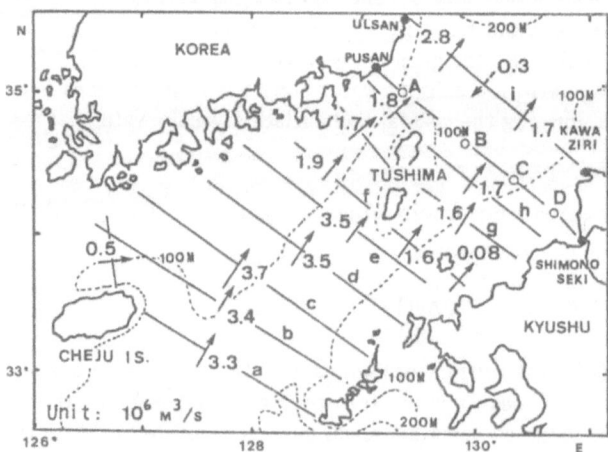

Fig. 2 Volume transport across the Tsushima Strait, based on the direct current measurements in summer, 1924–74, after Miita(1976).

ita and Ogawa, 1984). The volume transport through the channels northwest and southeast of Tsushima Island are about 1.8 and 1.7 $\times 10^6$ m^3 s^{-1}, respectively. In winter the transports decrease by about 50%. The vertical section of the current in the northeastern end (section i, Figure 2) in August is shown in Fig. 3 with the geostrophic currents referenced to the bottom. The vertical shear obtained by both methods suggests that the barotropic component is as strong as the baroclinic one, especially in the southeast channel. Counter currents appearing in the lee of the Tsushima Islands are thought to be associated with a local anticyclonic eddy (Isoda, 1989).

Figures 4a and 4b show mean water temperature and salinity in the northeastern section of the strait in August and February, respectively (Ogawa, 1983). The water is stratified in summer due to the increase of surface heating and the fresh water inflow. During winter, however, it is well mixed by surface cooling. The relatively warm, saline water flows into the Japan Sea along the Japanese coast (Inoue et al., 1985; Miita and Ogawa, 1984).

Short period fluctuations were investigated by Tawara et al. (1984). The surface water temperature across the Tsushima Strait between Pusan and Shimonoseki was monitored over a year from a ferry boat. Figure 5 shows temporal variations in the water temperature at fixed stations A–D (Fig. 2) indicating periodic warming events from April to August. These events are predominant in the eastern channel of the Strait because of the intermittent warm water intrusions from the Kuroshio front associated with onshore/offshore fluctuation of the Kuroshio west of Kyushu (see Fig. 17a and section 3.1).

Beside the seasonal and short term fluctuations, long term variations in water temperature are also significant. According to Toba et al. (1982) and Miita and Tawara (1984), secular variations of 2–3 years and six-year periods are predominant. Their generation processes are not well understood and are now under investigation (Iwasaka et al., 1988).

2.2. THE TSUGARU STRAIT

The Tsugaru Strait (Fig. 6) is relatively narrow compared with the Tsushima Strait (Fig. 4). The width of the narrowest section is about 20 km. The mean depth of the strait as a whole is about 200 m, but the maximum sill depth at the shallowest section near the entrance (western part) of the Strait is about 130 m.

From the difference between the geostrophic volume transport at the southern section a and the northern section b off the entrance of the Strait in Fig. 6, the volume transport into the Tsugaru Strait can be estimated. The transport (Q in Fig. 7) undergoes large seasonal changes, fluctuating from a maximum of about 3.0 $\times 10^6$ m^3 s^{-1} in September to a minimum of about 1.0 $\times 10^6$ m^3 s^{-1} in March.

Figure 8 shows the seasonal variation in the thermosteric anomaly at the eastern end of the Tsugaru Strait (section c), facing in the downstream direction of the Tsugaru Warm Current. Density stratification becomes much stronger in summer but very weak in winter and spring. Figure 7 shows seasonal variations in the inflow volume transport, Q, and geostrophic volume transport Q_g referenced to the bottom at the eastern end of the Strait (at section c). Also shown is the ratio of the barotropic component of the volume transport Q_b/Q, where Q_b is given by $Q - Q_g$. The figure indicates that this ratio becomes as much as 0.7 during winter and spring (Sugimoto and Kawasaki, 1984).

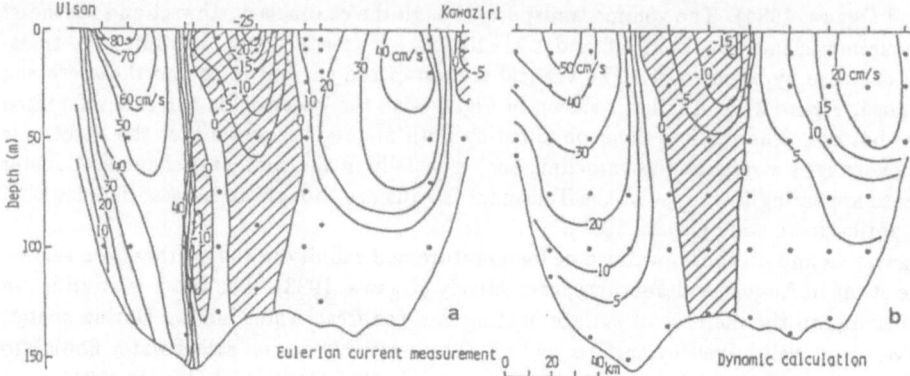

Fig. 3 The tidal residual currents in the exit section of the Tsushima Strait looking downstream in August, 1942—43 obtained by direct current measurements (a), and mean geostrophic currents referenced to the bottom (b), after Miita(1976).

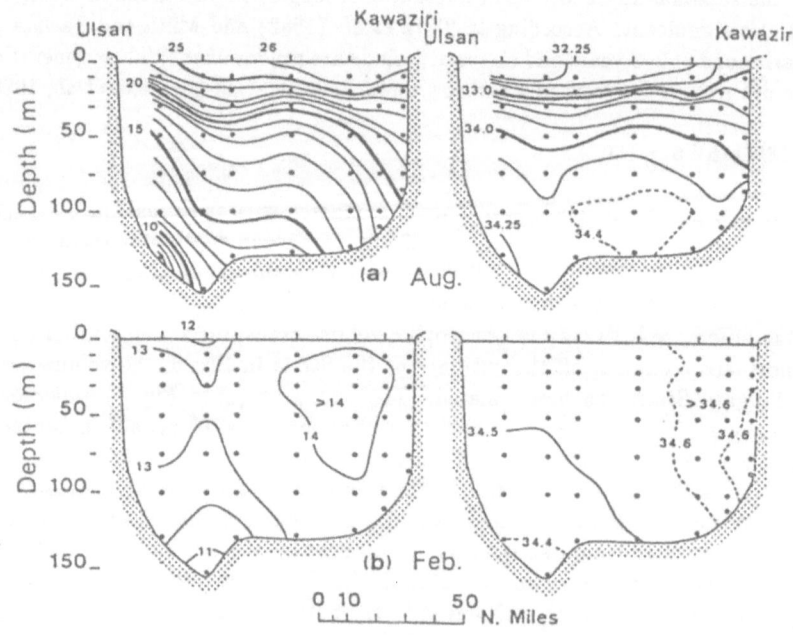

Fig. 4 Mean water temperature (left) and salinity (right) in the exit section of the Tsushima Strait in August (a) and February (b), after Ogawa (1983).

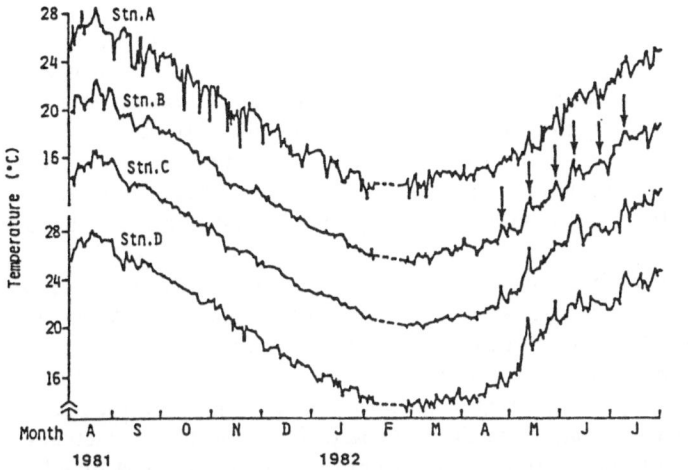

Fig. 5 Day-to-day variations in the sea surface temperature at stations A—D in Fig. 2 from August, 1981 to July, 1982 obtained by Tawara et al. (1984). The arrows indicate warming events.

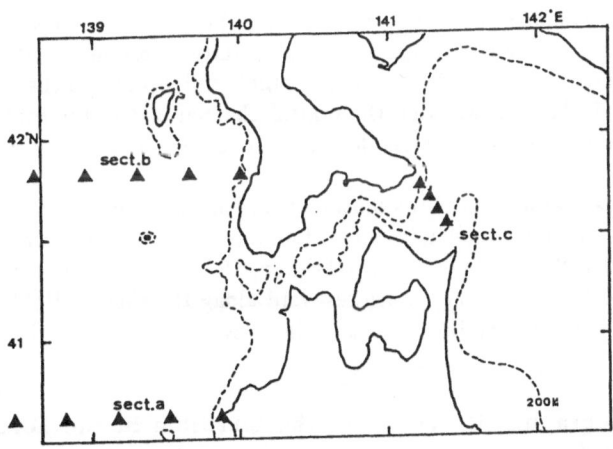

Fig. 6 Bathymetry in the Tsugaru Strait and hydrographic stations (solid triangles) monthly observed by the prefectural fisheries laboratory.

Figures 9a and 9b show the month-to-month variations in the outer edge of the Tsugaru Warm Gyre in a typical year, 1982. The outflowing Tsugaru Warm Water extends offshore, forming an anticyclonic gyre during summer and autumn, but flows southward along the coast as a coastal boundary current farther south. However, this summer–autumn development of the anticyclonic gyre is sometimes disrupted by approaching warm-core rings pinched off from the Kuroshio Extension. The gyre is also modulated by the secular variations in the outflow volume transport. Figure 10 shows year-to-year variations in the seasonality of the outflow patterns, symbol W indicating that Kuroshio warm-core rings are present.

The shape of the Tsugaru Warm Gyre during summer and autumn also varies on a shorter period, about 25 days, as shown in Fig. 11 (Yasuda *et al.*, 1988). Increases and decreases in the outflow transport induce an offshore extension (a), cold water intrusions from the outside (b), and then a southward extension (c). Sometimes solitary frontal disturbances are formed by the splitting or modification of the gyre (d, e), and the disturbances propagate southward along the Sanriku coast. These short term fluctuations are thought to be associated mainly with those of the outflow volume transport from Tsugaru Strait inferred from the daily mean sea-level difference through the strait. The features were simulated in a rotating hydraulic experiment with time-dependent flow rates (Kawasaki and Sugimoto, 1988), as schematically shown in Fig. 12.

2.3. THE SOYA STRAIT

The Soya Strait is wide but shallow and short, about 40 km in width and less than 20 km in length, and its maximum sill depth is about 70 m, as shown in Fig. 13. The tidal residual currents in summer show velocities of more than 50 cm s^{-1} over the whole depth (Aota, 1984). Sections of water temperature and salinity across the strait in August and April are shown in Figs. 14a and 14b. The warm, highly saline water of the Tsushima Warm Current flows into Okhotsk Sea, along the central channel and in the southern part of the strait. The volume transport through the strait in summer is about 1×10^6 m^3 s^{-1}, but almost vanishes in winter.

The Soya Warm Water flows southeastward along the Hokkaido Islands as a shelf-trapped barotropic coastal boundary current, forming a sharp meandering front with the offshore Okhotsk Sea Water (Aota, 1984; Ohshima, 1987). The current flows out into the Pacific Ocean and further extends southwestward along the shelf of Hokkaido Island as a coastal boundary current containing warm, saline water.

3. Hydrographic Structures of the Kuroshio in the Tokara Strait and Over the Izu Ridge

Although the Tsushima Warm Water originates from the East China Sea as mentioned above, the main part of the Kuroshio water in the East China Sea passes through the Tokara Strait and flows along the southern coast of Japan, and then crosses over the Izu Ridge. The hydrographic structure of the flow in these passages is described now.

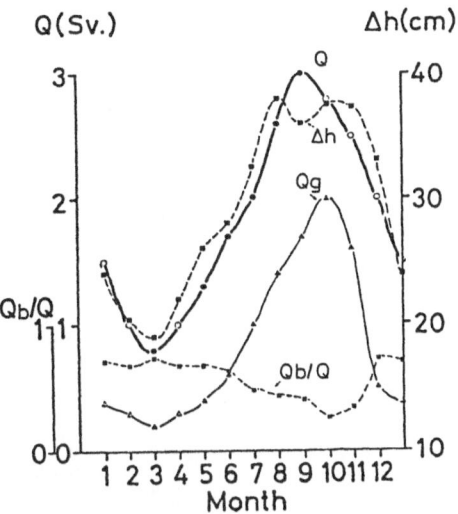

Fig. 7 Seasonal variations in the inflow volume transport Q, volume transport of the geostrophic current referenced to the bottom and the ratio of the barotropic component of the volume transport Qb (=Q- Qg) to the total transport Q.

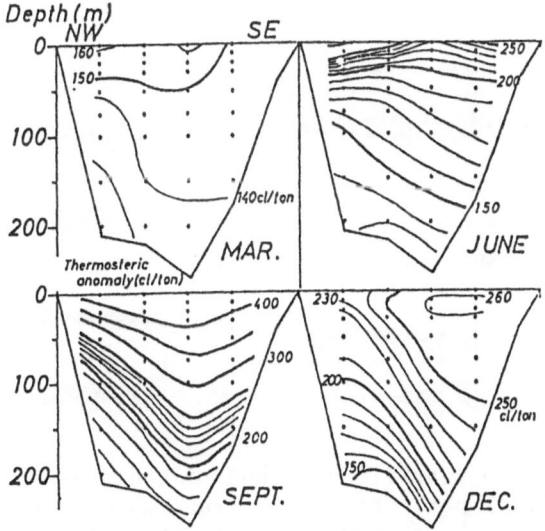

Fig. 8 Seasonal variation in the thermosteric anomaly at the eastern end section (exit) of the Tsugaru Strait, after Kawasaki and Sugimoto (1984).

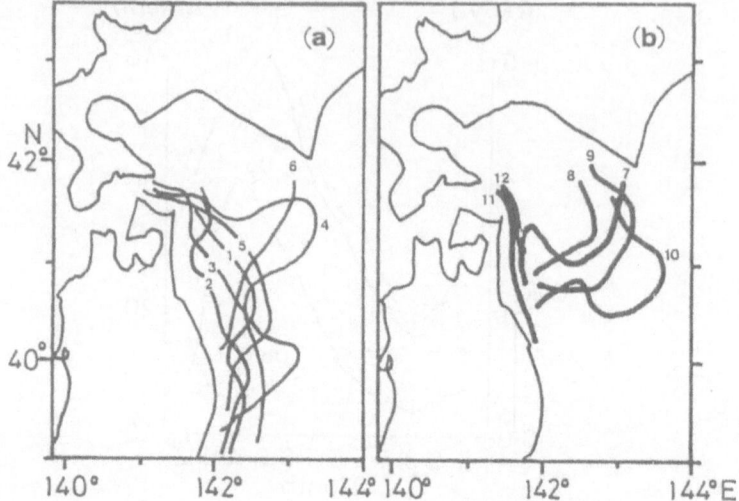

Fig. 9 Seasonal variation of the offshore boundary of the Tsugaru Warm Water during January—June (a) and July—December (b) in 1982, after Sugimoto and Kawasaki (1984).

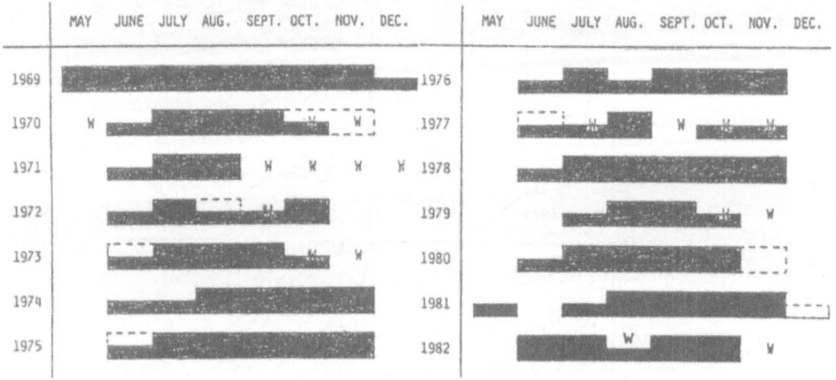

Fig. 10 Year-to-year fluctuations of the seasonal variation in the outflow patterns of the Tsugaru Warm Water. Thick bars indicate the gyre mode, thin bars the loop mode and mark W the possible effect of the Kuroshio warm-core ring, after Sugimoto and Kawasaki (1984).

3.1. THE TOKARA STRAIT

The bathymetry and the currents in the Tokara Strait and upstream are shown in Fig. 15. Figure 16 shows a typical lateral section of the geostrophic current referenced to the surface current velocity shown in the left bottom corner as measured with an acoustic Doppler current profiler (Nakano et al., 1989). One of the current cores is in the northern shallower part of the slope which corresponds to the Kuroshio axis indicated by the isotherm of 15°C at 200 m depth. Another deep core lies in the middle of Tokara Strait. A weak counter current appears at the bottom of the continental slope and deeper than about 800 m, coincident with the results obtained by Ishii et al. (1985) with moored current meters. These features are quite similar to those on the East China Sea slope (Sugimoto et al., 1988).

The path and hydrographic structures of the Kuroshio around Tokara Strait are complicated, with seasonal and 10–30 day fluctuations dominating. The Kuroshio path forms an anticyclonic meander to the northwest of the Strait and then a cyclonic meander in the Strait near the shelf-break. The amplitude of the frontal disturbance is also exaggerated in this area. Figures 17a and 17b show examples of the time series of the distance of the Kuroshio front from the southern tip of Kyushu Island observed during 1982 and December 1984 – May 1985 from a daily ferry boat (Takeshita, 1983; Miyaji, 1989). The path of the Kuroshio in Tokara Strait fluctuates onshore/offshore with a predominant period of 20–25 days, although it varies from year to year (Nagata and Takeshita, 1985). These short term fluctuations of the Kuroshio path have significant influence on the coastal circulation and water exchange between the Kuroshio and coastal waters in the East China Sea and south of Kyushu (Sugimoto et al., 1988; Miyaji, 1989). Satellite thermal images of sea surface temperature often show the entrainment of the coastal waters into the Kuroshio.

The current velocity of the Kuroshio is usually intensified in winter, causing the cyclonic meander of the Kuroshio path in the east of Tokara Strait, which subsequently propagates downstream. Cyclonic meanders with relatively large amplitude are produced once every 2–3 years and some develop into stationary meanders of large amplitude to the west of the Izu Ridge.

3.2. THE IZU RIDGE

South of Japan the path of the Kuroshio has two or three stable modes, as shown in Fig. 18 (Otsuka, 1985; Kawabe, 1985). One is the straight path which flows eastward south of Honshu and passes the Izu Ridge north of Hachijo Island (type N). The second passes south of Hachijo Island (type C). The third is the large meandering path with a cyclonic eddy to the west of the Izu Ridge (type A).

Otsuka (1985) found also that the vertical inclination of the frontal surface of the Kuroshio decreases as the Kuroshio approaches the Izu Ridge, which means that the Kuroshio current becomes weak and broad. The lateral departure of the 10°C contours at 500 m depth in Fig. 18 is consistent with this broadening. The paths of the drifting buoys which cross over the northern part of the Izu Ridge (Hirano and Hasunuma, personal communication) also indicate that the surface currents diverge horizontally and decrease their speed.

Tomosada (1977) made STD and GEK (geomagnetic electrokinetograph) observations in the northern part of the Izu Ridge and showed the presence of small local eddies and lee

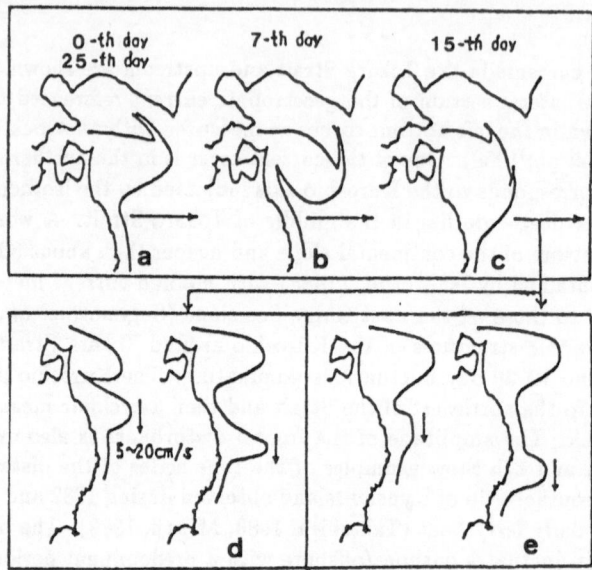

Fig. 11 Short term fluctuation patterns of the Tsugaru Warm Gyre in summer and autumn ; offshore projection (a), detaching (b), southward extension (c), splitting (d) and modification (e), after Yasuda et al., (1988).

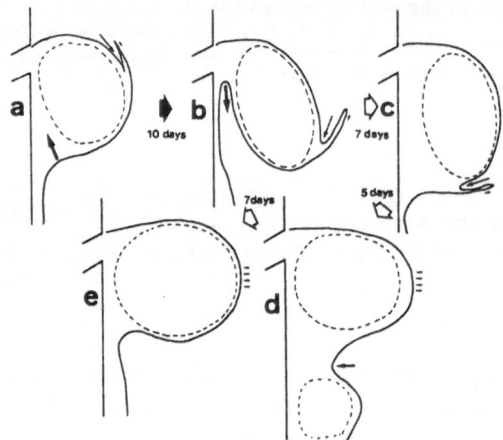

Fig. 12 Response of the gyre to the decreasing flow rate (closed arrow) b: intrusion of offshore water; c: response to increasing flow rate (open arrows); e: offshore extension; or d: splitting of the gyre; after Kawasaki and Sugimoto (1988).

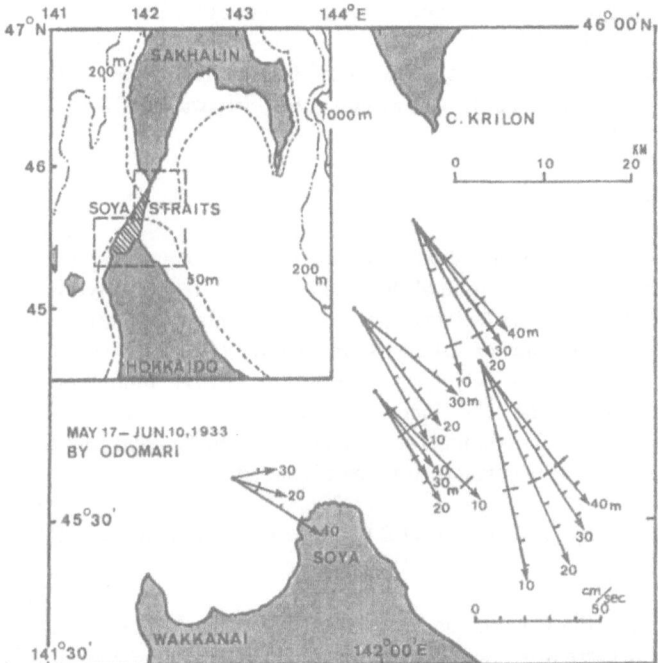

Fig. 13 Bathymetry and the current profiles measured in the Soya Strait in summer, after Aota (1984).

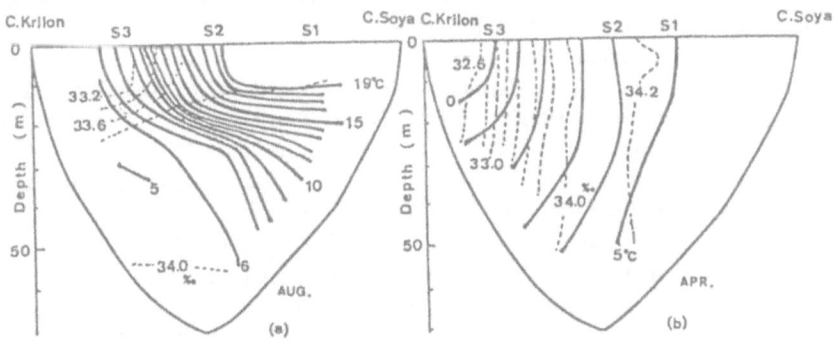

Fig. 14 Lateral sections of the water temperature and salinity across the Soya Strait in August (a) and in April (b), after Aota (1984).

waves in the temperature and salinity maps. The day-to-day fluctuations of the Kuroshio path over the Izu Ridge can be monitored by measuring the daily mean sea-levels and the sea surface temperature along the Izu Island Chain. The results indicate the predominant periods of 20–30 and 70–90 days (Kawabe, 1987; Segawa, 1988).

4. Discussion

Sea straits can be classified according to the type of current they contain. In many applications the currents fall into one of three categories: tidal, density-driven, and geostrophic. Some overlap between the last two can occur. In this paper the physical features of the currents around the Japanese Islands were described, dividing the straits into two further types. One includes relatively deep gaps where western boundary currents such as the Kuroshio cross over. The other includes relatively shallow straits where coastal boundary currents, such as the Tsushima–Tsugaru–Soya Warm Current, cross through. Some dynamical aspects of these currents are now discussed.

4.1. CONTROLLING DYNAMICS OF THE FLOW PASSING THROUGH THE TSUGARU AND SOYA STRAITS

It is thought that the Kuroshio is wind driven and it has been suggested that the Tsushima Warm Current system is created by the associated mean sea-level difference between the East China Sea and the exit of the Tsugaru and Soya Straits (Minato and Kimura, 1980). Seasonal changes of the volume transport, Q, through these straits such as in Fig. 7 corresponds well with the sea-level difference $\Delta\eta$ and dynamic depth anomaly through the straits (Conlon, 1981; Toba et al., 1982).

The volume transport through the straits is given by $(gh/f)\Delta\eta$ in the rotating barotropic frictionless condition or by $g' h_1^2/2f$ in the baroclinic condition, where g is the gravitational acceleration, $g' (= g\Delta\rho/\rho)$ reduced gravity, f the Coriolis parameter, h the mean depth of the strait, h_1 the thickness of the upper layer and $\Delta\rho$ the density difference between the upper and lower layers. In the former case Q is proportional to $\Delta\eta$ and h and independent of the width of the strait (Sugimoto and Kawasaki, 1984). The values 0.2 m for $\Delta\eta$ and 120 m for mean depth over the sill give a barotropic flow $Q = 2.4 \times 10^6$ m^3 s^{-1}, which roughly corresponds to the actual values in the field. However, the physical reason for the seasonal variation of the mean sea-level along the coastal boundary regions of the Kuroshio is not yet completely understood, as discussed by Ichiye (1984).

The unidirectional flow in the Tsushima, Tsugaru and Soya Straits resembles that associated with a salt wedge near a river mouth sill (Sugimoto and Taniya, 1978). To prevent the intrusion of the downstream heavier water crossing over the sill, the sill depth must be shallower than the critical thickness of the upper outflow h defined by the unity of the Froude number as $(Q^2/g' b^2)^{1/3}$, where b is the width of the strait. If we use the following values for the western part of Tsugaru Strait (Conlon, 1981),

$$Q = 2 - 3 \times 10^6 \text{ m}^3 \text{ s}^{-1}, \quad g' = 0.7 \times 10^{-2} \text{ m s}^{-2}, \quad b = 20 \text{ km},$$

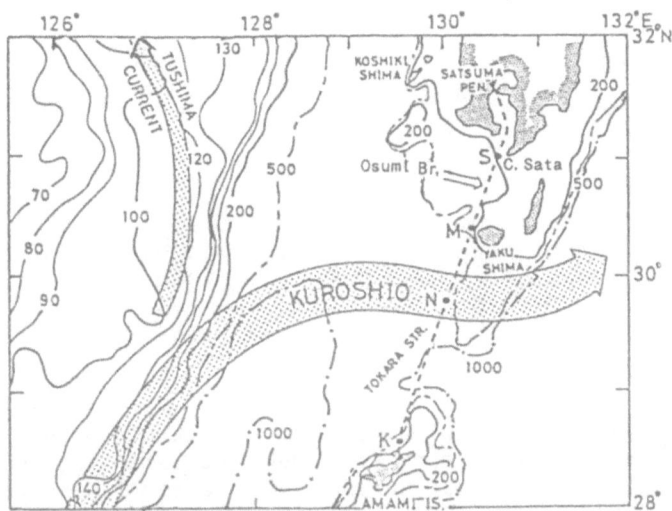

Fig. 15 Bathymetry around the Tokara Strait and the course of the ferry boat (broken line), after Miyaji (1989).

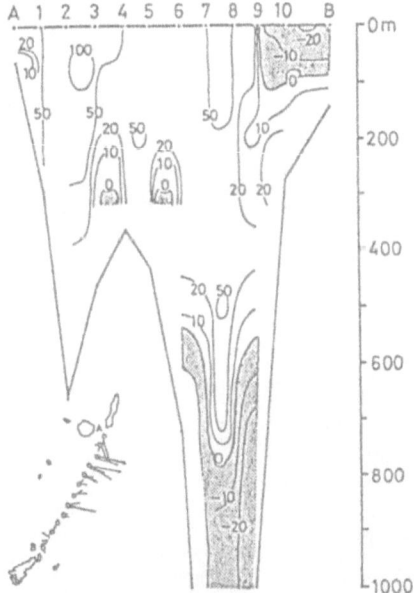

Fig. 16 Lateral distribution of the surface current measured with an acoustic doppler current meter and the geostrophic current velocity referenced to the surface current, obtained by Nakano et al. (1989).

204

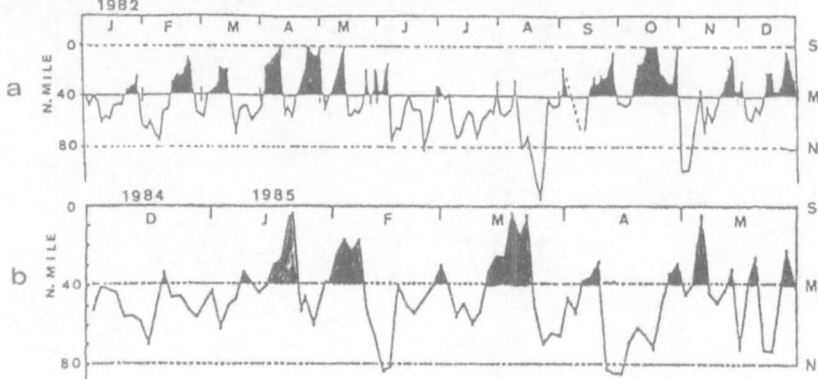

Fig. 17 Temporal variations in the offshore distance of
the Kuroshio front from the Cape Sata, the southern tip
of Kyushu ;   (a) during 1982,   after Takeshita (1983) ;
(b) December 1984—May 1985, after Miyaji(1989).

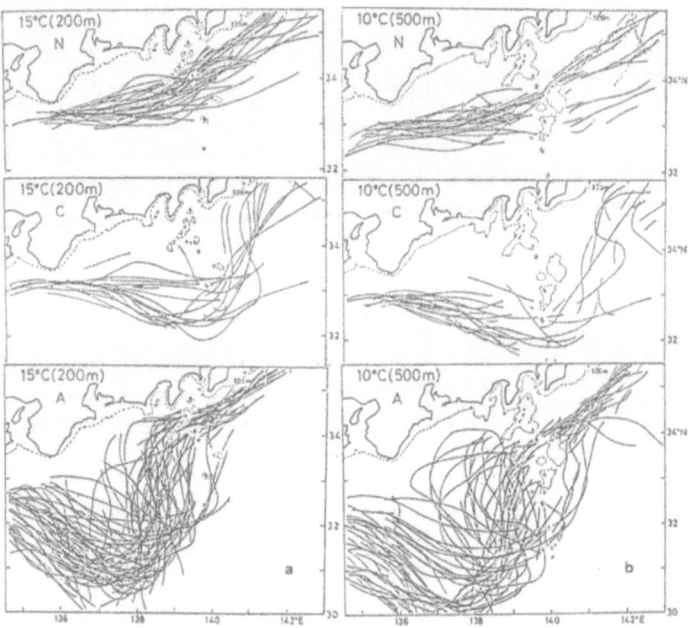

Fig. 18 The three modes of the Kuroshio path around
the Izu Ridge indicated by the isotherms of 15°C at
200m and 10°C at 500m, after Otsuka (1985).

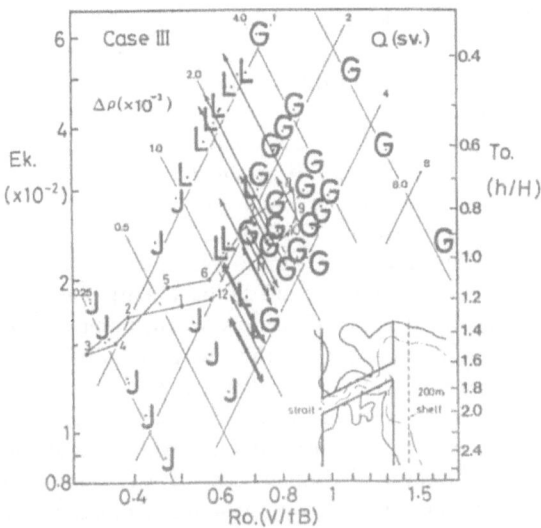

Fig. 19 Seasonal variations in the volume transport Q, the density difference between the upper and lower layers $\Delta\rho$ and the estimated upper layer thickness divided by the channel depth (200m) in the field shown by small dots and thin solid line in the oblique axes. The outflow patterns in the model for Case Ⅲ with constant flow rates are shown by G: growing gyre, L: loop current, J: coastal jet. Thin arrows indicate the cases for the temporal change of the flow rate, after Kawasaki and Sugimoto (1988).

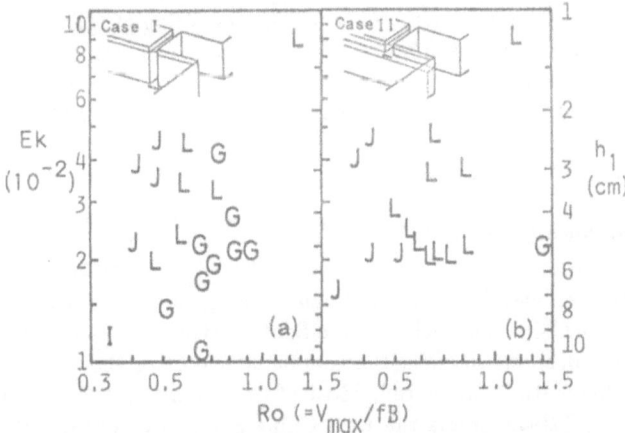

Fig. 20 Values of the Rossby number and the type of the outflow patterns in the model for Case I without shelf (a), Case II with shelf (b). The channel direction is right angle to the coast, after Kawasaki and Sugimoto (1984).

h_c becomes 110–140 m. The addition of the tidal current might make this value larger. As the maximum sill depth of Tsugaru Strait is about 130 m, the volume transport of the Tsugaru Warm Current is just above the critical value needed to prevent the intrusion of Pacific Water into the Japan Sea.

Although the flow has a tendency to become barotropic over the sill near the entrance when the upper layer is thick enough, it gradually becomes baroclinic near the exit, especially in summer and autumn. The dynamics of this baroclinic adjustment process must be made clearer in the near future.

The outflows of the Tsushima and Tsugaru Warm Currents form anticyclonic gyres during summer and autumn. Whitehead and Miller (1979) classified the outflow circulation patterns into coastal jet and gyre modes using a Rossby number R_d/r_g, where R_d is Rossby's internal radius of deformation and r_g is the radius of curvature of the right side coastal boundary. This scheme was applied by Conlon (1981) to the Tsugaru Warm gyre. Nof (1978) pointed out the importance of the non-dimensional parameter F/R, where F is the internal Froude number and R the Rossby number, which was applied to the Tsugaru Warm Gyre by Ichiye (1982). However, the actual seasonal variation in this Froude number in the field and also in the model experiment by Kawasaki and Sugimoto (1984) is small, meaning $\Delta\rho/\rho$ is roughly proportional to Q^2, as indicated in Fig. 19. The seasonal changes of the values of the Rossby number and Ekman number obtained using the same data are also shown.

The physical mechanism of the seasonal variation was also investigated by Kawasaki and Sugimoto (1984) using hydraulic models and direct current observations in the field. They found that the bottom control is as important as the effect of inertia and the angle of the outflow (about 27° from the east to the north), especially in winter and spring. Comparing the experimental results on the transition of the circulation patterns in the rotating model between Case I and Case II (Figs. 20a and b) and also with the results of Case III (Fig. 19), we notice that topographic control and the angle of the outflow are both effective in shifting the critical values of the Rossby number for the transition of the circulation patterns.

In exiting the Soya Strait, the outflow does not form a gyre throughout the year, possibly because of the small value of the Rossby number (due to the small volume transport). A second contributing factor may be topographic steering (Aota, 1984).

4.2. EFFECT OF THE TOKARA STRAIT AND THE IZU RIDGE ON THE PATH OF THE KUROSHIO

The Kuroshio flows through the strait northeast of Taiwan, the Tokara Strait and the gaps in the Izu Ridge. The maximum sill depth of the first two straits is in the range 700–800 m. Hence, intrusion of the less saline intermediate water into the Okinawa Basin between the East China Sea shelf and the Southwest Islands Chain is fairly limited. However, neither strait seems to affect the volume transport or the thickness of the main flow although the flow path is influenced. Most of the volume transport of the Kuroshio along the east coast of Taiwan enters the East China Sea and exits through Tokara Strait. No steady northeastward flow has been found even in the deep layer along the southeast coast of the Southwest Islands Chain (Kaneko et al., 1989). Hence both straits are considered to be deep and wide enough for the main flow of the Kuroshio to pass through freely.

In the Tokara Strait the core of the Kuroshio is laterally separated into two parts. The shallower part (less than 400 m) flows along the southern coast of Yakushima Island with weak topographic influence. Part of it flows through Ohsumi Strait, between Yakushima Island and Cape Sata, when the Kuroshio front is close to Cape Sata (see Fig. 15). The deeper part flows along the deepest gap in the Strait.

Over the Izu Ridge, the Kuroshio shifts north and south, according to the type of the Kuroshio path in the Shikoku basin, as shown in Fig. 18. During the period of type A (large meandered path) the water in the upper layer of the Kuroshio, indicated by 15°C, at 200 m crosses over the Izu Ridge in the northern part. However, waters in the deeper layer of the Kuroshio, indicated by 10°C at 500 m and 4°C at 1000 m, can cross only the deep gap of the maximum sill depth of about 1100 m just to the north of Hachijo Island. During the period of type C, the Kuroshio passes to the south of Hachijo Island and extends from the surface to the bottom.

It seems that the upper part of the Kuroshio can cross over the relatively shallow northern gaps of the Izu Ridge. This vertical incoherence or lateral branching of the current and the northward expansion of the surface current must be investigated more precisely and clearly.

It is thought that the bi-modal nature of the Kuroshio path is caused by the forcing due to coastal geometries. The path may run to the north or south of Hachijo Island, according to whether the wave length of the standing Rossby wave is larger than the distance between Kyushu and the Izu Ridge (Nitani, 1975; White and McCreary, 1976; Chao, 1984; Yoon and Yasuda, 1986). This bi-modality is thought to be associated with a regime of multiple solutions (Masuda, 1982; Yasuda et al., 1985), or the effect of the deep recirculations associated with the large meander before its spindown (Sekine et al., 1985). However, the role of the Izu Ridge in the transition and conservation of the patterns should be examined further.

Acknowledgment. The author would like to express his heartfelt thanks to Dr. L. Pratt for correcting his English and editing the manuscript.

References

Aota, M. (1984) Oceanographic structure of the Soya Warm Current. *Bull. Coastal Oceanogr.*, **22**, 30–39 (in Japanese).

Chao, S. Y. (1984) Bimodality of the Kuroshio. *J. Phys. Oceanogr.*, **14**, 92–103.

Conlon, D. M. (1981) Dynamics of flow in the region of Tsushima Strait. Coastal Studies Inst., Louisiana State University, Tech. Rep. No. 312, 62 pp.

Ichiye, T. (1982) A commentary note on the paper "On the outflow modes of the Tsugaru Warm Current" by D. M. Conlon. *La mer*, **20**, 125–128.

Ichiye, T. (1984) Some problems of circulation and hydrography of the Japan Sea and the Tsushima Current. In *Ocean Hydrodynamics of the Japan and East China Seas*, T. Ichiye (ed.), Elsevier Sci. Pub., Amsterdam, pp. 15–54.

Inoue, N., T. Miita and S. Tawara (1985) Tsushima Strait. In *Coastal Oceanography Around the Japanese Islands*, Coastal Oceanogr. Res. Committee of the Oceanogr. Soc. Japan (ed.), pp. 901–933 (in Japanese).

Ishii, H., Y. Seto and Y. Michida (1985) Current measurement in the Tokara Strait. KER II Rep., Japanese Sci. and Tech. Agency, **8**, 73–79 (in Japanese).

Isoda, Y. (1989) Topographic effects of the Tsushima Islands on the Tsushima Warm Currents. *Bull. Coastal Oceanogr.*, **27**, 76–84 (in Japanese).

Iwasaka, N., K. Hanawa and Y. Toba (1988) Partition of the North Pacific Ocean based on similarity in temporal variations of the SST anomaly. *J. Met. Soc. Japan*, **66**, 433–443.

Kaneko, I., K. Kimura and S. Takahama (1989) Does the Kuroshio undercurrent exist on the continental slope along the Nansei Islands? KER II Rep., Japanese Sci. and Tech. Agency, 107–119 (in Japanese).

Kawabe, M. (1985) Sea level variations at the Izu Islands and typical stable path of the Kuroshio. *J. Oceanogr. Soc. Japan*, **41**, 307–325.

Kawabe, M. (1987) Spectral properties of sea level and time scales of Kuroshio path variations. *J. Oceanogr. Soc. Japan*, **43**, 111–123.

Kawasaki, Y. and T. Sugimoto (1984) Experimental studies on the formation and degeneration processes of the Tsugaru Warm Gyre. In *Ocean Hydrodynamics of the Japan and East China Seas*, T. Ichiye (ed.), Elsevier Sci. Pub., Amsterdam, 225–238.

Kawasaki, Y. and T. Sugimoto (1988) A laboratory study of the short-term variation of the outflow pattern of the Tsugaru Warm Water with a change in its volume transport. *Bull. Tohoku Reg. Fish. Lab.*, **50**, 203–215.

Masuda, A. (1982) An interpretation of the bimodal character of the stable Kuroshio path. *Deep-Sea Res.*, **29**, 471–484.

Miita, T. (1976) Current characteristics measured with current meter at fixed stations. *Bull. Japanese Soc. Fish. Oceanogr.*, **28**, 33–58 (in Japanese).

Miita, T. and Y. Ogawa (1984) Tsushima Currents measured with current meters and drifters. In *Ocean Hydrodynamics of the Japan and East China Seas*, T. Ichiye (ed.), Elsevier Sci. Pub., Amsterdam, pp. 67–76.

Miita, T. and S. Tawara (1984) Seasonal and secular variations of water temperature in the East Tsushima Strait. *J. Oceanogr. Soc. Japan*, **40**, 91–97.

Minato, S. and R. Kimura (1980) Volume transport of the western boundary current penetrating into a marginal sea. *J. Oceanogr. Soc. Japan*, **36**, 185–195.

Miyaji, K. (1989) The main environmental factor related to the formation of sardine spawning ground in the waters south of Kyushu. *Bull. Coastal Oceanogr.*, **27**, 57–66 (in Japanese).

Nagata, Y. and K. Takeshita (1985) Variation of sea surface temperature distribution across the Kuroshio in the Tokara Strait. *J. Oceanogr. Soc. Japan*, **41**, 244–258.

Nakano, T., N. Ishikawa and Y. Takatsuki (1989) Volume transport of the Kuroshio through the Straits of Tokara. KER II Report, Japanese Sci. and Tech. Agency, **1**, 96–106 (in Japanese).

Nitani, H. (1975) Variation of the Kuroshio south of Japan. *J. Oceanogr. Soc. Japan*, **31**, 154–173.

Nof, D. (1978) On geostrophic adjustment in the sea straits and wide estuaries; theory and laboratory experiment, Part II, Two-layer system. *J. Phys. Oceanogr.*, **8**, 861–872.

Ogawa, Y. (1983) Seasonal changes in temperature and salinity of water flowing into the Japan Sea through the Tsushima Straits. *Bull. Japanese Soc. Fish. Oceanogr.*, **43**, 1–8 (in Japanese).

Ohshima, K. (1987) On the stability of the Soya Warm Current. *J. Oceanogr. Soc. Japan*, **43**, 61–67.

Otsuka, K. (1985) Characteristics of the Kuroshio in the vicinity of the Izu Ridge. *J. Oceanogr. Soc. Japan*, **41**, 441–451.

Segawa, K. (1988) Variation in surface water temperature over the Izu-Ogasawara Ridge. Abstract of Fall meeting of Oceanogr. Soc. Japan, 208.

Sekine, Y., H. Ishii and Y. Toba (1985) Spin up and spin down processes of the large cold water mass of the Kuroshio south of Japan. *J. Oceanogr. Soc. Japan*, **41**, 207–212.

Sugimoto, T. and M. Taniya (1978) Effects of boundary geometries on the intrusion of salt-wedge (1) Numerical experiment and field observations. Sci. Rep. Tohoku Univ., Ser. 5, **25**, 71–82.

Sugimoto, T. and Y. Kawasaki (1984) Seasonal and year-to-year variations of the Tsugaru Warm Current and their dynamical interpretation. *Bull. Coastal Oceanogr.*, **22**, 1–11 (in Japanese).

Sugimoto, T., S. Kimura and K. Miyaji (1988) Meander of the Kuroshio front and current variability in the East China Sea. *J. Oceanogr. Soc. Japan*, **44**, 125–135.

Takeshita, K. (1983) Onshore–offshore fluctuations of the Kuroshio IV. KER Rep., Japanese Sci. and Tech. Agency, **6**, 221–232 (in Japanese).

Tawara, S., T. Mita and T. Fujiwara (1984) Oceanographic structure and its variabilities of the Tsushima Strait. *Bull. Coastal Oceanogr.*, **22**, 50–58 (in Japanese).

Toba, Y., K. Tomizawa, Y. Kurasawa and K. Hanawa (1982) Seasonal and year-to-year variability of the Tsushima–Tsugaru Warm Current system with its possible cause. *La mer*, **20**, 41–51.

Tomosada, A. (1977) Observations of the Kuroshio disturbed by bottom topography near Izu Islands. *Bull. Tokai Reg. Fish. Lab.*, **89**, 17–42 (in Japanese).

White, W. B. and J. P. McCreary (1976) The Kuroshio meander and its relationship to the large scale ocean circulation. *Deep-Sea Res.*, **23**, 33–47.

Whitehead, J. A. and A. R. Miller (1979) Laboratory simulation of the gyre in the Alboran Sea. *J. Geophys. Res.*, **84**(C7), 3733–3742.

Yasuda, I., J. H. Yoon and N. Suginohara (1985) Dynamics of the Kuroshio large meander barotropic model. *J. Oceanogr. Soc. Japan*, **41**, 259–273.

Yasuda, I., K. Okuda, M. Hirai, Y. Ogawa, H. Kudo, S. Fukushima and K. Mizuno (1988) The short term variation of the Tsugaru Warm Current in autumn. *Bull. Tohoku Reg. Fish. Lab.*, **50**, 153–191.

Yoon, J. H. and I. Yasuda (1986) Dynamics of the Kuroshio large meander. *J. Phys. Oceanogr.*, **17**, 66–81.

INFLUENCE OF THE CLIMATIC CONDITIONS ON THE WINTER FLUXES IN THE CORSICAN CHANNEL

M. Astraldi and G. P. Gasparini
Consiglio Nazionale delle Ricerche
Stazione Oceanografica
c/o ENEA, P.O.   Box 316
19100 La Spezia
Italy

ABSTRACT. Currents along the eastern and western sides of the northern part of Corsica flow northward all year long. However, whereas on the western side the flow does not show any substantial seasonal change, the flux coming from the Tyrrhenian Sea has the highest values in winter and the lowest in summer. The winter rise takes place at the end of December and is very fast. The comparison between the flow variability in the Corsican Channel with several meteo-climatic parameters measured in the Ligurian Sea has permitted us to show that a correlation exists between the winter flux rise and the relevant heat loss which affects the Ligurian basin in this period. In this way, a link can be established between the intensification of the winter circulation and the deep water formation processes occurring in the northwestern Mediterranean basin.

1.   INTRODUCTION

A number of hydrographic campaigns carried out in the Ligurian Sea in the past years (e.g. Stocchino and Testoni, 1977; Nyffeler et al., 1980) have shown that the general circulation of the basin is basically cyclonic and fed by two different fluxes flowing along both sides of Corsica (Fig. 1). Direct current measurements (Taupier-Letage and Millot, 1986; Astraldi et al., 1990) revealed that, whereas the western flow is fairly steady all year round, the current flowing east of Corsica varies seasonally, having its highest values in winter. Even though this behaviour has been observed in several occasions, the mechanisms generating it are not yet explained.

In the following, a description will be made of the main characteristics of the circulation around the northern tip of Corsica. Special attention will be paid to the Corsican Channel current with the purpose of establishing a connection between its seasonal variability and the thermal balance of the Ligurian basin and evaluating the role of the deep water formation processes in modulating the current.

211

L. J. Pratt (ed.), The Physical Oceanography of Sea Straits, 211–224.
© 1990 *Kluwer Academic Publishers.*

212

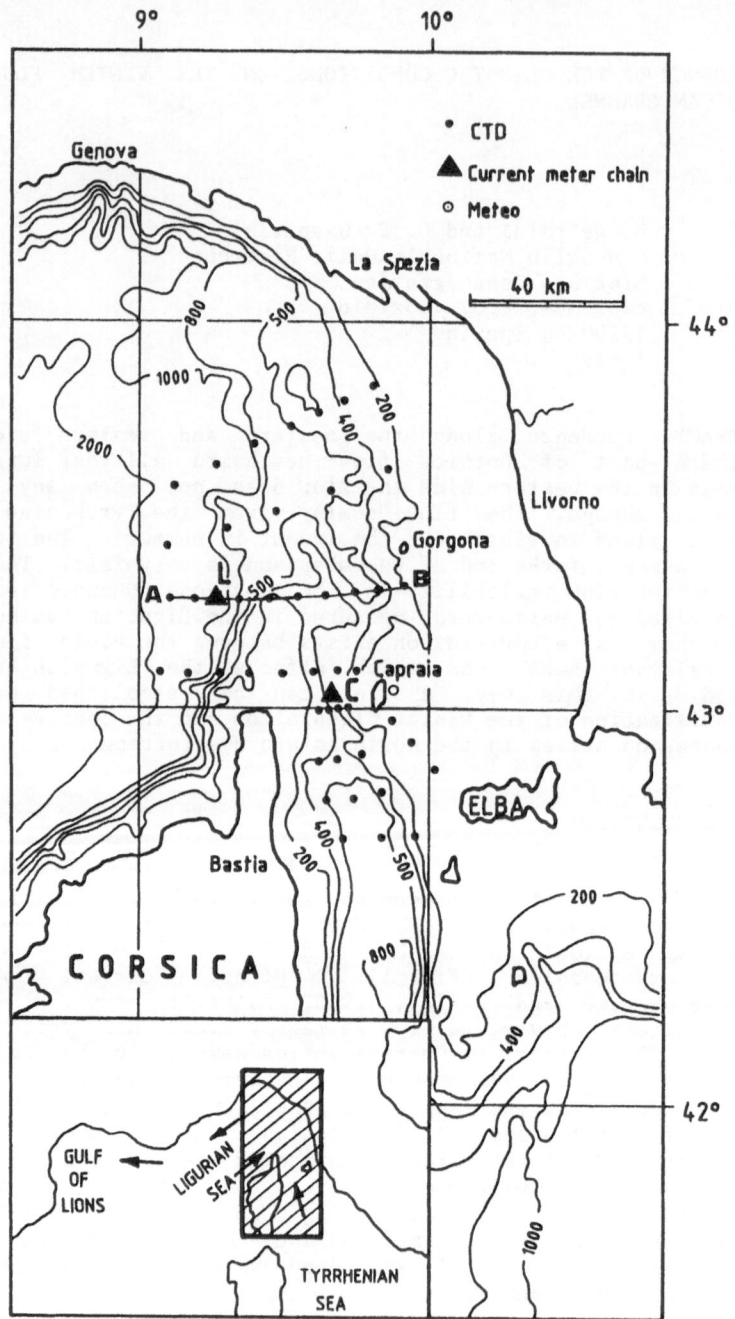

Figure 1. The east Ligurian sea and the sampling stations.   Arrows indicate the principal paths of the current in the region.

2. MEASUREMENTS AND METHODS

An experiment in the Corsican Channel has been conducted as a part of a regional environmental investigation aimed at monitoring the annual and interannual variability of the exchange between the Tyrrhenian and the Ligurian Sea. Oceanographic measurements lasted from July 1985 to September 1988 (Fig. 2). In this work we report on the results of the one-year period from October 1986 to September 1987. During this period, two moorings C and L, located northwest and northeast of Corsica respectively, measured the velocity and direction of the current as well as the pressure and temperature at 30-minute intervals at 5 levels in the Corsican Channel and at 4 levels outside of it, in 460 and 1250 m of water, respectively. The instruments used on these moorings were Aanderaa RCM-4 and RCM-7 current meters. The hydrographic observations were conducted in December 1986, March 1987, and May 1987 using a Neil Brown Mark III CTD probe, while meteorological measurements were effected on a fixed station on the top of the Capraia island.

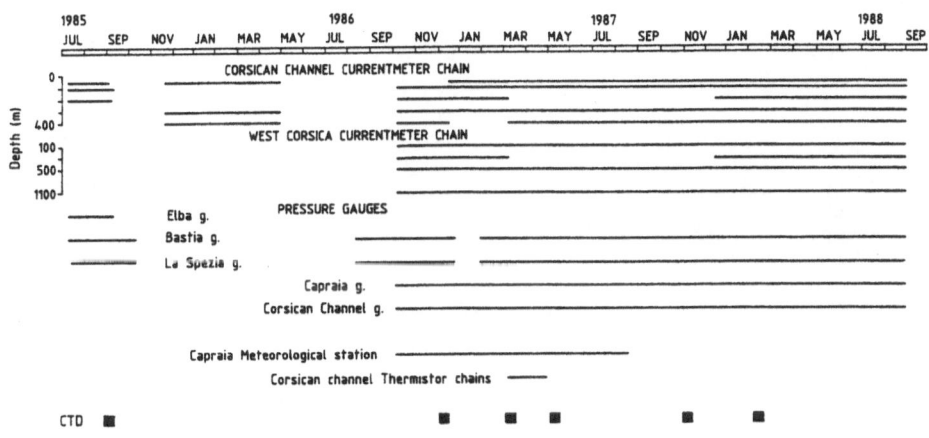

Figure 2. Overall schedule of the measurements.

3. SEASONAL CHARACTERISTICS OF THE CURRENTS

The data of the current meter chain C show that the flow in the channel is essentially northward at all depths, with a seasonally varying intensity (Fig. 3 and Tab. 1). From July to December the flow

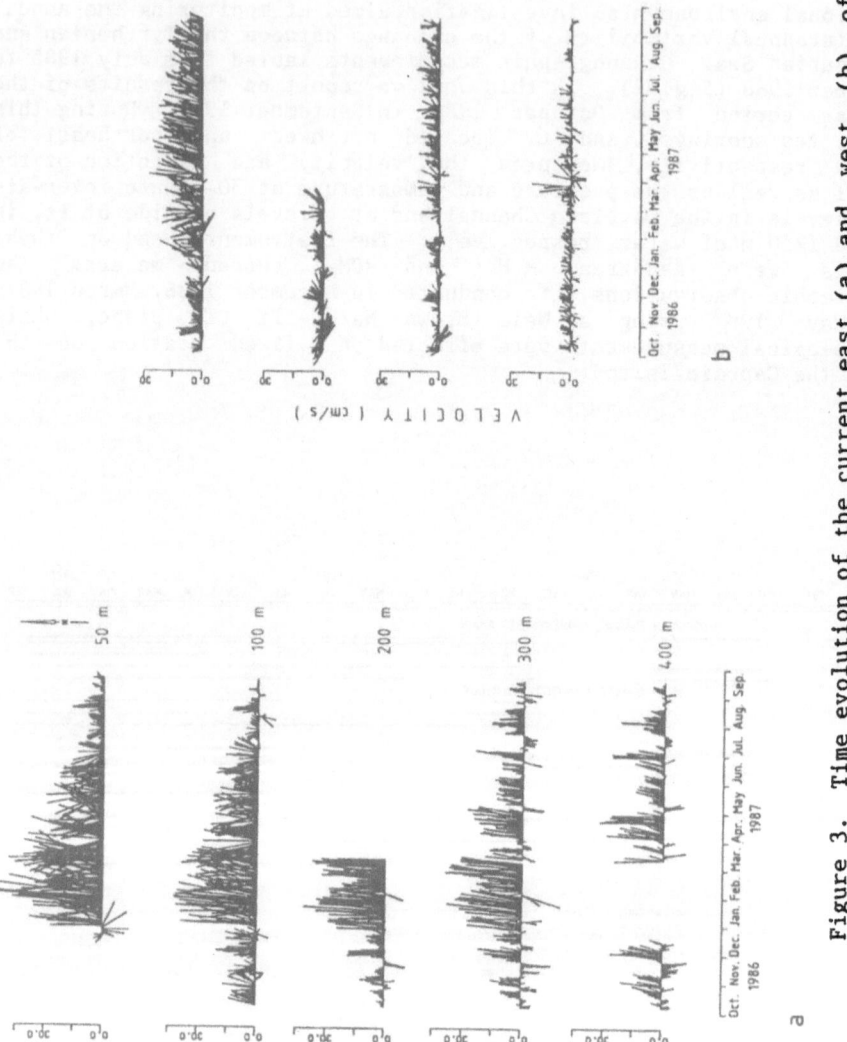

Figure 3. Time evolution of the current east (a) and west (b) of the Corsican island. The original data have been low-pass filtered to remove oscillations below 48h and daily time series have been reconstructed.

TABLE 1. Statistics of the N-S components of the current on both sides of Corsica.

---

	depth (m)	autumn mean+s.d. (cm/s)	winter mean+s.d. (cm/s)	spring mean+s.d. (cm/s)	summer mean+s.d. (cm/s)	year mean+s.d. (cm/s)
	50	--- ---	31.1+10.7	14.6+6.3	5.0+3.7	17.5+13.6
Mooring C	100	7.3+5.7	27.1 +8.4	13.3+5.4	2.6+4.0	13.7+11.3
East of	200	3.9+5.8	22.7 +9.9	---- ---	--- ---	13.0+12.4
Corsica	300	2.8+7.9	16.2+13.3	9.3+8.1	3.0+5.8	8.4+11.0
	400	3.4+9.1	---- ----	9.2+9.2	3.0+5.6	5.6 +8.9
	100	5.2+3.4	9.8 +6.5	15.5+4.4	10.5+3.6	10.4 +6.0
Mooring L	300	2.0+3.5	3.6 +2.9	---- ---	---- ---	2.7 +3.3
West of	500	1.3+3.3	2.5 +2.8	4.2+2.9	4.2+3.1	3.1 +3.2
Corsica	1100	1.6+3.2	1.7 +3.6	1.2+3.0	5.5+1.4	1.4 +3.1

TABLE 2. Transport in the Corsican Channel and percentage contribution to the total value. Units in Sv.

---

	autumn	winter	spring	summer	year
Surface water (0-200m)	0.20 (9%)	1.00 (58%)	0.51 (27%)	0.14 (6%)	0.50
Interm.te water (200-450m)	0.05 (8%)	0.30 (60%)	0.14 (25%)	0.05 (7%)	0.15

---

is generally weak; starting at the end of December it grows rapidly to reach the maximum surface value of 64 cm/s in January. From February to July the flow decreases progressively to the summer values of 2-5 cm/s. The seasonal variability affects both the surface layer of Atlantic water and the deep layer below (Intermediate water).

The current west of Corsica shows different features. Even though northward, the flow does not seem affected by any seasonal variation, the spring rise being mainly connected with a lateral shift of the Ligurian frontal system. These results agree with previous observations (Taupier-Letage and Millot, 1986; Bethoux et al., 1982), all indicating that the Ligurian current along the Nice-Calvi section is nearly steady on the Corsican side and varies seasonally off Nice.

The two currents at C and L appear better correlated in summer, when circulation assumes local features. A comparison with the Ligurian shelf indicates that the current in this location retains characteristics similar to the current coming from the Tyrrhenian Sea; this agrees with the results of the satellite images, which show that the Tyrrhenian current, once it has entered into the Ligurian Sea, follows the Italian coast, while the West Corsican current makes a more restricted loop inside the basin.

The hydrographic data along section A-B (Fig. 1) confirm that the path of the warmer Tyrrhenian water develops along the continental shelf, whereas the West Corsican current remains offshore of it, and a thermal front separates the two water masses (Fig. 4). The θ-S diagrams of individual cruises (Fig. 5) show that the differences in temperature between the two waters remain all year long. Since only a slight difference can be observed between their salinities, an overall major density can be ascribed to the Ligurian water compared to the Tyrrhenian water. This means a lower steric height for the Ligurian basin, thus the existence of a steady flow directed towards this basin.

From the current data inside the channel, the seasonal values of the transport have been computed both for the surface layer (0-200m) and the deeper one (200-460) (Tab. 2); the annual mean value of the surface water transport has been about 0.5 Sv (15.8×10^{12} m^3/year), while that of the LIW 0.15 Sv (4.7×10^{12} m^3/year). The winter transport alone is about 60% of the annual transport and this percentage increases to 85% if the spring transport is included; therefore the exchange between the Ligurian and the Tyrrhenian basin mostly occurs during the months of the weakly stratified season.

4. CURRENT VARIATIONS AND THE WINTER HEAT BALANCE

If we consider the N-S component (v) of the current at 100m (Fig. 6a) and compare it with the air temperature T of the Capraia island, we observe that all the principal events affecting the temperature curve are found in the current time series. In particular, for any drop in the air temperature there is a corresponding increment in the current component and vice-versa. The temperature variations anticipate the current variations by about 4-7 days.

With such behaviour, the flux variability in the channel can be associated with the regional meteo-climatic conditions through the influence they exert on the thermal content of the two basins,

217

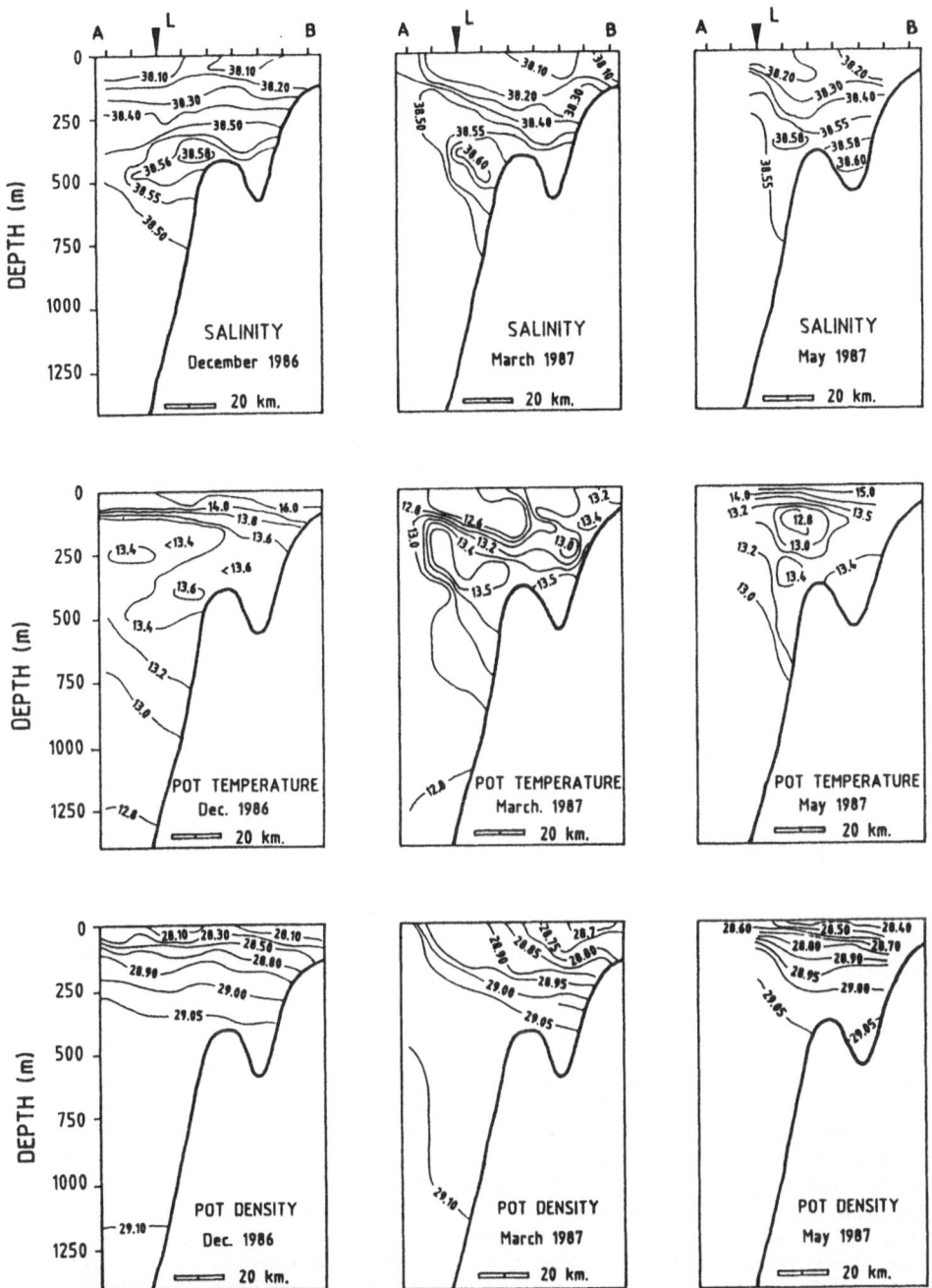

Figure 4. Vertical sections of salinity, potential temperature and density along the transect A-B of Figure 1. The vertical arrow indicates the position of current meter chain L.

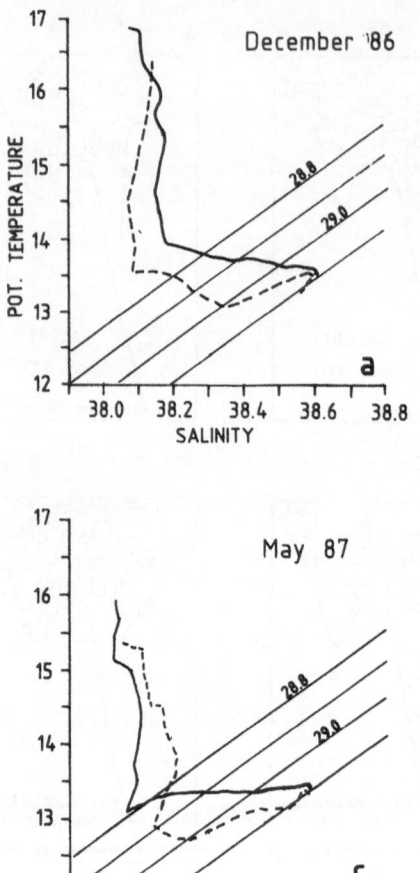
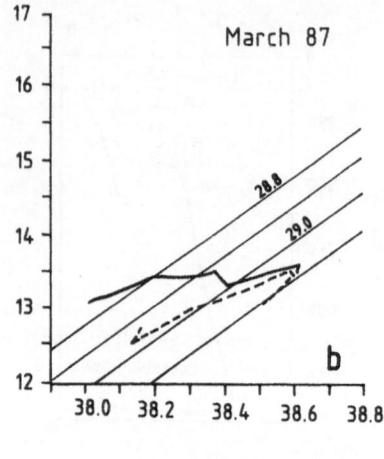

Figure 5. θ-S diagrams east (solid line) and west (dashed line) of Corsica.

particularly the Ligurian basin. We have already noted that the flux along the channel is mainly driven by the gradient in the steric level between the Ligurian and Tyrrhenian basin; now we can state that a link exists between the air-sea heat exchange, the hydrographic conditions of the water body and the current in the channel. Among other events, the first relevant rise in the flow at the end of December (Figs. 3 and 6a) is related to a sudden diminution of more than 4 degrees in the air temperature, associated with a consistent heat loss from the basin. The evolution of the thermal conditions of the surface layer in this period can be followed with the aid of the satellite images (Direction de la Meteorologie Nationale, 1986, 1987) which show (Fig. 7) that the winter rise of the current (at the end of

December) occurs in correspondence to a remarkable decrease in the surface temperature.

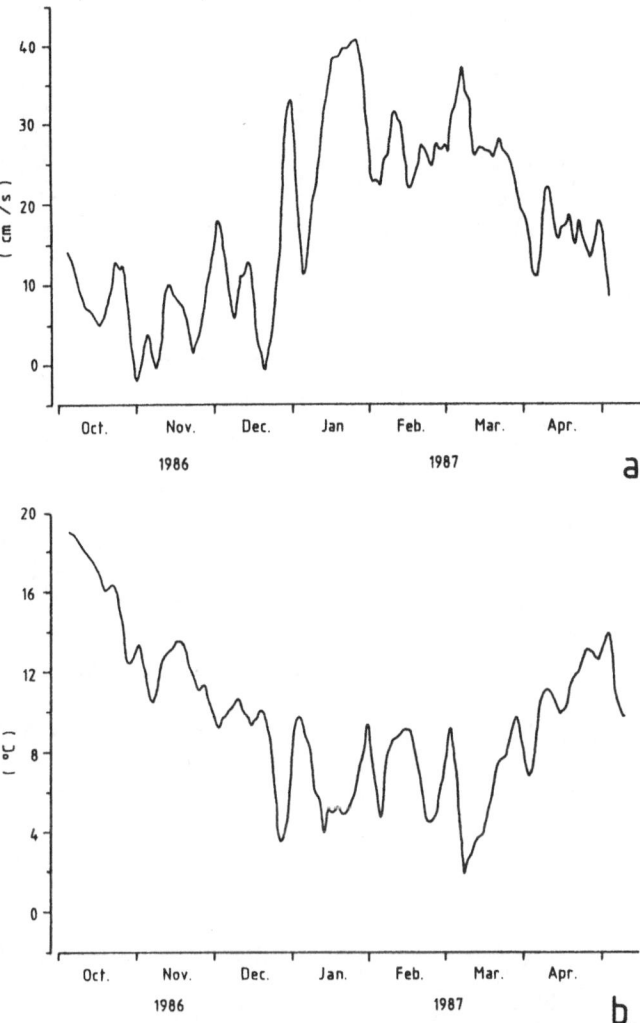

Figure 6. Time evolution of the N-S component of the current at 100m of depth in the channel (a) and of the air temperature at Capraia Island (b).

In order to give a quantitative indication of the influence of atmospheric conditions, we have computed the heat exchange rate between sea and atmosphere by summing up individual contributions:

220

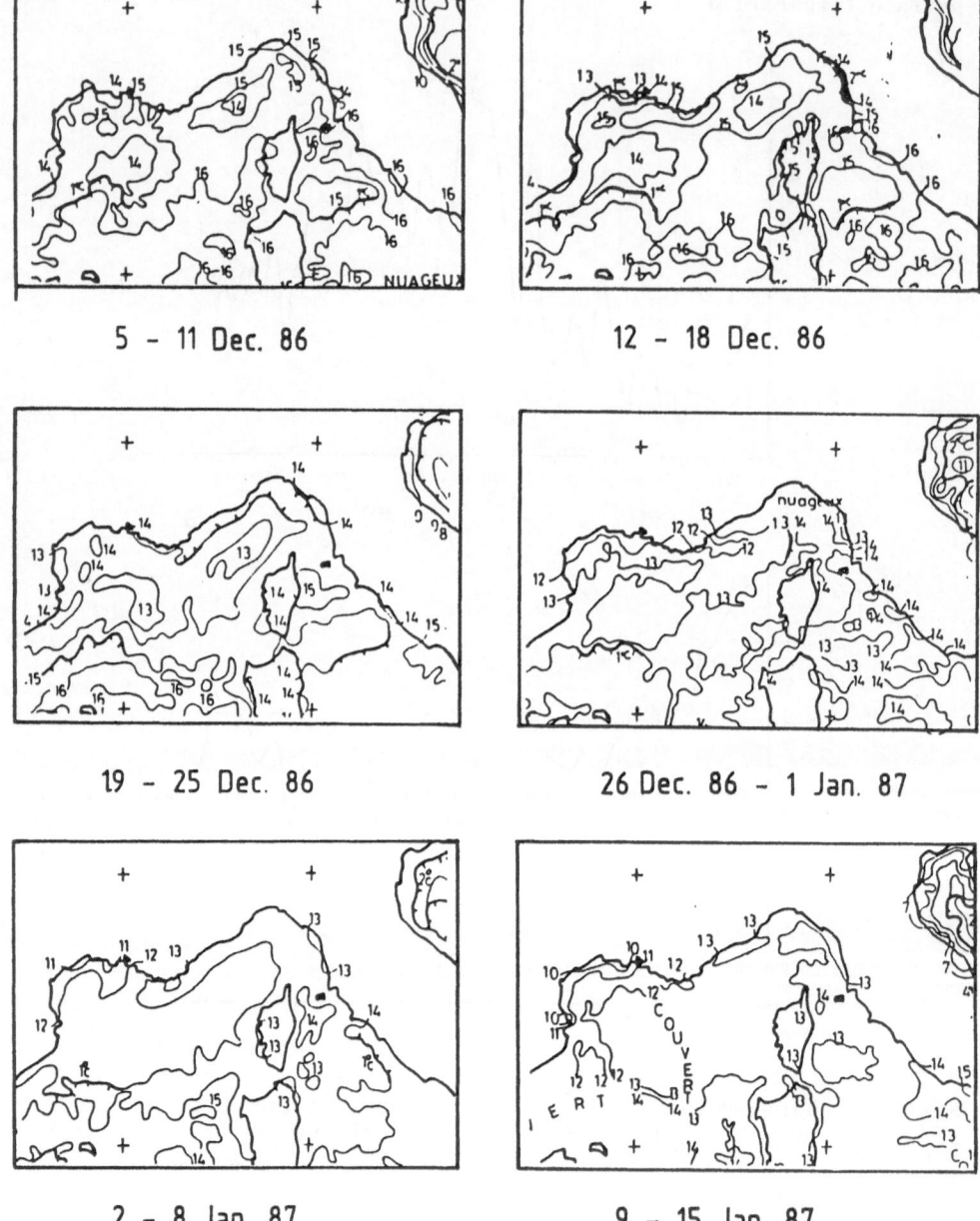

Figure 7. Weekly mean values of the surface isotherms in the northwestern Mediterranean sea computed in the periods between 5/11-12-1986 to 9/15-1-1987 (from Direction de la Meteorologie Nationale, 1986, 1987).

$$Q = Q_s + Q_l + Q_r - Q_i$$

where the subscripts s, l, r, and i refer to the sensible heat exchange, latent heat of evaporation, radiative heat loss and irradiative heating, respectively. The estimate of Q has been made following already known methods (Roll, 1965; Kraus, 1977).

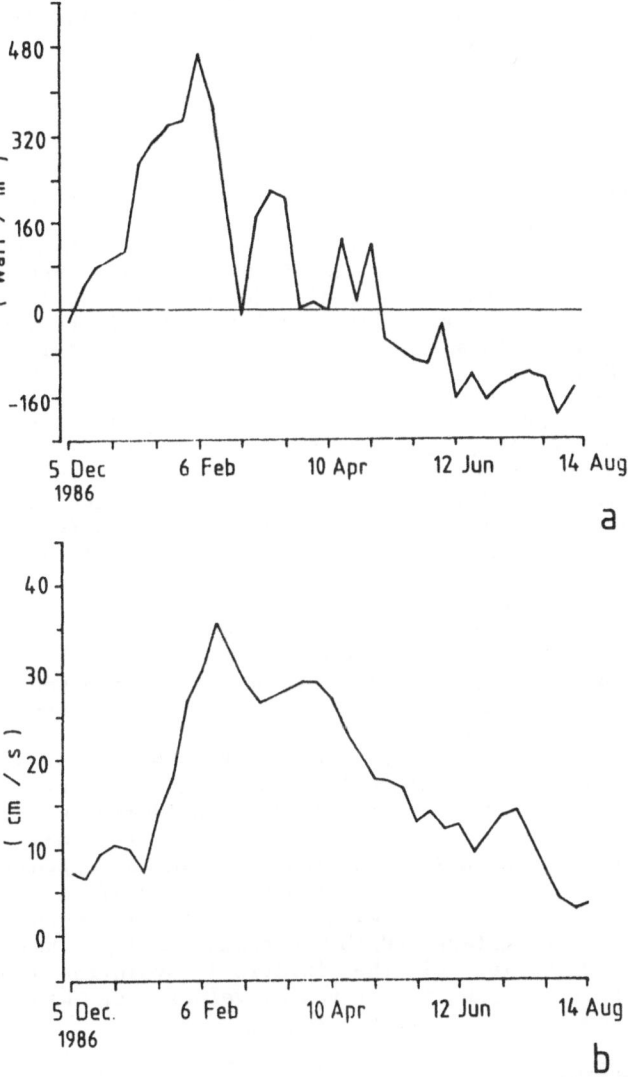

Figure 8. Time evolution of the weekly mean values of the total heat exchange Q between sea and atmosphere (positive Q values correspond to a flux from ocean to atmosphere) (a) and the current in the channel at 100m of depth (b). Ticks are three weeks apart.

In our computations, we have used the coastal meteorological parameters of the Ligurian Sea, while the surface temperature has been derived from the satellite data. Fig. 8a shows the evolution of the mean weekly value of Q from December 1986 to July 1987. Up to December, there is a progressive increase of the heat from the sea to the atmosphere, with the maximum heat flux occurring in February.

A comparison between the behaviour of the heat balance and the weekly mean values of the current at 100m of depth (Fig. 8b) shows that the two curves have the same trend and that the highest mean values of current occur about 7 days later than the maximum ocean heat loss.

5. DEEP WATER FORMATION AND FLUXES IN THE CHANNEL

It is well known that the so-called preconditioning-phase for the deep water formation processes begins in autumn in the Liguro-Provencal basin (Gascard, 1978). During this phase, the combined effects of cooling and evaporation at the air-sea interface cause the surface density to grow, thus producing the conditions for convective movements in the water body; the intensity of the vertical motion increases as the surface density rises and the internal stratification vanishes. Convective processes produce a turbulent vertical mixing creating a surface homogeneous layer whose thickness grows with time. In a cyclonic circulation like that of the Ligurian Sea, the vertical mixing erodes the stratification of the central part of the basin more rapidly than the external borders. Starting from the water and salt balances, Bethoux (1980) has estimated that the deep water formation processes involve about $13.53 \; 10^{12}$ m^3/year of surface water. It is interesting to note that this volume is about 65% of the total volume that crosses the channel during the year, in good agreement with the winter transport (Tab. 2).

Gill et al. (1979) showed that processes like this can induce an intensification of the surface cyclonic circulation around the involved areas. This intensification is a consequence of the geostrophic adjustment following the increase in the doming of the interface between the surface and the intermediate layer. Keeping this in mind, Crepon and Boukthir (1987) have constructed a simple model linking the winter intensification of the Ligurian circulation to the deep water formation processes. Considering a two-layer flow, consisting of two homogeneous layers at a pre-determined density, and assuming the hydrostatic and the Boussinesq approximations and linearity, the problem can be treated by decomposition into barotropic and baroclinic modes. Assuming as a forcing mechanism the vertical flux w between the surface and the intermediate layer below, the model allows the evaluation of the interface evolution between the two layers and the corresponding horizontal velocities, hence giving an estimate of the horizontal flux

$$F = r_i^2 \, f w t$$

where r_i is the baroclinic radius of deformation, f the Coriolis parameter and t the time. On the contrary, if the flux is known, one can evaluate the corresponding vertical velocity w. In our case, assuming the winter flux off Nice equivalent to that of the Corsican Channel: F=1.3Sv (Tab. 2) and r=9km (Crepon and Boukthir, 1987), we obtain an interface elevation wt of about 160m, in fair agreement with the experimental values (Campagnes Hydrokor, 1975).

We can say therefore that the deep water formation processes are likely to play a significant role in modulating the flow through the Corsican Channel.

6. CONCLUSION AND DISCUSSION

Currents east and west of northern Corsica are always northward. However, whereas the West Corsican current does not show any sensible seasonal variation, the eastern flow is highest in winter and lower in summer. The rise is very fast and takes place at the end of December.

A winter rise in the current has been observed in the Strait of Sicily (Manzella et al., 1988), and Manzella and LaViolette (1990) have recently proposed an explanation of this by coupling the variations in the fluxes entering the Western Mediterranean through the Gibraltar and Sicilian straits.

The models available in the literature indicate that we cannot explain the dynamics of the Ligurian basin through the direct action of the wind only (Heburn, 1987), nor with the influence of the fluxes in the Sicilian and Gibraltar straits (Loth and Crepon, 1984). The comparison between the winter flux rise in the Corsican Channel and the meteo-climatic conditions around the Ligurian basin indicates that a correlation can be established between the winter rise of the flux and the cooling of the surface water: a result of the heat loss which takes place during winter. As a consequence it is likely that the deep water formation processes play a significant role in shaping the circulation characteristics of the northwestern Mediterranean basin.

7. References

Astraldi, M., Gasparini, G. P., Manzella, G. M. R. and Hopkins, T. S. (1990) 'Temporal variability of currents in the Eastern Ligurian Sea', J. Geophys. Res, 95, 1515-1522.

Bethoux, J. P. (1980) 'Mean water fluxes across sections in the Mediterranean Sea, evaluated on the basis of water and salt budgets and of observed salinities, Oceanologica Acta 3, 79-88.

Bethoux, J. P., Prieur, L. and Nyffeler, F. (1982) 'The water circulation in the North-Western Mediterranean Sea; its relation with wind and atmospheric pressure,' in: J. C. Nihoul (ed.), Hydrodynamics of Semi-Enclosed Seas, Elsevier, Amsterdam, 129-148.

Campagnes Hydrokor (1975) 'Resultas des campagnes du N.O. Korotneff (1972-73)', Fasc. 16, Centre de Recherches Oceanographiques de Villefranche-sur-Mer.

Crepon, M. and Boukthir, M. (1987) ' Effect of deep water formation on the circulation of the Ligurian Sea', Annales Geophysicae 1B, 43-48.

Direction de la Meteorologie Nationale (1986), Satmer, 51.

Direction de la Meteorologie Nationale (1987), Satmer, 52.

Gascard, J. C. (1978) 'Mediterranean deep water formation, baroclinic instability and oceanic eddies', Oceanologica Acta 3, 315-330.

Gill, A., Smith, J. M., Cleaver, R. P., Hide, R. and Jonas, P. R. (1979) 'The vortex created by mass transfer between layers of a rotating fluid', Geophys. Astrophys. Fluid Dyn. 12, 195-220.

Heburn, G.W. (1987) 'The dynamics of the western Mediterranean Sea: a wind forced case study', Annales Geophysicae 5B, 61-74.

Kraus, E.B. (1977) 'Modelling and prediction of the upper layer of the ocean', Pergamon Press, Oxford.

Loth, L. and Crepon, M. (1984) 'A quasi-geostrophic model of the circulation of the Mediterranean Sea', in: J.C. Nihoul (ed.), Remote Sensing of the Shelf Seas Hydrodynamics, Elsevier, Amsterdam, 277-285.

Manzella, G.M.R., Gasparini, G.P. and Astraldi, M. (1988) 'Water exchange between the eastern and western Mediterranean through the Strait of Sicily', Deep Sea Res., 35, 1021-1035.

Manzella, G.M.R. and LaViolette, P.E. (1990) 'The relation of the transport through the Strait of Gibraltar and the seasonal transport of LIW through the Strait of Sicily', J. Geophys. Res., 95, 1623-1626.

Nyffeler, F., Raillard, J. and Prieur, L. (1980) 'Le bassin Liguro-Provencal, Etude statistique des donnes hydrologiques 1950-1973', CNEXO Rapp. No. 42.

Roll, H.U. (1965) 'Physics of Marine Atmosphere', Academic Press, New York.

Stocchino, C. and Testoni, A. (1977) 'Nuove osservazioni sulla circolazione delle correnti nel M. Ligure', Istituto Idrografico della Marina, Genova, Rapp. F.C. 1076.

Taupier-Letage, I. and Millot, C. (1986) 'General hydrodynamical features in the Ligurian sea inferred from the DYOME experiment', Oceanologica Acta, 9, 119-131.

LONG TERM CURRENT AND SEA LEVEL MEASUREMENTS CONDUCTED AT BOSPHORUS

Yalçın ARISOY, Ph.D. Student
Adnan AKYARLI, Prof.Dr.
Institute of Marine Science and Technology
Izmir, TURKEY.

ABSTRACT. Institute of Marine Science and Technology (IMST) has conducted "Meteo - oceanographic Surveys" at Küçükçekmece, Tuzla, Büyükada, Küçüksu, Baltalimanı, Paşabahçe and Büyükdere marine outfall regions. The latter four of these locations are located along the Bosphorus. In the meantime, IMST has also carried out the marine surveys foreseen within the scope of "Feasibility Studies and Preliminary Designs of Bosphorus Railroad Tunnel and Istanbul Metro System Project". As a result of this overlap, probably the most extensive data base of the long term current and sea level measurements has been created. This paper outlines the program and the methodology of the surveys and explains some features of the data evaluation.

1. INTRODUCTION

IMST has conducted "Meteo - oceanographic Surveys" at the Küçükçekmece, Tuzla, Büyükada, Küçüksu, Baltalimanı, Paşabahçe and Büyükdere regions (Figure 1) to collect data for the reliable design of the marine outfall systems. All these locations have previously been selected within the content of "Istanbul Sewerage Project", and the last four of them are located along the Bosphorus. IMST has also been involved in the "Feasibility Study and Preliminary Designs of Bosphorus Railroad Tunnel and Istanbul Metro System Project". The task was to gather the oceanographic data required for the environmental impact assessment of the railroad tunnel.

The accepted theories for the two layer flow in the Bosphorus assume that the flow is controlled by the critical sections at both ends. Since the location of the proposed tunnel may coincide with the critical section at the southern entrance, it may significantly affect the flow system of the Bosphorus. Among them, the most important influence may be on blocking phenomena. As it was pointed out in METU's recent report (METU, 1989), there have been various concerns about the probability of long term blocking of the Bosphorus underflow at the northern sill (i.e. Black Sea entrance). Furthermore, it has

L. J. Pratt (ed.), The Physical Oceanography of Sea Straits, 225–236.
© 1990 Kluwer Academic Publishers.

226

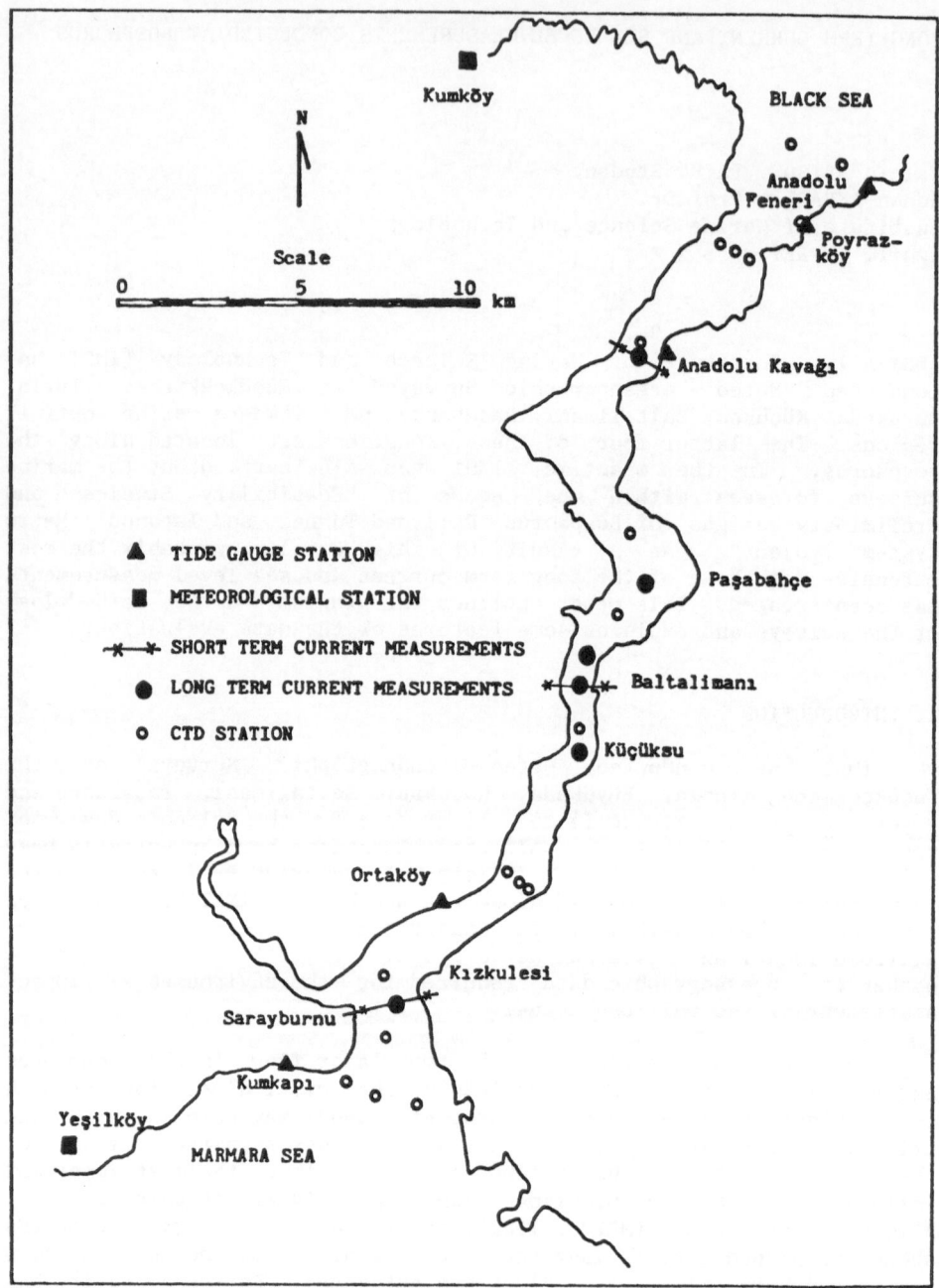

Figure 1. Measurement stations of the IMST's data gathering program.

also been speculated that the pollutants discharged through the
İstanbul Sewerage System into the lower layer would be mixed with the
flow of the upper layer and transported back to the Sea of Marmara.
According to such opinions, the construction of an underwater railway
tunnel will reduce the cross section, and will affect the underflow
system in an unfavorable way. In other words, it may cause an increase
in the period of the blocking.

Regarding the strong interrelations between the hydrodynamical
features of the Bosphorus and these two engineering applications, IMST
suggested to the owners of both projects that a joint data gathering
program of wider scope be conducted. The proposal was accepted by the
parties involved, and then, two independent measuring campaigns were
combined into a single program. The result of this cooperation has
been the establishment of a reliable, systematic and also an extensive
data base comprising the current and sea level data recorded at the
Bosphorus. The jointly determined program of the surveys appears in
Table 1.

Table 1: Time table of IMST's data gathering program.

Stations	Month 1984 1985 1986 123456789012123456789012123456789012
Anadolu Kavağı (1)	XXXXX XXXX
Paşabahçe	XX
Baltalimanı	XXXX XXXXXX
Küçüksu	XXXXX XX
Kızkulesi	XXXX XXXX
Sarayburnu	XXXXX XXX
Anadolu Kavağı (2)	X X
Baltalimanı	X X XX
Sarayburnu	X X
Anadolu Feneri (3)	XXXXXX
Poyrazköy	XXXXXXXXX
Anadolu Kavağı	XXXXXXXXXXXXXXX XXXXXXXX
Ortaköy	XXXXXXXXXXXXXX XXXXXXXX
Kumkapı	XXXXXXXXXXXXXXX
Kumköy (4)	XXXXXXXXXXX
Yeşilköy	XXXXXXXXXXX

(1) Long term current measurements   (3) Sea level measurem.
(2) Short term current measurements (4) Meteorological data

This paper outlines the program and the methodology of the surveys
and explains some features of the data evaluations.

2. CONTENT AND METHODOLOGY

It is a very well known fact that the density flow in the Bosphorus is a function of the following parameters:

- The water level difference between the adjacent seas,

- The density difference between these seas,

- Meteorological conditions,

- The geometry of the Bosphorus.

The main objective of the IMST's data gathering program is to collect simultaneous information on these parameters. Therefore, the content and the program of the measurements have been determined accordingly. All the measurements were carried out by a team comprising the relevant scientific and technical personnel of IMST. Also, the instrumental capacity of IMST and its versatile research vessel K. Piri Reis were utilized.

The instruments used for this purpose are:

- "Decca Trisponder System" and "Racal-Decca Autocarta II System",

- "Endeco 174A" and "Interocean 135M" type recording currentmeters,

- "Endeco 1032" type tide gauges,

- "Endeco 900" type acoustic releases and transmitter unit,

- "Interocean OSEAS" system for CTD measurements.

2.1. Long Term Current Measurements

In order to conduct long term current measurements, currentmeter moorings were deployed at the locations shown in Figure 1. The general schemes and properties of the mooring systems are given in Figure 2.

The positions of the moorings were determined either by the "Racal-Decca Radar Navigation System" in geographical coordinates or by the combination of "Decca Trisponder System" and "Racal-Decca Autocarta II System" in local or UTM coordinates. Meanwhile, water depths at these locations were measured by an "Atlas-Deso 10" echosounder.

Currentmeter moorings are released into the sea starting from the upper buoy while there is a certain distance between the ship and theoretical deployment point. Then, the vessel sails towards this point along a predetermined route and when she arrives at the point, the concrete block of the system is freely dropped.

229

Explanations:

1- The numbers and depths of the currentmeters were adjusted according to the average position of the interface.

2- The depth of upper buoy permits the safe navigation of the most deeply drafted ships.

3- Endeco currentmeters are trimmed so that they become neutrally buoyant in the sea water.

4- The mooring system is kept vertical by aluminium or polyester buoys with 50 cm diameters. The buoyancy provided by these buoys are 43 kg and 52 kg, respectively.

5- The upper buoy is attached to a steel frame which is 5 kg in the water.

6- Elements are connected with 8 mm steel wires and linked with stainless steel shackles and swivels.

7- The mooring system is kept fixed at a certain location with a 200 kg (in the water) concrete block.

8- The system is released from the concrete block by activating the Endeco 900 subsurface unit at the end of the deployment period.

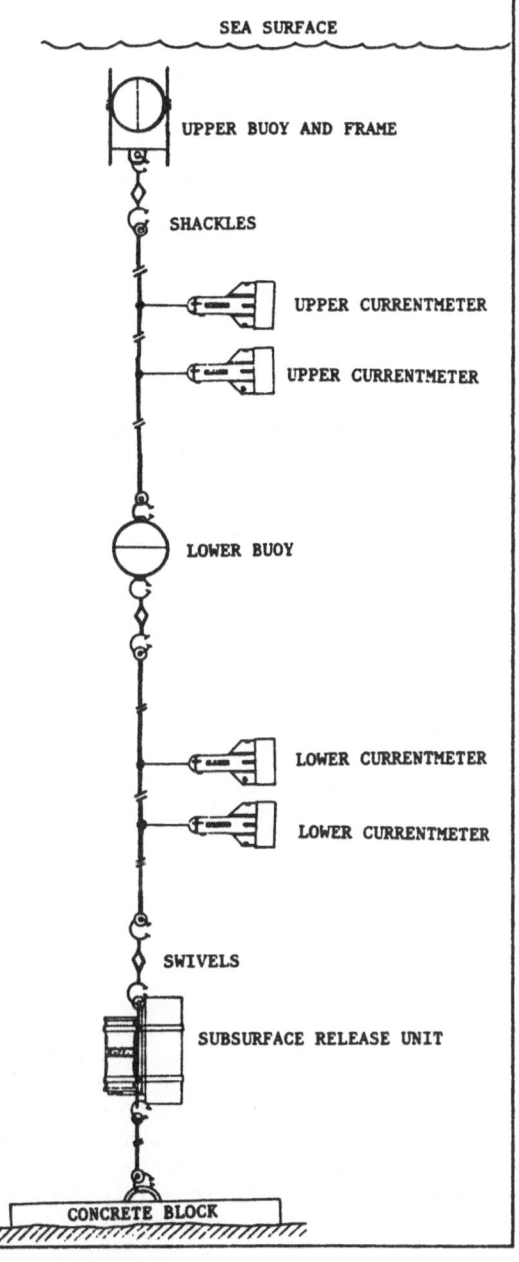

Figure 2. Long term currentmeter moorings.

2.2. Short Term Current Measurements

One of the most important contributions of this study has been the determination of the discharges of upper and lower layers from direct measurements (see Figure 1). In this respect, short term current measurements were carried out through the three cross sections to obtain the spatial and temporal variations of current velocities and directions. During the short term current measurements, moorings were deployed at five different locations on the same cross section in a sequential order. The general schemes and properties of the mooring systems are shown in Figure 3.

As every mooring system includes six currentmeters, five deployment points provided current data measured at 30 regularly spaced locations over the cross section. Each deployment lasted a minimum of 30 minutes and therefore, at every point, approximately 15 measurements with intervals of two minutes were recorded. Due to the methodology applied, there are time lags between the measurements conducted at different deployment points of the same cross section. The effect of this factor was controlled by examining the temporal variations in the simultaneous long term current data recorded in the vicinity of the section. The short term current measurements were carried out in a total of 12 days, with 6 days in winter and 6 days in summer period. In each term, measurements were repeated 3 days in Baltalimanı, 2 days in Sarayburnu and one day at Anadolu Kavağı.

2.3. Sea Level Measurements

The water level difference between the Black Sea and Marmara is one of the major parameters which affect the formation of the two layer flow. Consequently, a priority was given to the sea level measurements. First, the analog data recorded at Anadolu Kavağı and Ortaköy for many years were obtained and digitized. Then, IMST was requested to perform additional sea level measurements at Anadolu Feneri and Kumkapı (see Figure 1). Therefore, one tide gauge was mounted at Anadolu Feneri to determine the water level in the Black Sea and another was installed at Kumkapı to monitor the same parameter in the Sea of Marmara.

Apart from these measurements, the consortium preparing the "Feasibility Studies and Preliminary Designs of Bosphorus Railroad Tunnel and Istanbul Metro System Project" charged a private company to conduct a first order levelling along the Bosphorus to determine the relative elevations of the above mentioned stations. As a combination of both studies, it became possible to calculate the absolute differences between the sea levels of the Black Sea and Marmara within the accuracy of first order levelling.

Explanations:

1- The depths of the currentmeters were adjusted according to the actual position of the interface which was determined from CTD measurements or analog echosounder records.

2- As the upper buoy was at the sea surface, serious precautions were taken for safe navigation. Apart from them, R/V Piri Reis always sailed in the vicinity during the measurements.

3- The mooring arrangements were made considering the geometrical characteristics of the cross sections.

4- The lengths of connecting elements have been selected very carefully to shorten the preparation time of the moorings between the sequential deployments as much as possible.

5- Maximum effort was made to determine the horizontal and vertical positions of the deployment points as accurately as possible.

6- Items 3, 4, 6, 7, 8 mentioned for long term currentmeter moorings are also valid in this case.

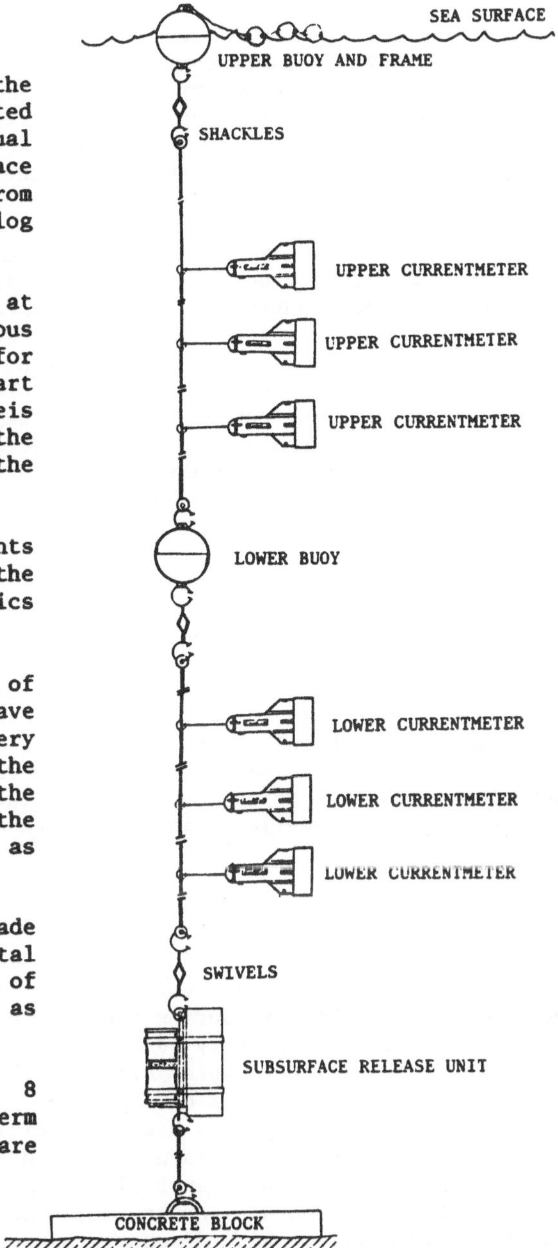

SEA SURFACE

UPPER BUOY AND FRAME

SHACKLES

UPPER CURRENTMETER

UPPER CURRENTMETER

UPPER CURRENTMETER

LOWER BUOY

LOWER CURRENTMETER

LOWER CURRENTMETER

LOWER CURRENTMETER

SWIVELS

SUBSURFACE RELEASE UNIT

CONCRETE BLOCK

Figure 3. Short term currentmeter moorings.

2.4. Conductivity, Temperature, Depth (CTD) Measurements

The consortium mentioned above organized a team to conduct simultaneous CTD measurements on the days when short term current measurements were performed by IMST. The measurements were made from a small boat at preselected stations situated between the Black Sea and Marmara along the Bosphorus (see Figure 1). This effort has provided an important data source to determine the position of the interface, another major parameter affecting the two layer flow system.

At the same time, IMST carried out similar measurements along the cross section to detect the transversal shape of the interface. However in some circumstances, such measurements could not be conducted due to some compulsory reasons (intensive sea traffic, improper weather conditions). Instead, this information is deduced from the analog records of the echosounder "Atlas Deso 10". As there is a density difference between the upper and lower layers, the interface behaves like a reflection surface and can be traced on the records. The same methodology has also been utilized to determine the position of the interface along the strait during the navigation from one end to another.

2.5. Meteorological Data

Meteorological conditions may have significant influence on the flow conditions in the Bosphorus. Due to this fact, some meteorological parameters were recorded at Kumköy (on the Black Sea coast at the northern end of Bosphorus) and Yeşilköy (on the Marmara coast at the southern end of Bosphorus). Collected data include air pressure, temperature, wind direction and velocity measured at standard elevations and according to WMO regulations. All information was transferred onto magnetic media for future computer processing.

3. FINDINGS

The measurements being performed in this study have added significantly to the knowledge about the flow system in the Bosphorus. The long term current measuring program has revealed that there is a two layer system with a heavy salty layer flowing from Marmara to the Black Sea and lighter more brackish layer flowing in the reverse direction for more than 95 % of the time. In other words, blocking is not a frequent phenomenon as has been claimed (Ilgaz, 1944; Ullyott, 1953). Morever, definitive identification has been made of the periods in which there is only one layer flow from the Black Sea to the Sea of Marmara. By fortuituous circumstance, one layer flow was observed on two of the 12 days in which discharge measurements and longitidunal CTD profiling were carried out. All current and CTD data recorded on these days have clearly indicated the existence of the blocking phenomenon. On the other hand, it was also determined that the strong south-westerly winds slow down the surface outflow, and on certain occasions may be the reason for the blockage of the upper layer flow. During such an event, the surface salinity increases in the southern half of the Bosphorus.

Another picture emerging from the examination of long term current data shows that the currents in the Bosphorus are subject to considerable temporal and spatial variations depending on the varying discharge rates, the meteorological conditions in the adjacent seas, the geometry, and the transience in such forcing. In other words, current velocity is generally in a transitory state. Even very small variations in the water levels at both ends create relatively large changes in the current and also, probably, in the position of the interface (Figure 4).

The longitidunal profiles of the interface have been determined by "Atlas Deso 10" echosounder during and just after the blocking phenomenon. Although the time difference between the records was on the order of few hours, the two profiles were completely different. These observations have also indicated that the tip of the lower layer progresses with a comparatively high speed towards the Black Sea exit.

The evaluation of the long term current data recorded at Kızkulesi and Sarayburnu indicates that a velocity exceeding 1.00 m/sec was never recorded in the lower layer whereas a maximum of 1.66 m/sec was recorded in the upper layer. Furthermore, it should be noted that the shear pin of the currentmeter (which was deployed at Kızkulesi as always to be in the upper layer) was broken after 5 days from deployment during the winter 1985 cruise. This pin which connects the propeller to the shaft is designed to withstand 2.50 m/sec current. Of course it is possible to explain this failure with the impact of floating debris. But, it can be the result of a very strong current which may reach up to 350 cm/sec in the Marmara junction (Black Sea Pilot, 1955), as well.

Previous sea level measurements in the Bosphorus are reported in some publications (Gunnerson and Özturgut, 1974; De Filippi, Iovenitti, Akyarlı, 1986; Yüce, 1986). According to these authors, the mean sea level difference between the two ends of the Bosphorus is 30-40 cm. Also, it was stated in both of the first two papers that there is nonlinearity in the slope of the free surface due to the variation of the slopes in the northern and southern halves. In the second publication, the dominant periods of the sea level signals are given as 5 days, 24 hr and 12 hr. The measurements of IMST have generally been in agreement with the previous findings. However, the peak monthly average level at the Black Sea end was reached in March of 1986. This is a somewhat surprising finding because the peak river discharge into the Black Sea is reached in the April-May period. This finding may be a random event or in fact may be the norm due to the meteorological conditions encountered in the winter time.

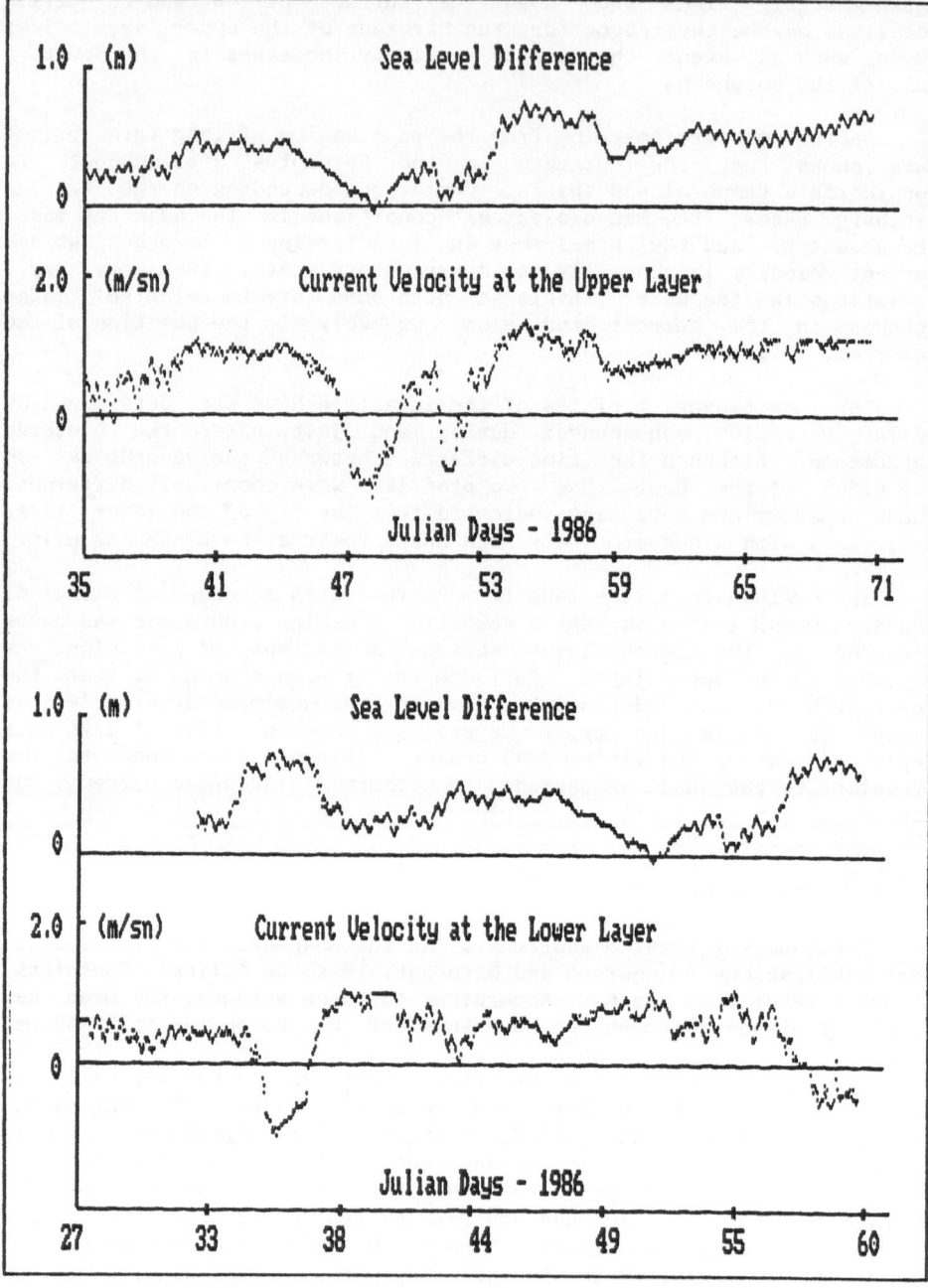

Figure 4. The interaction between sea level difference and current velocities at upper and lower layers.

The recent CTD data have confirmed the same picture as all previous CTD profiling efforts, namely that the water column in the Bosphorus can be divided into 3 layers as follows:

1. A top layer of Black Sea water which is nearly homogeneous in the winter and stratified due to temperature variations in the summer time. The salinity of this layer varies between 17-18 ppt at the northern half of the strait and increases progressively towards the southern exit.

2. A middle layer (the interface) which is vertically stratified with properties continually varying and with increasing density downward. It was also evident that the thickness of this layer decreases from the southern to the northern end.

3. A bottom layer of heavy saline Marmara sea water with more uniform temperature. However, the salinity of the lower layer attains a maximum value of 38 ppt at the southern entrance and decreases towards the northern exit at a rate depending on the intensity of vertical mixing.

4. CONCLUSIONS

The data collection program described above has provided a reliable, systematic and probably the most extensive data base on the current and sea level recorded in the Bosphorus. Up to now, this information has only been used in the design of marine outfall systems mentioned above and in the preparation of the relevant parts of the "Feasibility Studies and Preliminary Designs of Bosphorus Railroad Tunnel and Istanbul Metro System Project". Therefore, the findings submitted in this paper are mostly based on the data evaluation performed in this scope (Akyarlı, 1987 a; Akyarlı, 1987 b). However, the authors have recently started a study covering the subject in a much wider scope as the PhD thesis of the first author under the supervision of the second author.

5. ACKNOWLEDMENTS

This study has been conducted with the financial support of the "General Directorate of Istanbul Water Supply and Sewerage Administration" and the "General Directorate of Railways, Harbours and Airports Construction".
The marine surveys have been successfully carried out under the supervision of the authors by the capable academicians and experienced technical personnel of IMST. Also, the crew of R/V K. Piri Reis have performed their job in an excellent way although they worked in the Bosphorus which is one of the most complex marine environments not only from the hydrodynamical but also from the navigational point of view.

The authors greatly acknowledge everyone involved in this project for any reason. Among them, they appreciate the contributions of Prof. Dr. Erol İZDAR who is the director of the IMST and Mech. Eng. Nihat YAZICI who was the coordinator of R/V K. Piri Reis in that period, with their special thanks.

REFERENCES

Akyarlı, A.(1987 a): Geomorphological and Oceanographical Surveys for the Marine Sewerage Outfall Areas of the Istanbul Sewerage Project-Data Reports of the Oceanographical Surveys Conducted at Anadolu Kavağı, Paşabahçe, Baltalimanı, Küçüksu and Sarayburnu Regions (Five volumes). Institute of Marine Science and Technology, Dokuz Eylül University, Izmir, Turkey.

Akyarlı, A.(1987 b): Geomorphological and Oceanographical Surveys for the Marine Sewerage Outfall Areas of the Istanbul Sewerage Project-Data Reports of the Sea Level Measurements Conducted at Anadolu Feneri, Poyrazköy and Kumkapı Regions (Two volumes). Institute of Marine Science and Technology, Dokuz Eylül University, Izmir, Turkey.

Black Sea Pilot (1955): Hydrographic Department of the Admiralty, Her Majesty's Service Office, London, UK.

De Filippi, G.L.; L. Iovenitti and A. Akyarlı (1986): Current Analysis in the Marmara Sea-Bosphorus Junction, 1 st AIOM (Associazione di Ingegneria Offshore e Marina) Congress, Venice, Italy, pp.5-25.

Gunnerson, C. G. and E. Özturgut (1974): The Bosphorus. "The Black Sea-Geology, Chemistry and Biology", Am. Assoc. Pet. Geol. Memoir 20, Tulsa, Oklahoma, U.S.A., pp.99-113.

Ilgaz, O. (1944): Notes on the entry of water from the Black Sea into the Bosphorus. Journal of Türk Kongr., V.6.

METU (1989): Oceanographic Characteristics of the Region Surrounding the Northern Entrance of the Bosphorus as Related to Planned Sewage Outfalls. First Annual Report, Institute of Marine Sciences, Middle East Technical University, Ankara, Turkey, 111 p.

Ullyott, P. (1953): Conditions of flow in the Bosphorus. Publication of the Hydrobiological Research Institute, Faculty of Science, University of İstanbul, Volume B-3, pp.199-214.

Yüce, H. (1986): The Investigation of the Water Level Variations in the Bosphorus (in Turkish). Journal of the Institute of Marine and Geographical Sciences, Volume 2, Nr. 3, Istanbul University, Istanbul, Turkey, pp.67-78.

WOCE AND THE GIBRALTAR EXPERIMENT THIRD OBJECTIVE

G. PARRILLA
Instituto Español de Oceanografía
Avda del Brasil, 31
28020 - MADRID
SPAIN

ABSTRACT.- Some ideas are proposed about how some of the recommendations outlined during the workshop on the Physical Oceanography of the Strait of Gibraltar (WPOSC) held last October at Madrid, can be adapted to serve both: the third objective of the Gibraltar Experiment and the WOCE goals.

1. INTRODUCTION

In the WOCE Implementation Plan (1988), the Strait of Gibraltar is one of the sites which has been selected within the Project Moorings (Heat Flux and Boundary Currents) with the designation ACM9 (fig 1). It has been proposed that the Strait should be monitored throughout all the field work of WOCE. It has been noted that the results of the Gibraltar Experiment (GE) will help to design an effective array for the WOCE action.

This proposed array is directly related to WOCE Core Projects 1 and 3, and specifically to the following issues:

a) Direct currents and transport measurements at crucial choke points.

b) The Atlantic Ocean is the most saline of all the World Oceans. It is of great importance to determine whether high salinity inflows are sufficient to create the salinity excess or if there is a greater excess of evaporation over precipitation over the Atlantic than over the rest of the global Ocean. This question makes it indispensable to measure the inflow of Mediterranean water in the Atlantic.

These two points, a) & b) are included in the Core Project 1 (the Global Description).

The following, c), belongs to the Core Project 3 (the Gyre Dynamics Experiment):

c) Exchange with marginal seas, which is significant in the Atlantic Ocean. The mass and tracer fluxes need to be determined, at least in the statistical analysis mode.

The third goal of the Gibraltar Experiment is (Bryden and Kinder, 1986): "to define an efficient method for long term measurement of the Strait flows so that interannual variability of the Atlantic -

237

L. J. Pratt (ed.), The Physical Oceanography of Sea Straits, 237–242.

Mediterranean exchange can be monitored".

Thus, the confluence of purposes in both experiments is more than clear.

Keeping in mind the facts mentioned above and the results and papers related to the GE already published, I shall try to lay out a field work plan. Of course, this is only a preliminary step to be refined during this meeting and subsequent work.

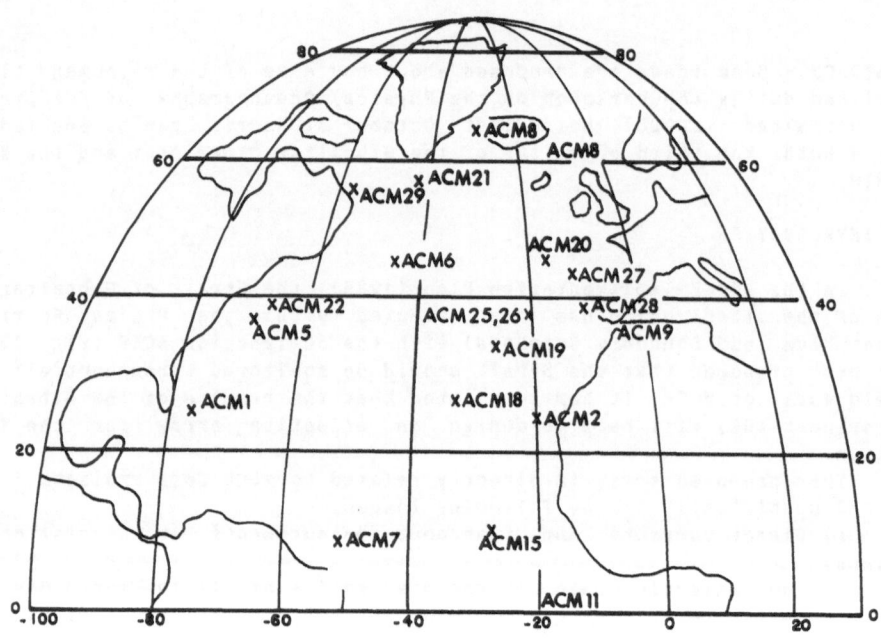

Fig.1 Moored instrumentation sites in the North Atlantic Ocean (WOCE)

2. SELECTION OF SITES

From the results of the WPOSG held last October in Madrid, two places stood out as the most appropriate for long term monitoring, for scientific and practical reasons: Espartel and Tangier region and the Tarifa Narrows (fig 2).

It should be relatively easy to measure the Mediterranean outflow at the first site, the Atlantic inflow at the second one, and control conditions at both sites.

2.1. Espartel

In this shallow region, with bottom depth less than 500 m, the outflow runs always west (Lacombe and Richez, 1982) with a velocity of about 1.4m/s and a tidal variation of less than 20% at Espartel West (Armi and Farmer, 1988). The Mediterranean water is confined to a relatively thin bottom layer (about 100m) with salinity and temperature maxima less than 38.10 and 13.10 °C respectively (Kennelly et al, 1989). During the GE three places within this region were customarily surveyed: the so-called Espartel West, Trafalgar and Tanger sections (fig 2).

It seems that the best choice for the survey is Espartel West for two reasons. First, according to Armi and Farmer (1988) there "the Strait is always bound at its western end by a supercritical flow". Second, it seems as good a place as the other two sections to monitor the volume and salt fluxes of Mediterranean water in the Atlantic, although a stronger Mediterranean signal is detected in the Tangier section.

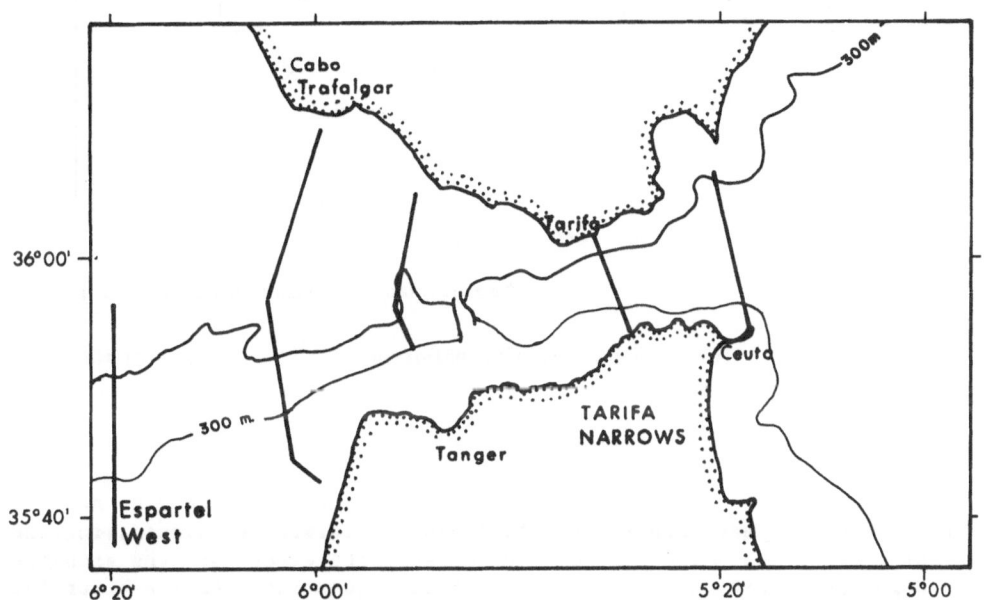

Fig. 2 Situation of Espartel- Tanger region and Tarifa Narrows.

2.2. Tarifa narrows

This region has a maximum depth of about 800 m and a width of some 14 Km, the narrowest in the Strait (fig 2). The inflow, although influenced by tides, always run to the east with velocities of the order

of 1 m/s. According to Armi and Farmer (1988), the flow is continuously
supercritical at the exit of this region (Froude number ranges from 1 to
3) and this section, with the West Spartel section, bounds the Strait
supercritically.

Furthermore, the distribution of the Mediterranean water in this
area is very similar to that in the eastern confines of the Alboran Sea,
a fact that, as one shall see later, expands the range of our
objectives.

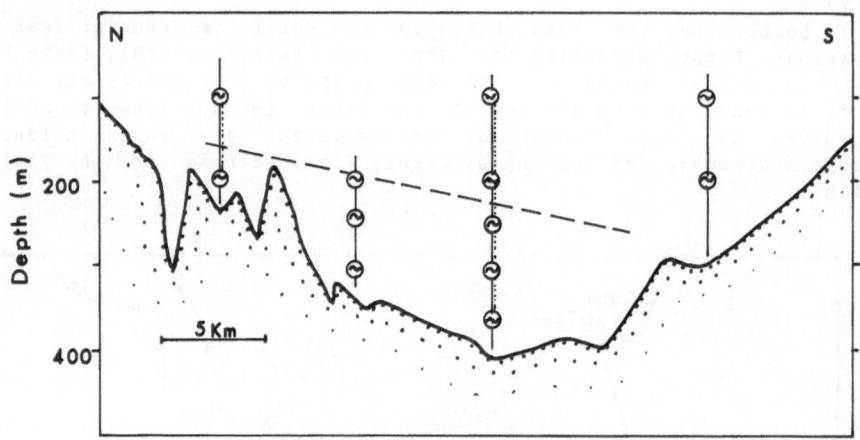

Fig. 3   Espartel west section (6° 20' W).

Adapted from Armi and Farmer (1988)

_ _ _ _ _ Interface   ⊚ Currentmeters   .... T/C chains

In conclusion, measurements in these two sections should help to
understand if these supercritical flows are permanent, elucidating the
problem about the maximal or submaximal character of the exchange
through the Strait, which is one of the main questions pending after the
WPOSG.

The monitoring of the outflow at Espartel will also contribute to
the WOCE goals.

The work in Tarifa Narrows will also help to observe changes in the
Mediterranean waters directly related to the climate changes in this
basin; these will probably be better detected in this area than in the
Mediterranean itself.

3. TENTATIVE FIELD WORK

It seems clear that several of the recommendations made during the WPOSG are most adequate for this joint - WOCE/GE - programming.

Among the recommendations made we emphasize the following:

The deployment in Espartel West of an array of up to six moorings of currentmeters and thermistor/conductivity (T/C) chains, possibly reduced to one or two later on, during the field work of WOCE.

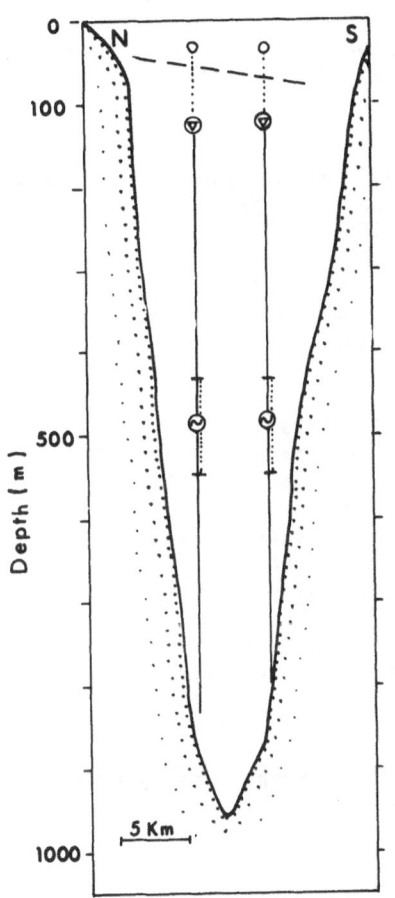

Fig. 4 Tarifa Narrows.

⊗ ADCP
⊘ Currentmeter
.... T/C chain

Moorings of acoustic Doppler currentmeters and T/C chains at three stations between Tarifa Narrows and the Gibraltar - Ceuta section, to measure conditions from the surface to 150 m at 10 m intervals for at least one year.

Although it would be desirable to have most of the equipment in the water simultaneously, I do not think that we shall be able to install that many moorings. I think some adjustments could be made without jeopardizing our goals.

Probably, the Espartel West section could be appropriately surveyed with four moorings (fig 3).

The one in the deepest part would be similar to Armi and Farmer's in their April cruise (Armi and Farmer, 1988) except for the addition of a shallower meter, about 100 m deep. Two moorings could be deployed to the North of the latter one, where the interface slopes upwards: The one nearer the coast with two meters at both sides of the interface with a separation of about 50m between them, and the one in between with 3 currentmeters distributed from about 100 m to the bottom. Finally another mooring could be set in the southern part with 2 meters, one about 25 m from the bottom and the shallower one at 100 m depth.

T/C chains should be arranged along the deepest mooring and across the interface in the northern one.

With respect to the Tarifa Narrows site, it seems more convenient to substitute the middle mooring, of the proposed three along the Strait, by two analogous across it (fig 4), so that slope of the interface can be monitored. In this part of the Strait, Levantine Intermediate Water and Western Mediterranean Deep Water are clearly distinguished and they are placed almost side by side (Parrilla et al, 1989), so if two current meters with the correspondent T/C chains are added at those moorings, at about 500 m, we shall be able to survey both water masses.

Obviously, it is not my purpose to limit the recommendations of the Madrid meeting - which I think should be followed for the accomplishment of the third GE objective - to the ones exposed above. What I have intended is to modify two of them to make WOCE and GE objectives compatible.

Furthermore I bring this to your attention not as a "fait acompli" but as a starting point for discussion during this meeting.

4. REFERENCES

Armi, L. and D.M. Farmer (1988). "The flow of Atlantic (Mediterranean) water through the Strait of Gibraltar". Progress in Oceanography, 21, 1 - 105.

Bryden, H. and T.H. Kinder (1986)."Gibraltar Experiment: A plan for dynamic and kinematic investigations of strait mixing, exchange and turbulence". Techn. Rep. WHOI - 86 - 29, 82 pp. Woods Hole Oceanographic Institution.

Kennelly, M.A., T.B. Sanford, and T.W. Lehman (1989)."CDT, Data from the Gulf of Cadiz Expedition: R/V Oceanus cruise 202". Techn. Rep. APL - UW TR 8917. University of Washington.

Lacombe, H. and C. Richez (1982). "The regime of the Strait of Gibraltar" in: Hydrodynamics of semi-enclosed seas, J.C.J. Nihoul, ed Elsevier Oceanogr. Ser. 34, 13 - 73.

Parrilla, G., T.H. Kinder and N. Bray (1989). "Hidrologia del agua mediterranea en el Estrecho de Gibraltar durante el Experimento Gibraltar". Seminario de Oceanografia Fisica del Estrecho de Gibraltar. Madrid 24 - 28, Oct., 1988: 95 - 122.

WOCE Implementation Plan (1988) Vol I&II, WCRP - 12, WMO/TD No. 243. WOCE I.P.O., Wormley, England.

II. Hydraulics

HYDRAULIC MODELS OF DEEP STRATIFIED FLOWS OVER TOPOGRAPHY

PETER G. BAINES and HENRY GRANEK
CSIRO Division of Atmospheric Research
Private Bag No.1
Mordialloc
Victoria
Australia

ABSTRACT. The flow of deep stratified fluids with an upper radiation condition (or absorbing upper boundary) over topographic ridges is composed mainly of internal lee waves for small Nh_m/U (where N is buoyancy frequency, h_m the maximum obstacle height and U the fluid speed), but as Nh_m/U is increased, a value is reached where these waves steepen and break. Observational and numerical studies show that for larger values of Nh_m/U the flow has a hydraulic character, with most of the topographic effects concentrated at relatively low levels. On the basis of one or two plausible assumptions we describe a quantitative model which provides a description of the flow properties for hydrostatic flow for values of Nh_m/U from zero up to 1.5, where upstream blocking first occurs in this model. The model is essentially a combination and unification of Long's model (Long 1953) for constant N and U, Smith's (1985) solutions to Long's model with a free upper surface, and the hydraulic model of Baines (1988) for finite depth systems. The stability criteria for Long's model determine the point (Nh_m/U value) at which the flow evolves to a hydraulic state, and hydraulic conditions then determine the depth of the hydraulic layer and its changing properties as Nh_m/U increases further.

1. Introduction

The nature of stratified flow over topography is very sensitive to the stratification (and to shear in the mean velocity, although we will omit this here). We may identify two stratification archetypes. In the first, the stratification is concentrated at low levels, so that internal wave energy generated by the topography propagates horizontally in discrete modes with well-defined vertical structure; if the fluid layer is deep (relative to other scales) there is little or no motion at the upper levels. This may be termed a *finite depth* situation. In the second type, internal wave energy generated by the flow over topography may propagate to great heights where it is "lost" or dissipated; energy may therefore be lost through the top of the system, and the vertical spectrum of internal wave motion is continuous rather than discrete. This may be termed an *infinite depth* system. It applies to the atmosphere when the stratification is approximately uniform, and to small (i.e. on the scale of hundreds of metres) bumps or ridges on the bottom of the deep ocean. These stratification types represent two extremes, and hybrid combinations of them are quite common in the atmosphere, for example. Stratified flows over subterranean ridges where the presence of the upper surface seems unimportant are also not uncommon in the ocean. Two examples are the Vema Channel (Hogg 1983) and the Denmark Strait.

When it comes to understanding the properties of the non-linear flow over large obstacles or ridges, the second type of flow has proved to be the more difficult of the two (Baines 1987). The properties of non-linear flows of the first type, having a discrete vertical spectrum, may be computed using (relatively) simple physically based models because the flow is controlled by one mode at a time (Baines 1988, Baines and Guest 1988). For the infinitely deep flows this is not

245

L. J. Pratt (ed.), The Physical Oceanography of Sea Straits, 245–269.
© 1990 *Kluwer Academic Publishers.*

possible because, in general, the vertical wave spectrum is not discrete. Numerical simulations (Peltier and Clark 1979), Pierrehumbert and Wyman 1985) and laboratory observations (Baines and Hoinka 1985) have shown that the flow is dominated by lee waves when Nh_m/U is small, but that for larger Nh_m/U it appears to have a hydraulic character where the perturbations to the flow are mainly concentrated at low levels. In this latter case, the flow resembles that for finite depth flows, so that the presence or absence of an upper boundary seems to be of little importance. In this paper we describe a new physically based model which may be used to describe and compute the properties of the second, infinitely deep type of flow, where the initial fluid velocity and buoyancy frequency N are assumed to be constant with height and there is no upper reflecting boundary. We make the simplifying assumption that the flow is hydrostatic, which is equivalent to saying that the horizontal length scales of the topography are long compared with the other length scales in the problem.

Note that we have carefully avoided calling Nh_m/U or its inverse a *Froude number*. Froude numbers are normally defined in such a way that they may be expressed as $F = U/C$, where C is a wave speed, and an example is given later in the paper. Physical significance may then be attached to the condition $F = 1$ where the flow is critical, and the behaviour associated with changing F values is common to many systems. The significance of Nh_m/U (or its reciprocal), on the other hand, is quite different.

The model is based on some dynamical assumptions which are detailed below in section 4. Assuming that these are valid, the model may be used for the following purposes. Firstly, it provides a description of the general character of the disturbance caused by given two-dimensional topography, up to and including the properties of blocked upstream layers. Secondly, it permits the computation of flow properties such as the form drag over given topography, and the regions where this drag is felt in the fluid; these are quantities of considerable importance when assessing the impact of topographic features on larger scale flows. The model takes as its starting point the Long's model solutions derived by Lilly and Klemp (1979) for infinitely deep flows over finite topography. These solutions are valid up to the point where the internal waves become unstable. It then uses the work of Smith (1985) to relate these infinitely deep flows at the point of instability to flows with a free upper surface. For yet greater topographic heights the flow may then be described by the hydraulic model of Baines (1988) with a free upper surface. Although the results presented here only cover obstacle heights up to the point where upstream blocking occurs, flow properties for larger topographic heights may be calculated if desired.

2. The Long's Model solution for infinitely deep fluids and its limits

The equations for the two-dimensional motion of an incompressible Boussinesq fluid with the hydrostatic approximation are

$$
\begin{aligned}
\frac{\partial u}{\partial t} + u\frac{\partial u}{\partial x} + w\frac{\partial u}{\partial z} + \frac{1}{\rho_0}\frac{\partial p}{\partial x} &= 0, \\
\frac{\partial p}{\partial z} + g\rho &= 0, \\
\frac{\partial u}{\partial x} + \frac{\partial w}{\partial z} &= 0, \\
\frac{\partial \rho}{\partial t} + u\frac{\partial \rho}{\partial x} + w\frac{\partial \rho}{\partial z} &= 0,
\end{aligned}
\tag{2.1}
$$

where u and w are horizontal and vertical velocity components respectively, p is pressure and ρ is fluid density, with ρ_0 a reference mean density. Two important parameters here are the upstream velocity profile $U(z)$ and the buoyancy frequency N, defined by

$$N^2 = -\frac{g}{\rho_0}\frac{d\rho_0}{dz}. \tag{2.2}$$

Long (1953) showed that if the upstream vertical profiles of kinetic energy $\frac{1}{2}\rho_0 U^2$ and buoyancy frequency N are independent of height z (effectively, this means that U and N are constant for a Boussinesq fluid), solutions of the equation

$$\frac{\partial^2 \delta}{\partial z^2} + \frac{N^2}{U^2}\delta = 0, \tag{2.3}$$

are also steady-state solutions of equations 2.1. Here the variable δ is defined to be the vertical displacement of a fluid particle from its level far upstream, and is given by

$$\delta(x, z) = z - z_0, \tag{2.4}$$

where z_0 is the upstream height of the particle. The density and velocity fields are related to δ by

$$\rho = \rho_0\left(1 - \frac{N^2 z_0}{g}\right), \quad u = U\left(1 - \frac{\partial \delta}{\partial z}\right), \quad w = U\frac{\partial \delta}{\partial x}, \tag{2.5}$$

so that the lines of constant z_0 are also streamlines and lines of constant density. For flow over topography of height $h(x)$, the lower boundary condition is

$$\delta(x, h) = h(x). \tag{2.6}$$

Solutions to equation 2.3 which also satisfy equation 2.6 take the form

$$\begin{aligned}\delta(x, z) &= h(x)\cos(N(z - h)/U) + f(x)\sin(N(z - h)/U), \\ &= \mathrm{Re}\left[He^{iN(z-h)/U}\right],\end{aligned} \tag{2.7}$$

where $H = h + if$. The real function $f(x)$ determines the whole flow field and must be obtained from an upper radiation condition of no incoming energy. Lilly and Klemp (1979) showed that this yields an integral equation for f which may be written in the form

$$H(x)e^{iNh/U} = \frac{i}{\pi} \int\limits_{-\infty}^{\infty} \frac{H(x')e^{iNh(x')/U}}{x' - x} dx' \tag{2.8}$$

They further showed that this equation may be solved numerically for various topographic profiles $h(x)$ by an iterative process which takes the Hilbert transform of $h(x)$ as its starting point.

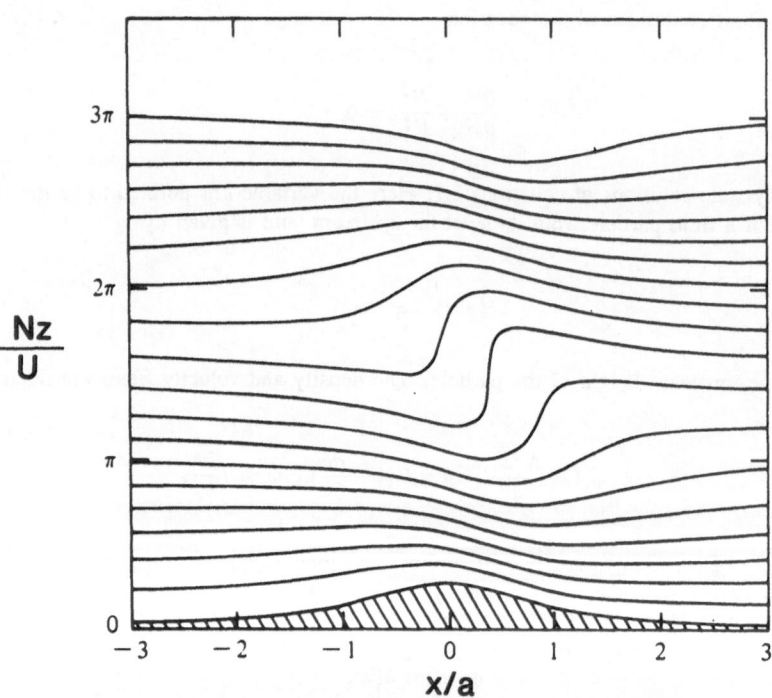

Figure 1. Hydrostatic flow over the symmetric *Witch of Agnesi* shape at the point of instability, $Nh_m/U = 0.85$ (from Lilly and Klemp 1979).

A number of solutions to this system, in both hydrostatic and non-hydrostatic forms, have been obtained, and a hydrostatic example is shown in Figure 1. It is well-known that the lee waves in the associated flow fields become steeper as Nh_m/U increases, where h_m is the maximum value of $h(x)$, until a transition value $Nh_m/U = (Nh_m/U)_t$ is reached where the streamlines become vertical somewhere (e.g. Miles 1968). Further increase in Nh_m/U results in flows which are statically unstable. Consequently, solutions from Long's model with Nh_m/U in excess of the transition value do not represent physically realisable flow fields, even, it seems, when local adjustments are made for statically unstable regions.

We have examined the stability criteria for flow over obstacles of the form

$$z = h(x) = h_m/(1 + (x/a_u)^2), \quad x < 0,$$
$$= h_m/(1 + (x/a_d)^2), \quad x > 0,$$
(2.9)

for the whole range of values of a_u/a_d from 0 to infinity (effectively). Various values of a_u and a_d have been tried for the same ratio but the variation has been small, so that the ratio seems to be the dominant parameter. The results are given in Figure 2, showing that the transition value for Nh_m/U lies in the range

$$0.5 < (Nh_m/U)_t < 1. \tag{2.10}$$

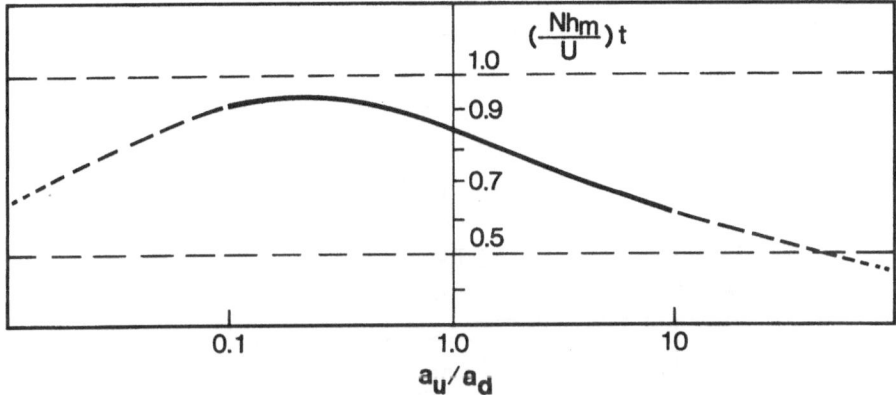

Figure 2. Values of $(Nh_m/U)_t$ as a function of the asymmetry of the topography, a_u/a_d. The topography has the shape $h = h_m/(1 + (x/a)^2)$, where $a = a_u$ on the left-hand (upstream) side and $a = a_d$ on the right.

Note that $(Nh_m/U)_t = 0.85$ for symmetric obstacles with this *Witch of Agnesi* shape ($a_u/a_d = 1$), as expected. If $h(x)$ is permitted to become negative, values of $(Nh_m/U)_t$ less than 0.5 may be obtained (Lilly and Klemp give an example with $(Nh_m/U)_t = 0.34$), but no cases with $(Nh_m/U)_t < 1$ have yet been found for hydrostatic flow (although they have for non-hydrostatic flow --- see for example Miles 1968).

We therefore conclude that obstacles with steep forward faces and gently sloping rear sides have critical values of Nh_m/U near 0.6, and those with flat forward faces and steep rear sides have $(Nh_m/U)_t$ close to but less than unity, although the precise value will depend on the actual shape. Since the solutions become unstable at this point, the flows must have a different character for larger values of Nh_m/U.

3. Long's Model solutions with a free upper surface

Smith (1985) has proposed a simple model for infinitely deep flows above the transition value, which was suggested by the results obtained from numerical and laboratory models. It assumes a free upper surface (above which the disturbances are weak), and further assumes that the fluid below the interface has uniform U and N upstream, so that it is governed by Long's Model (see Figure 3). Hence equations 2.1 to 2.6 still apply. With the free surface at a height D upstream, the boundary condition on this streamline (from the continuity of vertical displacement and pressure) is

$$u = U, \quad \frac{\partial \delta}{\partial z} = 0, \quad at \ z = D + \delta_T, \tag{3.1}$$

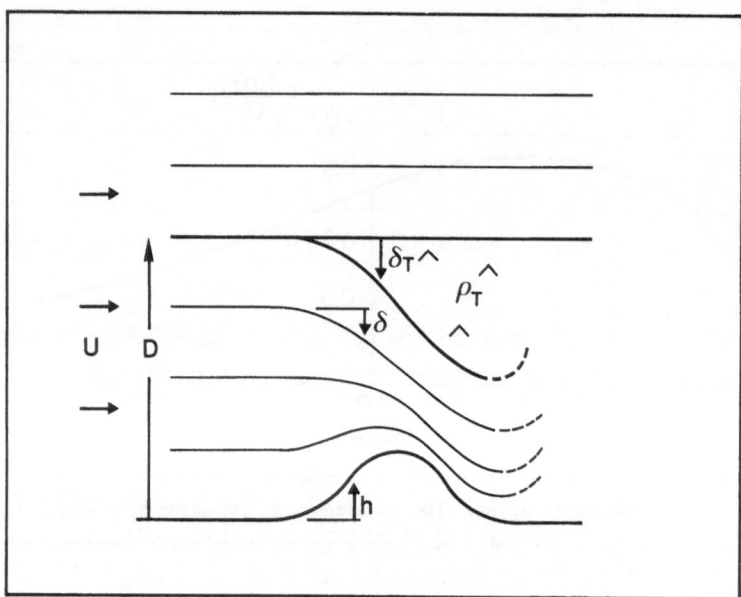

Figure 3. Schematic diagram of the flow for the hydraulic solution to the Long's model equations (adapted from Smith (1985)).

where δ_T is the vertical displacement of this streamline from its initial level.

A uniformly stratified layer of depth D at rest with a free upper surface has normal modes of the form

$$\delta = A e^{ik(x - C_n t)} sin(n + 1/2)\pi z / D, \quad C_n = \frac{\pm N D}{(n + 1/2)\pi}, \tag{3.2}$$

where the C_n are the wave speeds and $n = 0, +1, +2, ...$ If the layer is in motion, we may define a Froude number F for this layer by

$$F = U/C_0 = \pi U/2ND. \tag{3.3}$$

namely the mean flow speed divided by the speed of the fastest wave mode. When $F = 1$, therefore, the propagation speed of the fastest internal wave mode against the stream is zero.

The general solutions to equation 2.3 for flow over topography with a free upper surface may be written in the form

$$\delta(x, z) = \delta_T(x)\cos N(z - D - \delta_T(x))/U, \tag{3.4}$$

where δ_T is given by

$$h(x) = \delta_T(x)\cos N(h - D - \delta_T(x))/U. \tag{3.5}$$

These solutions represent realistic flows provided that h_m/D is less than a maximum which depends on the Froude number F. This maximum value of h_m/D is shown in Figure 4 for both positive and negative topography (i.e. depressions). For h_m/D less than the maximum, the solution is single-valued in terms of the height $h(x)$, so that the downstream flow is the same as the upstream flow if $h(-\infty) = h(+\infty)$.

The limiting curve for h_m/D in terms of F shown in Figure 4 takes two forms. The first form, which occurs for $h > 0$ for $F > 1/3$ and is marked by a heavy solid line, was described by Smith and denotes a hydraulic limit. For $h < 0$, it occurs for $0.27 < F < 1$, and for smaller F, in regions with F centred on $F = 1/(2n + 1)$, where n is an even integer for $h > 0$, and an odd integer for $h < 0$. These sections of the solid curve (for hydraulic solutions) are interspersed with sections where the limit to allowable values of h_m/D takes the second form, which is discussed below. For parameter values on these heavy solid curves, there are two possible flow solutions. The first is the same solution as that obtained for h_m/D less than the maximum, where the downstream flow state is the same as the upstream flow, as described above. The second type of solution is unchanged on the upstream side of the topography but is different on the downstream side, where the stream descends to a narrower, faster form for $h > 0, F > 1/3$. An example is given in Figure 5. From the general theory of stratified hydraulics (Baines 1988), we can see that this implies that the flow undergoes a transition from sub- to super-critical flow at the obstacle crest, where the relevant internal wave mode must be critical, i.e. have zero propagation speed against the stream. Where these hydraulic limits apply, the model does not have solutions for larger values of h_m/D. For values of F less than 1/3, similar behaviour occurs in corresponding parameter ranges, but higher modes are involved. An example of such a solution for $F = 0.3$ is given in a layered form in Figure 10a.

The second form of the limiting curve for solutions on Figure 4 is marked by the dot-dashed line, and this is found in between the regions bounded by the heavy solid hydraulic curves. This second type of limit marks the onset of stagnant or unstable regions within the body of the fluid layer, as determined by equations 3.4 and 3.5. This is the behaviour which normally limits the validity of Long's model solutions in flows which have a rigid upper boundary (e.g. Baines and

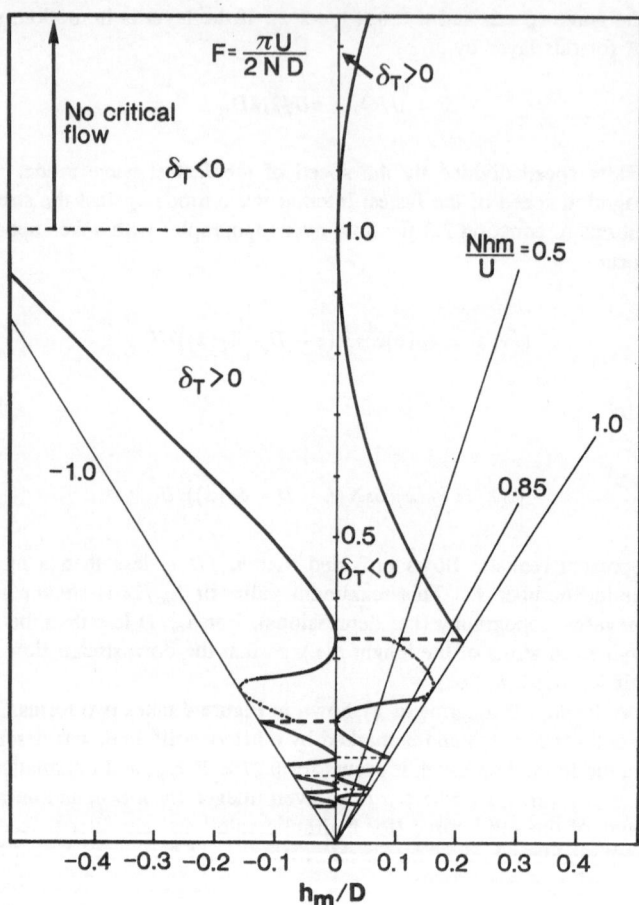

Figure 4. The boundaries to Long's model solutions for uniformly stratified flow with a free upper surface, as a function of the Froude number F, for both obstacles and depressions. Long's model solutions apply for smaller values of h_m/D.

Heavy solid lines --- denote the topographic height h_m (scaled by D) marking the limit of Long's model solutions due to critical flow, obtained by Smith (1985) but presented by him in a different manner. On this line the flow may take two forms: that obtained from Long's model, or a hydraulic transition with critical flow at the crest.

Dot-dashed lines --- denote the topographic height marking the limit of Long's model solutions due to static instability in the solution, for given F. These are interspersed between sections of the hydraulic curve, and only two sections (for $F > 0.2$) are shown. The pattern for $F < 0.2$ is similar.

Light solid lines --- denote lines of constant Nh_m/U.

Guest 1988). It also denotes the same type of boundary on physically realistic solutions as that described for the infinite depth system in section 2.

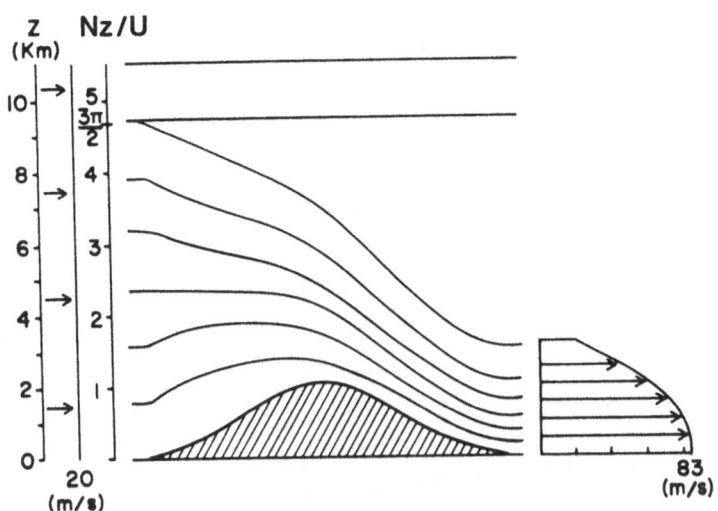

Figure 5. An example of a hydraulic solution from Long's model with $F = 1/3$, $h_m/D = 0.21$, the cut-off values for mode $n = 0$ (from Smith 1985).

Lines of constant Nh_m/U are marked in Figure 4, where it can be seen that the lines $N|h_m|/U = 1$ intersect cut-off points for the hydraulic flow regions for all the modes, for both positive and negative h. The bounding hydraulic curve for $h > 0$, $1/3 < F < 1$, in particular, corresponds to $1 > Nh_m/U > 0$.

4. Hydraulic models with a free upper surface

The flow properties of stratified flow with a free upper surface for obstacle heights greater than the limits described in the previous section may be calculated by using the layered hydraulic model described by Baines (1988). The equations and procedure are all described there in detail, and only a brief summary of them is given here.

If the stratified fluid is represented by $n + 1$ layers, as shown in Figure 6, the equations 2.1 for hydrostatic flow take the form

$$\frac{\partial d_i}{\partial t} + \frac{\partial(u_i d_i)}{\partial x} = 0,$$

$$\frac{\partial u_i}{\partial t} + \frac{\partial}{\partial x}\left[\frac{1}{2}u_i^2 + g\sum_{j=0}^{i} d_j + g\sum_{j=i+1}^{n} \frac{\rho_j}{\rho_i} d_j + \frac{p_s}{\rho_i}\right] = 0,\qquad (4.1)$$

$$i = 1, \ldots$$

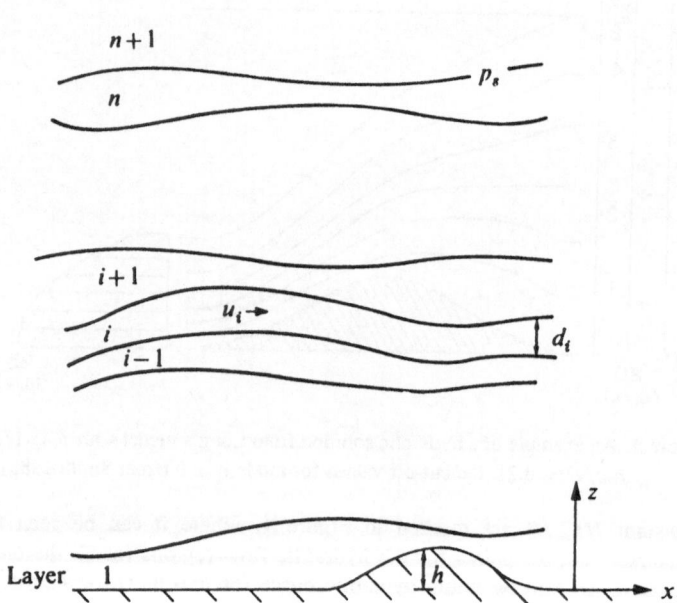

Figure 6. Schematic diagram for the layered model with a free upper surface.

where d_i, u_i and ρ_i denote the thickness, velocity and density of the ith layer respectively, and p_s is the pressure at the top of the nth layer. In the steady state these equations integrate to

$$d_i u_i = D_i U_i,$$

$$\frac{1}{2}(u_i^2 - U_i^2) + g\sum_{j=0}^{n} \rho_{ij}(d_j - D_j) + (p_s - P_s)/\rho_i = 0, \qquad (4.2)$$

where

$$\rho_{ij} = 1, \qquad j = i,$$

$$= \rho_j/\rho_i, \quad j > i, \qquad (4.3)$$

and $d_i = D_i$, $u_i = U_i$, $p_s = P_s$ far upstream with

$$p_s = P_s - g\rho_{n+1}(\sum_{j=0}^{n} d_j - D), \quad D = \sum_{j=0}^{n} D_j. \quad (4.4)$$

This system has $2n$ internal wave modes, n of which propagate against the stream, and n which propagate with it. The set of equations has solutions for flow over topography up to some maximum height, at which point some internal wave mode propagating against the stream becomes critical (i.e. has zero propagation speed relative to the ground) at the obstacle crest. Solutions for flow over higher obstacles are obtained in this model by calculating the changes which occur as the obstacle height is progressively increased by small increments. This may be imagined as occurring at successive times, with the flow passing through a sequence of steady states in the vicinity of the obstacle. The principal factor determining the flow is the "hydraulic alternative", namely that at the obstacle crest (where $dh/dx = 0$) we must have either

$$dd_i/dx = 0, \quad for\ all\ i = 1, 2, ..., n \quad or \quad (4.5a)$$

$$c_j = 0, \quad for\ some\ j, \quad (4.5b)$$

where c_j is the speed of internal wave mode j relative to the topography. If $c_j = 0$ at the obstacle crest, a small increase in obstacle height will normally cause the flow to adjust locally to keep $c_j = 0$, and this will cause a small disturbance of the form of a "columnar disturbance mode" to propagate upstream and change the upstream conditions. This change has the horizontal velocity and density structure of mode j. The relationship between the amplitude and sign of this change and the increase in obstacle height is determined numerically, and when these amplitudes are known the consequent downstream changes may be determined (at least in principle). This process may be repeated for successive small increases in obstacle height, causing progressive changes in the upstream velocity and density profiles.

With regard to the upstream columnar disturbances, there are two possible situations which can arise. Firstly, the speed of successive small-amplitude upstream disturbances (relative to the reference frame of the obstacle) may be constant or decrease, so that the upstream disturbances become "smeared out" as they propagate upstream. The total upstream disturbance is termed a "rarefaction". This implies that no resulting disturbances will be transmitted back to the region of the obstacle, and the procedure of calculating the flow changes due to successive increments in upstream disturbances and obstacle height, as described above, is valid. Alternatively, it is possible that the successive small-amplitude upstream disturbances travel faster, so that the overall upstream disturbance is an increasing function of its amplitude. This means that disturbances generated later will catch up to those generated earlier, and the result will be an internal bore or hydraulic jump, for that particular mode. A hydraulic jump imposes its own conditions on the change in the flow across it, and this information is transmitted back to the flow over the obstacle. Information may then be transmitted back to the jump, and an equilibration between the two is set up. When a jump is present, a different numerical procedure is required from that for the rarefaction situation. However, it turns out that provided the upstream disturbances are not too large, the flow profile obtained immediately upstream of the obstacle is very similar in both cases. The details are given by Baines (1988). This implies that the simpler rarefaction procedure may be used as an approximation to the jump procedure, and this has been done in this paper.

The rarefaction procedure may be employed as described above, until one of three things occurs. The first of these possibilities is that the upstream propagation speed of mode j decreases to zero, because of the altered propagation characteristics of the background flow. For higher obstacle heights, condition 4.5a applies at the obstacle crest since 4.5b now cannot, and the fluid makes an adjustment on the downstream side with a hydraulic drop or some equivalent flow structure. As h_m increases, this condition may continue until another mode (probably $j - 1$) becomes critical at the obstacle crest. The second possibility occurs when the increasing amplitude of a given mode upstream causes the next slowest mode to propagate faster, so that it eventually becomes critical in the upstream flow. Further increase in the height of the topography then results in the two modes increasing together. This phenomenon occurs readily with a free upper surface and is discussed below. The third possibility is that one of the upstream layers (normally the lowest) becomes blocked. With further increases in the obstacle height, the nature of the flow may still be much the same as before, but the upstream disturbances will be more complex and involve two or more modes because the blocked layer (or layers) must have zero velocity at the obstacle, although its thickness may change (generally it decreases).

These various steps are embodied in a set of computer programs which enable the calculation of the steady-state flow properties for all obstacles in a given initial stratified flow up to quite large heights. Results for uniformly stratified flow with a rigid upper boundary are described by Baines and Guest (1988). An example with a free upper surface with $n = 3$, where the initial layer thicknesses, density increments and velocities are all uniform, is shown in Figure 7. The rarefaction procedure has been used to obtain the flow up to the point where the lowest layer becomes blocked. Flow states are shown in terms of the two dimensionless parameters F, the Froude number, and h_m/D. Here $F = U/C_0$, as in equation 3.3, where C_0 is the propagation speed of the fastest mode in fluid at rest. Note that, as with Figure 4, this diagram gives the critical height as a function of F, but it also gives information about the upstream disturbances produced by the obstacle. This diagram is similar to Fig. 9 of Baines and Guest (1988) for four layers with a rigid upper boundary, with some notable exceptions. These are, firstly, that here the lowest mode ($n = 0$) behaves like the mode in a single layer system (at least as far as this diagram is concerned), and secondly, mode 1 does not affect the flow (i.e. become critical at $h = h_m$) until h_m reaches a finite value for any F.

We now apply this model to a uniformly stratified layer of depth D with uniform initial velocity and a free upper surface, with $n = 64$ layers. Again, the rarefaction procedure has been used. As in the previous section, we specify the system by the two parameters $F = \pi U/2ND$, the Froude number based on the lowest mode, and h_m/D. The properties of the system in terms of these parameters are shown in Figure 8 for $F < 1$ and $h > 0$. For $F > 1/3$, the model gives the same critical curve marking the limit of symmetric (Long's model) solutions as that shown in Figure 4 in the previous section. This indicates that 64 layers gives a good approximation to continuous stratification, at least in this parameter range. The section of the curve centred on $F = 0.2$ also agrees quite well. For the intervening region in Figure 4, where the Long's model solution is limited by static instability, the hydraulic model indicates that the flow becomes critical at the same value of h_m/D (or very slightly greater). This is consistent with the results with a rigid

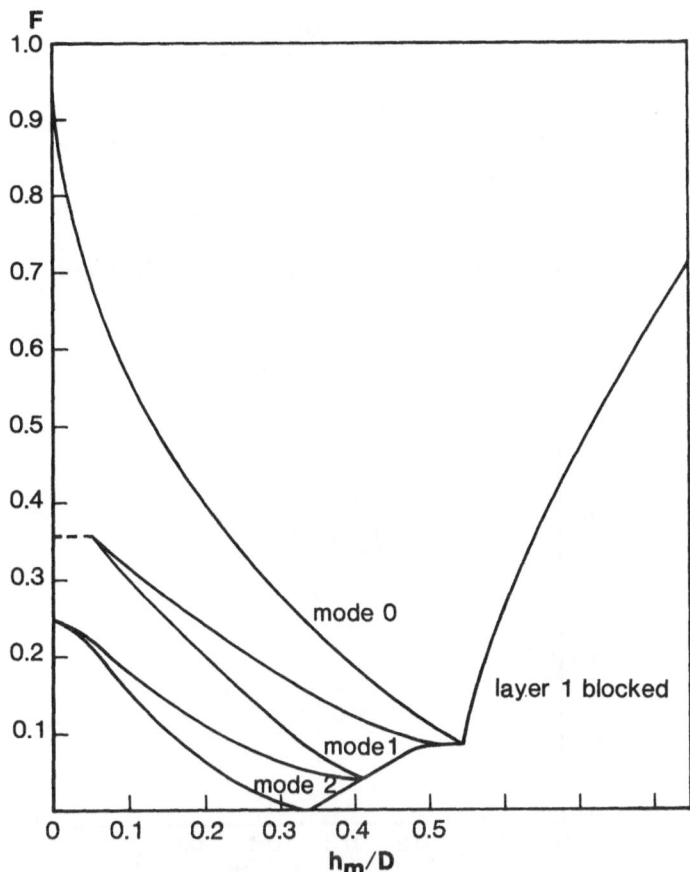

Figure 7. F -- h_m/D diagram for three layers with initially uniform velocity and thickness, with a free upper surface, calculated with the rarefaction procedure up to the point where the lowest layer becomes blocked. Regions where modes 0, 1 and 2 are critical at the obstacle crest are indicated.

258

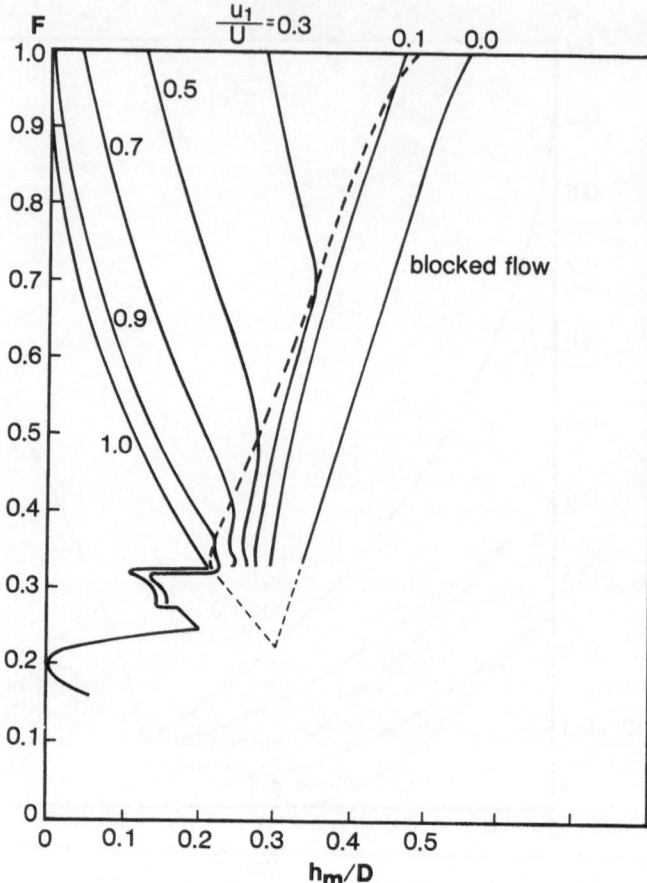

Figure 8. The $F - h_m/D$ diagram for the hydraulic 64-layer model with a free upper surface, showing contours of constant values of u_1/U. The dashed line marks the location where mode 1 (the second mode) begins to affect the upstream flow.

lid described by Baines and Guest, where the Long's model stability boundary was very close to (but lower than) the critical flow boundary of the 64-layer model for most of the parameter range.

Curves showing the relative upstream velocity at the ground, u_i/U, are plotted in Figure 8 for $F > 0.3$, up to the point of blocking for the lowest layer ($u_i = 0$). A change in character in these curves occurs at the heavy dashed line, which marks the point where mode $n = 1$ becomes critical upstream because its propagation speed has been increased by the presence of mode $n = 0$ (the "second possibility" mentioned above). When this point is reached, if a further increment in mode $n = 0$ is added to the upstream flow (in accordance with the rarefaction computation procedure), it renders mode $n = 1$ subcritical there (i.e. its propagation speed is reduced so that it can propagate upstream); a small increase in mode $n = 1$ upstream, on the other hand, renders this mode supercritical. The numerical procedure described in Baines (1988) causes the computation to alternate between the two modes as h_m is increased, and the nett result in practice is that they both increase together. This phenomenon may be expected whenever the increasing amplitude upstream of a particular mode causes another slower mode to propagate faster and become subcritical upstream of the obstacle. It has yet to be verified experimentally.

Some flows obtained from the 64-layer model are illustrated in Figure 9, up to the point of blocking of the lowest layer. In particular, we show velocity profiles upstream, over the obstacle crest, and immediately downstream of the isolated topography, for a number of F values in the range $0.35 < F < 0.9$, and a range of h_m/D values. In all of these the lowest mode ($n = 0$) is critical at the obstacle crest, and the flow undergoes a transition from subcritical upstream to supercritical downstream. For each value of F, the first frame shows the same flow as that described by the Long's model solution obtained by Smith (1985). The flow states are qualitatively similar for all F in this range, with the total depth decreasing monotonically from the upstream to the downstream side. The effect of mode $n = 1$ on the upstream velocity profiles may be seen in the change in curvature evident in the blocked flow profiles (Figures 9c,f,i, and l) and in Figure 9k. When blocking occurs, the depth of the blocked layer is a little less than half the obstacle height.

In this paper we are mostly concerned with results for $1/3 < F < 1$, but some results for F in the range $0.25 < F < 0.33$ are included in Figure 10. As expected, the flows found near $h_m/D = 0.15$ for $F = 0.3$ are dominated by the second mode, which becomes critical upstream at a small amplitude (Figure 10b). The flow then becomes critical with respect to the lowest mode at somewhat larger h_m/D.

5. Application of hydraulic models to infinite-depth flows

We may now combine the results of the preceding two sections to produce a model for infinitely deep hydrostatic flow over topography of finite height. In order to do this, we must make three assumptions. The first assumption is that the infinitely deep flow eventually reaches a steady state which is unique, in that it does not depend on the way in which the flow is established. In some circumstances, a discrete number of distinctly different flow states may be possible (as, for example, supercritical single-layer flow where two different flow states may occur depending on the initial conditions), but we will assume uniqueness for present purposes. There is in fact theoretical (Grimshaw and Smyth 1986) and experimental (Castro et al. 1989) evidence that inviscid finite-depth non-hydrostatic flows may be perpetually unsteady, depending on their past history, but with dissipation in the system they eventually reach a steady state (Smyth 1988).

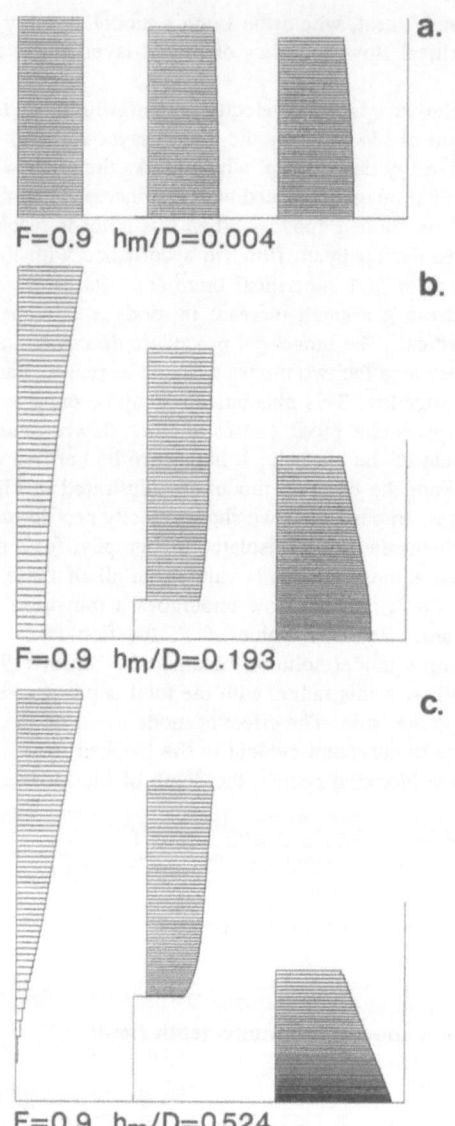

Figure 9. Velocity profiles upstream, over the obstacle crest and downstream, obtained from the 64-layer model with a free upper surface.
a) $F = 0.9$, $h_m/D = 0.0041$, critical height for the undisturbed flow;
b) $F = 0.9$, $h_m/D = 0.193$, mode 0 only upstream;
c) $F = 0.9$, $h_m/D = 0.524$, modes 0 and 1 upstream, with the lowest layer blocked;

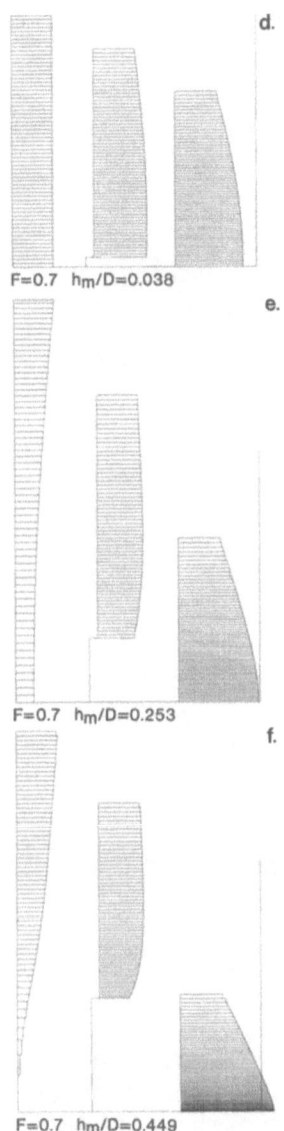

F=0.7 $h_m/D=0.038$

d.

F=0.7 $h_m/D=0.253$

e.

F=0.7 $h_m/D=0.449$

f.

d) $F = 0.7$, $h_m/D = 0.0385$, critical height with upstream flow undisturbed;
e) $F = 0.7$, $h_m/D = 0.253$, mode 0 only upstream;
f) $F = 0.7$, $h_m/D = 0.440$, modes 0 and 1 upstream, with the lowest layer blocked;

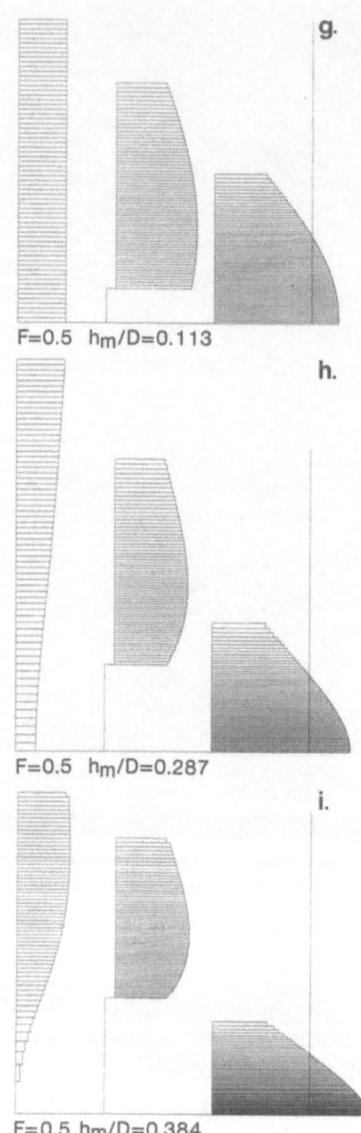

F=0.5 h$_m$/D=0.113

g.

F=0.5 h$_m$/D=0.287

h.

F=0.5 h$_m$/D=0.384

i.

g) $F = 0.5$, $h_m/D = 0.113$, critical height with upstream flow undisturbed;
h) $F = 0.5$, $h_m/D = 0.287$; modes 0 and 1 (at small amplitude) upstream;
i) $F = 0.5$, $h_m/D = 0.384$, lowest layer blocked upstream;

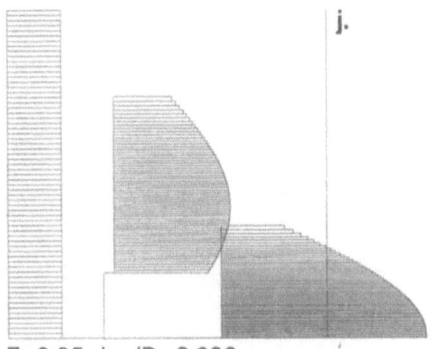

F=0.35 $h_m/D=0.200$

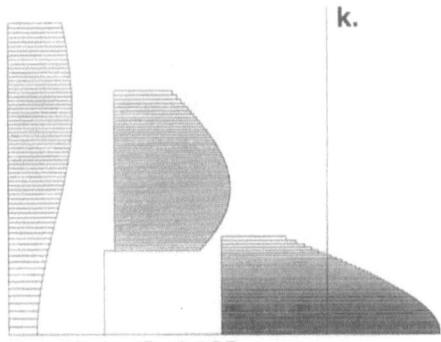

F=0.35 $h_m/D=0.255$

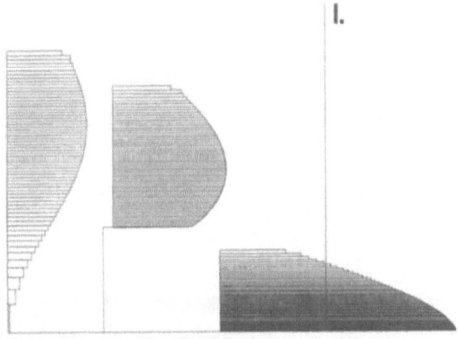

F=0.35 $h_m/D=0.320$

j) $F = 0.35$, $h_m/D = 0.200$, critical height with upstream flow undisturbed;
k) $F = 0.35$, $h_m/D = 0.255$; modes 0 and 1 upstream;
l) $F = 0.35$, $h_m/D = 0.320$, lowest layer nearly blocked.

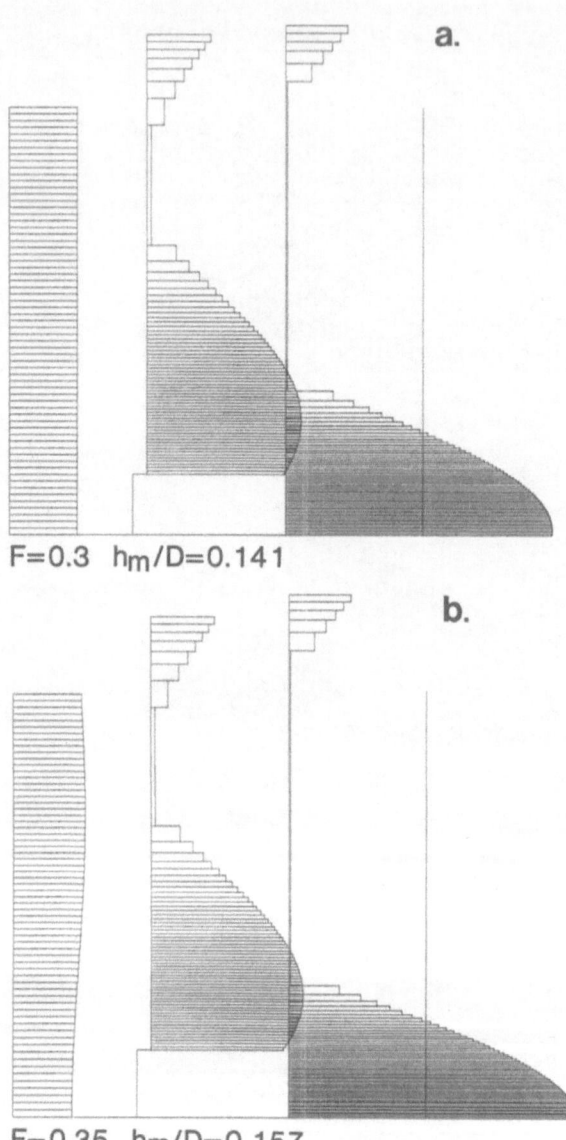

F=0.3 $h_m/D = 0.141$

F=0.35 $h_m/D = 0.157$

Figure 10. Velocity profiles from the 64-layer model with a free upper surface as for Figure
8, but for $F = 0.3$ with
a) $h_m/D = 0.141$ --- critical height for mode 1 with the upstream flow undisturbed;
b) $h_m/D = 0.157$ --- mode 1 critical upstream

As shown in Section 2, infinitely deep hydrostatic flows become statically unstable when Nh_m/U reaches a value $(Nh_m/U)_t$ between 0.5 and 1, which depends on the obstacle shape. The work of Laprise and Peltier (1989a,b) and others shows that when this point is reached the flow evolves towards a hydraulic state with a deep stagnant region over and on the lee side of the obstacle. An example is shown in Figure 11. These flows resemble the flow states in the model proposed by Smith (1985), where the disturbances to the flow caused by the topography are mainly confined to lower levels, and disturbances in the upper parts of the fluid are weak or non-existent. This situation is also familiar in laboratory experiments (Baines and Hoinka 1985). Our second assumption is therefore, following Smith (1985), that deep flows with Nh_m/U above the point of instability may be described by a model with a free upper surface. This is equivalent to neglecting the upper level vertical momentum flux in this parameter range. Whilst this wave flux is certainly greatly reduced relative to that in the *lee wave* regime at smaller Nh_m/U, it is not certain that its effects on the low-level flow are negligible.

With this assumption, the values of $(Nh_m/U)_t$ may be transferred to the hydraulic diagram (Figure 4) to obtain the depth D of the active hydraulic stratified layer, and hence the appropriate value of the Froude number, $F = \pi U/2ND$. this gives values of F in the range

$$0.333 < F < 0.4 \qquad (5.1)$$

$$(a_u/a_d \to 0) \quad (a_u/a_d \to \infty)$$

The values of D so obtained are found to be less than the corresponding values of z_s, the upstream height of the first streamline to become statically unstable in the Long's model solutions. This is consistent with the numerical solution shown by Laprise and Peltier (1989b) and reproduced in Figure 11.

In order to be able to calculate the flow properties for larger values of Nh_m/U (i.e. greater than $(Nh_m/U)_t$ for the particular obstacle shape), we make the third major assumption, namely that the streamline (or constant density surface) for the upper free surface (of the hydraulically active region) remains unchanged as h_m increases. This implies that F (the Froude number for the initial flow) remains unchanged as h_m increases, and the resulting flow states may be obtained from the model described in Section 4. This assumption is convenient, but it is also supported by the following physical considerations. Firstly, if we adopt the approach of the hydraulic model of Section 4, in which the flow is calculated by finding the change resulting from a succession of small increases in h_m, the initial response of the flow to the latter is to cause a local adjustment of the flow in the vicinity of the obstacle crest to keep the flow critical there (with respect to the lowest mode in this case). Since this local adjustment occurs to a dynamically active layer below a thick stagnant layer which tends to isolate it from the fluid above, one expects the process to have very little effect on this upper fluid. In particular, it seems intuitively unlikely that it will cause upper level layers to be incorporated into the hydraulically active region. Secondly, this local adjustment or altered flow state propagates upstream (and downstream) by means of a columnar disturbance mode. A consideration of simple prototype problems (the calculations are not reproduced here) shows that, as the wave propagates upstream to where the thick stagnant layer thins, the wave energy begins to leak into the upper level stratification and causes some changes to the flow there. However, this leakage is only partial and does not affect the eventual change which is felt at a given location upstream, which is the same as that which would be

266

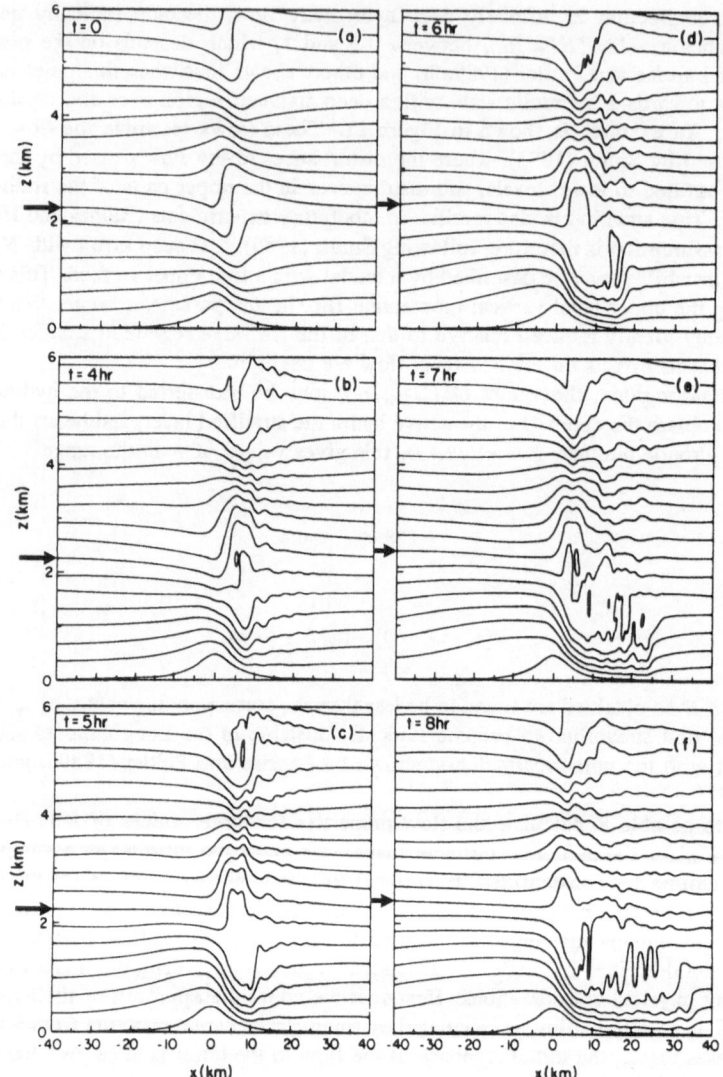

Figure 11. Time evolution of the stream function for a numerical experiment with $Nh_m/U =$ 0.95. The flow has been initialised with the Long's model solution (from Laprise and Peltier 1989b). The initially overturning streamline is marked in each frame with an arrow on the left-hand side.

obtained if the fluid were not stratified above the initial level D. The effect of upper level stratification on the lower level upstream flow, therefore, is to cause a time delay in the flow at a given point upstream reaching a steady state after an increase is made to the obstacle height, relative to that for a homogeneous upper layer.

These considerations give confidence in the third assumption above but by no means completely justify it. It is possible that the assumption breaks down or becomes progressively less valid as an approximation as the obstacle height h_m becomes large.

The flows predicted by this model for infinite depth flows with $N h_m/U > (N h_m/U)_t$ are therefore qualitatively given by Figures 9g, 9h and 9i (although these are slightly outside the range of equation 5.1) and Figures 9j, 9k and 9l. Flow fields for any particular value of a_u/a_d and of $N h_m/U$ (at least up to the point of blocking) may be readily obtained. From Figure 9 we may also see that, according to this model, upstream blocking begins at $N h_m/U \approx 1.5$, although the precise value depends on the value of a_u/a_d. It should be noted that Figure 8 has been computed by assuming that the upstream disturbances are all rarefactions. As noted above, mode 0 has hydraulic jump character upstream (which is more difficult to handle numerically) paralleling the behaviour for single layer flow. Hence the rarefaction procedure is an approximation in this case, and it is expected to become less accurate in describing the steady-state flow properties as the amplitude of the upstream disturbance increases. In particular, it may account for the fact that the model predicts upstream blocking when $N h_m/U = 1.5$, whereas the experimental value is about 2 (Baines 1979, Baines and Hoinka 1985, Pierrehumbert and Wyman 1985).

6. Conclusions and discussion

We have described an analytical/numerical model which provides predictions of the major properties of the flow resulting from the introduction of a two-dimensional ridge into a known flow where U and N are uniform with height. Given that the flow is essentially hydrostatic, the model depends on the following assumptions: (i) the flow reaches a steady state which is determined by the initial conditions; (ii) when $N h_m/U$ has increased to the point where the flow becomes statically unstable, the flow may be described by a model with a free upper surface, and (iii) when $N h_m/U$ increases further, the constant-density surface for the free upper surface remains unchanged. Assumptions (i) and (ii) were made by Smith (1985), who hypothesised that the flow would adjust to changes in external conditions by varying the depth of the hydraulically active layer (i.e. the height of the free surface). Such possible variations are limited to lie on the boundary curve (the solid curve with $F > 1/3$ in, for example, Figure 4). However, numerical (e.g. Pierrehumbert and Wyman 1985) and laboratory studies (Baines and Hoinka 1985) show that these flows adjust to changing external conditions by, in large measure, altering the upstream flow state to satisfy conditions over the obstacle. The third assumption made here is that, in the free surface model, this adjustment in the upstream flow is the total adjustment, and no variation in the free surface height takes place. This is plausible on physical grounds, and is supported by existing evidence, but it needs to be tested by more comparisons with results from high resolution numerical models. Very few of the published details of integrations from these models to date are suitable for making the comparisons which can determine in detail which proportion of the two processes are involved in the adjustment to the flows, and we await the results of further studies which can help resolve this question.

Acknowledgements

The authors are grateful for discussions with Ronald B. Smith on the subject of this paper.

References

Baines, P.G., 1979. Observations of stratified flow past three-dimensional barriers. J. Geophys. Res. **84**, 7834-7838.

Baines, P.G., 1987. Upstream blocking and airflow over mountains. Ann. Rev. Fluid Mech. **19**, 75-97.

Baines, P.G., 1988. A general method for determining upstream effects in stratified flow of finite depth over long two-dimensional obstacles. J. Fluid Mech. **188**, 1-22.

Baines, P.G. and Hoinka, K.P., 1985. Stratified flow over two-dimensional topography in fluid of infinite depth: a laboratory simulation. J. Atmos. Sci. **42**, 1614-1630.

Baines, P.G. and Guest, F. 1988. The nature of upstream blocking in uniformly stratified flow over long obstacles. J. Fluid Mech. **188**, 23-45.

Castro, I.P., Snyder, W.H. and Baines, P.G., 1989. Obstacle drag in stratified flow. Submitted to Proc. Roy. Soc. A.

Grimshaw, R.H.J. and Smyth, N. 1986. Resonant flow of a stratified fluid over topography. J. Fluid Mech. **169**, 429-464.

Hogg, N.G., 1983. Hydraulic control and flow separation in a multi-layered fluid with application to the Vema channel. J. Phys. Oceanog. **13**, 695-708.

Laprise, R. and Peltier, W.R., 1989a. The linear stability of non-linear mountain waves: implications for the understanding of severe downslope windstorms. J. Atmos. Sci. **46**, 545-564.

Laprise, R. and Peltier, W.R., 1989b. The structure and energetics of transient eddies in a numerical simulation of breaking mountain waves. J. Atmos. Sci. **46**, 565-585.

Lilly, D.K. and Klemp, J.B., 1979. The effects of terrain shape on non-linear hydrostatic mountain waves. J. Fluid Mech. **95**, 241-261.

Long, R.R., 1953. Some aspects of the flow of stratified fluids. I. A theoretical investigation. Tellus **5**, 42-58.

Miles, J.W., 1968. Lee waves in a stratified flow. Part 2. Semi-circular obstacle. J. Fluid Mech. **33**, 803-814.

Peltier, W.R. and Clark, T.L., 1979. The evolution and stability of finite amplitude mountain waves. Part II: Surface wave drag and severe downslope windstorms. J. Atmos. Sci. **36**, 1498-1529.

Pierrehumbert, R.T. and Wyman, B., 1985. Upstream effects of mesoscale mountains. J. Atmos. Sci. **42**, 977-1003.

Smith, R.B., 1985. On severe downslope winds. J. Atmos. Sci. **42**, 2597-2603.

Smyth, N.F., 1988. Dissipative effects on the resonant flow of a stratified fluid over topography. J. Fluid Mech. **192**, 287-312.

IS THE EXCHANGE THROUGH THE STRAIT OF GIBRALTAR MAXIMAL OR SUBMAXIMAL?

Chris GARRETT, Myriam BORMANS*, Keith THOMPSON
Department of Oceanography
Dalhousie University
Halifax, Nova Scotia
B3H 4J1
Canada

ABSTRACT. The Mediterranean Sea and Strait of Gibraltar can take either of two possible states. In the first the Mediterranean is overmixed, with minimum salinity difference between it and the Atlantic and maximal exchange through the Strait. In the second the Mediterranean is not over-mixed, the salinity difference is greater than for the overmixed case and the exchange is submaximal. Interannual and longer term variability in air–sea interaction over the whole Mediterranean will be much more readily detectable at the Strait of Gibraltar if the exchange there is submaximal than if it is maximal, so it is important to establish the present state of the exchange. We review indirect evidence from a variety of observations and theories and find support for both interpretations. We conclude that the system is close to maximal exchange and may flip from one state to the other on various time scales up to many years, but with a tendency for maximal exchange early in the year, and submaximal later in the year.

1. Introduction

The Strait of Gibraltar is a narrow and shallow connection between the Atlantic Ocean and Mediterranean Sea, with a minimum width near Tarifa of about $15 km$ and a sill depth of less than $300 m$ (Figure 1). It is an exciting place oceanographically with an internal hydraulic jump to the west of the sill and internal bores that can have a range from crest to trough of more than $100 m$ (e.g. Wesson and Gregg, 1988; Armi and Farmer, 1988). The dominant oceanographic feature, however, is a two-layer inverse estuarine circulation that is driven by an excess of evaporation over precipitation and river discharge for the Mediterranean.

In this circulation Atlantic Water, with a salinity of a little more than 36 ‰, flows in at the surface and spreads through the Mediterranean becoming progressively saltier as it moves eastwards (e.g. Wüst, 1961). Deep convection occurs at a number of locations, but it appears that the most important is in the Eastern Mediterranean, where so-called Levantine Intermediate Water (LIW), with a salinity of about 39 ‰, is formed in the wintertime. This water mass makes its way back towards the Strait of Gibraltar, experiencing some

---

* Now at Research School of Earth Sciences, Australian National University, Canberra, ACT 2601, Australia

L. J. Pratt (ed.), The Physical Oceanography of Sea Straits, 271–294.
© 1990 Kluwer Academic Publishers.

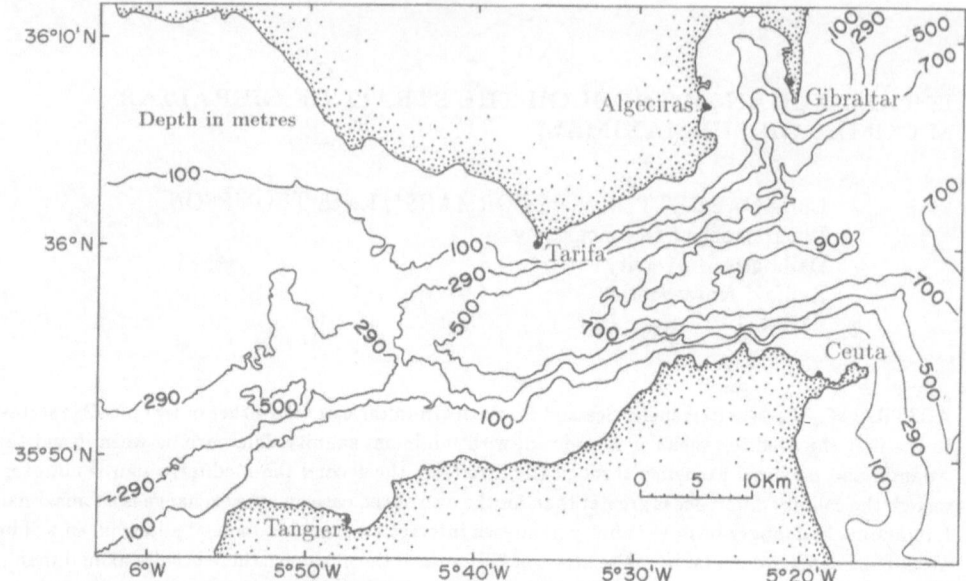

Figure 1. The Strait of Gibraltar.

mixing en route, and has a salinity of a little more than 38 ‰ as it begins to flow back
out to the Atlantic through the Strait of Gibraltar (e.g. Wüst, 1961).

Accounting for the 2 ‰ salinity difference between inflow and outflow is one of the key
oceanographic problems associated with the Mediterranean Sea and Strait of Gibraltar. If
the flow is regarded as a two-layer exchange, with flow rates Q_1, Q_2 and salinities S_1, S_2
(Figure 2), then volume and salt conservation require

$$Q_1 - Q_2 = \int_{\text{Med.}} (E - P)dA, \qquad Q_1 S_1 - Q_2 S_2 = 0 \qquad (1.1)$$

whence

$$Q_1 = \frac{S_2}{(S_2 - S_1)} \int_{\text{Med.}} (E - P)dA, \qquad Q_2 = \frac{S_1}{(S_2 - S_1)} \int_{\text{Med.}} (E - P)dA, \qquad (1.2)$$

where $E - P$ is evaporation minus precipitation (and river discharge). We have assumed
constant sea level and salt content of the Mediterranean (as seems appropriate for a time
scale of more than a year) and ignored unimportant subtleties to do with the use of volume
rather than mass fluxes. Consistency of these Knudsen formulae with $S_2 - S_1 \simeq 2$ ‰
and a yearly and spatial average of $E - P$ of about $1 m\,a^{-1}$ then requires $Q_1 = 1.68 Sv$
and $Q_2 = 1.60 Sv$ (Bethoux, 1979). These flow rates are reasonably consistent with direct

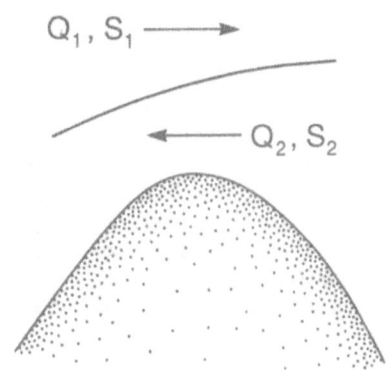

Figure 2. Schematic of a two–layer exchange over a sill.

measurements, though none of Q_1, Q_2 or $E - P$ appear to be known to better than $\pm 50\%$ or so.

It is important to recognise, though, that (1.2) does not determine $S_2 - S_1$. There is an infinity of possible solutions, with large flows corresponding to small salinity differences and vice versa. Further constraints are necessary to obtain a unique answer and account for the observed salinity difference of about 2 ‰.

Whitehead *et al.* (1974) seem to have been the first to suggest that one constraint might be hydraulic control at the sill, as also invoked by Bryden and Stommel (1984). For a steady two-layer flow through a rectangular cross-section, the hydraulic control condition is

$$G^2 = \frac{u_1^2}{g'h_1} + \frac{u_2^2}{g'h_2} = 1, \quad g' = g(\rho_2 - \rho_1)/\rho_2 \tag{1.3}$$

where u_i, h_i, ρ_i are the current speed, thickness and density of the upper ($i = 1$) and lower ($i = 2$) layer. There is still, however, a "semi-infinity" of solutions. As discussed by Bryden and Stommel (1984), an arbitrarily large salinity difference is still possible if h_2 becomes very small. There is then a weak exchange with very salty water trickling slowly over the sill. Bryden and Stommel (1984) point out that formation of very salty water in the Mediterranean would require that the inflowing Atlantic Water be kept at the sea surface, exposed to evaporation, for a long time and that it then sink below the surface without much mixing with the inflowing surface Atlantic Water. Increasing the mixing would reduce the salinity, increase the required exchange rate through the Strait and also, to satisfy (1.3), require a larger h_2.

As the mixing is increased still further, the solution can reach a so-called "overmixed" limit with a minimum salinity difference between the two layers, a maximum exchange and the interface at mid-depth ($h_2 = h_1$). Raising the interface above mid-depth would increase the salinity difference and reduce the exchange again. Using estimates of $E - P$ and the width of the Strait, Bryden and Stommel (1984) find that this overmixed limit corresponds to $S_2 - S_1 \simeq 1.7 \permil$, sufficiently close to the observed $2 \permil$ that they suggest that the limit has in fact been achieved. The inflow and outflow found by Bryden and Stommel (1984) for the overmixed limit were close to those of Bethoux (1979) cited above and agreed with the estimates of Whitehead *et al.* (1974) who also maximised the exchange after allowing for the additional minor effect of the earth's rotation.

Farmer and Armi (1986) showed that the Bryden and Stommel (1984) solution, with the interface at mid-depth at the sill, could not be connected hydraulically (using the continuity and Bernoulli equations for each layer) to the Atlantic and Mediterranean reservoirs at opposite ends of the Strait, and that the maximal exchange solution should have a somewhat deeper interface at the sill, less exchange and a bigger $S_2 - S_1$ than the Bryden and Stommel (1984) solution. For a reasonable approximation to the width and depth variations of the Strait of Gibraltar, though still with rectangular cross-sections, they find a 28% increase in $S_2 - S_1$, bringing it even closer to the observed $2 \permil$. The maximal exchange solution of Farmer and Armi (1986) has hydraulic control at both the sill and the narrows, with supercritical flow away from the Strait beyond these critical sections (Figure 3).

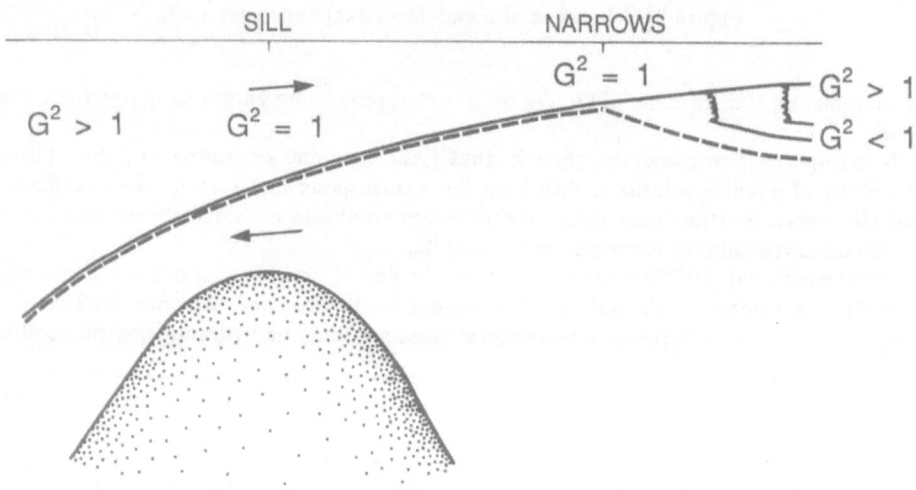

Figure 3. Schematic of the two–layer exchange through a strait with a sill and a contraction showing the interface depth for maximal exchange (solid lines) and for marginally submaximal exchange (dashed line). The square of the internal Froude number, from (1.3), is also shown for different regions.

The inflow must eventually match conditions in the Mediterranean. If the interface there is deeper than for the marginally submaximal inflow then the exchange will be submaximal. If, however, the interface in the Mediterranean is shallower than for the marginally submaximal inflow, the exchange will be maximal with a hydraulic jump or smoother dissipative transition somewhere east of the control section at Tarifa Narrows (as sketched in Figure 3). This transition could be between the Narrows and the Gibraltar/Ceuta section, but

Bormans and Garrett (1989b) have argued on energetic grounds that the transition will be further east if the interface depth to which the inflow must be matched is more than $10m$ or so shallower than that for the marginally submaximal solution. We return to this issue in Section 10.

The solutions of Bryden and Stommel (1984) and Farmer and Armi (1986) ignore friction, time-dependence, the earth's rotation and realistic cross-sections. One effect, friction, was included earlier by Assaf and Hecht (1974) in a simple model that approximated the Strait of Gibraltar by a flat-bottomed trapezoidal channel. They found that the interface then tilts along the channel and, in the overmixed limit, with maximal exchange, has hydraulic control at the ends of the Strait (which thus correspond to the sill and narrows in the models with more realistic geometry but no friction).

Other features, such as time-dependence, realistic cross-sections and the effect of the earth's rotation can be added to these two-layer models, but the over-riding assumption that would be carried over from the work of the above authors is that the Mediterranean is overmixed, with a maximum exchange rate through the Strait of Gibraltar and minimum salinity difference between the two layers there (Armi and Farmer, 1987). Some support for this view comes from its apparent success in accounting for the observed salinity difference, but the possibility must be considered that the system is close to, but not at, this limit.

It does seem likely that the outflow is hydraulically controlled at the sill (or perhaps some more western sill in a time-dependent flow, see Armi and Farmer (1988)) as it plunges down into the Atlantic, but we should also consider the submaximal exchange solutions of the Farmer and Armi (1986) model, with subcritical flow at the narrows and east of them; the corresponding solution of the Assaf and Hecht (1974) model would have subcritical flow at the eastern end of the Strait. A marginally submaximal exchange solution (Figure 3) would have $S_2 - S_1$ almost equal to ΔS_{min} as for the maximal exchange solution but, as the flow became increasingly submaximal with a deeper interface and less exchange, $S_2 - S_1$ would be larger than ΔS_{min}.

In summary, it seems that the Mediterranean Sea/Strait of Gibraltar system can have two possible states (Figure 4): 1) overmixed with maximal exchange, minimum salinity difference and a shallow, supercritical inflow at the eastern end of the Strait, or 2) not overmixed, with submaximal exchange, a bigger salinity difference and a thicker subcritical inflow. (Submaximal exchange is thus a continuum of possible states rather than a unique solution).

The purpose of this paper is to discuss the evidence for one or the other of these two states. We start in Section 2 by discussing the importance of resolving the question when considering the Strait of Gibraltar as a convenient climatic "choke point" at which time-dependent changes over the whole Mediterranean could be recorded. In Section 3 we discuss whether there is some fundamental theoretical reason why an inverse estuary, such as the Mediterranean, should be overmixed and in Section 4 describe some theoretical models that can be used to interpret data that might help us resolve the issue of maximal or submaximal exchange. Section 5 reviews data on the interface depth at the eastern end of the Strait and Section 6 summarizes the conclusions reached by Bormans et al. (1986) in their analysis of seasonal variability. Section 7 summarizes the interpretation of MEDALPEX sea level data by Garrett et al. (1989), Section 8 discusses evidence from geodetic levelling along the Strait, and Section 9 discusses the relevance of the study by Bormans and Garrett (1989c) of the Alboran Gyre just inside the Strait of Gibraltar.

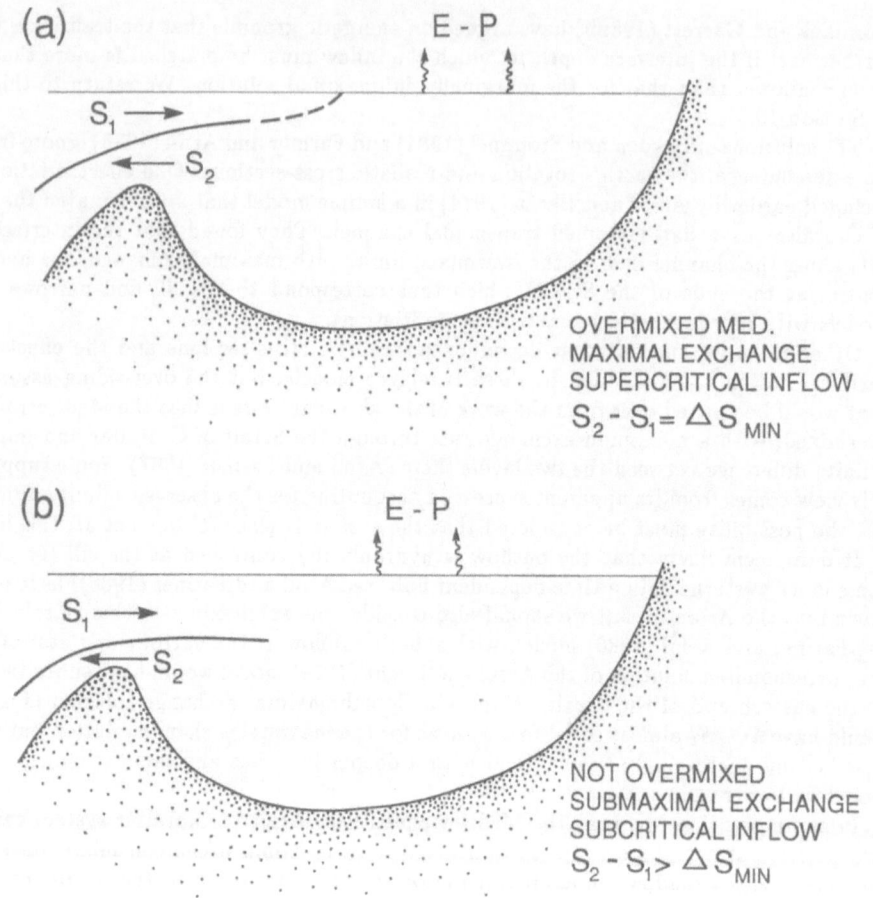

Figure 4. Summary of the two basic configurations of the Mediterranean and the Strait of Gibraltar.

We shall see that there is evidence in favour of both maximal and submaximal exchange at different times; Section 10 summarizes the evidence and suggests what further work is necessary to resolve this important issue.

2. Relevance of the Question

The Strait of Gibraltar would be a convenient point at which to monitor changes in the Mediterranean if some measurable quantity in the Strait were to respond quickly and significantly to changes in, say, evaporation minus precipitation $(E - P)$. As an extreme example we consider what would happen in three possible scenarios (Figure 5) if $E - P$ were suddenly put equal to zero.

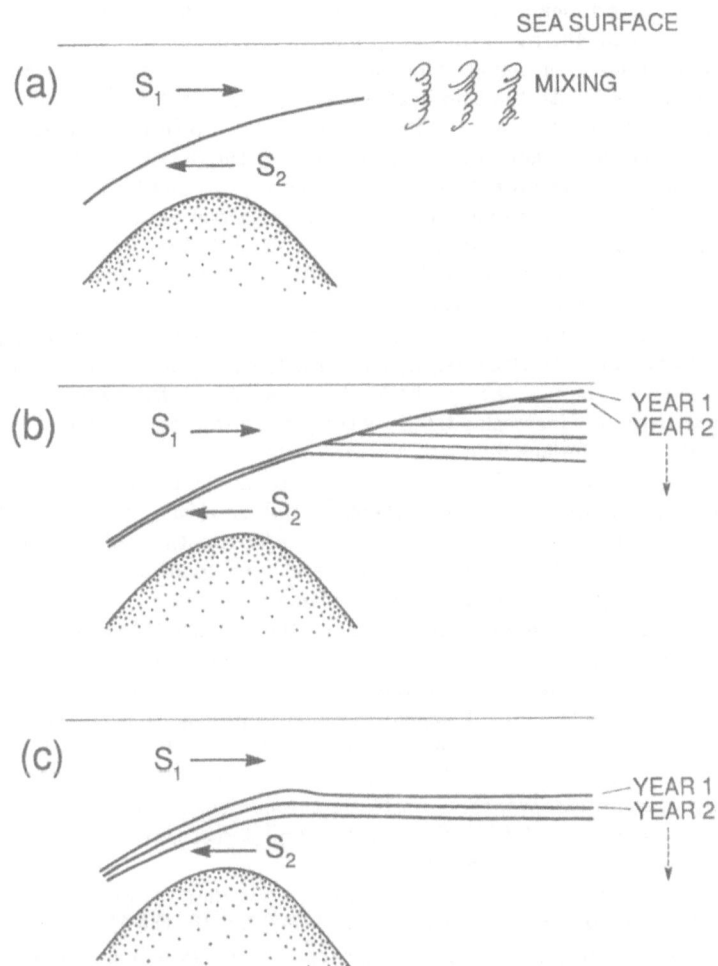

Figure 5. Schematic of the response of the interface to a cessation of $E - P$ for three different scenarios: a) Maximal exchange initially, mixing maintained, b) maximal exchange initially, mixing not maintained, c) submaximal initially.

In the first two cases, we assume that, up until that time, the exchange was maximal, corresponding to the lock exchange between reservoirs of salinity S_1 and S_2 and associated densities (e.g. Armi and Farmer, 1987). In the first case (Figure 5a) we assume that vigorous mixing continues after $E - P$ is put to zero. This maintains the lock exchange nature of the flow, though the Mediterranean salinity S_2 slowly decreases as Atlantic Water is mixed in with the Mediterranean Water. The exchange rate through the Strait of Gibraltar would be reduced on the same time scale but, to a first approximation, the shape of the interface would remain the same. The time scale for changes in $S_2 - S_1$ and the currents would be comparable with the time scale for the outflow to drain a significant part of the Mediterranean volume of about $4 \times 10^{15} m^3$. At about $1.5 Sv$ this is nearly 100 years. In this scenario, therefore, readily observable changes at the Strait would not be apparent for many years and so one could not recommend it as a good monitoring point.

In the second case (Figure 5b) we also assume maximal exchange initially, but assume that the mixing stops at the same time as $E - P$. Atlantic Water then forms a thicker and thicker layer over the top of the Mediterranean Water, with the interface sinking at a rate given by inflow rate divided by surface area, i.e. $1.5 \times 10^6 m^3 s^{-1} / 2.5 \times 10^{12} m^2 = 19 m\, a^{-1}$. The exchange through the Strait would remain critical at first with the same exchange, salinity difference and interface depth, but with the interface at the eastern end of the Strait connected to the interface in the Mediterranean by a hydraulic jump, or smoother dissipative transition of increasing intensity (not shown in Figure 5b but sketched in Figure 3). Eventually the interface depth would drop below that of the marginally submaximal exchange at the eastern end of the Strait, the control at the narrows would be lost and the interface throughout the Strait would therefore sink at a rate that would start at about $19 m\, a^{-1}$. As will be discussed in Section 4, the interface depth required to flood the control in this way is thought to be close to $90 m$, so several years could pass before a drastic change in $E - P$ and mixing in the Mediterranean became apparent.

The third scenario is the most interesting. In this (Figure 5c) the exchange is assumed to be submaximal from the start. Cessation of $E - P$ means that no more dense water is being made, so the interface in the Strait starts to sink at about $19 m\, a^{-1}$, possibly after a delay of the few weeks it would take for a signal to propagate from the former region of formation of LIW. One way of looking at this scenario, or the latter part of the second case, is that, after the narrows control is flooded, information about the changing interface depth in the Mediterranean can propagate upstream through the Strait. As will be discussed in Section 6, Bormans et al. (1986) did in fact interpret seasonal changes of surface inflow, inferred from sea level data, in these terms.

The main purpose of this discussion is to point out that the Strait of Gibraltar would show a much more rapid response to changing conditions in the Mediterranean, and so is a more suitable monitoring point, if the exchange is submaximal rather than maximal. It thus seems important to determine the nature of the present day exchange.

3. Is an Inverse Estuary Overmixed?

In Section 1 we discussed, following Bryden and Stommel (1984), how an inverse estuary could contain very salty water, and hence have submaximal exchange with the external ocean, if the inflowing water could be retained for long enough in a shallow shelf area with intense evaporation and then descend into the basin without excessive further mixing. These conditions seem plausible, though requiring further investigation.

It might be argued that, in reality, Mediterranean Intermediate Water and Deep Water are formed in the open sea by deep convection and so automatically set up conditions that correspond to lock exchange with its accompanying maximal exchange. A limited, but instructive, model has been proposed by Phillips (1966). He considered a one-dimensional sea of length L (Figure 6), with a uniform surface buoyancy loss rate B_0, separated from the external ocean by a sill of depth d. He obtained similarity solutions for the buoyancy driven circulation U (Figure 6) and buoyancy $b = -g(\rho - \rho_0)/\rho_0$ of the form

$$U = (B_0 x)^{1/3} F(z/d), \quad b = (B_0 x)^{2/3} d^{-1} G(z/d) \tag{3.1}$$

where F and G are functions which would depend on the detailed behavior of the vertical momentum and buoyancy fluxes in the water.

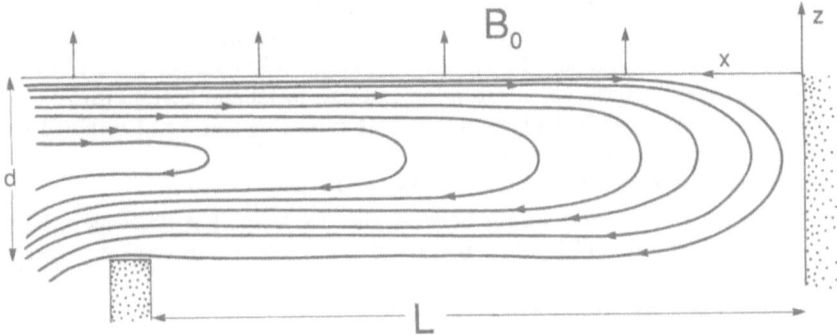

Figure 6. Schematic of the streamlines for the inverse estuarine circulation in an inland sea of uniform width separated from the exterior ocean by a sill (redrawn from Phillips, 1966).

As pointed out by Phillips (1966), the profile of the Richardson number $Ri = (\partial b/\partial z)/(\partial U/\partial z)^2$ is independent of the position x. If it is $O(1)$, then exchange with an exterior ocean through a narrow and shallow strait would lead to a Richardson number much less than one in the strait (or equivalently an internal Froude number much greater than one) if the flux in the top half and bottom half of the flow is the same in the strait as in the marginal sea. To avoid this violation of the hydraulic constraint in the strait, the flux in each layer in the strait must be reduced, presumably requiring significant recirculation within the sea of the flow shown in Figure 6, and perhaps penetration of the flow below sill depth.

The conclusion is that a marginal sea, separated from the external ocean by a strait and driven by a spatially and temporally uniform buoyancy loss rate, is likely to be overmixed, with a bigger buoyancy difference between the two layers in the strait than longitudinally in the sea. This would appear to require a rapid decrease in the surface buoyancy just inside the strait, which does not seem to be the case in the Mediterranean (Wüst, 1961). Moreover, Manzella et al. (1988) have found pronounced seasonal variability in the exchange through the Strait of Sicily, showing that the Eastern Mediterranean is not overmixed. We provisionally attribute this evidence against an overmixed limit to the temporal variability in the surface buoyancy loss, though further modelling and laboratory simulation is required.

4. Ingredients and Results of Recent Models

The hydraulic state of the flow at the eastern end of the Strait of Gibraltar is a key factor in determining whether the exchange is maximal or submaximal. The lower layer is deep and thus fairly stagnant there, so we need only consider the upper layer. In principle we require measurements of layer thickness and speed to determine the internal Froude number; in practice simultaneous data on these are not readily available, at least not with sufficient accuracy to discriminate clearly between supercritical and subcritical flow. We thus require dynamical models to aid in the interpretation of data on the interface depth alone or on sea level gradients.

We have already mentioned the Farmer and Armi (1986) model which has been a very useful guide in spite of the limiting assumptions of rectangular cross-sections, no friction, no rotation and no time-dependence other than quasi-steady barotropic fluctuations. Bormans and Garrett (1989a, b) extended the model to allow for triangular cross-sections (a better approximation to reality), the earth's rotation and interfacial and bottom friction, but they also largely neglected the effect of fluctuations at tidal frequency and retained the same simplifying assumption of two-layer flow.

The main results of these two papers can be summarized as follows:

1) The earth's rotation produces a tilt between the two layers but it seems to be valid to calculate this tilt from models which ignore rotation, rather than having to include it *ab initio*.

2) A supercritical surface layer inflow is caused by rotation to separate from the north shore of the Strait between Tarifa Narrows and the Gibraltar–Ceuta section at the eastern end of the Strait. The predicted separation point is rather insensitive to the upper layer inflow Q_1, and hence to barotropic flow fluctuations, but is shifted slightly eastwards by interfacial friction between the two layers.

3) Interfacial friction between the two layers shifts the control section eastward from Tarifa Narrows and brings the supercritical and subcritical solutions for the interface depth, in the eastern Strait, closer together. The effects are small, however, if the value of the interfacial friction coefficient C_i is chosen to be not much more than 10^{-4} on the basis of the microstructure measurements of Wesson and Gregg (1988).

4) Bottom friction on the sloping sides of the Strait has little effect on the position of the control section but tends to bring the supercritical and subcritical solutions slightly closer together at the eastern end of the Strait.

5) Taking realistic triangular, rather than rectangular, cross-sections reduces the interface depth in the Strait by 40 to $50m$.

6) Low-frequency barotropic fluctuations driven, for example, by changing atmospheric pressure over the Mediterranean, change the interface depth and sea level gradients. In particular, the ratio of sea level gradient fluctuations across and along the eastern end of the Strait is different for supercritical and subcritical flows.

7) Tidal fluctuations occur at too high a frequency to be included adequately in a steady model and may cause the mean exchange through the Strait to be different from that estimated from a model that neglects them. On the other hand, sea level gradient data do not show any significant fluctuations on the 15 and 28 day periods of tidal modulation

(Garrett *et al.*, 1989). Moreover, a one-layer reduced-gravity analysis of the flow at the eastern end of the Strait (Bormans and Garrett, 1989b) suggests that the upper layer depth varies fairly linearly with the transport, so that a solution based on the average flow will be very nearly the same as the average solution of a model with fluctuating flow. Bormans and Garrett (1989b) concluded that a model which neglects the tidal variability may be reasonably appropriate.

More quantitative results from the models will be discussed in subsequent sections while reviewing various pieces of evidence for maximal or submaximal exchange.

5. The Interface Depth at the Eastern End of the Strait

In reality the interface between the Atlantic and Mediterranean Water is spread over several tens of meters, but with the 37.0 and 37.5 isohalines representing the middle of the interface. Historical data, averaged over at least one tidal cycle, are scarce, but Figure 7 shows results from Lacombe and Richez (1982), based on data collected on 25-26 May, 1961, and a 4-day mid-strait mean obtained in September, 1971, by Cavanié (1973).

Figure 7 also shows the cross-strait interface depth, at the Gibraltar–Ceuta section, for a one-layer reduced gravity rotating hydraulics model with control at Tarifa Narrows. The cross-strait slope, and hence geostrophic velocity, found by Lacombe and Richez (1982) is close to that predicted, though the interface is rather deeper than predicted for supercritical flow. In fact the interface slope implies a geostrophic speed of about $1.7ms^{-1}$, giving a squared Froude number, based on a mid-strait thickness of $70m$, of about 2. This suggests that the flow was supercritical. Possibly, as suggested by Bormans and Garrett (1989b), lateral friction on the south shore of the Strait has reduced the current and flattened the interface there, effectively leading to a lateral shift of the interface shape to conserve the volume flux.

The interface depth reported by Cavanié (1973) seems too deep to be compatible with the predictions for supercritical inflow, at least with an inflow of $1.5Sv$. The interface also seems too shallow to match the subcritical solution, but this was obtained from a rotating hydraulics calculation with rectangular cross-sections which are inappropriate. As these data, and others we shall cite later, were all obtained in mid-strait we move next to comparison with the model discussed in Section 4, which ignores rotation but allows for triangular cross-sections and friction.

Figure 8 shows the predicted interface depth along mid-strait from the favoured model of Bormans and Garrett (1989b). It uses an interfacial drag coefficient of 10^{-4}, a bottom drag coefficient of 3×10^{-3} and has triangular cross-sections. The model has also been run with a net barotropic transport Q_0 that varies over the plausible range of -1 to $+1Sv$ (positive transport being into the Mediterranean). We see that Cavanié's (1973) observation of an interface depth of 100 to $120m$ is easily reconciled with a subcritical solution. The data of Lacombe and Richez (1982), with a mid-strait depth of 60 to $80m$, could have been associated with a significant negative Q_0 and subcritical flow, but, as described above, are more likely to represent a modification by friction of the supercritical solution. (The predictions shown in Figure 8 for the supercritical flow ignore all friction after the separation point).

In both cases it is possible that the flow was critical at Tarifa Narrows but underwent an internal hydraulic jump back to subcritical flow further east but before the Gibraltar–Ceuta section. However, as mentioned earlier, Bormans and Garrett (1989b) have discussed this

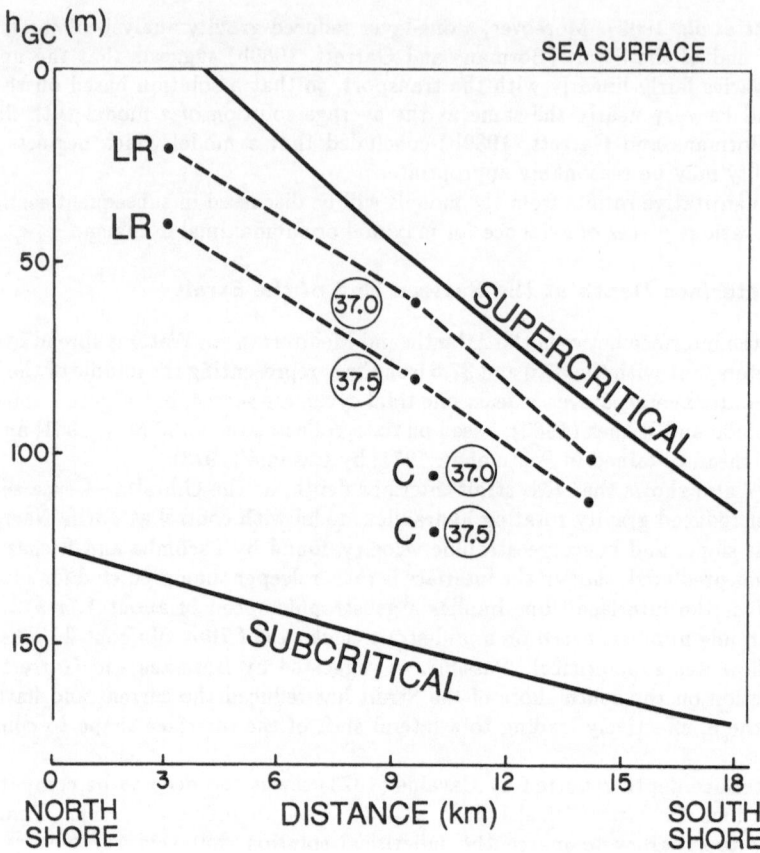

Figure 7. Predicted cross-strait profiles, at the Gibraltar–Ceuta section, of the upper layer depth for reduced gravity flow that is critical at Tarifa Narrows, ignoring friction, with vertical side walls and with a particular choice of potential vorticity (from Bormans and Garrett, 1989a). Observed depths of the 37.0 and 37.5 ‰ isohalines are also shown, LR for Lacombe and Richez (1982) and C for Cavanié (1973).

possibility and found that an internal head loss of more than $10m$ or so is unlikely to occur in the Strait itself.

Our conclusion is that Cavanié's data represent evidence for submaximal exchange in September 1971 and that the data of Lacombe and Richez (1982) correspond to maximal exchange in late May 1961.

More recent data have been obtained in the 1985–86 Gibraltar Experiment (Bryden and Kinder, 1988), though it failed to resolve the maximal/submaximal issue with any degree

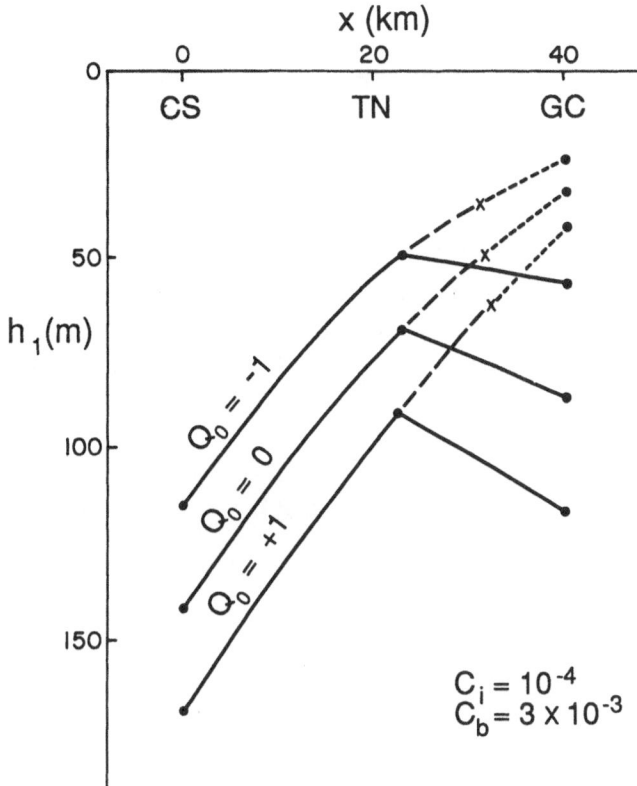

Figure 8. Predicted interface depth along the Strait from Camarinal Sill (CS) to Tarifa Narrows (TN) and the Gibraltar-Ceuta section (GC). C_i is the interfacial drag coefficient, C_b the bottom drag coefficient and Q_0 the barotropic transport (positive to the right). Long dashes indicate supercritical flow with finite upper layer depth at the north shore, short dashes flow that has separated from the north shore (from Bormans and Garrett, 1989b).

of certainty. Figure 9 shows the density structure, along the Strait, obtained by Wesson and Gregg (1988) in October 1985 and May 1986. Tidal effects have largely been averaged out, but some low-frequency variability may have been aliased into the spatial patterns.

We note the marked change from October 1985 to May 1986 in the depth of the 28.0 σ_θ isopleth which Wesson and Gregg took to represent the interface. The May 1986 inflow seems to have been supercritical, as also shown by the separation of the flow from the north shore of the Strait in April (Farmer and Armi, 1988), but the interface depth in October 1988 seems incompatible with supercritical flow.

A more continuous record of interface depths in mid-strait at the eastern end of the Strait (Figure 10) has been inferred by Candela (personal communication) from temperature

284

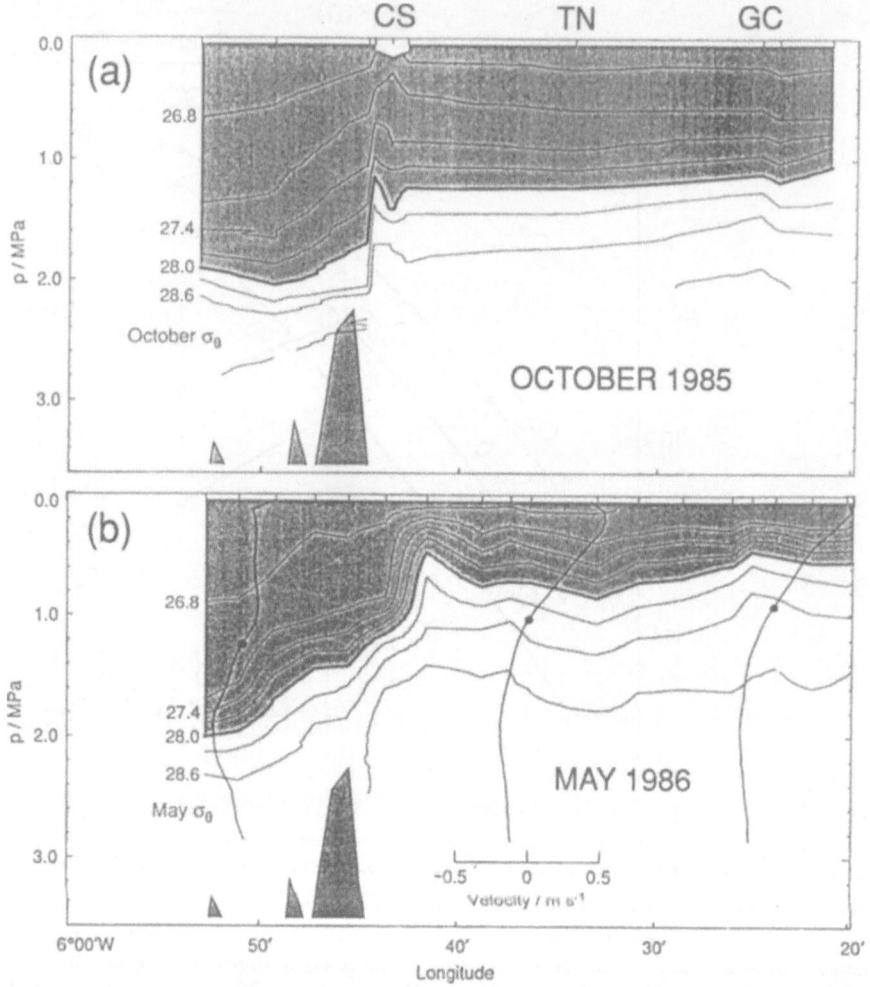

Figure 9. Tidally–averaged depth of the interface (represented by $\sigma_\theta = 28.0$), along the centerline of the Strait of Gibraltar, in October 1985 and May 1986 (from Wesson and Gregg, 1988).

sensors on a current meter mooring there during the Gibraltar Experiment. There are substantial low-frequency fluctuations, but there appears to have been a significant decrease in the mean depth, from more than $120m$ to less than $80m$, in late December, 1985. A similar, but smaller, change is discernible in the mean interface depth at the sill. These

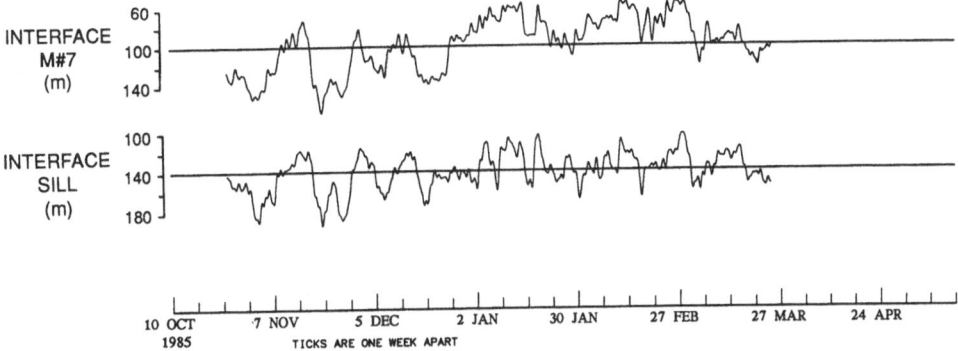

Figure 10. The low–passed depth of the interface (37.5% isohaline) at the eastern end (M7) of the Strait of Gibraltar and at Camarinal Sill (Candela, personal communication, 1989).

changes are compatible with the data of Wesson and Gregg (1988), suggesting a switch from predominantly submaximal exchange to predominantly maximal at the end of 1985.

Before leaving Figure 10 we note that the mean interface depth at Camarinal Sill, in early 1986 when it seems that the exchange was mainly maximal, was about $130m$, not much different from the prediction of $143m$ for $Q_0 = 0$ shown in Figure 8.

Our conclusion at this point, based on interface depth data from the eastern part of the Strait, is that the exchange has a tendency to be maximal in the early part of the year, submaximal in the later part of the year.

6. The Seasonal Cycle from Sea Level Data

Bormans et al. (1986) examined records of monthly mean sea level from the vicinity of the Strait of Gibraltar. In particular, they extracted the seasonal cycle for Ceuta (from data for 1950 to 1964) and Gibraltar (from data for 1962 to 1975). The data had been adjusted for air pressure and so the difference represented the seasonal cycle in total pressure across the Strait or, assuming geostrophy, the surface inflow through the Strait. Bormans et al. (1986) found a definite seasonal pattern, with more surface inflow than average in the first half of the year, less in the second half. They were unable to account for this except by assuming subcritical flow and allowing the interface depth at the Gibraltar–Ceuta section to vary.

Using the subcritical solutions of the Farmer and Armi (1986) model, and assuming marginally submaximal exchange when the upper layer was thinnest, they found a roughly sinusoidal pattern (shown in Figure 11, but using the Bormans and Garrett (1989b) model rather than the Farmer and Armi (1986) model), with the upper layer thinnest in March, thickest in August. After removing the effect of seasonally-varying wind stress by regression analysis (and checking that the coefficients were dynamically plausible) they found a

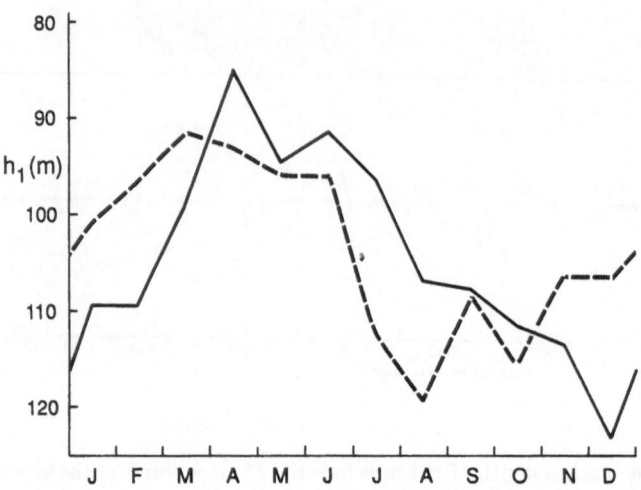

Figure 11. The seasonal cycle in the interface depth in mid–strait at the Gibraltar–Ceuta section assuming submaximal exchange (marginally submaximal in March). The dashed line is the prediction using the seasonal cycle in sea level difference across the strait and the seasonal density cycle (Bormans et al., 1986) in the model of Bormans and Garrett (1989b); the solid line is the residual after removing the effect of wind.

residual sawtooth pattern (Figure 11). Their interpretation of this was that the rise of the interface in late winter represents replenishment of the reservoir of Mediterranean Water by deep convection, and that the subsequent gradual fall-off showed draining of this dense water at the times of the year when it is not being formed. The magnitude of this sawtooth is nearly $40m$, somewhat greater than the change of $13m$ that would occur in 8 months by draining a surface area of $2.5 \times 10^{12} m^2$ at a rate of 1.5Sv, but not completely out of line; perhaps only part of the surface area is involved. The physical plausibility of this result seemed to lend support to the assumption of submaximal exchange, with the seasonal cycle following the scenario illustrated in Figure 5c.

More recently, Manzella and LaViolette (1989) have used hydrographic data from the Western Mediterranean to determine the seasonal pattern in the depth of the interface, defined by a salinity of 38.45 ‰, between Modified Atlantic Water and Mediterranean Water. The mean depth of about $300m$ would perhaps suggest submaximal exchange, but the seasonal cycle they find is out of phase with that deduced by Bormans et al. (1986). This is puzzling.

7. MEDALPEX Sea Level Data

Garrett et al. (1989) have analysed thirteen months of hourly heights collected at Ceuta, Gibraltar, Algeciras and Tarifa (Figure 1) from September 1981 through September 1982 as part of MEDALPEX (The Mediterranean-Alpine Experiment). In particular, after making necessary corrections for air pressure, they compared the low frequency fluctuations in the pressure head T–G along the Strait from Tarifa to Gibraltar with the pressure head C–G across the Strait, from Ceuta to Gibraltar, which represents the surface inflow. These two series were positively correlated; after removal by regression of the effect of wind the correlation coefficient is 0.67 and the ratio of C–G to T–G for the first empirical orthogonal function is 1.7 (Figure 12).

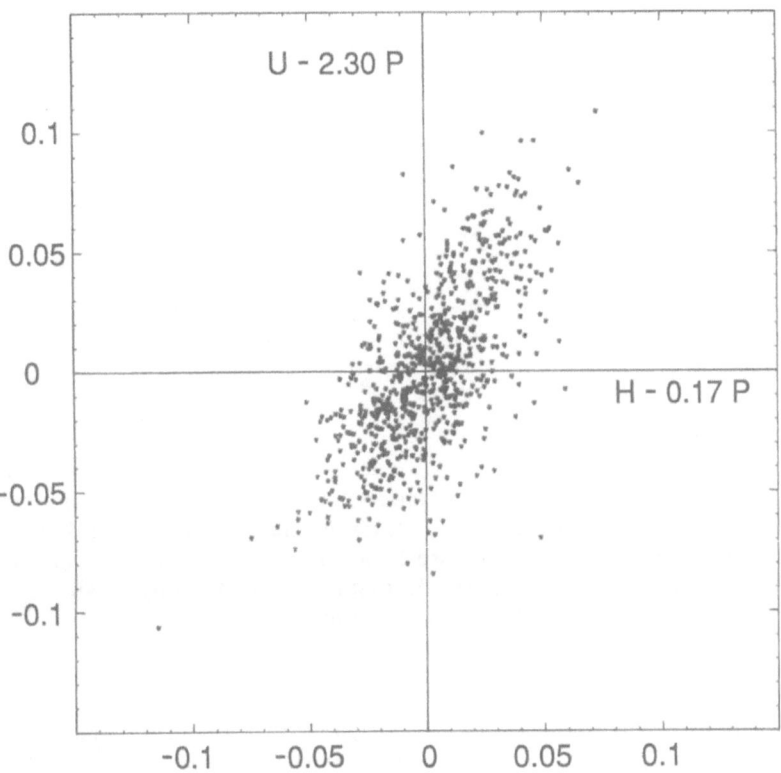

Figure 12. Scatter plot of the low passed residual sea level difference along the strait (x-axis) and across the strait (y-axis) after removing the effect of the along–strait atmospheric pressure gradient (Garrett et al., 1989).

288

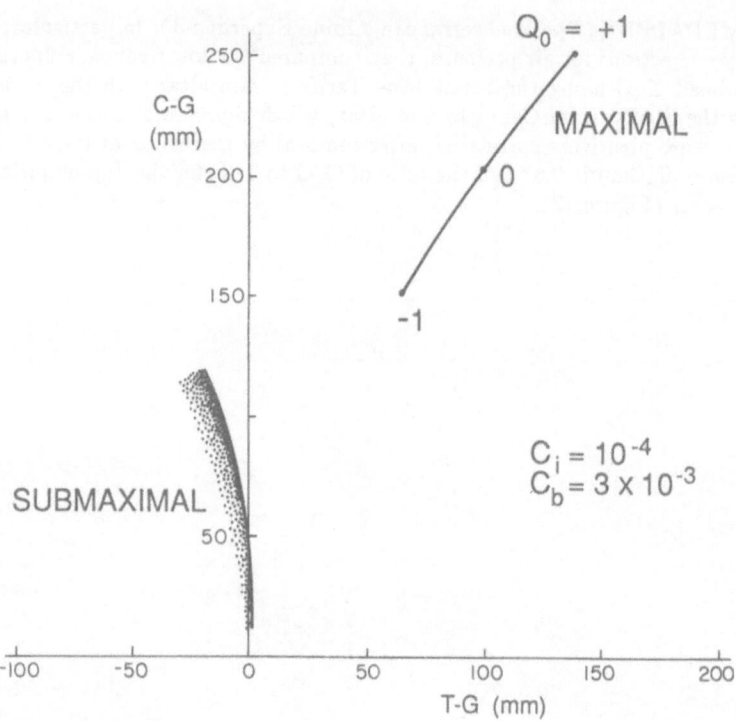

Figure 13. Predictions (Bormans and Garrett, 1989b) of the sea level difference along the strait (T–G) and across the strait (C–G) for maximal and submaximal exchange. For maximal exchange the changes are caused by a varying barotropic flow Q_0, with a range of $\pm 1 Sv$. For submaximal exchange the fluctuations are induced by varying Q_0 (over the range of $\pm 1 Sv$) or by varying the interface depths at the eastern end of the Strait, and hence occupy a region of the graph rather than just a line.

The points on this scatter plot may be compared with the model predictions of Bormans and Garrett (1989b) for maximal or submaximal exchange. Figure 13 shows that, for changes induced by fluctuations in the barotropic flow Q_0, the ratio of C–G to T–G fluctuations should be about 1.4 for maximal exchange (and would reach 1.7, in fact, if the internal drag coefficient C_i were increased from 10^{-4} to 5×10^{-4}).

For submaximal exchange, fluctuations may be induced either by changes in the barotropic flow Q_0 or by changes in the depth of the interface at the eastern end of the Strait. Figure 13 shows the theoretical predictions with $C_i = 10^{-4}$ and $C_b = 3 \times 10^{-3}$, giving a large negative slope to the ratio of C–G to T–G fluctuations (T–G may be negative due to the Bernoulli effect as the water slows down from Tarifa to Gibraltar in subcritical flow). If C_i is increased the envelope of submaximal solutions opens up and develops a tendency to have a positive slope. However, for reasonable values of friction, and knowing that a control section can exist, Bormans and Garrett (1989b) and Garrett *et al.* (1989) have argued that the observed ratio of $(C–G)/(T–G)$ provides evidence for maximal exchange at the time of MEDALPEX.

8. Geodetic Levelling

Bormans and Garrett (1989b) have used their dynamical model to estimate the sea level drop from Atlantic to Mediterranean for both maximal and submaximal exchange. For maximal exchange they find a drop of $100mm$, mostly due to the Bernoulli effect rather than friction, whereas for marginally submaximal exchange the drop is only $30mm$. These estimates were based on a one-dimensional model, so for the gradient along the north shore one should add half the value of C–G from Figure 13 and possibly subtract an allowance for a coastal current along the coast of Spain outside the Strait. Very roughly, then, one might expect to see a sea level drop, along the coast of Spain, from outside to inside the Strait, of about $150mm$ for maximal exchange, $50mm$ for submaximal.

First order levelling along the coast of Spain, reported by Levallois and Maillard (1970) does in fact show a drop of mean sea level (for which they took the annual mean for 1950) of about $150mm$ from Cadiz to Malaga (Figure 14), the closest reference points to the ends of the Strait. Figure 14 also shows significant changes in mean sea level elsewhere where there is no obvious dynamical reason, so the accuracy of the data is perhaps not entirely adequate. However, after smoothing the data there does seem to be evidence for a 150 to $200mm$ drop in sea level. This seems to be more in accord with maximal than submaximal exchange.

On the other hand, we have found that the monthly mean sea level difference from Cadiz to Malaga (with a small adjustment for atmospheric pressure) varies by about $100mm$ in the course of a year, typically being higher in the first half of the year and lower in the second though with considerable interannual variability. The geodetic levelling data is thus better summarized as indicating a drop from Atlantic to Mediterranean of order $200mm$ early in the year, $100mm$ later in the year. The difference of $100mm$ between these two values is the same as between the $150mm$ and $50mm$ estimated earlier for maximal and submaximal exchange.

We conclude that the geodetic levelling data, in combination with monthly mean sea level data, suggest that the system is close to a state of maximal exchange and can change from maximal exchange in the early part of the year to submaximal later, but with significant interannual variability.

9. The Alboran Gyre

The surface inflow through the Strait of Gibraltar tends to separate from both shores and form a large gyre in the Alboran Sea (e.g. Donde Va Group, 1984). It seems that this gyre can occasionally disappear, and perhaps be replaced by a coastal current which hugs the south shore (Cheney and Doblar, 1982; Kinder, 1984; Perkins *et al.*, 1989).

Bormans and Garrett (1989c) have suggested that the criterion for non-separation is that the inertial radius u/f of the inflowing current with speed u should be less than the radius of curvature of the boundary and that the relevant radius of curvature is that where the

290

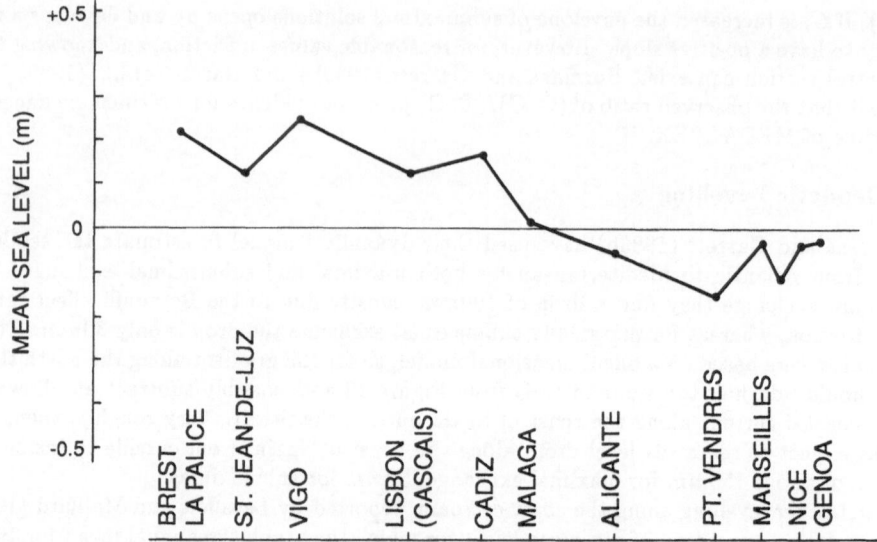

Figure 14. Mean sea level along the coast through the north shore of the Strait of Gibraltar as obtained by geodetic levelling (Levallois and Maillard, 1970).

interface between upper and lower layers intersects the sea floor. For the Strait of Gibraltar they estimate a critical speed of $0.4ms^{-1}$, too low for supercritical surface inflow even with a net barotropic outflow from the Mediterranean to the Atlantic. On the other hand, the surface current for a marginally submaximal exchange is $0.6ms^{-1}$ and, while this should still give a gyre, a speed of $0.4ms^{-1}$ could be achieved by a barotropic outflow of $0.5Sv$ or more or by a lowering of the interface at the eastern end of the Strait to more than $110m$ or so.

These ideas and numbers are still very speculative, but it does seem that the presence of a gyre could be associated with either maximal or submaximal exchange, but that the replacement of the gyre by a coastal current is an indication of submaximal exchange.

10. Summary

We have discussed a number of pieces of evidence that bear on the issue of maximal or submaximal exchange. Table 1 is an attempt to summarize them. We see that, as was clear from our earlier discussion, the evidence is mixed and even contradictory. It seems quite likely that the hydraulic nature of the flow through the Strait flips between the two states, but in a way that we do not yet understand.

One mechanism which might tend to push the system towards submaximal exchange is the draining of the reservoir of Mediterranean Water. We see, in fact, some evidence for maximal exchange early in the year, submaximal later. It may be that the average seasonal cycle discussed by Bormans et al. (1986) reflects a situation in which the system has a higher probability of maximal exchange early in the year, after wintertime replenishment of the reservoir of dense water, and a higher probability of submaximal exchange later in

Evidence	Reference	Maximal	Submaximal
Interface Depths	Lacombe and Richez (1982)	May 1961	September 1971
	Cavanié (1973)		October 1985
	Wesson and Gregg (1988)	May 1986	
	Candela (pers. comm.)	Jan.,Feb.,Mar. 1986	Oct.,Nov.,Dec. 1985
Seasonal cycle	Bormans et al. (1986)		1950–1975
MEDALPEX Sea Level Gradients	Garrett et al. (1989)	Sept. '81–Sept. '82	
Land Levelling	Levallois and Maillard (1970)	First half of year	Second half of year
Gyre	Kinder (1984)		12–22 Sept. 1982

Table 1. Summary of the evidence for maximal or submaximal exchange at different times.

the year after some of this water has drained away. It is unfortunate that the overlap of sea level records from Gibraltar and Ceuta is only 3 years so that the issue of interannual variability could not be addressed.

One would certainly expect significant variability in the amount of dense water formed in the winter. Quite possibly the MEDALPEX sea level data, which Garrett et al. (1989) claim to show evidence for year-long maximal exchange, occurred after a year or two of more than average dense water formation. It would be worthwhile to examine meteorological records for any evidence of this.

Our tentative conclusion is that the regime is close to maximal exchange; the second scenario in Figure 5 may be the most appropriate, but with only a year or two of no $E - P$ required for the narrows control to be flooded, so the implications for monitoring are more like those implied by the third scenario of Figure 5. It certainly seems worthwhile for efforts to be made to develop a suitable monitoring scheme for the Strait of Gibraltar, and possibly also for the Strait of Sicily. Monitoring schemes should go hand-in-hand with further investigation of the hydraulic nature of the exchange.

In fact, the conclusion reached in this paper, that the exchange typically switches between maximal and submaximal in the course of the year, implies that there will be periods when the exchange is maximal but the inflow jumps back from supercritical to subcritical within the Strait itself rather than east of Gibraltar. It would be exciting to discover such a transition, though it would probably be complicated by the tides. According to Pratt's (1987) investigation of hydraulic jumps in a rotating system, one signature might be the sudden reattachment of the upper layer to the north shore east of a point of separation.

In general, measurements of sea level gradients, upper layer depths and currents in the eastern part of the Strait would be valuable. A careful geodetic survey, along the north and south shores of the Strait, tied in to sea level measurements at permanent tide gauges, is also crucial.

Acknowledgements

We have benefited from discussions with many colleagues in the Gibraltar Experiment and from the financial support provided by the U.S. Office of Naval Research and Canada's Natural Sciences and Engineering Research Council.

References

Armi, L. and D. Farmer (1987) A generalization of the concept of maximal exchange in a strait. *J. Geophys. Res.*, *92*, 14,679-14,680.

Armi, L. and D. Farmer (1988) The flow of Mediterranean water through the Strait of Gibraltar. *Progress in Oceanography*, *21*, 1-105.

Assaf, G. and A. Hecht (1974) Sea straits: a dynamic model. *Deep-Sea Res.*, *21*, 947-958.

Bethoux, J.-P. (1979) Budgets of the Mediterranean Sea. Their dependence on the local climate and on the characteristics of the Atlantic waters. *Oceanologica Acta*, *2*, 2, 157-163.

Bormans, M., C. Garrett and K.R. Thompson (1986) Seasonal variability of the surface inflow through the Strait of Gibraltar. *Oceanologica Acta*, *9*, 403-414.

Bormans, M. and C. Garrett (1989a) The effect of rotation on the surface inflow through the Strait of Gibraltar. *J. Phys. Oceanogr.*, *19*, 1535-1542.

Bormans, M. and C. Garrett (1989b) The effects of non–rectangular cross–sections, friction and barotropic fluctuations on the exchange through the Strait of Gibraltar. *J. Phys. Oceanogr., 19,* 1543-1557.

Bormans, M. and C. Garrett (1989c) A simple criterion for gyre formation by the surface outflow from a strait, with application to the Alboran Sea. *J. Geophys. Res., 94,* 12,637-12,644.

Bryden, H. and T. Kinder (1988) Gibraltar experiment: a plan for dynamic and kinematic investigations of strait mixing, exchange and turbulence. *Oceanologica Acta,* SP, 29-40.

Bryden, H.L. and H.M. Stommel (1984) Limiting processes that determine basic features of the circulation in the Mediterranean Sea. *Oceanologica Acta, 7,* 289-296.

Cavanié, A. (1973) Observations océanographiques dans le détroit de Gibraltar pendant la Campagne Phygib (septembre-octobre 1971). *Ann. Hydrogr., 5e ser., 1,* 1, 75-84.

Cheney, R.E. and R.A. Doblar (1982) Structure and variability of the Alboran Sea frontal system. *J. Geophys. Res., 87,* 585-594.

Donde Va Group (1984) Donde Va? An oceanographic experiment in the Alboran Sea. *Eos Trans. AGU, 65*(36), 682-683.

Farmer, D.M. and L. Armi (1986) Maximal two–layer exchange over a sill and through the combination of a sill and contraction with barotropic flow. *J. Fluid Mech., 164,* 53-76.

Farmer, D. and L. Armi (1988) The flow of Atlantic water through the Strait of Gibraltar. *Progress in Oceanography, 21,* 1-106

Garrett, C., J. Akerley and K.R. Thompson (1989) Low frequency fluctuations in the Strait of Gibraltar from MEDALPEX sea level data. *J. Phys. Oceanogr., 19,* 1682-1696.

Kinder, T.H. (1984) Net mass transport by internal waves near the Strait of Gibraltar. *Geophysical Research Letters, 11,* 987-990.

Lacombe, H. and C. Richez (1982) The regime of the Strait of Gibraltar. In: *Hydrodynamics of Semi–enclosed Seas.* Ed. J.C.J. Nihoul, Elsevier, 13-73.

Levallois, J.J. and J. Maillard (1970) The New French 1st order levelling net. Practical and scientific consequences. Report on the Symposium on Coastal Geodesy, Munich 1970, 644 p.

Manzella, G.M.R., G.P. Gasparini and M. Astraldi (1988) Water exchange between the eastern and western Mediterranean through the Strait of Sicily. *Deep-Sea Res., 35,* 1021-1035.

Manzella, G.M.R. and P.E. LaViolette (1989) The relation of the transport through the Strait of Gibraltar and the seasonal transportation of LIW through the Strait of Sicily. *J. Geophys. Res.* (in press).

Perkins, H., T.H. Kinder and P.E. LaViolette (1989) The Atlantic inflow in the western Alboran Sea. *J. Phys. Oceanogr.,* (in press).

Pratt, L.J. (1987) Rotating shocks in a separated laboratory channel flow. *J. Phys. Oceanogr., 16,* 1970-1980.

Phillips, O.M. (1966) On turbulent convection currents and the circulation of the Red Sea. *Deep-Sea Res., 13,* 1149-1160.

Wesson, J.C. and M.C. Gregg (1988) Turbulent dissipation in the Strait of Gibraltar and associated mixing. In *Small–Scale Turbulence and Mixing in the Ocean,* Proceedings of the 19th International Liège Colloquium on Ocean Hydrodynamics. J.C.J. Nihoul and B.M. Jamart, eds., 201-212, Elsevier.

Whitehead, J.A., A. Leetmaa and R.A. Knox (1974) Rotating hydraulics of strait and sill flows. *Geophys. Fluid Dyn., 6,* 101-125.

Wüst, G. (1961) On the vertical circulation of the Mediterranean Sea. *J. Geophys. Res., 66,* 3261-3271.

ASPIRATION OF DEEP WATERS THROUGH STRAITS

THOMAS H. KINDER[1]
Department of Oceanography
United States Naval Academy
Annapolis, Maryland, 21402
United States of America

HARRY L. BRYDEN
Department of Physical Oceanography
Woods Hole Oceanographic Institution
Woods Hole, Massachusetts, 02543
United States of America

ABSTRACT. In 1973 Stommel, Bryden and Mangelsdorf conjectured that high speed shallow flow in the Strait of Gibraltar is capable of sucking deep Mediterranean Water from the adjacent Alboran Sea directly up and over the sill into the Atlantic Ocean. This mechanism, which we call Bernoulli aspiration, has since been demonstrated in laboratory models, and the direct outflow has been confirmed by field experiment. Laboratory modeling, numerical experiment, and field measurement also have shown that the upstream path of the outflowing deep water is constrained by the combination of rotation and topography to form a narrow boundary current against the African coast. Data that was recently acquired during the Gibraltar Experiment (1985–1986) is used to describe the direct outflow of deep water and its distribution upstream of the sill.

We hypothesize that the uplift and direct outflow of the deep water from the Mediterranean Sea through the Strait of Gibraltar is a mechanism which may be important in other strait and semi-enclosed sea systems. A key diagnostic for such flows is the upstream intensification of the deep flow as it approaches the sill.

1. Introduction

Straits influence the adjacent ocean through several mechanisms. One of the most interesting is aspiration in which the deep waters of a semi-enclosed sea flow directly up and over a shallow sill and then down into the ocean. This mechanism is inherently fascinating because of its paradoxical quality: the same geomorphic constriction that traps the deep water also permits its escape. Aspiration is more than a curiosity, however, as the direct outflow of the deep waters affects the abyssal circulation of the semi-enclosed sea as well as the circulation of the adjacent ocean. Such flows have obvious practical importance for such issues as the fate of long-lived pollutants.

Henry Stommel and colleagues (Stommel, Bryden and Mangelsdorf, 1973) proposed that such a mechanism operates at the Strait of Gibraltar. They conjectured that the deep water in the western Mediterranean could be sucked over the sill by the fast flowing shallow flow: Bernoulli aspiration. Over the past decade, several groups of oceanographers working in the

---

[1] Also with the Office of Naval Research, 800 North Quincy Street, Arlington, Virginia, 22217, USA.

L. J. Pratt (ed.), The Physical Oceanography of Sea Straits, 295–319.
© 1990 *Kluwer Academic Publishers.*

vicinity of the Strait of Gibraltar have confirmed the direct outflow, and laboratory models suggest that Bernoulli aspiration is the operative mechanism. The Strait of Gibraltar is a paradigm for an interaction between straits and semi-enclosed seas that could operate elsewhere in the world ocean.

Observations have shown that the outflowing deep water forms a narrow current on its left hand side (looking downstream) as it approaches the strait. Laboratory and numerical modeling explain this effect in terms of vorticity conservation of a water column forced over a shoaling bottom. This boundary current of the outflowing deep water may be a useful diagnostic for the detection of aspiration, as vigorous mixing and energetic temporal variations near sills make direct observations of outflow difficult. Kinder and Parrilla (1987) showed hydrographic data evidence for direct outflow using early Gibraltar Experiment results (Kinder and Bryden, 1987), and geochemical tracer measurements have also suggested a component of deep water in the Gibraltar outflow (Roether and Weiss, 1975; Measures and Edmond, 1988). In this paper, more complete data are presented, confirming the earlier results and suggesting that the outflow occurs regularly.

First, the oceanographic background of the Strait of Gibraltar and its approaches is discussed. We will then show the observational evidence for aspiration, and review relevant laboratory and numerical modeling results. Finally, we suggest that this mechanism may be important elsewhere.

2. Oceanographic Environment

The Strait of Gibraltar is a narrow (about 15 km) and shallow (about 300 m) constriction that separates the open Atlantic Ocean from the Mediterranean Sea (Fig. 1). The inflowing Atlantic Water is lower in salinity, and forms a surface layer of less than 200 m thickness in the Alboran Sea. Extending to the bottom beneath this layer are the two Mediterranean water types, Levantine Intermediate Water (LIW) and Western Mediterranean Deep Water (WMDW). While the distributions of properties are described in the literature (e.g. Lacombe and Tchernia, 1972; Lanoix, 1974; or Parrilla and Kinder, 1989), we review salient features of the water mass properties using recently acquired data (Figure 2). Careful examination of these properties provides subtle but unmistakable evidence of the direct outflow of the deep water (WMDW).

Two stations obtained in November 1985 show the contrasting properties within the strait. Station 503 was taken just inside the Alboran Sea (Fig. 3), and it shows the shallow Atlantic layer with salinities much less than 37.0, a shallow sharp pycnocline, and then a nearly uniform deeper layer. Station 108 was taken at the western end of the strait, and shows the fresher Atlantic layer with salinities below 36, and a deep Mediterranean layer with salinities above 38. Density and salinity contrasts are somewhat greater at this station than at 503.

The marked salinity minimum in the western station (108) is caused by North Atlantic Central Water (NACW; Lacombe and Tchernia, 1972) and it is usually prominent at neap tides in the western strait, but absent or barely discernible in the east at all stages of the tide. The upper-most water properties are highly variable: often the eastern stations have similar temperatures and salinities as the western stations, in contrast to these November

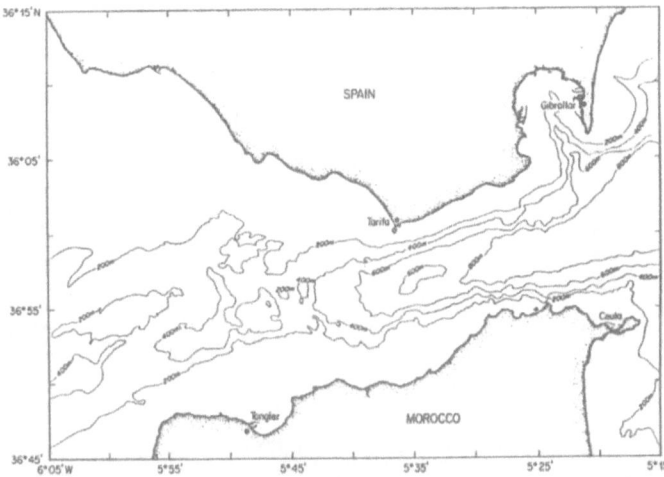

Figure 1: The Strait of Gibraltar, with bathymetry (m). To the east of the strait lies the western Alboran Sea, the western-most basin of the Mediterranean Sea. To the west of the strait lies the Gulf of Cadiz, an embayment of the North Atlantic Ocean. The shallowest sill is north of Point Malabata, and is slightly less than 300 m deep. The eastern approaches to the strait exceed 800 m depth with a U-shaped cross section, while the western approaches are generally shallower than 400 m except for a narrow channel with a V-shaped cross section. (Figure is after IGN–SECEG, 1988).

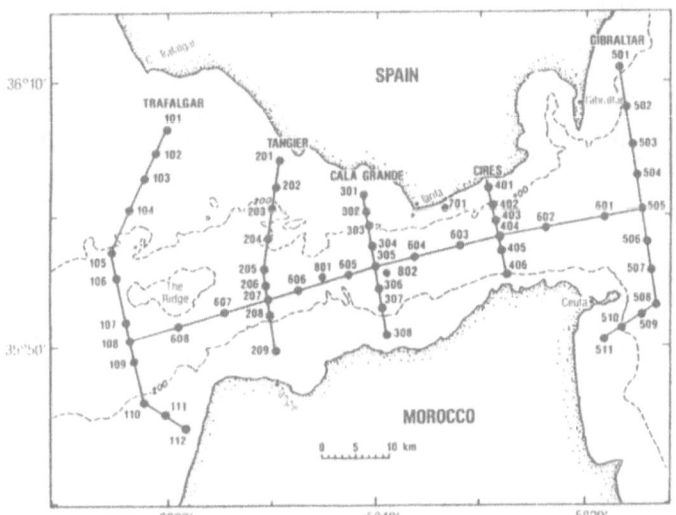

Figure 2: The station plan for CTD (conductivity-temperature-depth profiler) surveys during the 1985–1986 Gibraltar Experiment. The primary sill is between stations 606 and 801. Two major motivations for this station pattern were to provide high resolution in the cross-strait direction and to avoid the large instrument array which was moored in the strait concurrently.

298

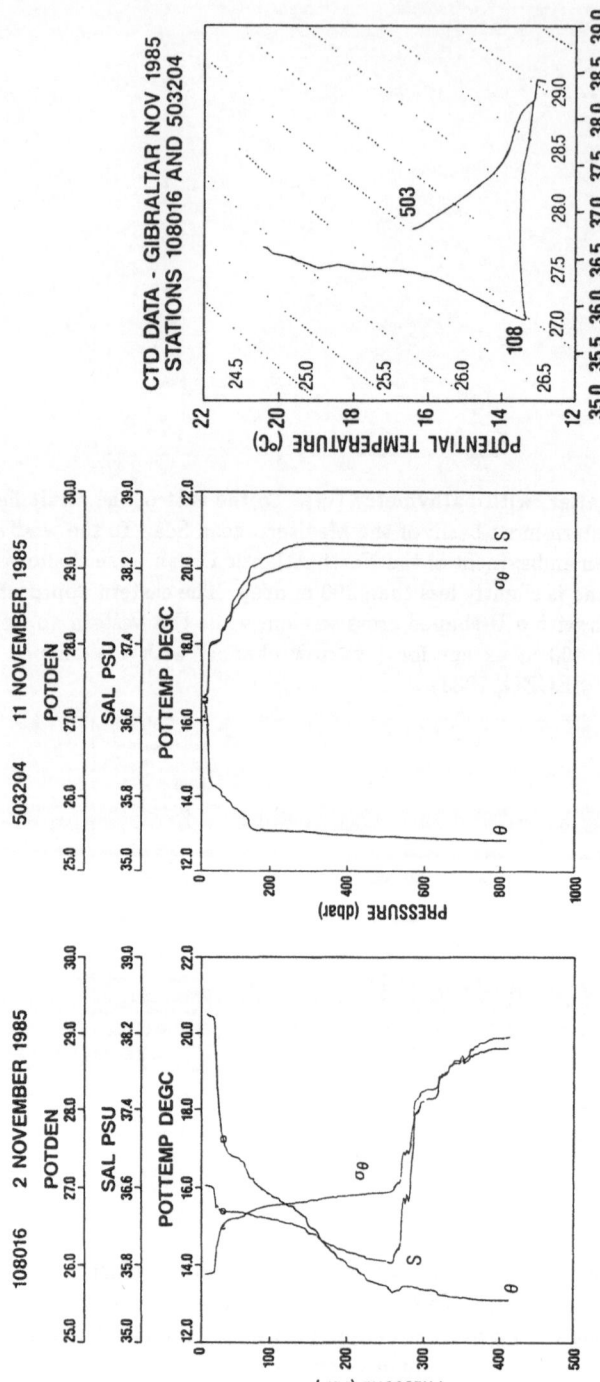

Figure 3: Hydrographic data illustrating the water mass properties along the strait axis. Vertical profiles of potential temperature, salinity and potential density are presented for stations 108 (a) and 503 (b). Station 108 was taken on the Trafalgar line in the western approaches, and station 503 on the Gibraltar line in the eastern approaches (Fig. 2). Station 108 was taken near neap tide, when the salinity minimum is strongly expressed. The potential temperature–salinity curves (c) for the two stations exhibit the water mass characteristics in the Strait.

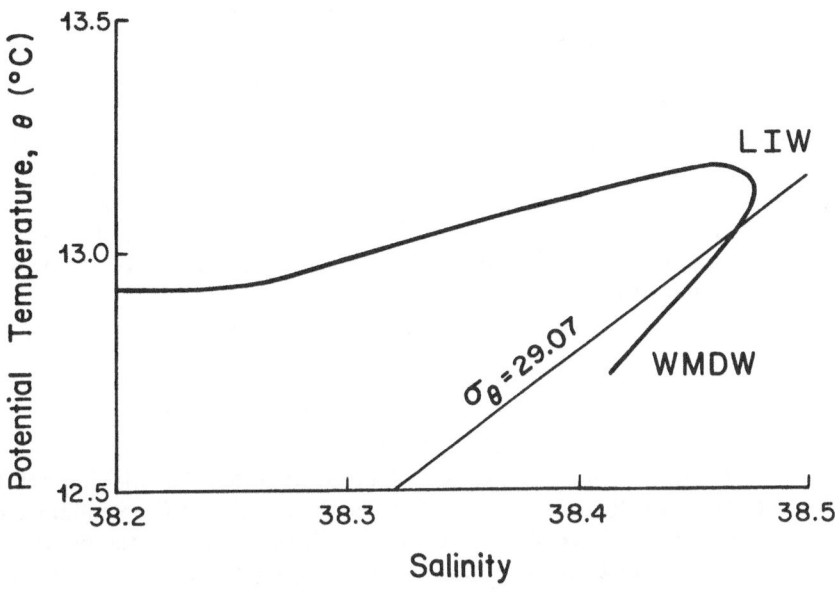

Figure 4: Schematic T–S correlation of the Mediterranean Water in the Alboran Sea. The Levantine Intermediate Water (LIW), formed in the eastern Mediterranean, is characterized by intermediate maxima in temperature at about 13.2°C and in salinity at about 38.5 with a potential density of about 29.0 kg m^{-3}. The Western Mediterranean Deep Water (WMDW), formed off the southern coast of France, has temperatures less than 12.9°C and salinities of 38.42 to 38.44 in the deep water. LIW is found typically at depths above the Gibraltar Sill, while the WMDW is typically below sill depth.

data. The deeper Mediterranean water, with salinities above 38, however, shows spatial differences that are robust with respect to time of year and tidal phase. In the east, salinities exceed 38.4, potential temperatures are below 13°C, and densities approach 29.1 kg/m^3. In the west, these values are never found together: there, the deep water is fresher, warmer and lighter as a result of mixing during its transit through the strait.

In order to examine the subtle differences within the Mediterranean Water, we focus on the deepest part of the water column. (Note: We use the modern salinity definition and equation of state, which lowers calculated values in the Mediterranean Water by 0.006–0.007 in salinity and by 0.02 kg/m^3 in density. See Parrilla, 1984, for a discussion.) The classical LIW and WMDW (as expressed in the western Alboran Sea) appear only in the right part of the T–S plane, at densities greater than 29 kg/m^3 (Fig. 4). The Mediterranean Water at station 108 contains no water of this density, although it was located just 30 km west of the Gibraltar sill. This rapid eradication of the subtle characteristics of the WMDW and LIW is the reason for the long delay from the publication of the hypothesis of a direct outflow of deep water in 1973 and its confirmation with 1985 data (Kinder and Parrilla, 1987; there was much circumstantial evidence published between these dates, however, as we will discuss later). The salinity and temperature maxima at about 38.48 and 13.15°C mark the LIW, and the monotonic decrease of both salinity and potential temperature toward the bottom marks the WMDW. Adopting the definition of Bryden and Stommel (1982), the water with potential temperature below 12.9°C and salinity greater than 38.4 is regarded as WMDW.

An along-strait section taken in June 1986 (Fig. 5) illustrates the continuum of change represented by the two earlier stations. In the Alboran Sea (eastern end of the strait), there are deep temperatures below 12.90°C with salinities exceeding 38.4 and densities exceeding 29.0 kg/m³: clearly WMDW. By the time that the Mediterranean water has reached the longitude of Trafalgar (station 108), the maximum salinity has been reduced to < 38.35 and the minimum potential temperature is above 12.95°C.

3. Bernoulli Aspiration

The model put forward by Stommel *et al.* (1973) was based on a very simple idealization of the Strait of Gibraltar and the Alboran Sea. They incorporated the following well-known facts: the WMDW normally resides below 700 m depth or so (12.90°C potential isotherm), whereas the sill depth is about 300 m; near the strait the speeds in the overlying LIW are high (e.g., Lacombe and Richez, 1982, summarize experimental data from the 1960's); and the density difference between the LIW and WMDW is small (about 0.02 kg/m³; see previous section). They then used a Bernoulli energy function and the hydrostatic relation to examine whether a deep streamline in the quiescent WMDW in the Alboran Sea would be uplifted to the sill depth (Fig. 6). Their argument shows that the WMDW could outflow ($v_1{}^2 > 0$) if:

$$v_1'^2 > 2g\,\delta(z_0 - z_1) - \mathcal{E}_0(z_0 - z_0') + \mathcal{E}_1(z_1 - z_1')$$

where

$$\delta = \frac{\rho_0 - \rho_0'}{\rho_0'}\ , \quad \mathcal{E}_0 = \frac{\overline{\rho}_0 - \rho_0'}{\rho_0 - \rho_0'}\ , \quad \mathcal{E}_1 = \frac{\overline{\rho}_1 - \rho_1'}{\rho_0 - \rho_0'}\ ;$$

that is, if the dynamic pressure reduction induced by the swift-flowing LIW at the strait overcomes the small stratification in the Mediterranean Water. $\mathcal{E}_0$ and $\mathcal{E}_1$ are of order 0.5–1.0 (as a function of density ratios); d is proportional to the relatively small density difference between the LIW and WMDM and is of order 10^{-5}; and v_1' is the speed of the LIW at the sill. Their estimates indicated that such a streamline exists, and they therefore argued that WMDW from about 1000 m depth might participate directly in the outflow.

The Bernoulli argument ignores three salient features of the physical regime in the strait. Their analysis assumes no friction, no rotation (and attendant meridional structure), and no temporal variability. In spite of the vigorous mixing within the strait (e.g. Wesson and Gregg, 1988), the short residence time of a parcel in the strait suggests that friction is not a zero-order consideration. Rotation does indeed play an important role in the path of the outflowing water (as we discuss in the section on boundary currents below), but does not affect the above energy argument. Finally, the vigorous temporal variability on very short scales might be expected to overwhelm the simple steady Bernoulli argument. It was known then that the currents vary on many time scales, and especially that they reverse at many locations with the semidiurnal tide, but this variability appears to be a secondary consideration. Next, the distribution of the Mediterranean Water within the Alboran Sea

301

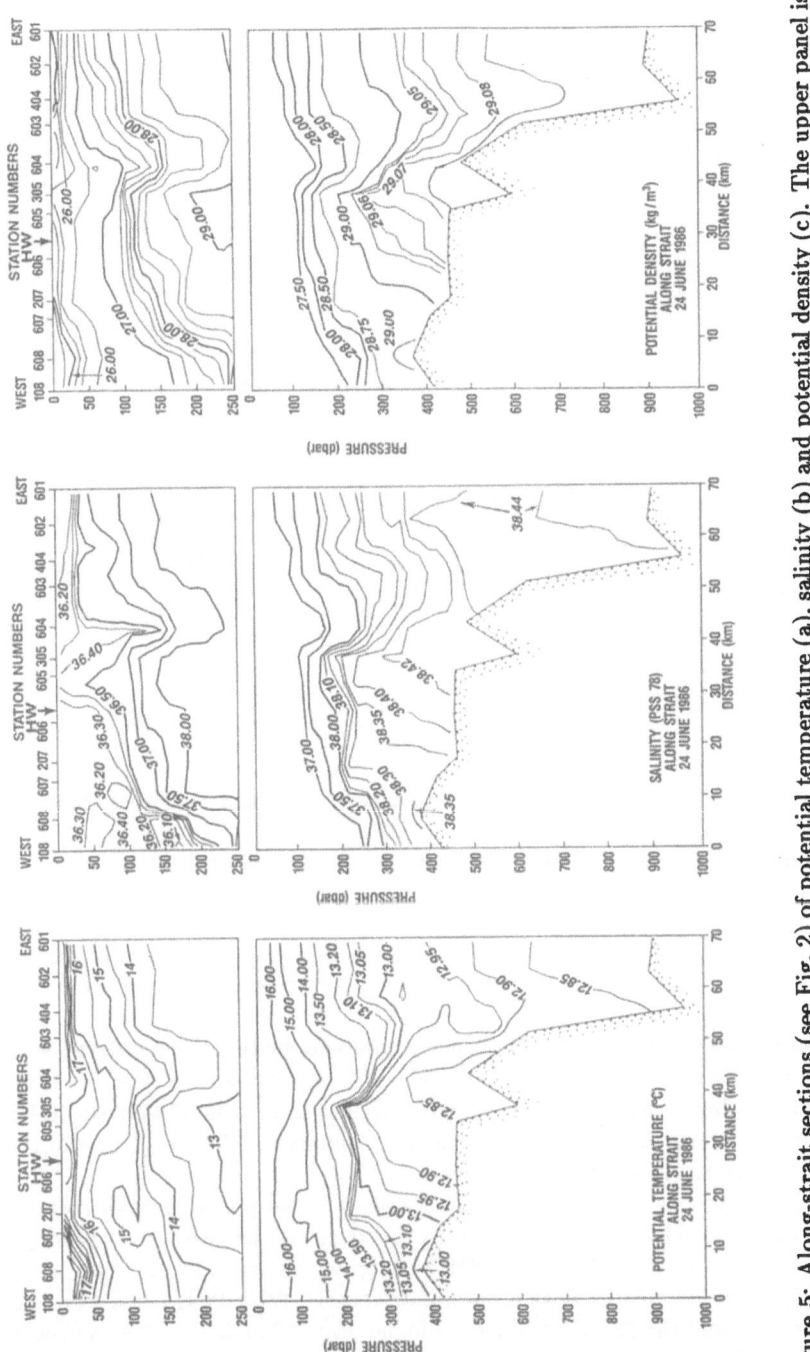

Figure 5: Along-strait sections (see Fig. 2) of potential temperature (a), salinity (b) and potential density (c). The upper panel is constructed to show the pycnocline and Atlantic Water clearly, while the lower panel is constructed to show the lower pycnocline and Mediterranean Water. Note that the choice of isolines is irregular in the lower panel in order to illustrate the subtleties of the Mediterranean water (e.g. Figs. 3 and 4). The bottom represents the depth observed at each station, and is *not* a typical along-strait bathymetric profile. The sill is between stations 605 and 606 at about 284 m depth. The section was occupied from west to east over a 10-hour period. The location of the ship at predicted high water is indicated by "HW".

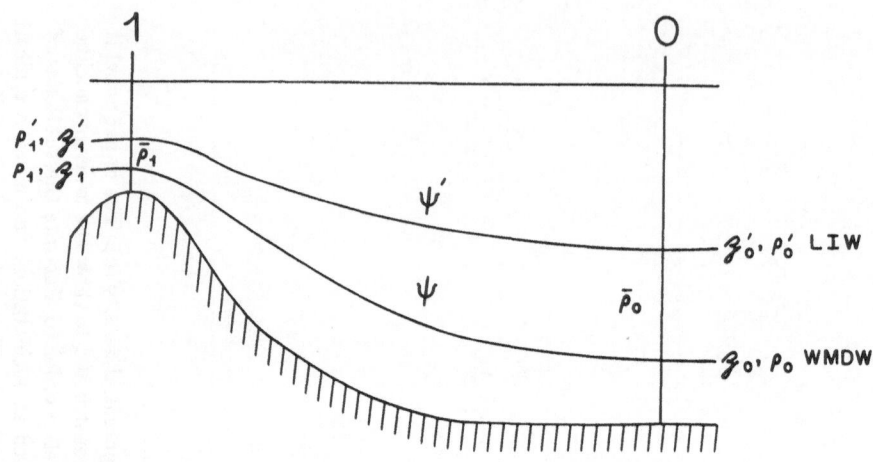

Figure 6: A schematic of Bernoulli aspiration in the Strait of Gibraltar and Alboran Sea. A zonal section is shown, with the upper streamline representing LIW (which can outflow directly) and the lower streamline representing WMDW (which must be 'sucked' over the sill from depth). If the speed along the upper streamline is great enough, the lower streamline will be able to rise above the depth of the sill according to equations 1–3.

and the hydrographic evidence for direct outflow are discussed.

4. Mediterranean Water in the Alboran Sea and Confirmation of Direct Outflow

In 1973, adequate data did not exist to rigorously test the notion of direct outflow, although there was some hydrographic data that lent support to the idea. Stommel, Bryden and Mangelsdorf (1973) noted a single 1961 station that showed WMDW at the sill. Both an understanding of the path of WMDW to the sill and observations of WMDW west of the sill were required. Three different groups designed field programs in the Alboran Sea that included attempts to examine the direct outflow hypothesis. Bryden and Stommel (1982) combined a 1975 CTD (conductivity, temperature and depth profiler) survey with a 1979–1980 current meter mooring to examine WMDW flow. Gascard and Richez (1985) used 1981 CTD and subsurface drifters, while Parrilla, Kinder and Preller (1986) used CTD and current data from 1982. Taken together, their work showed a coherent picture of the WMDW and LIW flow in the western Alboran Sea.

In the western Alboran Sea (Fig. 7) the LIW shows clearly in deep hydrographic sections as a salinity maximum which occupies the northern two-thirds of the basin in a layer from about 200 to 600 m depth (Fig. 8). The WMDW occupies the entire basin beneath the LIW, mostly below 600–700 depth. In the south, however, the WMDW is banked against the African slope such that the 12.90°C isotherm is found near 400 m depth. A plan view of the depth of the 12.85°C isotherm and a year long current record show its distribution and

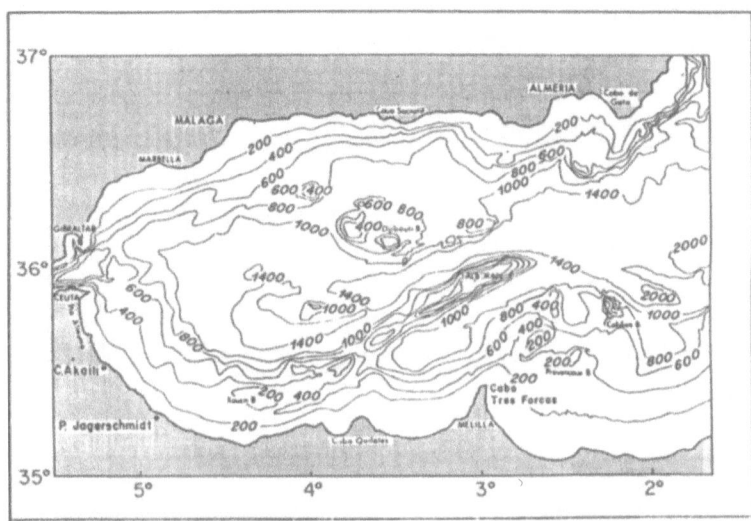

Figure 7: Bathymetry of the Alboran Sea, with depths in meters. North of Alboran Island, there is an unrestricted path for WMDW flowing westward from the adjacent Mediterranean Sea. LIW is free to flow toward the strait both north and south of the Island. (From Parrilla, Kinder and Preller, 1986.)

velocity (Fig. 9). The narrow vein of WMDW along the slope of Morocco is a westward flowing boundary current. The record had a mean velocity of 5 cm/s toward the strait entrance, and this pattern was confirmed by neutrally-buoyant floats deployed by Gascard and Richez (1985). A similar picture for the LIW shows the salinity maximum in the north, and three 110-day current records show flow toward the strait with a mean speed of 1–2 cm/s (Fig. 10). Pistek, de Strobel and Montanari (1985) also reported on subsurface drifter trajectories that fit into the inferred Mediterranean Water flow pattern, summarized in Figure 11.

During the Gibraltar Experiment, extensive hydrographic data were taken within the strait. Some of these stations (see Fig. 2) were taken to search for WMDW west of the sill using modern high quality CTD instruments and taking data within 10 m of the bottom. The cross-strait sections also showed that the meridional separation of the WMDW and LIW persists nearly to the sill.

A typical cross-strait section from the Cires line, about 20 km east of the sill, illustrates this separation (Fig. 12). Mediterranean Water is present throughout the section below 150 m depth, but the delineation between LIW and WMDW is still marked. Along the African side of the strait, low potential temperatures reveal WMDW banked steeply (vertical exaggeration is about 33:1 in the lower panel) against the slope. The salinity maximum, while less clearly delineated, is concentrated along the European slope, showing the preferred northern position of the LIW. All the meridional sections (Gibraltar, Cires and Cala Grande; see Fig. 2) obtained east of the sill show such WMDW–LIW meridional separation.

Kinder and Parrilla (1987) reported finding WMDW west of the sill at a single station in November 1985. Further analysis of this data, including a 25-hour CTD station taken about

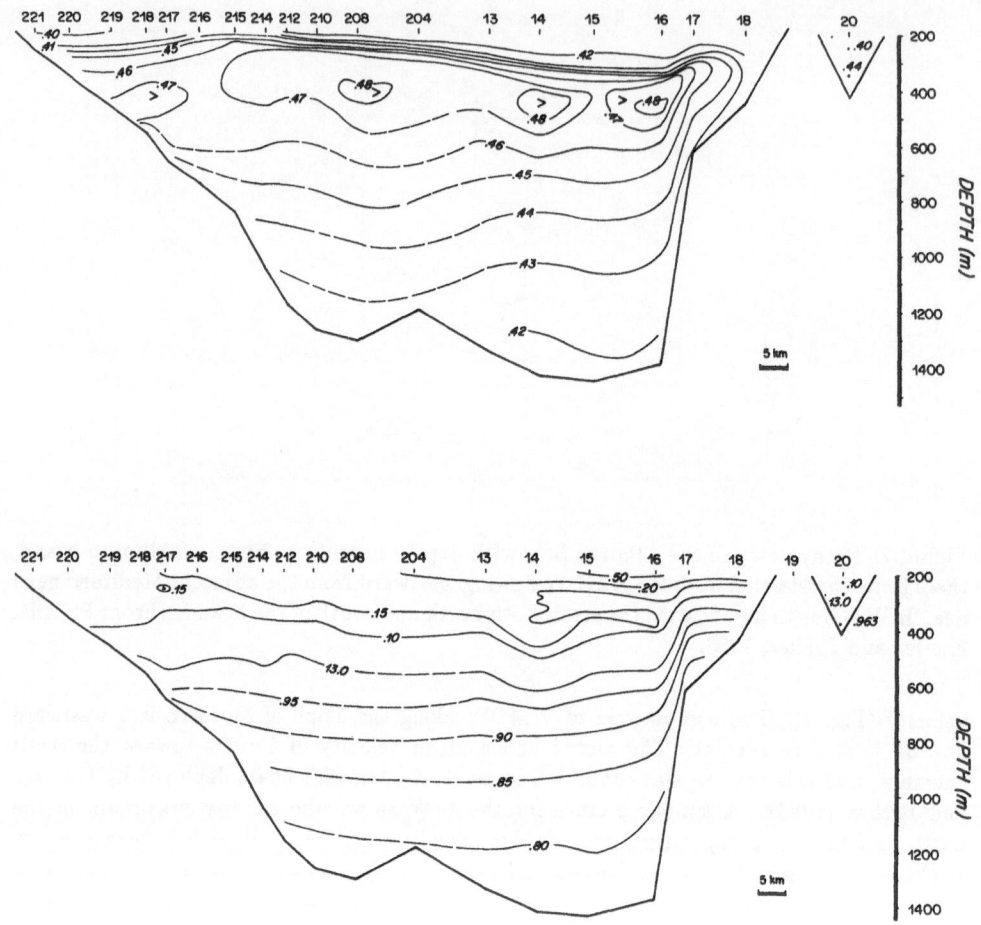

Figure 8: A 1982 meridional hydrographic section of salinity (a) and potential temperature (b) taken south of Málaga, showing the Mediterranean water (below 200 m). The salinity maximum reveals the LIW in a broad mid-depth plume over the northern three-quarters of the sea. Potential temperatures below 12.9°C show the WMDW at depth, but also banked against the Moroccan slope (where the LIW is absent). This meridional distribution of the LIW and WMDW is typical. (From Parrilla, Kinder and Preller, 1986.)

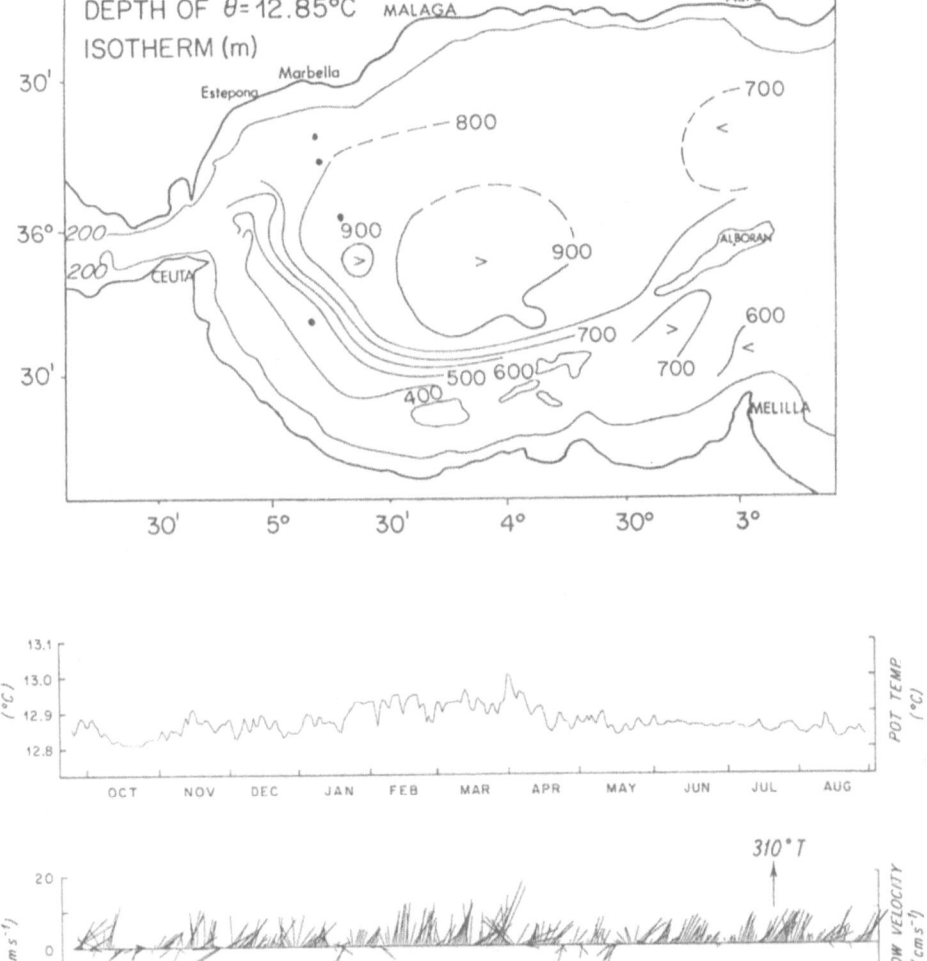

Figure 9: Distribution of the WMDW in the western Alboran Sea. (a) Depth (m) of the 12.85°C potential isotherm. Throughout most of the basin this isotherm lies below 800 m, but near the African coast it is uplifted to less than 400 m. The 12.90°C isotherm is about 50–90 m shallower. (From Parrilla, Kinder and Preller, 1986.) (b) Velocity and potential temperature from a 341-day current record at 500 m depth along the Moroccan slope (note large dot in a). The low-passed current vectors are rotated toward the strait (310°), and the temperature confirms that the instrument was in WMDW. (From Bryden and Stommel, 1982.)

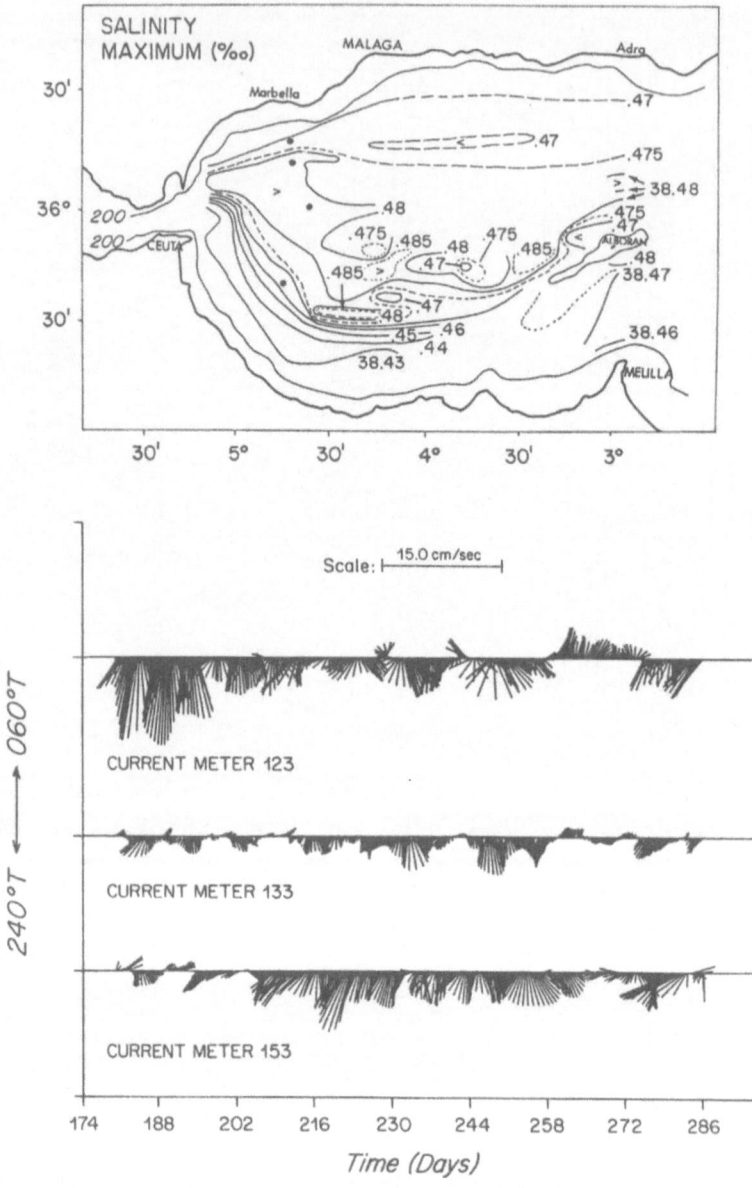

Figure 10: Distribution of the LIW in the western Alboran Sea. (a) The salinity maximum, showing the LIW in the northern part of the sea. (b) Three 110-day-long current meter records from 540 m depth south of Marbella (note three dots in a). The low-passed current vectors have been rotated toward the strait (about 240° for the middle mooring). (From Parrilla, Kinder and Preller, 1986.)

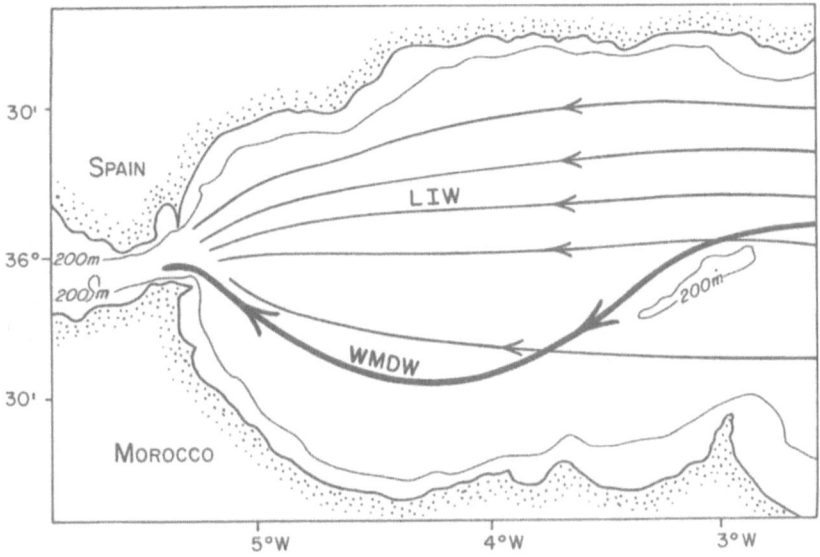

Figure 11: Schematic of the mean circulation of LIW and WMDW in the western Alboran Sea. The figure is a synthesis of the work by several teams of investigators.

1 km east of the south sill (Fig. 13), establishes WMDW at shallow depths adjacent to the sill. The strong semidiurnal and diurnal variability dominates the pycnocline movement. WMDW, as defined by the 12.90°C potential isotherm, is present on nearly every cast. During the time when the pycnocline is shallowest (e.g., slightly before predicted high water at Tarifa), the 12.90°C isotherm stands well above the greatest depth along the sill section (about 284 m). During the second high tide, the WMDW occupies over half the water column and extends nearly 150 m above the sill depth. Strong westward tidal flow occurs near high water (e.g. Lacombe and Richez, 1982), so this station suggests frequent WMDW outflow.

Because of the obvious tidal modulation of the vertical extent of the WMDW in the 1985 time series, finding significant WMDW west of the sill during springs seemed likely. Therefore, in June 1986 a time series station was taken about 5 km west of the sill during spring tides. The data (Fig. 14) show WMDW west of the sill, below sill depth, in a layer 100–200 m thick on both semidiurnal tides. The station was repeated over 4 semidiurnal cycles during neap tide (Fig. 15), and some WMDW is found during each cycle. The thickness and duration are both less than during neaps, but there is no ambiguity as to the presence of WMDW on the Atlantic side of the sill.

Near-bottom data taken during a pilot deployment in 1984 (Pettigrew, 1989; Fig. 16) showed a high correlation between westward flow and potential temperatures below 12.90°C. At the southern sill, nearly all cycles of westward outflow showed WMDW, whereas on the northern sill the potential temperature seldom dropped below 12.9°C. These data also suggest that some WMDW flows eastward across the sill when the tidal flow reverses,

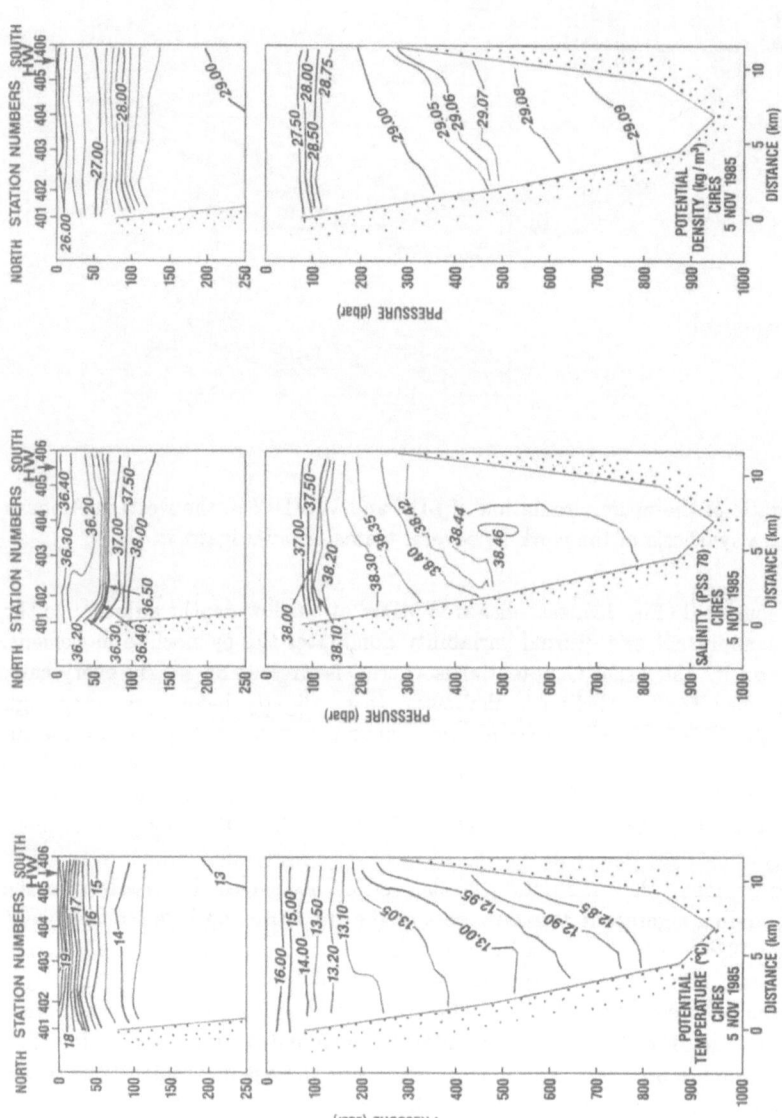

Figure 12: Cross-strait CTD section at Point Cires (see Fig. 2) of potential temperature (a), salinity (b) and potential density (c). The potential temperature shows the WMDW banked against the African slope, and the salinity shows the LIW in the northern part of the strait. All the 16 Cires sections obtained in November 1985 show a similar distribution, and the meridional separation is also clear in the Cala Grande section farther west. (See legend for Fig. 5 for explanation of plot.)

309

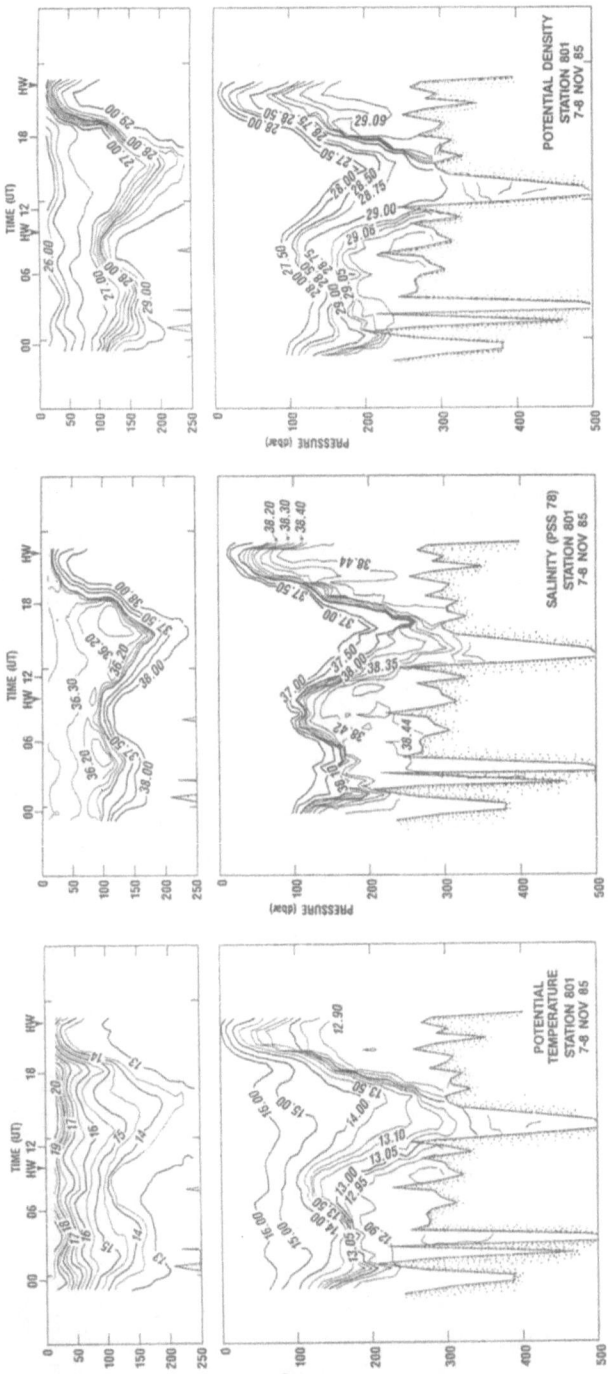

Figure 13: A 25-hour time series CTD station of potential temperature (a), salinity (b) and potential density (c), taken about 1 km east of the sill between neap and spring tides (Station 801; see Fig. 2). The WMDW stood more than 100 m above the sill depth of 284 m during both semidiurnal tides, as indicated by the 12.90°C isotherm. Note the large diurnal inequality. The ship remained within about 500 m of the designated station position for most of the casts, although some casts were about 1 km east of the designated position. The varying bottom depth indicates the positioning variability. (See Fig. 5 for further explanation of plot.)

310

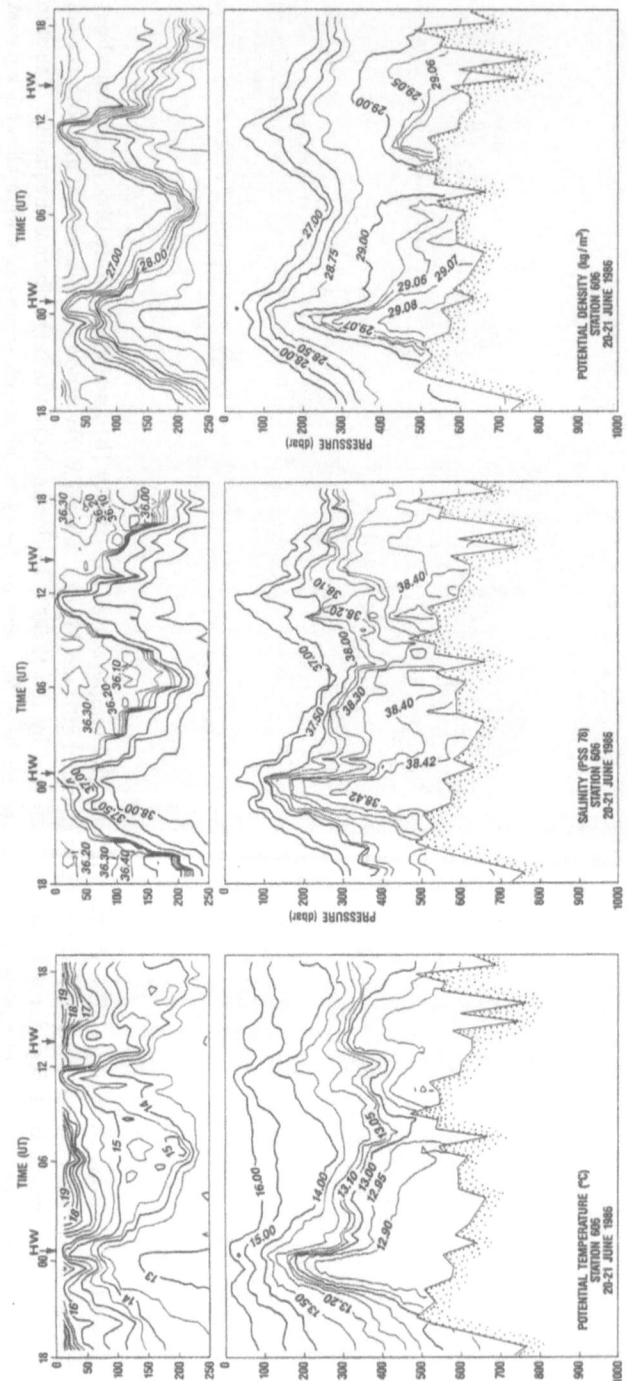

Figure 14: A 25-hour time series CTD station of potential temperature (a), salinity (b) and potential density (c) taken about 5 km west of the sill during spring tide (Station 606; see Fig. 2). The 12.90°C isotherm delineates WMDW in a layer over 100 m thick, west of the sill, and deeper than the sill depth. (See Figs. 5 and 13 for an explanation of plot.)

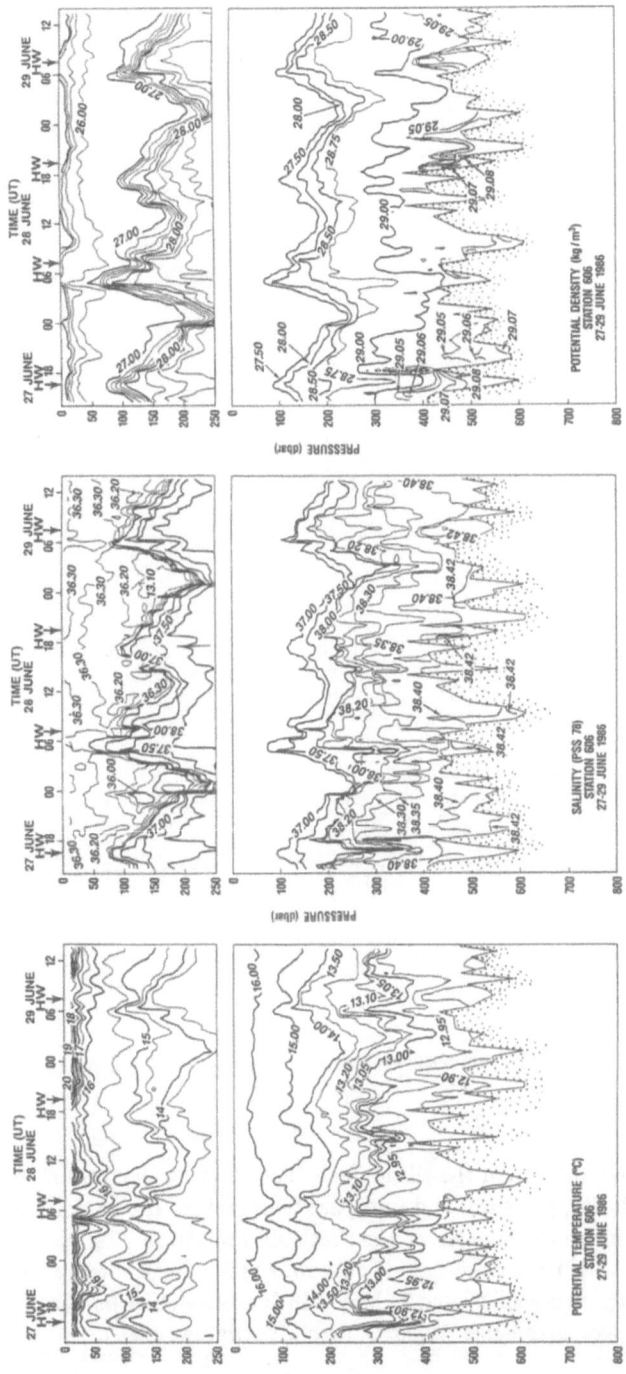

Figure 15: A 48-hour time series CTD station of potential temperature (a), salinity (b) and potential density (c), taken about 5 km west of the sill during neap tide (Station 606, see Fig. 2). The 12.90°C isotherm delineates WMDW in a thin layer near the bottom during all four semidiurnal tides (cf. Fig. 14). (See Figs. 5 and 13 for an explanation of plot.)

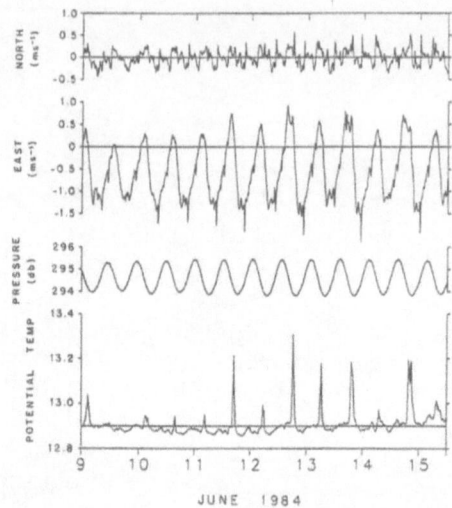

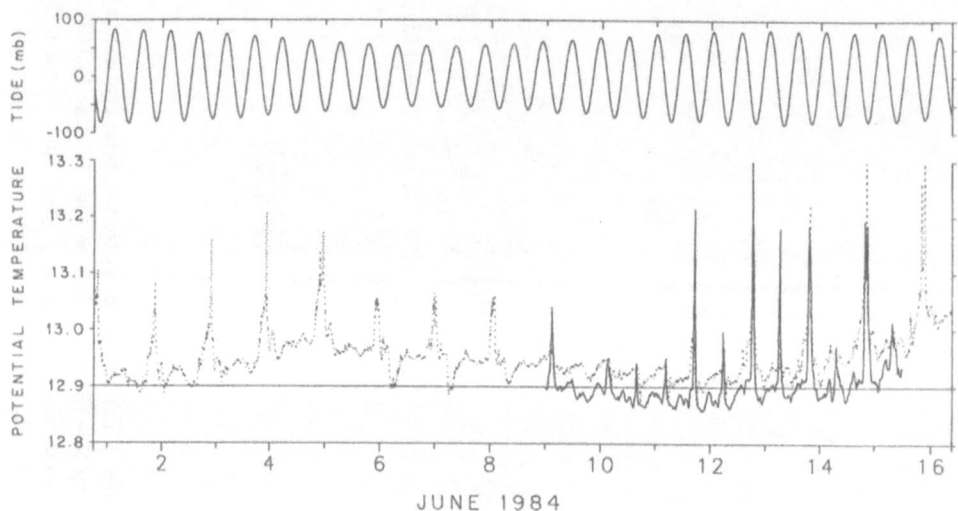

Figure 16: Pilot mooring data from the sill in June 1984 showing the outflow of WMDW. (a) North and east velocity, pressure, and potential temperature from a bottom mounted acoustic doppler profiling current meter at the south sill are shown. Temperatures below 12.90°C are correlated with strong (westward) outflow. (b) A potential temperature record from an instrument at the north sill during 31 May – 16 June 1984. The northern instrument showed almost no temperatures below 12.90°C. (From Pettigrew, 1989).

but both the generally warmer temperatures and the much lower speeds indicate that this reverse flow is minor compared to the westward flow.

The 1984–1987 data are a strong confirmation of the direct outflow of WMDW. Potential temperatures below 12.90°C are found west of the sill during both spring and neap tides. The CTD data demonstrated that the low temperatures are indeed WMDW, and not some of the cold but fresh Atlantic Water that is sometimes found near the strait. Current measurements during 1985–1986 show that neither the November 1985 nor the June 1986 data were taken during periods of unusual low frequency flow which could bias the results (Candela, Winant and Bryden, 1989).

The most straightforward extrapolation is that the WMDW outflows directly on nearly every semidiurnal tidal cycle. Data are presently inadequate to confirm such an idea unambiguously, but the outflow has been observed over several diurnal cycles during both spring and neap tides, during periods of moderate low frequency flow, and during three years. While there are several factors that could modify our interpretation, such as the idea that the volume of deep water in the western Mediterranean has a strong annual cycle (Bormans, Garrett and Thompson, 1986), we put forward this extrapolation as a new hypothesis that is consistent with the extensive Gibraltar data set.

Is it reasonable to assert that this outflow is caused by the Bernoulli mechanism? The WMDW is found preferentially on the southern sill, a continuation of its distribution in the Alboran Sea. What causes this current to keep to the left (looking downstream)? Next we discuss modeling results, first of the uplift and then of the boundary current, which help to answer these questions.

5. Laboratory and Numerical Models

Relevant laboratory and numerical experiments have addressed both key issues of the Gibraltar outflow: the uplift of the deep water, and the formation of the upstream boundary current. First, results pertinent to uplift are discussed.

Whitehead (1985) performed a series of rotating lock exchange runs, in which waters of different density were allowed to exchange between two reservoirs via a narrow channel. He scaled the ratio of the Mediterranean Rossby number to the strait width in order to correspond with the Gibraltar/Alboran Sea system. The denser water in the model Mediterranean was uplifted from a scaled depth near 700 m and flowed out over the sill; water from a scaled depth of 900 m did not outflow. The uplifted water formed a narrow current along the sloping bottom of the modeled African slope. The experiments confirm that the essence of the WMDW outflow is independent of the details of bottom bathymetry, but that the general trends (the sill and Moroccan slope) do play a role. The laboratory experiments were in near steady state, so that the strong temporal variability in Gibraltar does not appear critical. The simplest interpretation of the experiments is that of Bernoulli uplift, with "southward" intensification of the deep outflow upstream of the strait, as the result of rotation.

Preller (Parrilla, Kinder and Preller, 1986; Preller, 1986) used a numerical model of the Alboran Sea to examine the horizontal distribution of the LIW and WMDW as they flow toward the strait. The model had two layers and was forced by Gibraltar inflow in the upper layer, and by uniform inflow at the eastern boundary of the lower layer. Both of the outflow boundaries were open. When the model bottom was flat, the resulting flow resembled the broad westward LIW flow that is inferred from measurements (compare Figs. 17a and 11). The authors argued that this experiment simulated the LIW, since the WMDW insulates the LIW from direct contact with the bottom.

Their experiment with realistic bottom topography was intended to simulate the WMDW which is in direct contact with the bottom. This experiment showed a narrow outflow along the Moroccan slope, with a width and speed similar to the observed values (compare Figs. 17b and 11). This result was attributed to the dominant effect of bottom shoaling on the potential vorticity.

This notion of the key role of the shoaling bottom and rotation in forming a narrow outflow current was examined in more detail by Kinder, Chapman and Whitehead (1986). They hypothesized that if flow were forced across shoaling bathymetry, then the resulting flow would form a "western" boundary current. The bathymetry would dominate the potential vorticity gradient: changes in the water column thickness are normally orders of magnitude greater than changes in the Coriolis parameter. A rotating tank experiment with a sloping bottom tested this idea with a sink of fluid in the shallow part of a plane sloping bottom. A narrow boundary current developed on the left side with rotation in the northern hemisphere sense. A simple linear numerical model replicated this experiment. The model included only pressure gradient (sea level), rotation, and friction terms in the depth-averaged momentum equations over a basin with a linear bottom slope (Figure 18a). Streamlines from the numerical experiment with a large reservoir in the deep water, and a narrow sink (strait) in the shallow water, connected by a narrow boundary current, showed a similar behavior to the rotating tank experiment, and to the deep Alboran Sea (Figure 18b). The (meridional) variation in the Coriolis parameter in the Alboran is about $2 \times 10^{-13} \mathrm{cm}^{-1} \mathrm{s}^{-1}$. The topographic equivalent, $f h_x / h$, using the zonal change of depth from 700 to 300 m depth over 100 km, is two orders of magnitude greater than the planetary variation. Thus, the topographic beta effect dominates the planetary beta in the dynamics of the deep outflow.

While Gill (1977) predicted a boundary current on the left side upstream of a sill even in the absence of bottom topography due to the propagation characteristics of Kelvin waves, the marked differences in the paths of LIW and WMDW are more consistent with an explanation based on the effect of topographic beta on the outflow. In the rotating tank, in the numerical model, and in the Alboran Sea, an equivalent to a western boundary current arises. Instead of the planetary variation in the Coriolis parameter, however, it is the bathymetric gradient (in the direction of flow) that causes a variation in the potential vorticity.

6. Discussion and Summary

Recent research has confirmed the direct outflow of deep water from the Mediterranean, and supported the combination of Bernoulli aspiration and topographic beta as the forcing for the deep water uplift and upstream boundary current. Not all the evidence fits exactly into this scenario, but we think that the broad picture is correct.

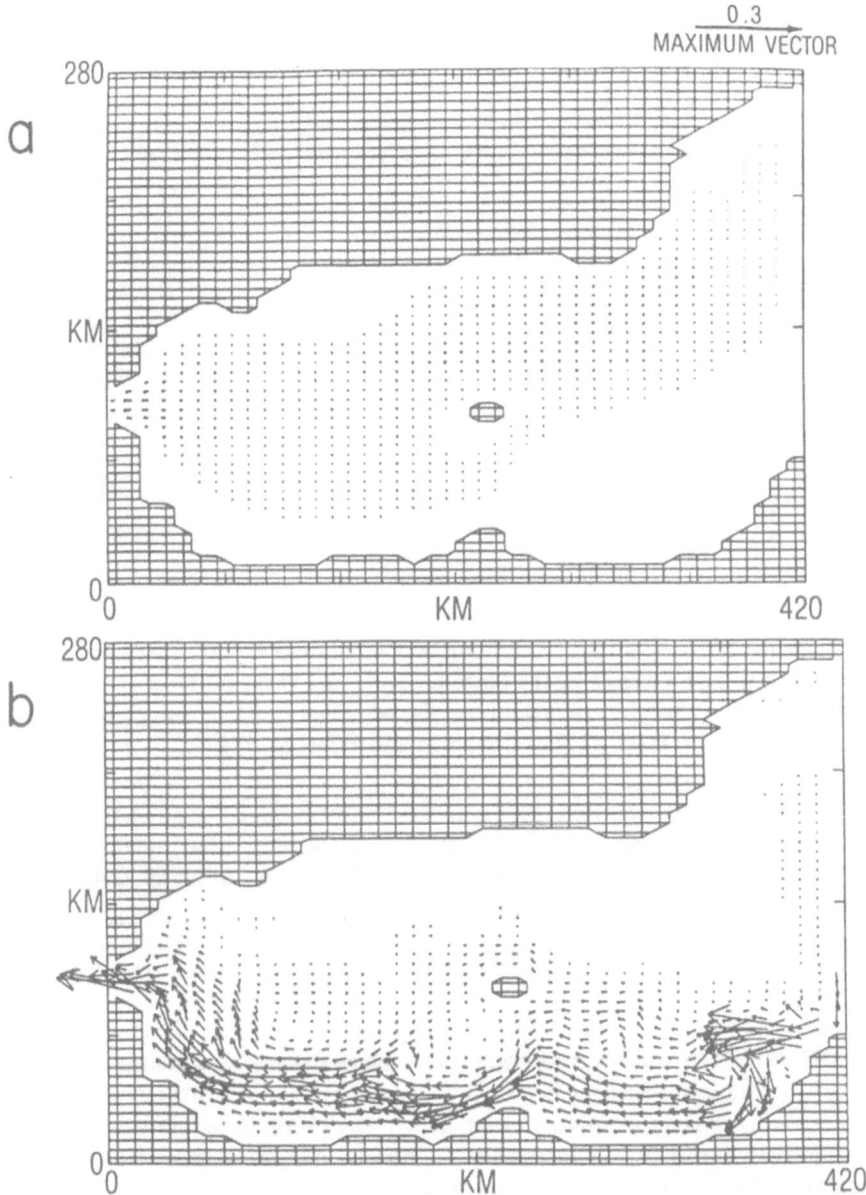

Figure 17: Lower layer velocities from two primitive equation numerical experiments. (a) With a flat bottom, intended to simulate LIW flow, isolated from the bottom topography by the WMDW. (b) With realistic bottom topography, meant to simulate the WMDW flow, which is in contact with the bottom. (cf. Fig. 11) (From Parrilla, Kinder and Preller, 1986).

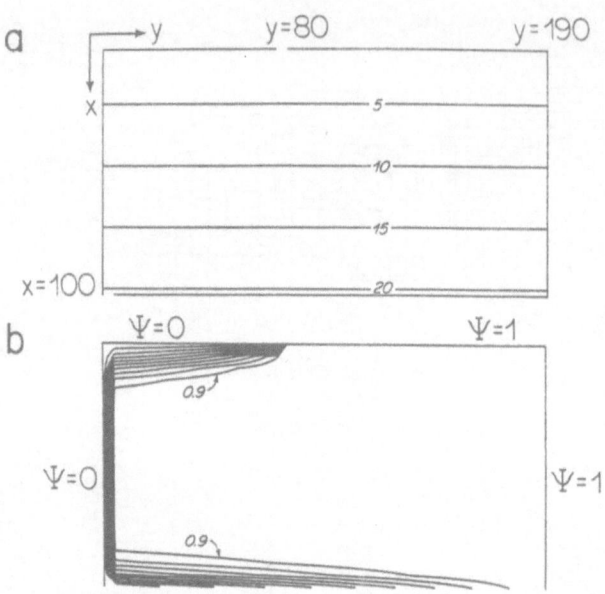

Figure 18: Numerical simulation of a rotating tank laboratory experiment, with a fluid sink (strait) at the top and a source at the bottom. (a) Model bathymetry in cm. (b) Stream-lines which illustrate the effect of flow that is forced across shoaling topography under the constraint of rotation. (From Kinder, Chapman and Whitehead, 1986).

Hydrographic observations have shown that WMDW is found west of the sill during spring and neap tides, in the spring and in the fall, and during three separate years. The flow of WMDW approaching the strait is in a narrow boundary current against the African slope, as revealed by hydrographic observations and direct current measurements. Rotating tank and both linear and primitive equation numerical experiments are consistent with the explanations of uplift by Bernoulli aspiration and boundary current formation by topographic beta.

The numerical simulations were incapable of including aspiration. The outflow was assumed and its upstream consequence examined. A numerical model capable of incorporating the abrupt topography and a deep initially isolated layer would be instructive (i.e., replicate the rotating lock exchange laboratory experiment). The ability to estimate momentum, energy and vorticity terms directly in a numerical model would reveal the extent to which the simple ideas of aspiration and topographic beta hold.

The models and ideas are all steady, but flow in the Strait of Gibraltar is notably variable (e.g., Armi and Farmer, 1988; Candela, Winant and Bryden, 1989). At Gibraltar, this intense variability on scales from 20 minutes to 20 days seems to have only minor effect on the predictions of the simple steady flow ideas (Bryden and Kinder, 1990). The effect of temporal variability poses an interesting research issue (Armi and Farmer, 1988, discuss this issue in the context of applying hydraulic control theory to the strait dynamics.)

The practical importance of a direct outflow of deep water for ideas of the Mediterranean as a pollutant sink or nutrient trap is obvious. Direct outflow also modifies thoughts about abyssal circulation. Stommel and Arons (1960), for example, constructed a model of abyssal circulation using uniform upwelling of the deep waters as the forcing function. Significant direct outflow from a semi-isolated basin changes this forcing, at least quantitatively. Bryden and Stommel (1984) discuss the ramifications for the Mediterranean circulation.

Finally, the evidence is so convincing at Gibraltar that other systems should be examined for evidence of Bernoulli aspiration. Perhaps the Strait of Sicily may exhibit aspiration. There are many semi-isolated basins in the Caribbean Sea where intermittent strong flows may exist at the sills (e.g. Kinder, Heburn and Green, 1985). Oceanographers can all nominate their favorite candidate. The requirements are mild stratification and strong flow at the sill. The topographic beta effect provides a useful diagnostic: in many regions the concentrated upstream boundary current is much easier to detect than the direct outflow because of the intense mixing that normally accompanies the strong shallow currents. Careful examination of the appropriate data upstream of various sills might suggest other locations where Bernoulli aspiration occurs.

Acknowledgments. We were funded by the Coastal Sciences Program of the United States Office of Naval Research. We have enjoyed collaboration with many colleagues including Gregorio Parrilla, Jack Whitehead, Nelson Hogg, Neal Pettigrew, David Chapman, Ruth Preller, and Nancy Bray who have generously provided thoughts and data that we have used. Both anonymous reviewers provided cogent and useful criticism. We especially thank shipmates, scientists and seamen, who have shared many hours in the Strait of Gibraltar and Alboran Sea.

References

Armi, L. and D. M. Farmer (1988) The flow of Mediterranean Water through the Strait of Gibraltar. (also: Farmer, D. M. and L. Armi. The flow of Atlantic Water through the Strait of Gibraltar.) *Progress in Oceanography* 21(1):1–105.

Bormans, M., C. Garrett and K. R. Thompson (1986) Seasonal variability of the surface inflow through the Strait of Gibraltar. *Oceanologica Acta* 9:403–414.

Bryden, H. L. and T. H. Kinder (1990) Steady two-layer exchange through the Strait of Gibraltar. *Deep-Sea Research*, in press.

Bryden, H. L. and H. M. Stommel (1982) Origin of the Mediterranean outflow. *Journal of Marine Research* 40(suppl.):55–71.

Bryden, H. L. and H. M. Stommel (1984) Limiting processes that determine basic features of the circulation in the Mediterranean Sea. *Oceanologica Acta* 7:289–296.

Candela, J. C., C. D. Winant and H. L. Bryden (1989) Meteorologically forced subinertial flows though the Strait of Gibraltar. *Journal of Geophysical Research* 94:12,667–12,679.

Gascard, J. C. and C. Richez (1985) Water masses and circulation in the Western Alboran Sea and in the Straits of Gibraltar. *Progress in Oceanography* 15:157–216.

Gill, A. E. (1977) The hydraulics of rotating channel flow. *Journal of Fluid Mechanics* **80**:641–671.

IGN–SECEG (1988) Estrecho de Gibraltar: Mapa Fisico, Instituto Geográfico Nacional M.6.058–1988, Madrid.

Kinder, T. H. and H. L. Bryden (1987) The 1985–1986 Gibraltar Experiment: Data collection and preliminary results. *Eos, Transactions, American Geophysical Union* **68**(40):786–787, 793–795.

Kinder, T. H., D. C. Chapman and J. A. Whitehead, Jr. (1986) Westward intensification of the mean circulation on the Bering Sea Shelf. *Journal of Physical Oceanography* **16**(7):1218–1229.

Kinder, T. H., G. W. Heburn and A. W. Green (1985) Some aspects of the Caribbean Circulation. In *Benthic Ecology and Sedimentary Processes of the Venezuela Basin: Past and Present. Marine Geology* **68**:25–52.

Kinder, T. H. and G. Parrilla (1987) Yes, some of the Mediterranean outflow does come from great depth. *Journal of Geophysical Research* **92**(C3):2901–2906.

Lacombe, H. and C. Richez (1982) Regime of the Strait of Gibraltar and of its east and west approaches. In *Hydrodynamics of Semi-Enclosed Seas*, J. C. J. Nihoul, ed., Elsevier, Amsterdam, pp. 13–73.

Lacombe, H. and P. Tchernia (1972) Caracteres hydrologiques et circulation des eaux en Mediterranee. In *The Mediterranean Sea: a natural sedimentation laboratory*, D. J. Stanley, ed., Dowden, Hutchinson and Ross, Stroudsburg, Pennsylvania, pp. 26–36.

Lanoix, F. (1974) Projet Alboran: Etude Hydrologique et Dynamique de la Mer D'Alboran (Juillet et Aout 1962). NATO Report 66, Brussels, 39 pp. plus figures.

Measures, C. I. and J. M. Edmond (1988) Aluminium as a tracer of the deep outflow from the Mediterranean. *Journal of Geophysical Research* **93**(C1):591–595.

Parrilla, G. (1984) Comparison between salinities of the Alboran Sea obtained according to Fofonoff *et al.* (1974) and the new practical salinity scale. In *Preliminary Results of Donde Va?*, G. Parrilla, editor, Instituto Español de Oceanografia, Informes Tecnicos No. 24, pp. 153–155.

Parrilla, G. and T. H. Kinder (1989) The physical oceanography of the Alboran Sea. In *Atmospheric and Oceanic Circulation in the Mediterranean Basin*, H. Charnock, ed., in press.

Parrilla, G., T. H. Kinder and R. H. Preller (1986) Deep and Intermediate Mediterranean Water in the western Alboran Sea. *Deep Sea Research* **33**(1):55–88.

Pettigrew, N. (1989) Direct measurements of the flow of Western Mediterranean Deep Water over the Gibraltar sill. *Journal of Geophysical Research* **94**(C12):18,089–18,093.

Pistek, P., F. de Strobel and C. Montanari (1985) Deep-sea circulation in the Alboran Sea. *Journal of Geophysical Research* **90**(C3):4969–4976.

Preller, R. H. (1986) A numerical model of the Alboran Sea Gyre. *Progress in Oceanography* **16**:113–146.

Roether, W. and W. Weiss (1975) On the formation of the outflow through the Strait of Gibraltar. *Geophysical Research Letters* **2**(7):301–304.

Stommel, H. and A. B. Arons (1960) On the abyssal circulation of the World Ocean — II. An idealized model of the circulation pattern and amplitude in oceanic basins. *Deep Sea Research* **6**:217–233.

Stommel, H., H. Bryden and P. Mangelsdorf (1973) Does some of the Mediterranean outflow come from great depth? *Pure and Applied Geophysics* **105**:879–889.

Wesson, J. C. and M. C. Gregg (1988) Turbulent dissipation in the Strait of Gibraltar and associated mixing. In *Small-Scale Turbulence and Mixing in the Ocean*, J. C. J. Nihoul and B. Jamart, eds., Elsevier, Amsterdam, pp. 201–212.

Whitehead, J. A., Jr. (1985) A laboratory study of gyres and uplift near the Strait of Gibraltar. *Journal of Geophysical Research* **90**(C4):7045–7060. Correction, **90**(C6): 12,011–12,013.

A REVIEW OF ROTATING HYDRAULICS

K. M. BORENÄS
Department of Oceanography
Göteborg University
Box 4038
S-400 40 Göteborg, Sweden

and

L. J. PRATT
Department of Oceanography
Woods Hole Oceanographic Institution
Woods Hole, Massachusetts 02543

ABSTRACT. The concepts of hydraulics of rotating-channel flow are presented. The review includes work on uniform as well as nonuniform potential vorticity flow. A unified formalism is used when discussing the various contributions to the field. Rotating hydraulic jumps and bores are also considered.

1. Introduction

1.1. BACKGROUND

The classical problem of weir flow has been extensively examined within hydraulic engineering. Recently, the concepts of hydraulics have been applied to large-scale geophysical flows such as airflow over mountain ridges and deep-water currents through constrictions. An example of the latter is the discharge of Norwegian Sea Deep Water through the Faroe Bank Channel into the North Atlantic. Figure 1 shows a hydrographic section taken along the Faroe Bank Channel during the ICES-coordinated 'Overflow-60' experiment. The graph shows a marked drop in the height of the isotherms over the sill, a feature characteristic of weir flow. A similar behavior of the isotherms is found in the Windward Passage, Figure 2. On the lee side of the sill a hydraulic jump, another particularity of weir flow, seems to be present. Other examples of hydraulically-driven flow include the deep-water currents in the Denmark Strait, Vema Channel, Anegada Passage and Strait of Gibraltar.

Ocean strait flows can be strongly influenced by the earth's rotation when the Rossby radius of deformation [see (2.12)] is the same order of or smaller than the strait width. As will be discussed below, the effect of rotation is to induce boundary layer structure in the flow or to cause the flow to become banked against one of the sloping sides of the strait. In order to compute the flow field, one must consider the cross-strait structure (absent in classical hydraulics) in addition to the nonlinear along-strait structure. Because

321

L. J. Pratt (ed.), The Physical Oceanography of Sea Straits, 321–341.
© 1990 Kluwer Academic Publishers.

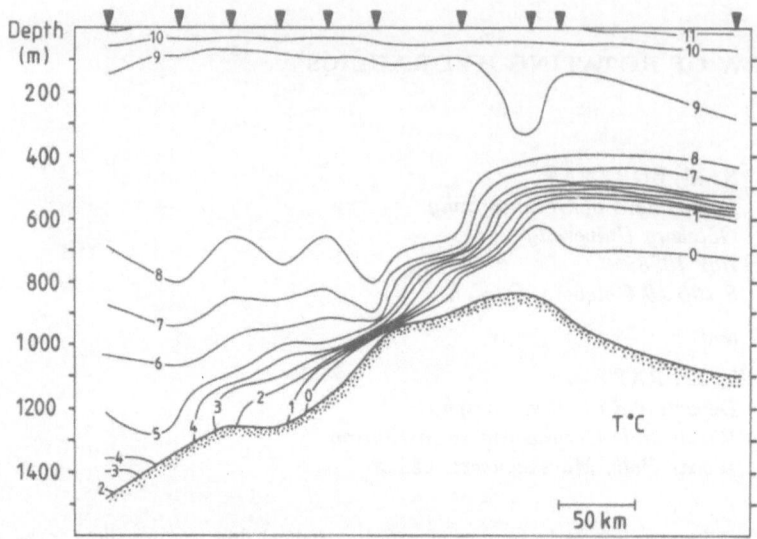

Figure 1.   Isotherms in a section along the Faroe Bank Channel (after Lee, 1967). The deep-water flow is from right to left.

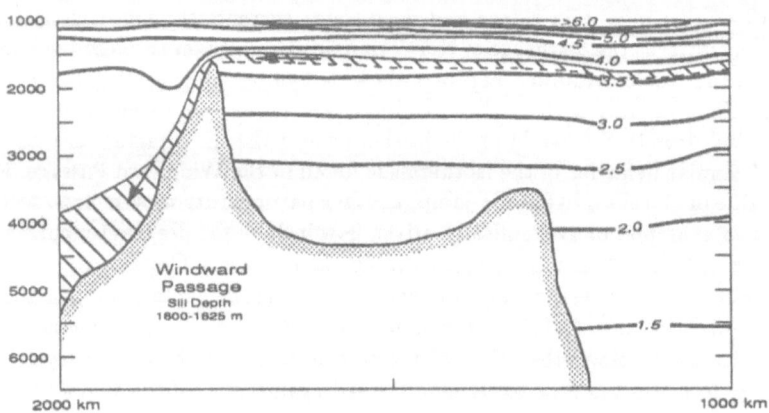

Figure 2.   Isotherms in a section along the Windward Passage (after Nof, 1986).

of these difficulties, most investigations of rotating strait and sill flows have been confined to homogeneous fluids. In the present review, the discussion will be restricted to systems with a single active layer. Investigations of multi-layer systems have been restricted to a few simple cases, and a review by Dalziel (1990) appears in this volume.

The outline of the paper is as follows: For the purpose of introducing certain basic ideas of hydraulic theory, the case of non-rotating, inviscid, one-layer channel flow is discussed first. In section 2 a uniform potential vorticity flow is considered, and in section 3 the theory is extended to encompass the case of nonuniform potential vorticity. The subsequent section is devoted to shocks and jumps, and a section discussing some unsolved problems concludes the review.

1.2. SOME BASIC IDEAS FROM NON-ROTATING HYDRAULIC THEORY

The analysis given in this section follows to a large degree the one presented by Long (1954) for a non-rotating, one-layer flow. Other discussions of basic concepts can be found in the text books of Chow (1959) and Rouse (1961).

Consider a flow of an inviscid fluid through a slowly varying channel. Let u and v be the x- and y- components of the horizontal velocity where the y- coordinate is aligned along the channel axis. The horizontal length scale is taken to be much larger than the depth scale, making the hydrostatic approximation valid. In Figure 3 the geometrical features of an arbitrarily shaped channel are shown. The reference level $z = 0$ is taken to be the elevation of the sill. The bottom elevation is given by $z = -h$, and the height of the free surface by $z = \eta$. Hence, the fluid depth becomes $D = \eta + h$. (All the results presented here are equally valid for a flow in a layer of density ρ underlying a passive layer of density $\rho - \Delta\rho$. The variable η then denotes the elevation of the interface between the two layers.) The geometry considered in this section is simplified in that the bottom depth is taken to be uniform across the channel and the width, w, is constant. If the total flow is denoted by Q, the continuity equation is given by

$$vwD = Q \quad . \tag{1.1}$$

The upstream reservoir is assumed to be large, rendering the velocities in this area negligible, and the energy equation takes the form

$$\frac{v^2}{2g} + D = (\eta_\infty + h) \quad . \tag{1.2}$$

Here n_∞ represents the upstream elevation of the surface and g the acceleration due to gravity. (For the two-layer case g is replaced by the reduced gravity $g = \frac{g\Delta\rho}{\rho}$). The left-hand side is called the specific energy, E, and by using equation (1.1), this quantity may be expressed as a function of D

$$\frac{Q^2}{w^2 D^2 2g} + D = E(D) \quad . \tag{1.3}$$

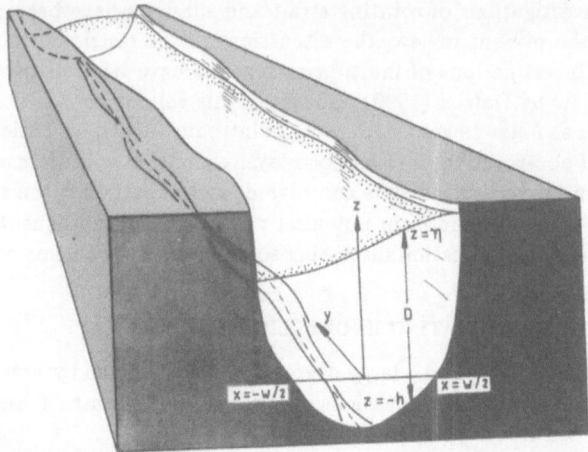

Figure 3. The geometry of the channel.

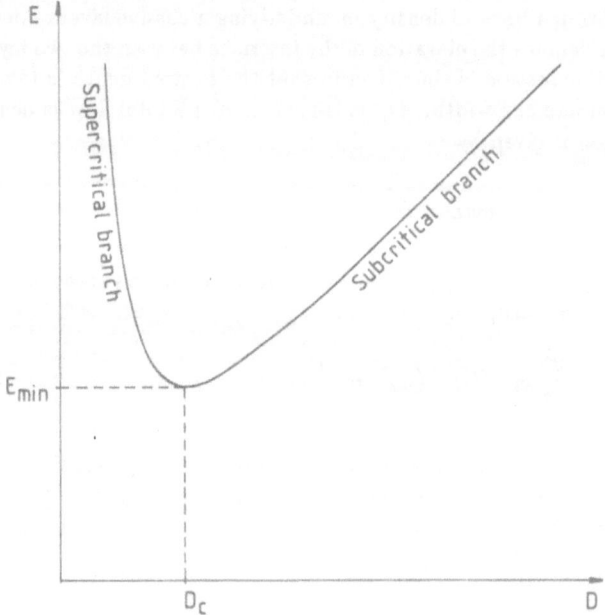

Figure 4. Relation (1.3) plotted for a fixed value of Q^2/gw^2.

Equation (1.2) is then restated as

$$E(D) \; = \; \eta_\infty \; + \; h \quad .$$

(1.4)

The relation (1.3) is plotted in Figure 4 for a given positive Q/w. The specific energy shows a minimum for $D_c = (\frac{Q^2}{gw^2})^{\frac{1}{3}}$ which is obtained from $\frac{\partial E}{\partial D} = 0$. The velocity at the specific energy, minimum is $v = (gD_c)^{\frac{1}{2}}$ and consequently the speed $v - (gD_c)^{\frac{1}{2}}$ of a small-amplitude long wave is zero. Disturbances are therefore unable to propagate upstream at the section of minimum specific energy, and the flow here is said to be critical. Conditions upstream of this 'control' section are hence insensitive to conditions downstream.

The right-hand branch of the solution curve in Figure 4 is associated with a comparatively slow flow while the left-hand branch describes a more rapid current. These flows are denoted sub- and supercritical, respectively. For a given upstream height η_∞, moving along the curve towards decreasing E ($E = h + \eta_\infty$) corresponds to moving along a rising bottom. In Figure 5 the elevation of the surface is drawn for a purely subcritical and a purely supercritical solution (the two solid lines). These solutions can be obtained by choosing the upstream value of $E(D)$ large enough so that the branch point ($E = E_{\min}$) of the Figure 5 curve is not encountered. Purely supercritical flows are relatively sensitive to frictional effects and generally do not exist in the upstream basin; hence attention will be restricted to flows with subcritical upstream states.

The minimum values of E and D in a given subcritical solution occur at the sill. If one fixes Q/w and reduces the upstream surface elevation η_∞, the sill values of E and D are also reduced. Eventually E is reduced to $E_{\min}$ at the sill, meaning that the branch point (or "critical" point) in the Figure 5 solution curve is encountered. Under these conditions the energy or Bernoulli head η_∞ has clearly reached a minimum value, say $\eta_{\infty c}$.

To investigate the behavior of the solution near the branch point, it is helpful to differentiate (1.4) with respect to y. Using the definition of $E(D)$ the result can be written

$$\frac{dD}{dy} = \frac{D^3}{[D^3 - Q^2/(gw^2)]} \frac{dh}{dy}$$

(1.5)

At the branch point ($D^3 = Q^2/gw^2$), $dD/dy \to \infty$ unless $dh/dy = 0$. In other words, a well-behaved critical solution can only occur at a topographic extremum. That this extremum corresponds to a sill and not a depression can be shown by differentiating equation (1.4) once again with respect to y. After evaluating at the branch point, the following expression is obtained

$$\left(\frac{dD}{dy}\right)^2 = \frac{d^2h}{dy^2} \frac{D_c}{3}.$$

(1.6)

For positive depth $\frac{d^2h}{dy^2} > 0$, implying a sill geometry. Furthermore dD/dy will be finite for finite d^2h/dy^2, so that the solution must change branches at the branch point. An example of the corresponding subcritical-to-supercritical solution is shown by the dashed line in Figure 5. This solution is often referred to as being "critically" or "hydraulically" controlled. The upstream flow is insensitive to conditions downstream of the sill, the Bernoulli head

326

is a minimum for all solutions having the same flow rate, and (conversely) the flow rate is maximized over all solutions having the same Bernoulli head. The transport for such a solution is given by

$$Q_{\max} = wg'/2 \left\{ \frac{2}{3} \eta_\infty \right\}^{3/2}. \tag{1.7}$$

[Here equation (1.2) has been used to express the depth at the critical section in terms of the upstream height; $D_C = 2\eta_\infty/3$.]

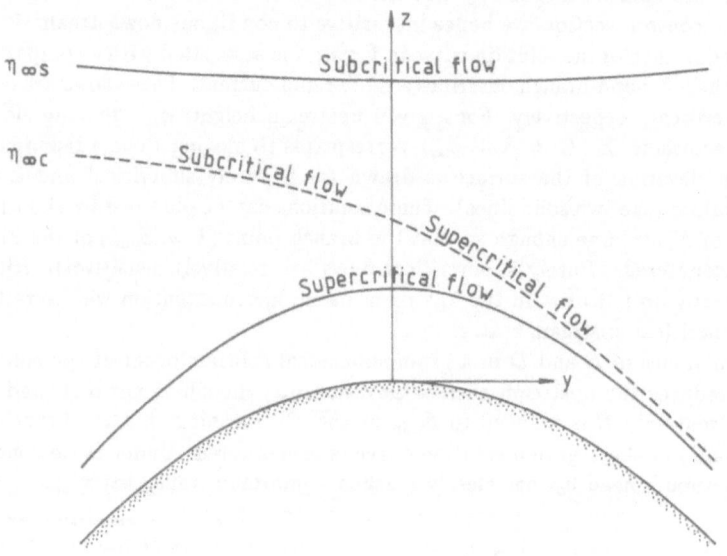

Figure 5. The form of the free surface of a fluid flowing in the right-hand direction along a parabolic shaped bottom. The upper and lower solid lines correspond to the sub- and supercritical branches, respectively. The upstream elevation of the subcritical solution is indicated by $n_\infty s$. The hatched line represents a flow changing branch from a sub- to a supercritical state. The upstream elevation is indicated by n_∞.

Gill (1977) identified a number of general features characterizing hydraulic control problems, rotating or non-rotating:

A. The flow can be expressed by a single dependent variable D whose dependence upon the along-channel coordinate y is parametrical. The relationship between D and the geometric variable h, w, etc. may be expressed by

$$F(h, w, \cdots; D) = constant. \tag{1.8}$$

For the flow considered here the corresponding expression is

$$F = E(D) - h = \eta_\infty. \tag{1.9}$$

B. As a result of the nonlinearity in the equations of motion, F is multivalued. Different solution branches coincide for

$$\frac{\partial F}{\partial D} = 0, \tag{1.10}$$

or for the present example $\frac{\partial E}{\partial D} = 0$.

C. The geometry involves some kind of constriction, defined by

$$G \equiv \frac{\partial F}{\partial h} \frac{dh}{dy} + \frac{\partial F}{\partial w} \frac{dw}{dy} + \cdots = 0. \tag{1.11}$$

From these statements it follows that $\frac{\partial F}{\partial D}(\frac{dD}{dy}) = -G$. For $\frac{dD}{dy}$ to remain finite when $\frac{\partial F}{\partial D} = 0$ requires that $G = 0$. This criterion determines the position of the critical section. In the case of constant width (1.10) reduces to $\frac{dh}{dy} = 0$ as already demonstrated in equation (1.5).

The prerequisites listed above are rather abstract and do not easily yield physical insight into the concept of hydraulic control, particularly for a reader new to the field. One difficulty is that hydraulic control is exercised through time-dependent processes (upstream wave propagation) and is therefore difficult to describe using steady theory. To motivate the concern about time dependence, consider the implications of equation (1.7) relating the flow rate Q and the upstream surface elevation η_∞. Ordinarily, one thinks of these quantities (which specify the upstream conditions) as being independent. However, when the flow is critical at some section, Q and η_∞ become linked. One might ask, therefore, what would be the outcome of a laboratory experiment in which critically-controlled steady flow initially existed and Q was suddenly altered at an upstream location? Pratt (1984b) has examined this problem numerically and shown that the violation of equation (1.7) exists only temporarily. When Q is increased, the upstream elevation changes from its initial value $\eta_{\infty 0}$ to $\eta_{\infty 1}$, generating a wave which propagates downstream, eventually reaching the critical section. A reflected wave is then created which propagates upstream and further alters the upstream elevation from $\eta_{\infty 1}$ to $\eta_{\infty 2}$ so as to reestablish equation (1.7). Thus the relationship between Q and η_∞ is controlled at the critical section by signals in the form of long gravity waves. When a critical section does not exist, equation (1.7) no longer holds and one is free to vary Q and η_∞ independently.

2. Uniform Potential Vorticity Theory For Steady Flow

2.1. A SUMMARY OF THE DEVELOPMENT OF ROTATING HYDRAULIC THEORY

In a paper by Stern (1972) the concept of criticality in a rotating channel flow was first discussed. A sloping bottom was introduced so as to compensate exactly for the free surface slope, rendering the depth and the velocity across the channel uniform. The flow is critical when no solution exists for an infinitesimal change of the geometry. (Motivation for this requirement can be given by reference to Figure 4. If the flow is critical, D lies at the extremum of the solution curve. Increasing the topographic elevation causes E to decrease into a range where no solutions exist.)

Whitehead, Leetmaa and Knox (1974) examined the more realistic case of zero potential vorticity flow through a channel of rectangular cross-section connecting two basins. (The potential vorticity is a measure of the total angular momentum of a fluid column depth.) A unique solution was obtained by requiring the flow rate to be maximal for fixed Bernoulli head. The transport in the channel is determined by the upstream height of the free surface. One complication which arises is that, when the rotation rate or the velocity becomes large, the surface will intersect the bottom rather than the left wall (looking downstream in the Northern Hemisphere). The width of the channel will then be of no importance for the transport capacity. It is therefore necessary to deal with these two cases separately. Whitehead *et al.* (1974) also conducted experiments showing the relationship between the flow rate and the upstream free surface elevation to be in good agreement with the theory.

Gill (1977) introduced a more complicated model in which the potential vorticity is uniform but finite. Gill assumed that the reservoir is very wide and showed that the flow is confined to boundary layers on each side wall, leaving the interior of the basin quiescent. The potential vorticity is then inversely proportional to the interior depth D_∞ in the upstream basin. In order to obtain the solution for maximum transport, D_∞ together with the distribution of volume flux between the upstream boundary layers must be prescribed. The possibility of separation at the left side wall necessitates two solution regimes. As already pointed out in section 1, Gill provided a general discussion of the hydraulic problem. He also showed that the principle of maximum transport is equivalent to linear Kelvin waves having zero phase speed at the control section. (In fact, Kelvin waves play a similar role in Gill's problem to that played by long gravity waves in classical hydraulics, at least when the fluid remains attached to both side walls.) Several examples of various flow types and channel geometries were presented.

A further extension of the theory of rotating hydraulics was made by Shen (1981) who examine three different configurations: subcritical flow subjected to different downstream conditions, critical flow in a channel of irregular cross section, and separated supercritical flow. Transport formulae were derived and careful experiments were carried out. Comparison between experiments and theoretical predictions based on zero potential vorticity were in general favorable. However, the zero potential vorticity transport formula not hold when the flow at the narrowest section became separated.

The studies outlined above all deal with the stationary situation. In order to examine how a critical flow is established, Pratt (1983) analyzed the adjustment problem for a flow to a sudden obtrusion of an obstacle. When the height of the obstacle is below a critical value, the Kelvin waves formed by the disturbance cause only a temporary change in the upstream flow. For obstacle heights exceeding the critical value, bores are generated and the upstream conditions are permanently altered. In this case the final steady solution has a flow rate which is less than the initial value (the flow is partially blocked). Pratt (1983) also found that the potential vorticity of the flow can be modified by the bores and hydraulic jumps appearing in the flow field.

In order to avoid the inconvenience of possible flow separation from a vertical side wall (and to model a more realistic topography), the flow in a channel of parabolic cross section was examined by Borenäs and Lundberg (1986). They considered a flow of constant, non-zero potential vorticity. As a consequence of vanishing depth at the boundaries, a new phenomenon is revealed, namely that the flow can become reversed over portions of the critical section. This feature is incompatible with the upstream pre-specification of potential vorticity along streamlines since it implies that streamlines extend back into the upstream reservoir or form closed trajectories. Elimination of solutions with reversed streamlines at the control section restricts the applicability of the hydraulic model to certain parameter regimes.

The case of a zero potential vorticity flow has also been analyzed for the parabolic shaped channel by Borenäs and Lundberg (1988). The formula obtained for the maximum transport is used in order to estimate the deep water discharge through the Faroe Bank Channel. The theoretically predicted value of the transport in the channel (whose cross section at the sill is very close to a parabola) is well in accordance with estimates based on hydrographic observations and current measurements.

A number of other studies of rotating, hydraulically driven flow with uniform potential vorticity have been carried out. For the most part this work has aimed at exploring new dynamical complications introduced by variations in the geometry, including, for example, the effect of large sill width-to-length ratio (Sambuco and Whitehead, 1976) and finite side-wall curvature (Reed, 1980). These studies will not be discussed here, though some explore novel and interesting dynamical departures from earlier work.

A related theory deserving mention involves the principle of "geostrophic control" put forth by Garrett and Toulany (1982) and also discussed by Whitehead (1986). This theory asserts that the free-surface level difference across a strait cannot exceed the sea level difference between the upstream and downstream reservoirs, placing an upper limit on the volume transport. The solutions of Gill (1977) can, however, experience larger cross-strait than mean basin-to-basin sea level changes. At present it is not known whether the solutions of rotating hydraulics experience geostrophic control in any limit.

2.2. RELEVANT EQUATIONS AND DIFFERENT TYPES OF SOLUTIONS

As mentioned in the beginning of this section, the along-channel flow is assumed to be in geostrophic balance, a consequence of neglecting the nonlinear terms in the x-component of the momentum equation. This simplification is suggested when changes along the channel are more gradual than those experienced across, implying that $v \gg u$ and $v_x \gg u_y$. The semi-geostrophic shallow water equations for a flow in a channel, rotating at the angular velocity $\omega = f/2$, are hence given by

$$fv = \frac{g\partial n}{\partial x} \qquad (2.1)$$

$$\frac{u\partial v}{\partial x} + \frac{v\partial v}{\partial y} + fu = \frac{g\partial n}{\partial y} \qquad (2.2)$$

$$\frac{\partial(uD)}{\partial x} + \frac{\partial(vD)}{\partial y} = 0 \qquad (2.3)$$

From these equations it can be shown that the potential vorticity is conserved along stream-lines:

$$\frac{\partial v/\partial x + f}{D} = G(\psi). \qquad (2.4)$$

The Bernoulli function, B, is related to the potential vorticity by

$$\frac{dB}{D\psi} = G(\psi), \qquad (2.5)$$

where the stream function ψ is defined by

$$\frac{\partial \psi}{\partial x} = Dv \qquad (2.6a)$$

$$\frac{\partial \psi}{\partial y} = -Du. \qquad (2.6b)$$

In this section only the case of a constant (zero or non-zero) potential vorticity is considered, and, hence,

$$B = \psi G + C \qquad (2.7)$$

where C is a constant to be determined. Combining equations (2.1) and (2.4) yields a differential equation for η:

$$\frac{\partial^2 \eta}{\partial x^2} - \frac{fG}{g}(\eta + h) + \frac{f^2}{g} = 0. \qquad (2.8)$$

If D_∞ is the depth at which the relative vorticity vanishes (sometimes called the "potential depth"), then the potential vorticity is given by $G = f/D_\infty$. For a very deep upstream basin the potential vorticity is approximately zero, and the Bernoulli function is constant. This constant will, as in the nonrotating case, be given by η_∞, the upstream elevation of the free surface. The solution to (2.8) for η is quadratic in x, and the two integration "constants" may be linked to D_∞, η_∞, and the channel geometry (using the Bernoulli continuity equations). After eliminating one of the constants, an algebraic equation of the form $F = $ constant remains to be solved. This equation exhibits all the features (listed in section 1) characterizing the hydraulic flow problem. In the case of a channel of rectangular cross-section (the geometry considered by Whitehead *et al.*, 1974), two regimes must be examined separately. If the channel width is limited by $w \leq (\frac{2g\eta_\infty}{f^2})^{1/2}$, the maximum transport becomes

$$Q_{\max} = \left\{\frac{2}{3}\right\}^{3/2} wg^{1/2} \left\{\eta_\infty - \frac{f^2 w^2}{8g}\right\}^{3/2}. \tag{2.9}$$

By putting $f = 0$ the formula for a non-rotating rectangular channel is obtained [cf. equation (1.7)]. When the velocity increases for a given rotation rate, the tilt across the channel of the interface will increase until the active layer is no longer in contact with the left side wall. The flow is said to be separated and the width of the channel becomes unimportant. Separation occurs for $w > (2g\eta_\infty/f^2)^{1/2}$, and the maximum transport is now given by

$$Q_{\max} = \left\{\frac{g\eta_\infty^2}{2f}\right\}. \tag{2.10}$$

Borenäs and Lundberg (1988) examine the case of a parabolic cross section where the position of the bottom is given by $h = \beta - \alpha x^2$. (Here the width of the flow is unknown, and the boundary condition of vanishing depth is used.) The formula for the transport is found to be

$$Q_{\max} = \frac{\eta_\infty^2}{(2 + f^2/g\alpha)} \left\{\frac{3g}{2\alpha}\right\}^{1/2}. \tag{2.11}$$

This expression is only valid when the flow at the critical section is unidirectional. For $f^2/g\alpha > 2/3$ a reversed flow exists and the formula can no longer be applied.

When the upstream depth D_∞ is finite, the procedure of the analysis differs somewhat. It is now assumed that the reservoir is very wide and that the upstream flow takes place adjacent to the boundaries.

The boundary layer thickness is

$$\lambda = (gD_\infty)^{1/2}/f, \tag{2.12}$$

the Rossby radius of deformation based on the "potential" depth D_∞. (Note that λ is constant along the channel.) The constant in equation (2.7) can be determined by applying the Bernoulli equation to a streamline ψ_i emanating from the quiescent interior of the upstream basin. The solution for η is expressed in terms of hyperbolic functions. In the case of a flow in a channel of rectangular cross section (the geometry considered by Gill, 1977), it is convenient to express the integration "constants" in terms of the average and differential wall depth:

$$\overline{D} = \frac{1}{2}[D(w/2) + D(-w/2)] \qquad \hat{D} = \frac{1}{2}[D(w/2) - D(-w/2)]. \tag{2.13}$$

(For a channel of parabolic cross section the integration "constants" are given in terms of the intersection points between the surface and the bottom.) In order to determine $\overline{D}$ and

$\hat{D}$ the Bernoulli equation is applied on the streamlines along the boundaries. An alternative way is to apply the Bernoulli equation along one of the boundaries and combine it with the transport equation

$$Q = \int_{-w/2}^{w/2} vDdx = \frac{g}{2f} \left\{ D^2(w/2) - D^2(-w/2) \right\} = \frac{2g}{f} \overline{D}\hat{D}, \qquad (2.14)$$

in view of the geostrophic relation. When side wall Bernoulli equations (or Bernoulli and transport equations) are combined, the resulting equation is of the desired form (1.7). However, the functional relation $F = $ constant takes the form of a transcendental equation and no explicit formula for the maximum flow can be given.

The transport capacity for the different types of channel geometry has been discussed by Borenäs and Lundberg (1986, 1988). They found that in most situations the rectangular channel gave a larger maximum flow than a channel of parabolic cross section. [Borenäs and Lundberg (1986) chose to compare flows which occupied the same cross-sectional area.]

3. Hydraulics of Nonuniform Potential Vorticity Flow

In view of the complicated nature of overflow water mass formation and the strength of bottom friction (Pratt, 1986), there is little reason to believe that overflows should possess uniform potential vorticity. The primary dynamical complication resulting from this nonuniformity is the presence of additional wave modes associated with the potential-vorticity-gradient restoring mechanism. These potential vorticity waves (related to Rossby waves) give rise to new modes of criticality and critical control (Pratt and Armi, 1987).

The semigeostrophic equations (2.1) – (2.4) are once again used, and the differential equation (2.8) remains the same with G now a function of ψ. In the case of a rectangular cross section this equation may be restated in terms of the total depth:

$$\frac{\partial^2 D}{\partial x^2} - \frac{Df}{g} G(\psi) + \frac{f^2}{g} = 0. \qquad (3.1)$$

In order to express ψ in terms of D, the geostrophic equation (with η replaced by D) is multiplied by D and the relation (2.6a) is used, giving:

$$\psi = \frac{g}{2f} \left\{ D^2(x,y) - D^2\left(-w/2, y\right) \right\}. \qquad (3.2)$$

The value of ψ along the wall $x = -w/2$ has arbitrarily been chosen as zero.

The solution of equation (3.1) will contain two integration "constants," and, since the channel has a rectangular cross section, these are most conveniently expressed in term of $\overline{D}$ and $\hat{D}$. The next step is to construct the function F, and, as was demonstrated in section 2, this may be done in several ways, each form providing different physical insights. Again, the branch-point condition together with the requirement of finite $\partial D/\partial y$ determines the position of the critical section. If $w = $ constant, this section lies at a sill. For a channel of varying width but constant h, the branch-point solution is finite only where the width has an extremum or where $\partial D^-/\partial w$ and $\partial E^+/\partial w$ vanish. (Here E^- and E^+ are specific energy at $x = -w/2$ and $x = w/2$, respectively.) Pratt and Armi (1987) argue that the last condition can be met only for a flow separated from one of the side walls.

For a flow separated from the wall at $x = -w/2$ conservation of volume flux necessitates that the depth along $x = w/2$ is independent of y. The branch-point/critical condition then requires that a stagnation point ($v = 0$) exists at $x = w/2$ implying a recirculation upstream of the critical section. As was mentioned earlier, recirculations render the pre-specification of potential vorticity less convincing, and it is noteworthy that experimental verification of theories based on such specification (cf. Shen, 1981) has been poor when recirculations exist.

Pratt and Armi (1987) also demonstrate that the branch-point condition implies that small amplitude waves propagating in the basic potential vorticity distribution are stationary when the specific energy E^- and E^+ are minimized.

As an example, Pratt and Armi consider the special case of nonuniform potential vorticity flow in a nonrotating channel, so that $\frac{\partial D}{\partial x} = 0$. The flow supports a free surface gravity wave and a discrete spectrum of potential vorticity waves. For the potential vorticity distribution $G = G_0 - \alpha\psi$, with α and G_0 constant, the cross-channel structure of the velocity v is given by

$$\frac{\alpha^2 v}{\partial x^2} + \alpha D^2(y)v = 0 \qquad (3.3)$$

[This equation is obtained by inserting the expression for G in equation (2.4), differentiating the result, and finally making use of (2.6a)].

For $\alpha < 0$ the potential vorticity gradient is directed in the positive$-x$ direction and the potential vorticity waves (which tend to propagate with high potential vorticity to their right) move in the same direction as the mass transport (here the positive-y direction). The only wave which can become stationary in the flow is the gravity mode. Hence, for a given transport there is only one critical steady state.

For $\alpha > 0$ the direction of the potential vorticity gradient is reversed, allowing propagation of potential vorticity waves against the flow. Multiple critical states are now possible, and a unique specification of the steady controlled flow no longer exists for fixed positive transport and topography. This nonuniqueness is demonstrated schematically in Figure 6, where E is plotted as a function of D in a manner similar to Figure 4. Pratt and Armi suggest the solution controlled by the gravity mode, corresponding to the left-hand lobe in Figure 6, may be preferred on the basis of stability. Resolution of this problem will, however, require laboratory experiment or solution to an appropriate initial value problem.

From an observational point of view, the most striking feature distinguishing potential vorticity wave control from gravity wave control is the presence in the former of velocity reversals at the critical section. Figures 7 and 8 show examples of both kinds of flow. The solution controlled by the gravity wave (Figure 7) has unidirectional flow at the critical section. In Figure 8 the flow is controlled by the potential vorticity wave of the lowest horizontal mode and the along-channel velocity is reversed (i.e., flow towards the left) along one wall at the critical section. Velocity reversals have been observed in the Vema Channel (Hogg, et al., 1982 and Hogg, 1983) and in the Faroe Bank Channel (Saunders, 1990).

334

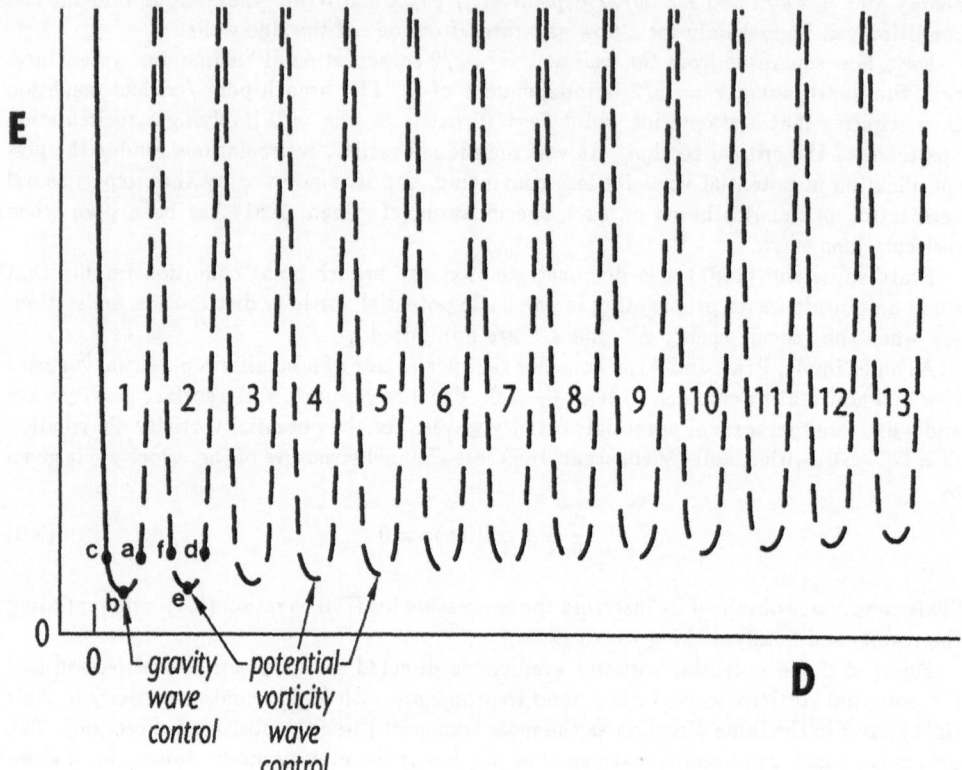

Figure 6. The Functional relation $F = $ constant for the case $\alpha > 0$.

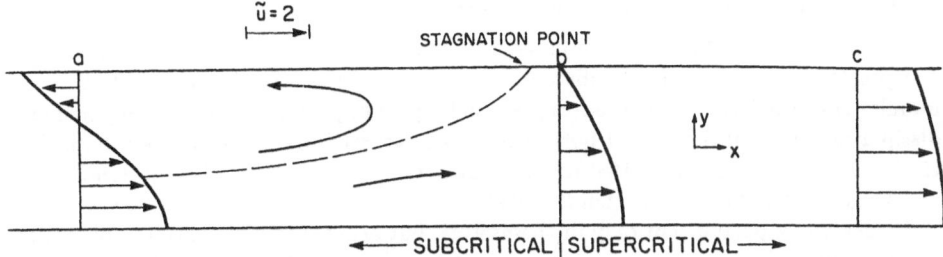

Figure 7. Plan view of a solution corresponding to lobe 1 of Figure 6. The velocity profiles at points a, b and c correspond to a, b, and c, in Figure 6.

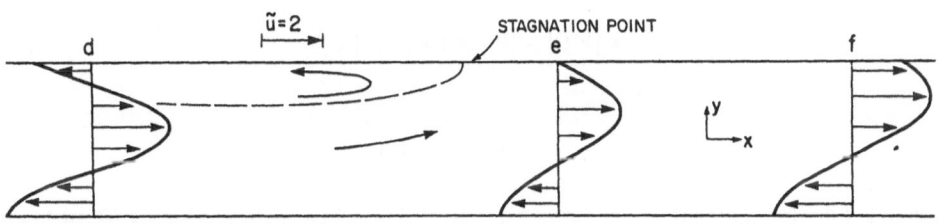

Figure 8. Same as Figure 7 except that the solution corresponds to lobe 2 of Figure 6.

4. Rotating Hydraulic Jumps and Bores

Hydraulic jumps and bores (shocks) arise naturally in straits, estuaries, and deep overflows. Ranging from strongly turbulent discontinuities in free surface or isopycnal depth to smooth, undular transitions, jumps and bores potentially play important roles in mixing and energy dissipation. Internal jumps and bores have been observed in a number of straits including the Windward Passage (Nof, 1986), the Strait of Gibraltar (La Violette and Arnone, 1988), and Knight Inlet (Farmer and Smith, 1980).

Shocks are important not only in themselves, but also for the computation of strait and sill flow. For example, it is possible to construct the time-dependent solution to various initial-value problems in which a simple steady flow is forced to adjust to an obstacle or side contraction suddenly placed in its path. The solution shows how controlled flows are established and is extremely useful in developing an intuition about hydraulic control and upstream influence. Furthermore, the solution gives the threshold obstacle height which must be exceeded in order for the flow to be critical, a result which might indicate the minimum sill depths or contraction widths necessary to control ocean currents. Baines and Davies (1980) have given a review of these problems for nonrotating flow.

The time-dependent solutions described above consist of segments of steady flow separated by (stationary) hydraulic jumps or (moving) bores. In order to piece together the steady components, it is necessary to employ shock-joining theory (Whitham, 1974). For prescribed values of the velocity and layer depth on one side of the shock, the theory specifies these quantities on the other side, provided that the shock speed also is given. In one-dimensional shallow flows the joining is based on conservation of the transport and the total momentum flux or "flow force." For a shock moving at speed c, these constraints result in the Rankine-Hugoniot conditions

$$c\,[VD] - [V^2 D + gD^2/2] = 0 \qquad (4.1)$$

$$c\,[D] - [vD] = 0 \qquad (4.2)$$

where [ ] denote jumps in the indicated quantity. For a hydraulic jump ($c = 0$), (4.1) and (4.2) specify the depth and velocity on one side given the depth and velocity on the other side.

In rotating-channel flows, bores and jumps have a two-dimensional structure which greatly complicates the shock-joining problem. Early investigations of rotating shocks (Yih et al., 1964; Houghton, 1969; Williams and Hori, 1970) utilized geometries which rendered the flow one-dimensional. However, the influence of side walls in rotating channels leads to two-dimensional effects in practically all cases of interest. As an example, consider the rotating hydraulic jump produced in the numerical solutions of Pratt (1983). As shown in Figure 9, the depth discontinuity is largest on the left (facing downstream) wall of the channel and decays over a distance of the order of the deformation radius. Pratt (1983) also attempted to formulate a shock-joining theory, with mass and momentum flux continuity applied in two dimensions. The two constraints (4.1) and (4.2) continue to hold, provided v is interpreted as the velocity normal to the shock at some value of (x, y). When these results

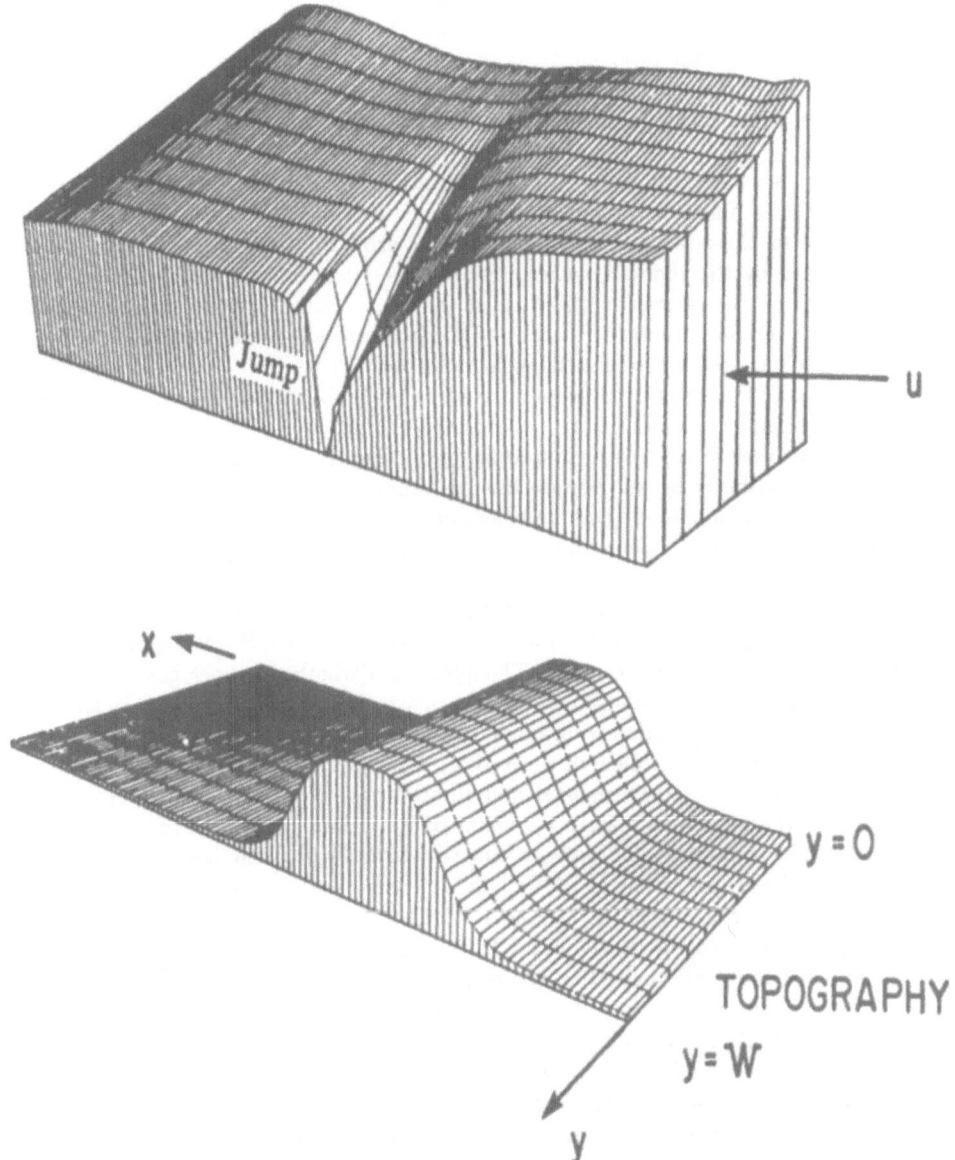

Figure 9.   Upper part shows the shape of the free surface for a rotating hydraulic jump produced from a numerical experiment of Pratt (1983). Lower part displays bottom topography.

are combined with the statement of momentum flux conservation tangent to the shock at (x, y), it follows that the tangential velocity must be continuous. These three constraints would appear to close the problem, however, an additional unknown exists — the position of the depth discontinuity (which may vary with x and t). A detailed analysis of the flow field near the discontinuity is thus required, precisely the type of calculation shock-joining seeks to avoid.

An alternative approach is to consider the implications of conservation of the cross-sectional integrals of mass and momentum flux on either side of the shock. These constraints are sufficient to close the shock-joining problem provided that the change in potential vorticity forced by the shock is known. Pratt (1983) has shown that the change at any point depends on the tangential structure of the depth discontinuity. Again, one requires foreknowledge of this structure in order to close the shock-joining problem, a difficulty comparable to that encountered in the previous approach.

Nof (1984) and Pratt (1987) have formulated shock-joining theories based on the assumption that the change of potential vorticity across the shock is zero everywhere. However, Nof (1986) has cast doubt on this approach by finding a special class of shocks in which the depth discontinuity occurs along a line perpendicular to the channel axis. This simplification makes it possible to compute the upstream and downstream end states and the potential vorticity jump. It is found that the jump in G can be as large as the upstream value of G itself.

Until recently all investigations of rotating jumps involved flows of finite depth. As discussed earlier, it is possible for the fluid to separate from the left side wall (facing downstream) of a rectangular channel. The question is whether a shock can exist in such a flow, particularly in view of the fact that shocks in attached flows have structural similarities with Kelvin waves. (Separation from the left wall removes one of the Kelvin waves.)

To investigate shock existence in a rotating, separated flow, Pratt (1987) set up a hydraulic jump in a laboratory channel and rotated the channel swiftly enough to cause separation of the flow immediately upstream of the jump. Although the upstream (supercritical) flow separated, the downstream (subcritical) flow did not, so that the fluid reattached at the original position of the jump. As the rotation rate was increased further, the supercritical flow narrowed, occupying a decreasing fraction of the channel width, while the subcritical flow remained attached. At the same time the depth jump decreased at all points along the depth discontinuity, so that the shock evolved into a lateral hydraulic jump; *i.e.*, the main discontinuity became one in stream width rather than depth.

Presently, little is known about the interior structure of rotating shocks. Pratt's experiment was rather crude and revealed little about the turbulent structures and energy dissipation in the interior. Experiments with better flow visualization and more complete velocity and depth measurements are needed to resolve these issues.

5. Some Outstanding Problems

In the course of discussion a number of timely research problems have been mentioned. We now summarize some of the more outstanding examples.

A number of the cited papers have reported on the possibility for the flow to become stagnant at the channel wall. Under this condition a reverse flow may be present upstream of the stagnation point. The stagnation of the flow may reach all the way upstream or form closed eddies. It is not known what determines the potential vorticity in such regions.

For a rectangular channel a stagnation point of a critical flow is accompanied by the free surface separating from the opposite wall. As was pointed out in section 2, a critical separated flow has never been observed, and it is possible that the recirculation area alters the initially prescribed potential vorticity. Shen (1981) suggested a pool of stagnant water separating the flow from the right hand side of the channel. Such a pool has been observed for a subcritical flow (Borenäs, 1983). The problem with reverse flows becomes even more troublesome if negative velocities are found at the critical section. This feature was reported by Borenäs and Lundberg (1986, 1988) for a flow in a parabolic channel within certain parameter ranges. Pratt and Armi (1987) found that this type of solution was also present for a nonrotating flow of nonuniform potential vorticity.

In relation to nonuniform potential vorticity flows, at least two major research problems presently exist. The first concerns the nonuniqueness question; for a given transport what selects the controlled steady solution? The simplest path toward resolution would probably be to investigate an initial-value problem based on the Pratt and Armi (1987) flow, which has relatively simple cross-channel structure. The second problem consists of finding a solution to (3.1) for nonzero f and nonconstant $dB/d\psi$, so that the effects of rotation and nonuniform potential vorticity can be considered simultaneously.

The most common assumption when working on hydraulic problems is that variations along the channel are small compared to those encountered across. The cross-channel momentum equation is then reduced to the geostrophic relation [cf. equation (2.1)]. The importance of the ageostrophic terms is measured by the parameter δ^2, where δ is the ratio of the Rossby radius of deformation based on the depth scale, to the scale of along-channel variations. In a study by Pratt (1984a) the effects of including the non-linear terms in the cross-channel momentum equation were examined for small but finite δ. The analysis, which was restricted to obstacles of small heights, showed that oscillatory behavior could be found in the along-channel direction. It was revealed that not only the height but also the shape of the obstacle could influence the upstream flow.

Finally, the general shock-joining problem for bores and jumps in a rotating-channel is still unsolved. Given the upstream flow and shock speed, there is no available means of computing the downstream flow without previous assumption or detailed analysis of the shock interior.

Acknowledgements

This work was supported by the Nordic Council for Physical Oceanography, the Swedish Natural Science Research Council under contract R-RA 9811-300, and the US Office of Naval Research under grant N00014-89-K-1182. Technical assistance from Agneta Malm, Veta Green, and Anne-Marie Michael is greatly appreciated.

References

Baines, P. G. and Davies, P. A., 1980. Laboratory studies of topographic effects in rotating and/or stratified fluids, in *Orographic Effects in Planetary Flows*, GARP Publ. Ser. No. 23, World Met. Org.

Borenäs, K., 1983. Subcritical rotating channel flow across a ridge, in H. M. Gade, A. Edwards and H. Svendsen (eds.), *Coastal Oceanography*, Plenum Press, New York, 363–372.

Borenäs, K. and Lundberg, P., 1986. Rotating hydraulics of flow in a parabolic channel, *J. Fluid Mech.*, **167**, 309–326.

Borenäs, K. M. and Lundberg, P. A., 1988. On the deep-water flow through the Faroe Bank Channel, *J. Geophys. Res.*, **93**, 1281–1292.

Chow, V. T., 1959. Open Channel Hydraulics, McGraw-Hill, New York, 680 pp.

Dalziel, S. B., 1990. Rotating two-layer sill flows, in L. J. Pratt (ed.), *The Physical Oceanography of Sea Straits*, Kluwer Academic Publishers Dordrecht.

Farmer, D. M. and Smith, J. D., 1980. Tidal interaction of stratified flow with a sill in Knight Inlet, *Deep Sea Research*, **27A**, 239–254.

Garrett, C. J. R. and Toulany, B., 1982. Sea level variability due to meteorological forcing in the northeast Gulf of St. Lawrence, *J. Geophys. Res.*, **87**, 1968–1978.

Gill, A. E., 1977. The hydraulics of rotating-channel flow, *J. Fluid Mech.*, **80**, 641–671.

Hogg, N. G., 1983. Hydraulic control and flow separation in a multi-layered fluid with application to the Vema Channel, *J. Phys. Oceanogr.*, **13**, 695–708.

Hogg, N. G., Biscaye, P., Gardner, W., and Schmitz, W. J., 1982. On the transport and modification of Antarctic Bottom Water in the Vema Channel. *J. Mar. Res.*, **40**, 231–283.

Houghton, D. D., 1969. Effect of rotation on the formation of hydraulic jumps, *J. Geophys. Res.*, **74**, 1351–1360.

La Violette, P. F. and Arnone, R. A., 1988. A tide-generated internal waveform in the western approaches to the Strait of Gibraltar. *J. Geophys. Res.*, **93**, 15653–15667.

Lee, A. J., 1967. Temperature and salinity distributions as shown by sections normal to the Iceland-Faroe Ridge, *Rapp. P. V. Reun. Cons. Int. Explor. Mer.*, **157**, 100–135.

Long, R. R., 1954. Some aspects of the flow of stratified fluids. II. Experiments with a two-fluid system, *Tellus*, **6**, 97–115.

Nof, D., 1984. Shock waves in currents and outflows, *J. Phys. Oceanogr.*, **14**, 1683–1702.

Nof, D. 1986. Geostrophic shock waves, *J. Phys. Oceanogr.*, **16**, 886–901.

Pratt, L. J., 1983. On inertial flow over topography. Part 1. Semigeostrophic adjustment to an obstacle, *J. Fluid Mech.*, **131**, 195–218.

Pratt, L. J., 1984a. On inertial flow over topography. Part 2. Rotating-channel flow near the critical speed, *J. Fluid Mech.*, **145**, 95–110.

Pratt, L. J., 1984b. A time-dependent aspect of hydraulic control in straits, *J. Phys. Oceanogr.*, **14**, 1414–1418.

Pratt, L. J., 1986. Hydraulic control of sill flow with bottom friction, *J. Phys. Oceanogr.*, **16**, 1970–1980.

Pratt, L. J., 1987. Rotating shocks in a separated laboratory channel flow, *J. Phys. Oceanogr.*, **17**, 483–491.

Pratt, L. J. and Armi, L., 1987. Hydraulic control of flows with nonuniform potential vorticity, *J. Phys. Oceanogr.*, **17**, 2016–2029.

Rouse, H., 1961. Fluid Mechanics for Hydraulics Engineers, Dover Publishing Company, New York, 422 pp.

Reed, L. P., 1980. Curvature effects on hydraulically driven inertial boundary currents, *J. Fluid Mech.*, **96**, 395–412.

Sambuco, E. and Whitehead, J. A., 1976. Hydraulic control by a wide weir in a rotating fluid, *J. Fluid Mechanics*, **73**, 521–528.

Saunders, P. M., 1990. Cold outflow from the Faroe Bank Channel. *J. Phys. Oceanogr.*, **20**, 29–43.

Shen, C. Y., 1981. The rotating hydraulics of open-channel flow between two basins, *J. Fluid Mech.*, **112**, 161–188.

Stern, M. E., 1972. Hydraulically critical rotating flow, *Phys. Fluids*, **15**, 2062–2065.

Whitehead, J. A., 1986. Flow of a homogeneous rotating fluid through straits, *Geophys. Astrophys. Fluid Dyn.*, **36**, 187–205.

Whitehead, J. A., Leetmaa, A. and Knox, R. A., 1974. Rotating hydraulics of strait and sill flows, *Geophys. Fluid Dyn.*, **6**, 101–125.

Whitham, G. B., 1974. *Linear and Nonlinear Waves*, Wiley Interscience, New York, 636 pp.

Williams, R. T. and Hori, A. M., 1970. Formation of hydraulic jumps in a rotating system, *J. Geophys. Res.*, **75**, 2813–2821.

Yih, C. S., Gascoigne, H. E. and Debler, W. R., 1964. Hydraulic jump in rotating fluid, *Phys. Fluids*, **7**, 638–642.

ROTATING TWO-LAYER SILL FLOWS

STUART B DALZIEL
Department of Applied Mathematics and Theoretical Physics
University of Cambridge
Silver Street
Cambridge CB3 9EW
England

ABSTRACT. Maximal exchange flow is investigated for a rotating channel containing a simple sill. Attention is restricted to situations where there is no net flow along the channel and the channel width is less than one Rossby radius at the sill crest. The channel cross-section is taken to be rectangular and the potential vorticity of the fluid to be zero. The hydraulic problem is formulated as a single functional with a number of special features. This formalism was introduced by Gill (1977) for single-layer flows and extended by Dalziel (1989) for two-layer nonrotating flows. The flow is found to be very sensitive to small departures from reflectional symmetry about a horizontal plane, as are introduced by the sill geometry. If the total channel depth away from the crest of the sill is greater than approximately 150% of that at the crest, the resulting flow is almost indistinguishable from the case where the depth goes to infinity. As found by Whitehead, Leetmaa & Knox (1974) for a simple channel of constant depth, rotation leads to a decrease in the exchange flow rate along the channel. For channels greater than approximately one local Rossby radius in width at the sill crest, the exchange flow rate is independent of the channel width. A second effect of rotation is to increase the communication between the two layers, reducing the influence of any asymmetry in the geometric forcing.

1. Introduction

Two-layer exchange flows arise in a wide variety of situations ranging from thermally driven airflow within buildings to salinity driven flow through oceanographic straits. In this paper we shall concentrate on the latter where the earth's rotation may play a significant role. The earliest studies with an oceanographic context are those by Marsigli (1681 – see *pp* 147-149, Deacon, 1971) and von Waitz (1755 – see Deacon, 1985) who studied the flow through the Bosporus Strait and the Strait of Gibraltar, respectively.

Progress in understanding two-layer exchange flows has been slow. Stommel & Farmer (1953) introduced some of the key ideas and demonstrated the relevance of two-layer hydraulics to oceanography. The work of Wood (1968, 1970) showed the concept of hydraulic control to be more complicated than for single layer flows. A number of basic aspects of these flows were missed until a series of papers by Armi and Farmer (Armi, 1986; Armi & Farmer 1986; Farmer & Armi,

343

L. J. Pratt (ed.), The Physical Oceanography of Sea Straits, 343–371.

1986). These two authors pointed out the fundamental difference between the flow through a contraction in width and that over a sill. Variations in the channel width are felt equally by both layers, whereas variations in the depth (unless such variations are symmetric in both top and bottom boundaries) are felt directly by only one of the two layers.

Armi (1986) demonstrated the possibility of formulating two-layer hydraulics as a quasi-linear hyperbolic problem, though chose to base his solution method on a description of the flow in terms of the layer Froude numbers. While this has the elegance associated with its nondimensionality, and may be extended to channels of nonrectangular cross-section without too much difficulty, it obscures some of the essential physics of the flow. Furthermore, the layer Froude number approach is not a convenient vehicle for investigating the effects of rotation where the fluid velocities vary across the channel.

There have been many studies of the effect of rotation on single-lay hydraulic flows. Following the introduction of a simple zero potential vorticity model by Whitehead, Leetmaa & Knox (1974), Gill (1977) outlined the underlying mathematical structure for uniform potential vorticity flows in channels of rectangular cross-sections. Shen (1981) demonstrated the effects of the downstream reservoir and nonrectangular cross-sections (for zero potential vorticity flows) as well as discussing the limitations to hydraulic theory in channels much wider than the local Rossby radius. The effect of uniform potential vorticity in parabolic cross-sections was studied by Borenas & Lundberg (1986), while nonuniform potential vorticity in rectangular cross-sections has been investigated by Pratt & Armi (1987). Bormans & Garrett (1989) have applied some of these ideas to the surface inflow through the Strait of Gibraltar.

Whitehead *et al.* (1974) made a preliminary attempt at analysing the effect of background rotation on two-layer exchange flows. They found that rotation played an important role when the local internal Rossby radius was of the same order as the channel width. To determine the flow they assumed the rate of increase in kinetic energy of the mean motion was equal to the rate at which potential energy was released. For the flow they looked at – zero potential vorticity along a channel of uniform rectangular cross-section – this assumption proves correct. However, for more general along-channel geometries the set-up process is dissipative (Dalziel, 1989) and for nonzero potential vorticity some of the energy is radiated away by Poincaré waves (Rossby, 1938; Gill, 1976; Dalziel, 1988).

In this paper we analyse the effects of rotation on the exchange flow over a sill of simple geometry. To simplify the problem we restrict our attention to situations where there is no net flow over the sill. In addition we shall assume the flow is of zero potential vorticity. Dalziel (1988) has shown that the effects of small, uniform, nonzero potential vorticity are very small for channels of constant depth.

The flow is analysed using a functional formulation based on Gill's (1977) approach to single-layer hydraulics and adapted by Dalziel (1988, 1989) for two-layer flows. This formulation may be extended readily to include more complex along-channel geometries, the addition of a net flow and nonzero potential vorticity (Dalziel, 1988).

2. Geometry and Equations

2.1. MODEL GEOMETRY

Figure 1 shows the typical along-channel geometry we shall use in this paper. The cross-section of the channel is taken to be rectangular. Depth variations are confined to regions of the

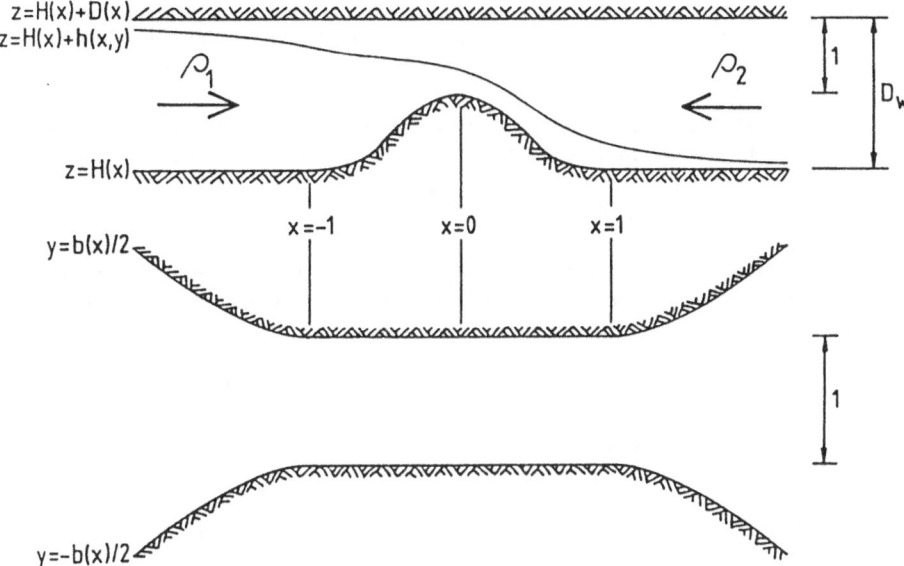

Figure 1: Schematic diagram of a simple sill. The x values are for identification purposes only: the along-channel length scale does not enter into the problem.

channel where the width is constant; width variations are confined to regions where the depth is constant. Variations in both width and depth occur over length scales much greater than the depth of the channel. The channel contains two homogeneous fluid layers flowing in opposite directions. The lower layer, denoted by subscript 1, has density ρ_1 and flows towards the right. The upper layer of density ρ_2 flows towards the left. We shall assume the fluids are Boussinesq. The sections marked $x=\pm1$, 0 are included as a means of identifying three important geometrical features within the channel. For steady flows the along-channel length scale will not enter the problem and so the numerical values shown have no physical significance.

For later convenience we shall nondimensionalise our system of equations as follows:

$$(x,y)^* = \frac{f\,(x,y)}{(D_m g')^{\frac{1}{2}}}, \qquad\qquad z^* = z/D_m,$$

$$(u,v)_i^* = \frac{(u,v)_i}{(D_m\,g')^{\frac{1}{2}}}, \qquad\qquad t^* = ft,$$

$$p_i^* = \frac{2p_i}{(\rho_1+\rho_2)D_m g'}, \qquad\qquad \rho_i^* = \frac{2\rho_i}{\rho_1+\rho_2}, \qquad (1)$$

where the superscript stars indicate dimensionless quantities and g' $(=2(\rho_1-\rho_2)/(\rho_1+\rho_2))$ is the reduced gravity. The vertical length scale D_m is the channel depth at the crest of the sill and the horizontal length scale $(D_m g')^{1/2}/f$ is the local Rossby radius of deformation at this section. The right-handed coordinate system is oriented with the x axis along the channel such that the lower layer velocity is positive. Rotation is about the vertical (z) axis at an angular velocity of f/2.

One characteristic of hydraulic flows is that the Reynolds number is sufficiently large for viscous effects to be ignored. In addition, if for some region of the channel the height of the interface changes over length scales which are large compared with the total depth of the channel, we may apply the shallow water equations using the hydrostatic approximation. Moreover, if the cross-channel component of the velocity remains small compared with the along-channel component (the conditions for this assumption to apply will be discussed in section 5), we may eliminate the pressure between the shallow water equations for the two layers to give the semigeostrophic equations

$$\frac{\partial}{\partial t}(u_1 - u_2) + \frac{1}{2}\frac{\partial}{\partial x}(u_1^2 - u_2^2) = -\frac{\partial}{\partial x}(H+h), \tag{2a}$$

$$0 = -\frac{\partial}{\partial y}(H+h) + (u_1 - u_2), \tag{2b}$$

$$\frac{\partial h}{\partial t} + \frac{\partial}{\partial x}(h u_1) = 0, \qquad\qquad \frac{\partial h}{\partial t} - \frac{\partial}{\partial x}[(D-h)u_2] = 0, \tag{2c}$$

where superscript stars have been dropped, $H=H(x)$ is the elevation of the channel floor above some datum, $D=D(x)$ is the total depth of the channel, and $h=h(x,y)$ the thickness of the lower layer.

Integrating equation (2a) yields the difference in the Bernoulli potentials G_i between the two layers, viz.

$$\Delta G = (\partial/\partial t)(\Phi_1 - \Phi_2) + \tfrac{1}{2}(u_1^2 - u_2^2) + H + h, \tag{3}$$

where Φ_i are velocity potentials ($u_i = \nabla\Phi_i$). Equation (2b) is the geostrophic balance for a *relatively straight* flow. The potential vorticities for the two layers, Π_1 and Π_2, are given by

$$\Pi_1 = \frac{1 - \partial u_1/\partial y}{h}, \qquad\qquad \Pi_2 = \frac{1 - \partial u_2/\partial y}{D-h}. \tag{4}$$

As we are considering zero potential vorticity flows, equations (4) require uniform shear in the two layers. Moreover, the relationship between the Bernoulli potentials and potential vorticities for steady flow, $\Pi_i = \partial G_i/\partial\psi_i$ (where the stream functions ψ_i are defined by $u = \nabla\wedge(k\,\psi_i)$, and k is the unit vector in the vertical direction), demonstrates that G_i are independent of position, so ΔG is a universal constant. [Steady flow with nonzero potential vorticity requires ΔG to be a function of y].

The cross-channel structure of the flow falls into two categories: *attached* and *separated*

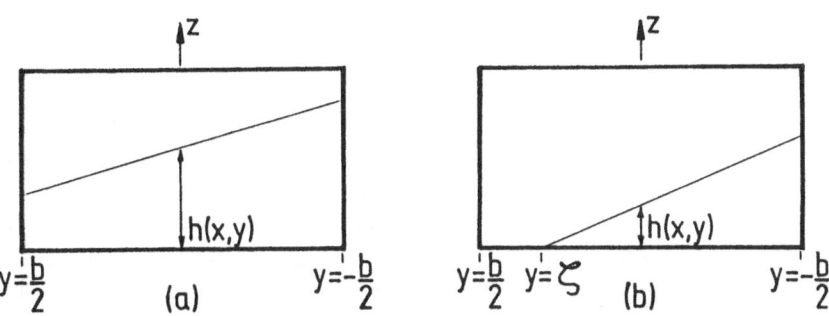

Figure 2: Typical cross-sections of hydraulic flow: (a) attached flow with two layers over entire channel width; (b) separated flow with two layers over only part of channel width.

flow. By *attached* flow we mean that, for a given cross-section x, there are two layers over the entire width of the channel. The flow is *separated* at a given section if for some region of the cross-section only one fluid layer is present. Examples of these two situations are illustrated in figure 2.

2.2. ATTACHED FLOW

For a section containing attached flow we may solve (2b) and (4) simultaneously to give the cross-channel structure of the interface height and layer velocities, *viz.*

$$h = D(\tfrac{1}{2} + A + 2By/b), \tag{5a}$$

$$u_1 = y + c - BD/b, \tag{5b}$$

$$u_2 = y + c + BD/b, \tag{5c}$$

where $b=b(x)$ is the dimensionless channel width (*ie.* the channel width in terms of the local Rossby radius at the shallowest section; $b=0$ implies the channel is not rotating). The functions of integration $A=A(x)$, $B=B(x)$ and $c=c(x)$ are unknown at this stage. They describe the along-channel structure of the flow. We shall term $A(x)$ the *interface height coefficient* and $B(x)$ the *interface slope coefficient*. For the class of flows we are considering $B\leq0$ (u_1 essentially positive) and the lower layer is banked up against the right-hand wall (looking downstream with respect to that layer) by the Coriolis force.

By integrating hu_1 and $(D-h)u_2$ over the channel width we may determine the layer volume flow rates, q_1 and q_2, in the two layers. Defining the exchange flow rate as

$$\overline{q} = q_1-q_2 = |q_1|+|q_2|, \tag{6}$$

(5) shows that for attached flows

$$\overline{q} = D(\tfrac{1}{3}Bb^2 - BD + \tfrac{1}{2}Acb). \tag{7}$$

Similarly, the net flow rate, defined as

$$Q = q_1 + q_2 = |q_1| - |q_2|, \tag{8}$$

is given by

$$Q = D(cb - 2ABD). \tag{9}$$

As we have restricted attention to situations where there is no net flow (Q=0), we may utilise (9) to eliminate the function of integration c from equations (5) and (7). For later use we note that equation (7) may be rearranged to give an expression for the interface slope coefficient B in terms of the interface height coefficient A and the exchange flow rate $\overline{q}$:

$$B = \frac{\overline{q}}{D[\frac{1}{3}b^2 - 4D(\frac{1}{4} + A^2)]}. \tag{10}$$

Much of the discussion in the following sections will be in terms of the *specific exchange flow rate*, $\overline{q}/b$, to show more clearly the dynamical effects of rotation (remember that the dimensionless channel width b is the width of the channel in terms of the local Rossby radius at the shallowest section).

2.3. SEPARATED FLOW

There are three possibilities for separated flow: the interface may intersect the floor of the channel, it may intersect the top of the channel, or it may intersect *both* the top and bottom of the channel at a given section. As the rotation rate is increased for a channel of given geometry, separation will occur first on the supercritical flows leading towards the reservoir. On the left-directed flow the interface will begin to intersect the top of the channel, while on the right-directed flow the interface will intersect the channel floor. In this section we shall analyse the intersection of the interface with the channel floor. Intersection with the top may be treated in a similar manner. Conditions giving rise to the interface intersecting both top and bottom at a given section are beyond the scope of this paper.

The interface configuration illustrated in figure 2b shows a two-layer region extending from y=-b/2 to y=ζ. The remainder of the cross-section, from y=ζ to y=b/2, is occupied by the upper layer only. Within the two-layer region the geostrophic balance condition (2b) must hold, allowing the y dependence of the flow to be determined from the conservation of potential vorticity. In the single-layer region Π_2 must still vanish. Moreover $\Pi_2 = 0$ requires the velocity in the upper layer to be a continuous function of y over the whole channel width so we may write

$$h = \begin{cases} D(\frac{1}{2} + A + 2By/b), & -b/2 \leq y \leq \zeta \\ 0, & \zeta < y \leq b/2 \end{cases} \tag{11a}$$

$$u_1 = \begin{cases} y + c - BD/b, & -b/2 \le y \le \zeta \\ \\ \text{undefined}, & \zeta < y \le b/2 \end{cases}$$ (11b)

$$u_2 = y + c + BD/b.$$ (11c)

Again the interface height coefficient $A=A(x)$, the interface slope coefficient $B=B(x)$ and the velocity coefficient $c=c(x)$ are functions of integration. As in the previous section $B \le 0$ for the class of flows we are investigating. From the geometry of the interface,

$$\zeta = - \frac{\frac{1}{2}+A}{2B} b,$$ (12)

which may be used to eliminate ζ.

Integrating hu_1 and $(D-h)u_2$ over the channel width again allows us to determine the exchange flow rate. For this separated section the exchange flow rate is

$$\bar{q} = D \left\{ \frac{B^3 - (\frac{1}{2}+A)^3}{6B^2} b^2 - \left[\frac{1}{2} + A + \frac{2Bc}{b} \right] \frac{B^2 - (\frac{1}{2}+A)^2}{4B^2} b^2 \right.$$

$$\left. - (\frac{1}{2}-A)cb - c\frac{(\frac{1}{2}+A)^2}{B}b - BD \right\}.$$ (13)

Evaluating the net flow rate and setting it to zero allows us to eliminate c using

$$c = \frac{Q/D + BD(\frac{1}{2}-A) + \frac{1}{2}D[B^2 + (\frac{1}{2}+A)^2]}{b}.$$ (14)

In contrast to the previous section, we are not able to rearrange (13) and (14) to give an explicit expression for B in terms of A and $\bar{q}$. This is of importance when calculating the complete solution, as the numerical determination of B for separated flows is very costly compared with attached flows. The problem is further hampered by the existence of multiple solutions, only one of which is accessible to the flow.

3. Formulation

3.1. HYDRAULIC FUNCTIONAL

Following the work of Gill (1977) for rotating single-layer flows and Dalziel (1989) for nonrotating two-layer flows, we formulate a hydraulic functional of the form

$$J(D, H, b, Q, \overline{q}, \mathcal{G}; A) = 0, \tag{15}$$

to describe the flow. Functions satisfying $J(\bullet;A)=0$ must conserve volume flow rates and Bernoulli potentials in the two layers. The single dependent function $A=A(x)$ (the interface height coefficient) depends on the along-channel coordinate x only through the geometric parameters D, H and b (ie. we may write $A=A(D,H,b)$). The net flow rate Q is a prescribed parameter; for this paper $Q=0$. Nonzero net flow rates are covered by Dalziel (1989) for nonrotating flows and Dalziel (1988) for rotating flows. The exchange flow rate $\overline{q}$ and constant $\mathcal{G}$ are parameters which select between the various conceivable flows. The constant $\mathcal{G}$ is similar to the constant on the right hand side of Gill's (1977) equation (3.1); the physical significance of $\mathcal{G}$ is described later in this section.

An important feature of hydraulic problems is that $J(\bullet;A)=0$ is multiple valued for some range of D, H, b, Q, $\overline{q}$, $\mathcal{G}$ in that there is more than one function A satisfying (15). In order that more than one solution branch is accessible to the flow, there must be a *constriction*, in the sense that

$$K = \frac{\partial J}{\partial D}\frac{dD}{dx} + \frac{\partial J}{\partial H}\frac{dH}{dx} + \frac{\partial J}{\partial b}\frac{db}{dx} = 0, \tag{16}$$

at some point along the channel. Different solution *sheets* of the surface J in D, H, b, Q, $\overline{q}$, $\mathcal{G}$, A space meet along lines defined by

$$\partial J/\partial A = 0. \tag{17}$$

Differentiation of (15) with respect to x shows that $(\partial J/\partial A)(dA/dx)=-K$ along such lines. Unless $K=0$, dA/dx must be infinite where sheets meet; this violates the fundamental assumptions of the flow. Further differentiation with respect to x reveals that $K=0$ must be a constriction and not an expansion.

As pointed out by Dalziel (1989), it is desirable for J to be defined uniquely in terms of its parameters. For this reason, plus the simple relationship between J and the composite Froude number for nonrotating channels, we adopt a functional of the same form as Dalziel (1989):

$$J(D, H, b, Q, \overline{q}, \mathcal{G}; A) = \mathcal{G} - \tfrac{1}{2}(u_1^2 - u_2^2) - (H+h). \tag{18}$$

The layer velocities and interface height may be written in terms of A, B and y using equations (5) and (11) for attached and separated flows respectively. For attached flows we may eliminate B to give

$$J(D,H,b,Q,\overline{q},\mathcal{G};A) = \mathcal{G} + \frac{4A\overline{q}^2}{b^2[b^2 - 4D(\tfrac{1}{4}-A^2)]^2} - H - D(\tfrac{1}{2}+A), \tag{19}$$

for no net flow $(Q=0)$. Unfortunately, as we are unable to write an explicit expression for B in terms of A and $\overline{q}$ for separated flows, we can not give J explicitly. In terms of A, B and c the functional is

$$J(D, H, b, Q, \overline{q}, \mathcal{G}; A) = \mathcal{G} + 2DBc/b - H - D(\tfrac{1}{2}+A), \tag{20}$$

where c is given by (14) and B must be determined numerically from equations (13) and (14).

The requirement $J(\bullet;A)=0$ is a statement of conservation of Bernoulli potential; $\mathcal{G}$ is simply the value of G_1-G_2 for the flow. Conservation of mass has been ensured by the method of calculating the interface slope coefficient B.

3.2. INFORMATION PROPAGATION

In hydraulics the propagation of long, small amplitude gravity waves play an important role. For two-layer problems such waves are supported on the density interface and at the free surface (should this exist). In this paper we shall assume the flow has a rigid lid (external Froude number much less than unity) and so we need not consider the surface modes.

Traditionally the composite Froude number F is used to characterise the propagation of long waves on the interface. Previous authors (*eg.* Armi, 1986) have shown that the composite Froude number may be written in terms of the Froude numbers F_1, F_2 of the two layers as

$$F^2 = F_1^2 + F_2^2, \tag{21}$$

where

$$F_1^2 = u_1^2/(hg'), \qquad\qquad F_2^2 = u_2^2/[(D-h)g']. \tag{22}$$

Dalziel (1989) has shown that (21) may be expressed in terms of the long wave phase velocities C_1, C_2, (relative to the geometry) as

$$F^2 = 1 + \frac{D}{(D-h)\ h} C_1 C_2. \tag{23}$$

We may analyse waves on the interface in terms of the hydraulic functional by differentiating J with respect to x for a channel of uniform cross-section (any variations in geometry occur over length scales large compared with the depth of the fluid and so this approach is valid),

$$\partial J / \partial x = (\partial J/\partial h)(\partial h/\partial x),$$

$$= -u_1\partial u_1/\partial x + u_2\partial u_2/\partial x - \partial(H+h)/\partial x, \tag{24}$$

which may be compared with equation (2a) to show that for a time dependent flow

$$(\partial J/\partial h)(\partial h/\partial x) = (\partial/\partial t)(u_1-u_2). \tag{25}$$

For small amplitude travelling wave solutions we may write $\partial/\partial t=-C_n\partial/\partial x$, where C_n, $n=1, 2$ are the phase velocities of such waves. Eliminating the layer velocities from (24) using (25) and the continuity equation (2c) allows us to obtain the relationship for the phase velocity of long, small amplitude gravity waves in terms of $\partial J/\partial h$:

$$(D-h)\ h\ \frac{\partial J}{\partial h}\ \frac{\partial h}{\partial x} - C_n[(D-h)u_1 + hu_2]\frac{\partial h}{\partial x} + DC_n^2\frac{\partial h}{\partial x} = 0 . \quad (26)$$

From equations (5a) and (26) we can see the composite Froude number is related to $\partial J/\partial A$ by

$$F^2 = 1 + \frac{1}{D[1+2(\partial B/\partial A)(y/b)]}\ \frac{\partial J}{\partial A}. \quad (27)$$

The effect of the cross-channel variations of the particle velocities, u_1, u_2, is to introduce some y dependence for the composite Froude number. Hence we must consider (21) to (27) as local Froude numbers which do not necessarily characterise the information propagation of the flow as a whole at a given section.

In section 5 we shall demonstrate that, for the range of channels we are considering, the denominator of (27) is always positive. Thus while the local Froude number varies across the channel, it remains finite, real *and* on the same side of unity. Using this observation we shall define a *channel-wide* or *section* composite Froude number,

$$F_s^2 = 1 + D^{-1}\partial J/\partial A = 1 + \frac{C_1 C_2}{D(\frac{1}{2}+A)(\frac{1}{2}-A)}, \quad (28)$$

which summarises the nature of the flow in terms of information propagation over the entire width of the channel. Note that F_s^2 is effectively the local Froude number at the centre of the channel.

The expression for the section composite Froude number, given by equation (28), facilitates the understanding of information propagation. If C_1 and C_2 are the same sign, both waves propagate in the same direction (relative to the geometry) and F_s^2 is greater than unity. The flow is therefore *supercritical* and is able to communicate information about disturbances only in the direction of *both* C_1 and C_2. When C_1 and C_2 are of opposite sign $F_s^2<1$. This *subcritical* flow is able to propagate information in both directions by the long wave mechanism. The flow is *critical* at sections where one or both of the phase velocities vanish.

Equations (27) and (28) show that supercritical flows correspond to roots of $J(\bullet;A)=0$ where $\partial J/\partial A>0$, and subcritical flows to roots of $J(\bullet;A)=0$ where $\partial J/\partial A<0$. Critical conditions, $\partial J/\partial A=0$, correspond to lines along which different sheets of the solution to $J(\bullet;A)=0$ meet. For any given section $J(\bullet;A)$ varies with A in a *cubic-like* manner, though the order of the functional is not integral. In general $J(\bullet;A)=0$ will have three real solutions. Two of these solutions will be supercritical ($\partial J/\partial A>0$) and one subcritical ($\partial J/\partial A<0$). The two supercritical solutions differ in that one has both phase velocities towards the left and the other has both phase velocities towards the right. The direction of the phase velocities is obtained from the second term of (26) as

$$C_1+C_2 = [(D-h)u_1 + hu_2]/D. \tag{29}$$

Typically C_1+C_2 will be in the direction of the thinner, faster moving layer. The value of A for the subcritical solution is intermediate between the values of A for the two supercritical solutions.

From the basic symmetry of the channel geometry shown in figure 1, the only way for the height of the interface to differ between the two reservoirs is for there to be at least one hydraulic transition, where the solution passes from one branch of the solution to another. To understand the flow properly, we must look at the properties of the various conceivable transitions. We can reduce the size of the set by noting that any transition will be between a supercritical flow and a subcritical flow. Under some circumstances the distance between two such transitions may be vanishingly small, but this does not affect the analysis. There remain essentially two cases: the phase velocities in the supercritical region adjoining the subcritical region may be either towards or away from the subcritical region.

Transitions where the phase velocities in the supercritical region are away from the subcritical region are stable in the sense that disturbances propagating towards the transition in the subcritical region are accelerated away from the transition on entering the supercritical region. The transition will be characterised by a smooth, continuous change in the flow variables.

In contrast, if the phase velocities in the supercritical region are towards the subcritical region, then the slower moving wave in the supercritical region will be brought to rest on passing through the transition. Such a situation is unstable as disturbances will accumulate at the transition. In real flows the flow will adjust itself so that this does not occur, possibly by the propagation of an internal bore or the formation of a hydraulic jump. This instability has been discussed in detail by Pratt (1984).

Hydraulic jumps will generally form on the supercritical flow towards the reservoirs. The resultant discontinuity in interface height, velocity and Bernoulli potential is important for matching supercritical channel flows on to subcritical reservoirs. Pratt (1983) has shown that the change in potential vorticity across a hydraulic jump is related to the rate of change of the jump amplitude along the line of depth discontinuity. If the jump forms a straight line,

$$_A[\Pi]_B = \frac{1}{4Vh} \frac{\partial}{\partial s} \frac{_A[h]_B^3}{h_A h_B}, \tag{30}$$

(equation (6.6) in Pratt, 1983) where $_A[\bullet]_B$ represents the jump in the quantity $\bullet$ over the hydraulic jump between locations A and B; s is an along-jump coordinate and V is the fluid velocity normal to the jump seen in a frame moving with the jump (Vh is conserved). This finding is supported by Pratt's (1983) numerical work. Nof (1986) analysed a special class of jumps where the cross-channel component of velocity vanishes everywhere and the jump takes the form of a straight line normal to the flow. If a weak jump of this form were to occur, the change in potential vorticity would vary with the jump amplitude as

$$_A[\Pi]_B \sim {}_A[h]_B^2. \tag{31}$$

Thus, provided the hydraulic jump is small, or the cross-channel structure dependent only weakly on y, the potential vorticity will remain approximately zero over the channel width.

Moreover, Wood & Simpson (1984) have shown that the change in the Bernoulli potential of the contracting layer is small for nonrotating channels. If this carries over to rotating jumps then we would expect the potential vorticity (which is related to the Bernoulli potential by $\Pi_i = dG_i/d\psi_i$) of the layer flowing out of the reservoir to be essentially unchanged by any jump.

The preceding arguments show that if the channel contains a subcritical region bounded on either side by supercritical regions (in which the phase velocities are directed *away* from the subcritical region), the flow is *hydraulically controlled* in the sense that small disturbances in the supercritical zones or the reservoirs are unable to affect the processes in the subcritical zone. Such flows are able to match on to a wide variety of reservoir conditions by the formation of hydraulic jumps. Further details are given in section 6.

If only one smooth hydraulic transition occurs along the channel the flow is *partially controlled*. One reservoir is connected to the channel by subcritical flow while the other is isolated by a region of supercritical flow. Partial control is the limiting form when the position of the interfaces within one or both of the reservoirs can no longer be matched on to by a hydraulic jump from the fully controlled flow. Further discussion on such flows is delayed until section 6.

Using Long's (1956) criterion for two-layer shear layer instabilities in a channel, it is possible to show shear instabilities are not able to occur within the subcritical region. Any instabilities in the bounding supercritical region will be carried away from the subcritical region before they have had time to grow to finite amplitude. Lawrence (1985) has shown similar arguments apply when both layers are flowing in the same direction (relative to the channel). In addition, as there are no gradients in potential vorticity in this problem, we do not expect any instabilities from this source. The subcritical region may thus be treated as stable.

4. Nonrotating Sills

In order to understand the use of the functional J, we shall turn briefly to the flow over a nonrotating sill. This flow is covered more fully in Dalziel (1989).

To completely specify the flow (*ie*. A(x)), using the roots of $J(\bullet;A)=0$, we need to determine the exchange flow rate $\overline{q}$ and the constant G. While it would be possible to formulate an algorithm to find the values of $\overline{q}$ and G which satisfy all the requirements for hydraulic control, it proves more convenient to search for the locations of the two control sections.

Suppose we know one of the controls is located at $x=x_c$, say. Solving $\partial J/\partial A=0$ for the exchange flow rate allows us to compute the value of $\overline{q}$ required to produce critical conditions at $x=x_c$ for a specified interface height coefficient. For nonrotating and attached flows $\overline{q}$ may be given explicitly in terms of the geometry and $A_c=A(x=x_c)$. For separated flows the exchange flow rate must be evaluated numerically (details may be found in Dalziel, 1988). Solution of $J(\bullet;A)=0$ then allows us to determine G in terms of A_c. At the second control, $x=x_v$ (say), we may determine $A_v=A(x=x_v)$ as a function of A_c from $J(\bullet;A_v)=0$. The interface height coefficient A_c may then be determined to give critical conditions ($\partial J/\partial A=0$) at x_v. The task is, therefore, to determine the positions of the two control sections such that the solution may be traced continuously between the two supercritical solution branches. Dalziel (1989) has shown that the exchange flow rate for the controlled solution is smaller than that associated with critical conditions at any other pair of sections (for such a pair the flow required to match the solution on to the reservoir conditions can not be traced throughout the channel). This feature may be used in the construction of a solution algorithm (Dalziel, 1989).

For simple geometries the constriction condition (16) may be used to reduce the number of sections which need to be sought to determine the control sections. For our simple sills, as H=*const*–D, we may write

$$K = \frac{\partial J}{\partial b} \frac{db}{dx} + \frac{\partial J}{\partial D} \frac{dD}{dx} - \frac{\partial J}{\partial H} \frac{dD}{dx} = 0 \, . \tag{32}$$

The geometric terms of (32) suggest x=±1, 0 as possible control sections (both dD/dx and db/dx vanish at these points; see figure 1). For regions of nonzero db/dx, K=0 requires $\partial J/\partial b$=0. For no net flow (Q=0) the solution to $\partial J/\partial b$=0 is A=0. This is equivalent to the condition of Armi & Farmer (1986) that u_1^2=u_2^2. Critical conditions occurring with A=0 (when Q=0) correspond to a triple root of J($\bullet$;A)=0. Such solutions yield the largest value of $\bar{q}$ for any solution with a control at the sill crest. Solutions where only one of the controls is at the sill crest give lower exchange flow rates, suggesting the A=0 solution is not realised. Regions of nonzero dD/dx are unable to support any control sections as there are no real solutions to $\partial J/\partial D$–$\partial J/\partial H$=0.

In the absence of net flow a complete analysis shows that the controls are positioned with one at the crest of the sill (the *crest control*; x_c=0) and one at the foot of the sill on the side of the dense reservoir (the *exit* or *foot control* x_v=-1), whether or not the channel is rotating. The positions are due to both db/dx and dD/dx vanishing at these points.

Figure 3 sketches how the hydraulic functional varies along a nonrotating channel containing a simple sill. The depth of the channel away from the sill crest is twice that at the crest (D_w=2), though the following discussion applies for all values of D_w>1. At section (a) near the dense reservoir the interface is near the top of the channel. The dependence of the hydraulic functional on A at this section is plotted above the channel. The appropriate supercritical root of J($\bullet$;A)=0 is indicated by an arrow. Moving towards the foot control at section (b) the supercritical and subcritical roots converge. At section (b) they coincide, allowing the flow to switch smoothly from the supercritical branch (with phase velocities towards the left) to the subcritical branch of the solution.

On the slopes of the sill the two solution branches again separate (figure 3c), the solution now following the subcritical branch. Closer to the sill crest the subcritical branch converges on the other supercritical branch, the two coinciding at the crest (section (d)). The solution is then able to switch to the supercritical branch with both phase velocities towards the right. To the right of the sill crest the solution branches move apart again. The shape of the functional at (e) is identical to that at (c) due to identical geometry at these two sections. The difference lies in which solution branch is being followed by the flow. Moving towards the light reservoir the solution remains on the supercritical branch with the smaller value of A. In general, a hydraulic jump will form somewhere in each of the supercritical regions to match on to the reservoir conditions.

Figures 4 and 5 are included to demonstrate why it is necessary for the two controls to be positioned at the crest and the foot of the sill. In figure 4 we have assumed that the two controls coincide at the sill crest. The triple root associated with such a situation requires A_c=A_v=0 and maximises the exchange flow rate in the sense that $\partial \bar{q}_{crit}/\partial A$=0, where $\bar{q}_{crit}$ is the value of the exchange flow rate required to give critical conditions at the sill crest in

356

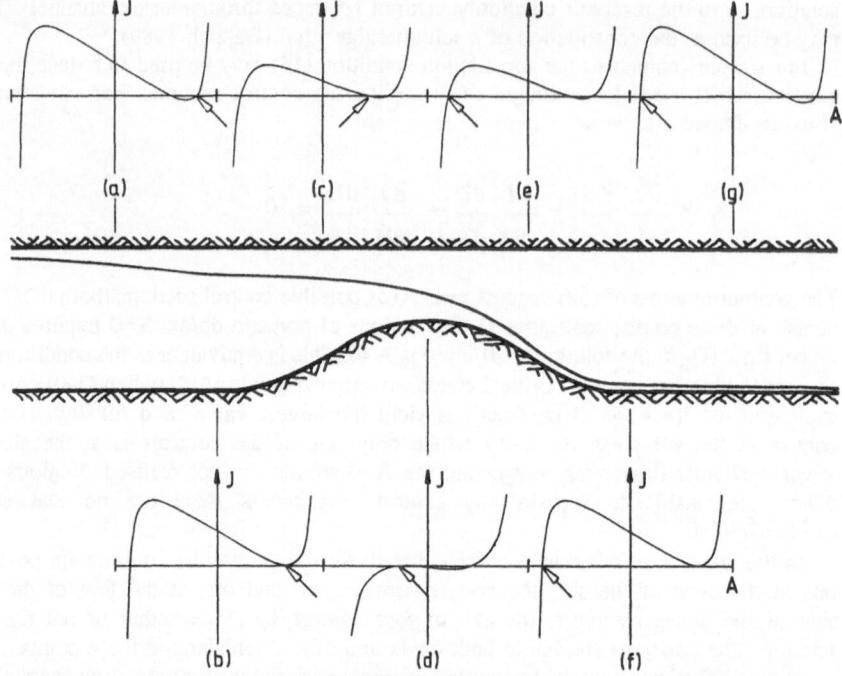

Figure 3: Variations in the hydraulic functional for a fully controlled flow over a simple sill. The interface profile along the channel is shown, along with plots of the hydraulic functional at seven locations along the channel. The root of J=0 giving the interface position shown is indicated by an arrow. Plots of J at sections (a) and (g) are identical due to identical geometry, although the indicated root differs. Similarly for sections (b) and (f), and sections (c) and (e). The two controls are positioned at sections (b) and (d).

isolation from the remainder of the channel. For values of $A>0$, solutions to $\partial J/\partial A=0$ give $\partial \overline{q}_{crit}/\partial A<0$ while those with $A<0$ give $\partial \overline{q}_{crit}/\partial A>0$. This feature shall be utilised in section 6 in demonstrating the maximal nature of these flows.

Tracing the supercritical root with the smaller value of A from the light reservoir (section (g)) towards the sill crest (section (d)) is straight forward. At the crest all three roots to $J(\bullet;A)=0$ coincide, allowing the solution to branch from the supercritical root with the phase velocities towards the right to either the subcritical root or directly to the other supercritical root. However, as we move down the sill towards the dense reservoir, the subcritical root and supercritical root with $\partial^2 J/\partial A^2>0$ become imaginary away from the crest. Due to critical conditions at the sill crest the remaining supercritical root (with $\partial^2 J/\partial A^2<0$) would not be stable if traced to the left of the crest. It is, therefore, not possible to trace the solution past the crest through sections (c), (b) and (a). We note however that these two roots are regained to the left of section (a). The inability to connect the flow smoothly to the dense reservoir shows the two controls can not coincide at the sill crest. Moreover, if the second control is anywhere except at the foot (on the dense reservoir side), an analysis will again show the solution may not be traced smoothly through two hydraulic transitions from one reservoir to the other.

Figure 5 shows a situation where the flow is partially controlled. The height of the

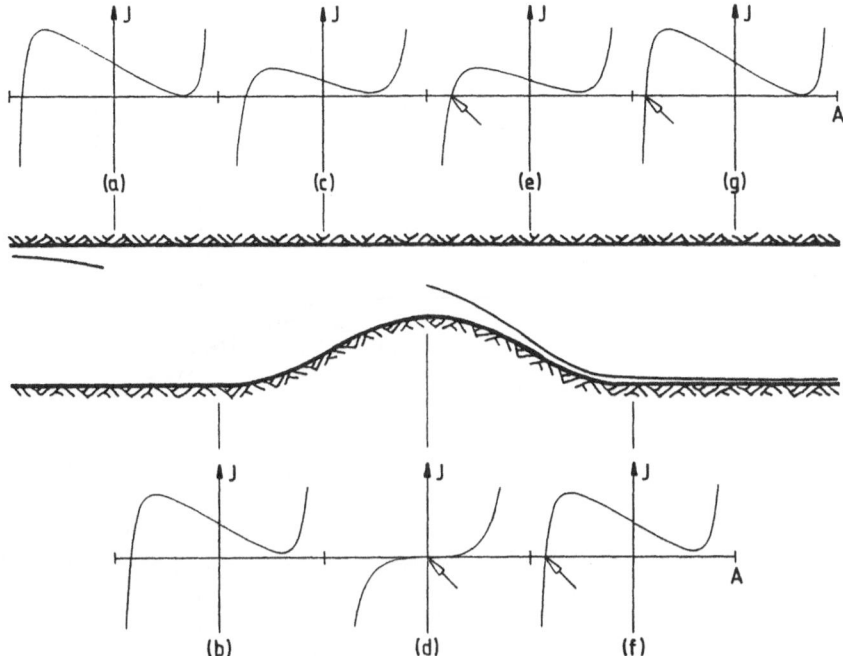

Figure 4: Variations in the hydraulic functional for an unrealisable flow over a simple sill. The calculation assumed both controls were positioned at the sill crest (section (d)). The channel geometry is identical to that for figure 3.

interface at the sill crest is lower than that associated with the fully controlled solution of figure 3. Tracing the solution towards the left from the light reservoir is achieved as before. At the sill crest two of the solution branches coincide, allowing a transition from supercritical to subcritical flow. To the left of the crest the solution remains on the subcritical branch as the geometry does not allow it to coincide with the supercritical branch with the larger value of A.

5. Rotating Sills

As we have shown in the previous section, the position of the control sections is known for a simple sill when no net flow is present. The crest control is at the sill crest (quantities denoted by subscript c) and the foot control is at the base of the sill on the dense reservoir side (quantities denoted by subscript v). Despite this it is not possible to determine the controlled flow analytically, except for channels of constant depth ($D_w=1$). We are, however, able to make some progress by considering the asymptotic limits $D_w \rightarrow 1$ and $D_w \rightarrow \infty$. Dalziel (1988) showed that asymptotic evaluation of the solution to $J(\bullet;A)=0$ and $\partial J/\partial A=0$ at the two controls (ie. four equations) for small sills ($D_w \rightarrow 1$) gives

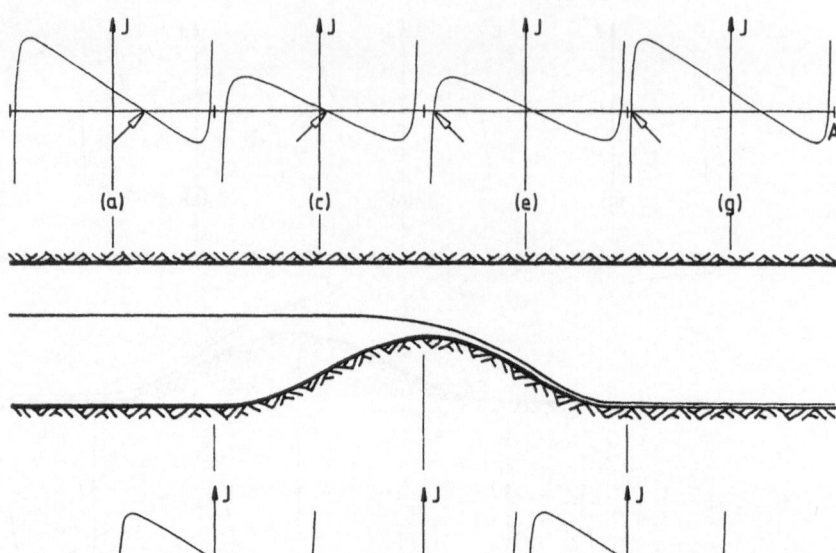

Figure 5: Variations in the hydraulic functional for a partially controlled flow over a simple sill. The geometry of the channel is identical to that in figures 3 and 4. Critical conditions occur only at the sill crest (section (d)). The flow is subcritical ($\partial J/\partial A<0$) everywhere to the left of the crest.

$$A_c = -\frac{1}{4}(1-b^2)^{\frac{1}{3}}\epsilon + \frac{9-b^2}{72(1-b^2)^{\frac{1}{3}}}\epsilon^2 + \frac{(13/6)-(37/27)b^2-(1/162)b^4}{4(1-\frac{1}{3}b^2)(1-b^2)}b^2\epsilon^3 + O(\epsilon^4),$$

$$A_v = \frac{1}{4}(1-b^2)^{\frac{1}{3}}\epsilon + \frac{9-b^2}{72(1-b^2)^{\frac{1}{3}}}\epsilon^2 - \frac{(13/6)-(37/27)b^2-(1/162)b^4}{4(1-\frac{1}{3}b^2)(1-b^2)}b^2\epsilon^3 + O(\epsilon^4),$$

$$\overline{q}/b = \frac{1}{2}(1-\frac{1}{3}b^2) - (3/8)(1-b^2)^{2/3}\epsilon^2 + (3/8)[1-(1/9)b^2]$$

$$+ \frac{(3/8) + \frac{1}{3}b^2 - (67/216)b^4 - (1/324)b^6}{4(1-b^2)^{2/3}(1-\frac{1}{3}b^2)}\epsilon^4 + O(\epsilon^5), \qquad (33)$$

where

$$D_v = D_w = 1+\epsilon^3, \qquad (34)$$

as $\epsilon \to 0$. The channel width b equals that for the two controls ($b=b_c=b_v$). Note that the $O(\epsilon^2)$ and

$O(\epsilon^3)$ terms for A_c and A_v become infinite as $b_c \to 1$, so the expansion is formally valid only for $\epsilon, b_c \ll 1$. In the limit of $\epsilon = 0$ we recover the solution for a channel of constant depth ($A_c = A_v = 0$) as found by Whitehead *et al.* (1974). When $b_c = 0$ we recover the nonrotating solutions of Armi & Farmer (1986) for $\epsilon = 0$ and Dalziel (1989) for $\epsilon \to 0$.

For a channel of constant depth the present assumption that the flow is *relatively straight* breaks down when $b_c > 1$. Dalziel (1988) has shown that for $b_c > 1$ the relative straight solution predicts a flow separated from both walls. The position of the two-layer region within the channel at the contraction can not be determined without considering the initial transients and the introduction of a nonzero cross-channel component to the velocity. Just away from the contraction, on the dense reservoir side, the two-layer region would be confined to the $y = \frac{1}{2}b$ wall; on the light reservoir side the two-layer region would be confined to the $y = -\frac{1}{2}b$ wall. Through the contraction itself (extending from $x = -1$ to $x = 1$) the two-layer region would have to cross from one side of the channel to the other. The geometry is unable to set the length scale of this process.

Dalziel (1988) used an adaptation of Gill's (1976) single-layer linear analysis of the dambreak problem for the two-layer problem and found that a crossing of the channel would occur over the local Rossby radius. The nonlinear single-layer analysis of Hermann, Rhines & Johnson (1989) suggests the length scale may be larger than the local Rossby radius for the channel crossing. We note, however, that the second layer alters the manner in which waves propagate making it unlikely that the analysis of Hermann *et al.* can be applied to two-layer flows. The experiments of Whitehead *et al.* (1974) and Dalziel (1988) confirm that a channel crossing does occur, but are unable to determine the length scale due to the local Rossby radius and geometric along-channel length scales being of comparable size.

Further discussion of this channel crossing is beyond the scope of this paper. For channels wider than $b_c = b_v \approx 1$ the cross-channel component of the velocity is no longer negligible, leading to the breakdown of the present *relatively straight* assumption.

For sills in which $D_w \to \infty$ we must expand the solution in terms of the channel width as well as the sill height. Even then the coefficients of the expansion must be evaluated numerically to show

$$A_c \sim -0.12544 + 0.03946b^2 + 0.03125b^4 + O(b^6) + O(D_w^{-1}),$$

$$A_v \sim \frac{1}{2} - (0.35104 + 0.017799b^2 + 0.010649b^4)/D_w + O(b^6) + O(D_w^{-2}),$$

$$(\overline{q}/b)^2 \sim 0.17304 - 0.009691b^2 + 0.002344b^4 + O(b^6) + O(D_w^{-1}). \tag{35}$$

Exact solution (with respect to b) of the $D_w \to \infty$ asymptotic limit reveals that the flow will be separated at the sill crest if $b_c > 0.89416$ (*cf.* separation when $b_c > 1$ for channel of constant depth). As the flow remains attached at the x_v control beyond $b_c = 0.89416$, numerical evaluation of the solution remains possible since the length scale associated with any cross-channel flow remains fixed on the length scale of variations in the channel geometry. In the nonrotating limit we recover the results of Farmer & Armi (1986) for $D_w = \infty$ and Dalziel (1989) for $D_w \to \infty$.

Figure 6 plots how the solutions to equations (33) vary as a function of both D_w and channel width. The height coefficients at the two controls, A_c and A_v, are shown in figure 6a. Agreement with numerical evaluation of the full solutions (which shall be presented shortly) is good for narrow channels even for $D_w \sim 1.5$. However the agreement is much more tightly confined to $D_w \to 1$ as the channel width is increased and the second and higher order terms become close to

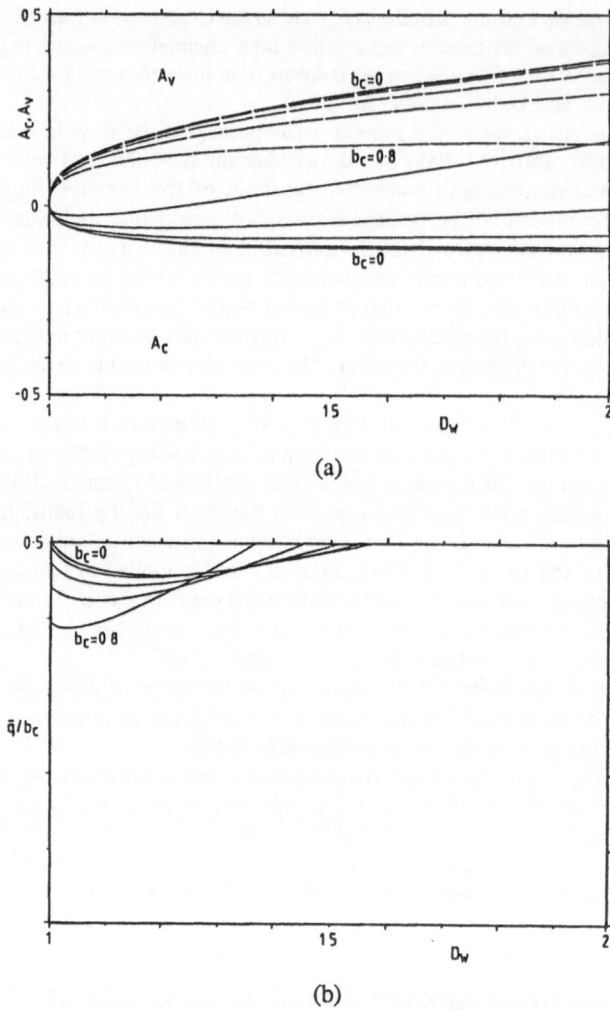

(a)

(b)

Figure 6: Variations with D_w in the small sill asymptotic limit. (a) Interface height coefficients at sill crest (solid lines) and foot control (dashed lines); (b) specific exchange flow rate. Channel widths as marked.

singular.

Variations in the specific exchange flow rate, $\bar{q}/b_c$, are plotted in figure 6b. The apparent turning point is the result of the solution breaking down due to the binomial expansion of a square root.

The asymptotic solutions for $D_w \rightarrow \infty$ are plotted in figure 7. The height coefficients at the two controls are given in figure 7a. The continuous line in figure 7a is the exact (with respect to b_c) solution of the $D_w \rightarrow \infty$ equations, while the dashed curve is the small b_c approximations given by equations (35). Note that A_c *increases* towards zero as the width of the channel increases. This suggests that in rotating channels the two layers respond more equally to the influence of

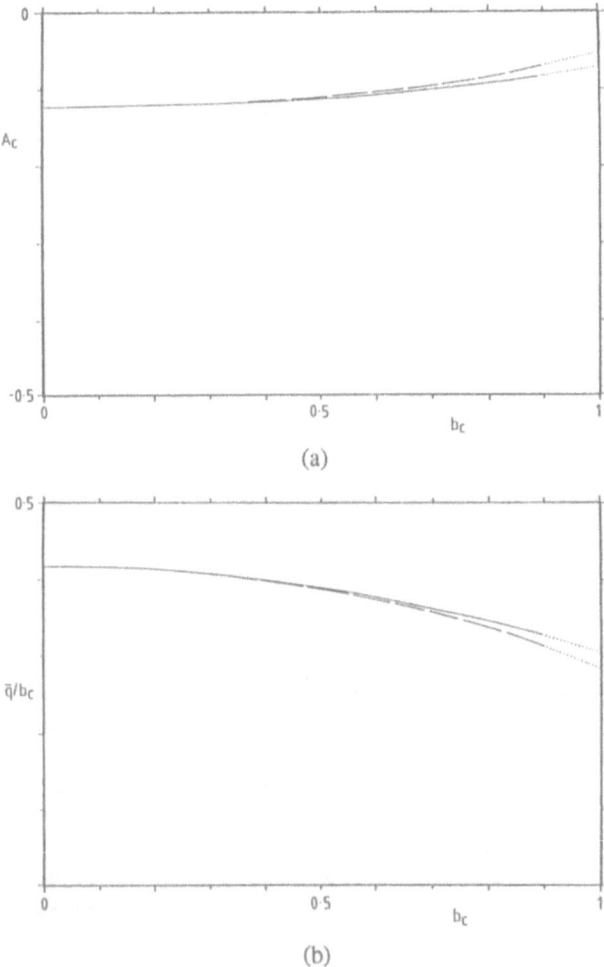

(a)

(b)

Figure 7: Variations with b_c in the large sill asymptotic limit. (a) Interface height coefficients at sill crest. (b) $D_v(\frac{1}{2}-A_v)$, a measure of the interface height at the channel exit. Solid lines are exact (with respect to b_c) solutions of the $D_v \to \infty$ equations; dashed lines are an expansion in b_c. Dotted lines indicate flow is separated at the sill crest (equations no longer valid).

a sill (protruding into only the lower layer) than is the case for their nonrotating counterparts. A possible explanation of this stems from the Taylor-Proudman theorem. For a continuously stratified fluid the scale of the penetration, parallel to the axis of rotation, of any disturbance increases with the rotation rate as fL/N (section 6.15, Pedlosky, 1979), where f is the Coriolis parameter, L the length scale of the disturbance and N the buoyancy frequency. It is reasonable for this to carry over to the two-layer case with N replaced by $(g'/D)^{\frac{1}{2}}$ and L by b. In our dimensionless system, the vertical penetration scales with the channel width. Increasing the rotation rate increases the communication between the two layers, reducing the asymmetry of the forcing associated with the presence of the sill. In contrast A_v

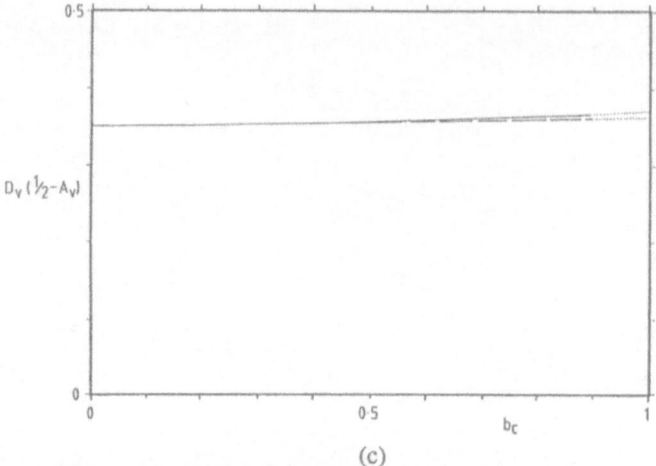

Figure 7: Variations with b_c in the large sill asymptotic limit. (c) The specific exchange flow rate. Solid, dashed and dotted lines as in 7 a & b.

(figure 7b) is relatively little affected by rotation (the penetration here scaling on $bD_w^{-\frac{1}{2}}$).

Figure 7c shows how the exchange flow rate per unit width is reduced as the width of the channel is increased, the two sets of curves having the same meaning as for the height coefficients.

In principle, asymptotic solutions could be found when the flow is separated from one of the side walls at the sill crest, but such a solution is beyond the scope of this paper. Instead we evaluate the exact solution numerically. Figures 8 to 15 plot variations in the controlled solution over the b_c–D_w plane. The plots extend past $b_c=1$ so that we may see the cause of any breakdown in the solution for wide channels.

The height coefficient at the sill crest is plotted in figure 8. The surface shading is used to indicate whether or not the flow is attached. Due to limited numerical resolution the boundaries between the regions may not be exact, particularly near $D_w=1$. Lines perpendicular to the b_c axis are plotted every 0.1 units; those perpendicular to the D_w axis are plotted every 0.25 units.

For channel widths less than approximately 0.9 Rossby radii (evaluated at the sill crest), the flow is attached (light gray regions of the surface) at both control sections and the length scale associated with any crossing of the channel remains determined by the geometry. The interface height coefficient A_c varies with the depth at the foot control $D_v=D_w$ in a manner very much like the nonrotating limit. Increasing the channel width within these bounds decreases the value of A_c slightly. Increasing the rotation rate (or equivalently the channel width) causes the flow to become separated first at the sill crest (unshaded regions of the surface). The interface height coefficient responds by increasing in magnitude rapidly as b_c is increased, reflecting the confinement of the two-layer zone to the y=-b/2 wall. Separation at the virtual control (dark gray regions) for still wider channels does little to modify this trend. The length scale of the channel crossing is still set by the geometry if the sill is sufficiently high.

The slope coefficient B_c, shown in figure 9, has comparatively little variation with D_w – the

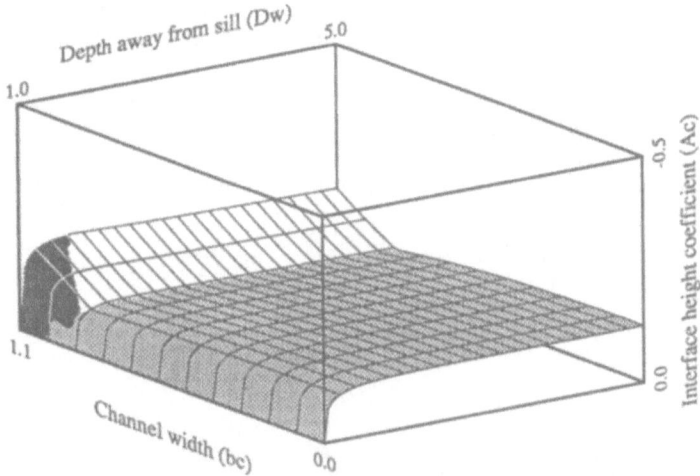

Figure 8: Variations in the interface height coefficient A_c at the sill crest with the channel width b_c and depth away from the sill at the foot control (D_y). Surface shading: light gray regions – flow attached at both controls; unshaded regions – flow separated at sill crest but attached at foot control; dark gray regions – flow separated at both controls.

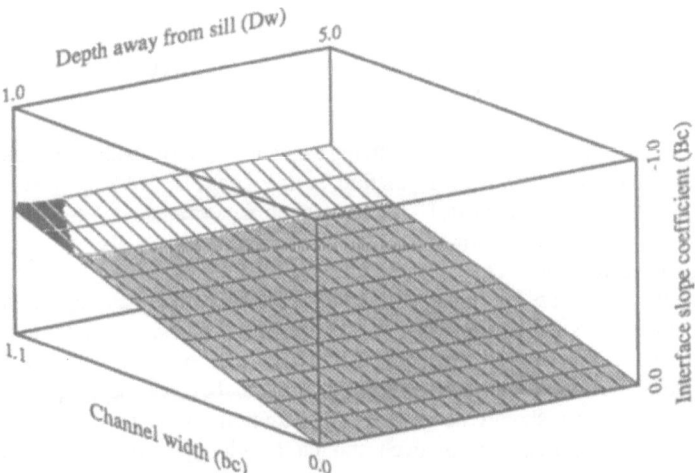

Figure 9: Variations in the interface slope coefficient B_c at the sill crest with the channel width b_c and depth D_w away from the sill. Surface shading as for figure 8.

variations that do exist are confined to $D_w\sim1$. The magnitude of the coefficient increases approximately linearly with the channel width (light gray regions; in dimensional terms this means that the slope of the interface is approximately the depth divided by the internal Rossby radius at the sill crest) up until the flow separates at the sill crest (unshaded regions of the surface). For wider channels, the variation is still nearly linear, though slightly more rapid.

Provided the flow is attached at at least one of the controls (*ie.* light gray and unshaded regions of figure 10), the interface height coefficient A_v behaves almost exactly like the nonrotating limit, approaching its asymptotic form for D_w greater than approximately 1.5. Separation at both controls results in the A_v surface increasing in both b_c and D_w.

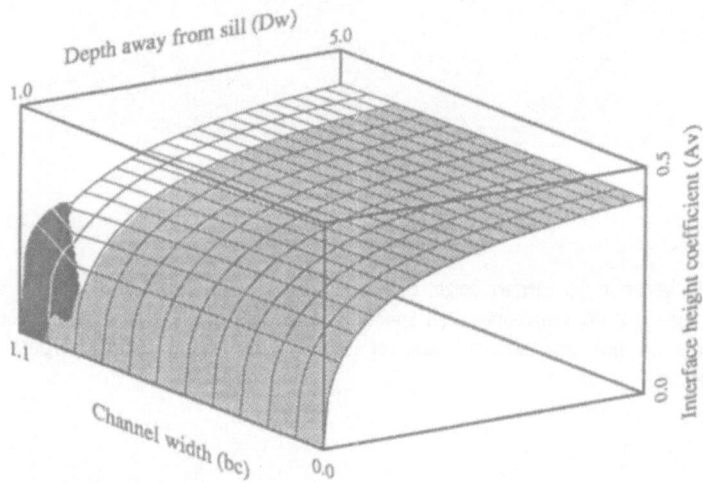

Figure 10: Variations in the interface height coefficient A_v at the foot control with the channel width b_c and depth D_w away from the sill. Surface shading as for figure 8.

Figure 11 also shows the A_v surface, though now the surface shading indicates features of the upper layer velocity rather than separation at the controls. Dark gray regions indicate the velocity is strictly unidirectional at both control sections, light gray indicates u_2 is unidirectional at the foot control, but undergoes a reversal at the sill crest, and unshaded regions indicate that the velocity undergoes reversal at both the foot control and the sill crest. In this last case it is not possible to assert that all upper layer streamlines originated in the less dense reservoir, and so the resulting solution may not be valid. As for $D_w=1$, Dalziel (1988) showed this paradox may be resolved by the introduction of a zone of stagnant fluid. Further discussion of this paradox is beyond the scope of this paper. The solution beyond $b_c=1$ should be treated with caution. Note that the lower layer remains unidirectional at the foot control even though a velocity reversal will occur at the crest control *before* separation.

The interface slope coefficient at the foot control is shown in figure 12, the surface shading indicates whether the flow is attached or separated as was used for figures 8 to 10. This plot shows that variations in the channel depth greatly reduce the cross-channel slope coefficient at the foot control (the interface slope $D_w B_v$ varies much less rapidly). The value of B_v varies approximately linearly with b_c regardless of separation, though the constant of proportionality is a strong function of D_w.

As the channel width increases, the specific exchange flow rate $\overline{q}/b_c$ (plotted in figure 13;

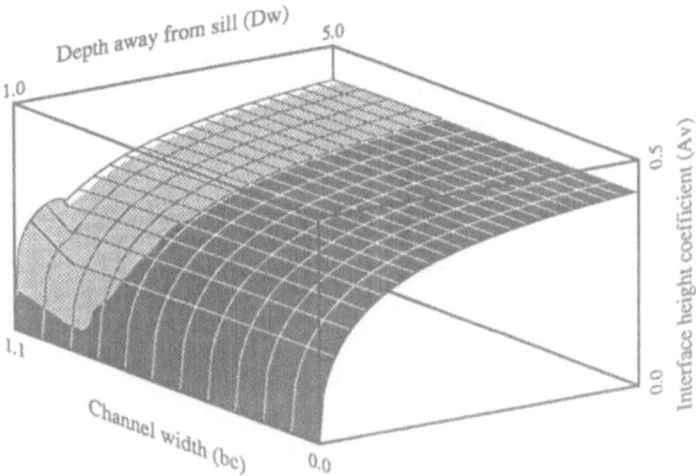

Figure 11: Surface as for figure 10, but surface shading based on the velocity of the upper layer: dark gray regions – velocity unidirectional at both controls; light gray regions – u_2 unidirectional at foot control but changes sign at sill crest; unshaded regions – u_2 changes sign at both controls.

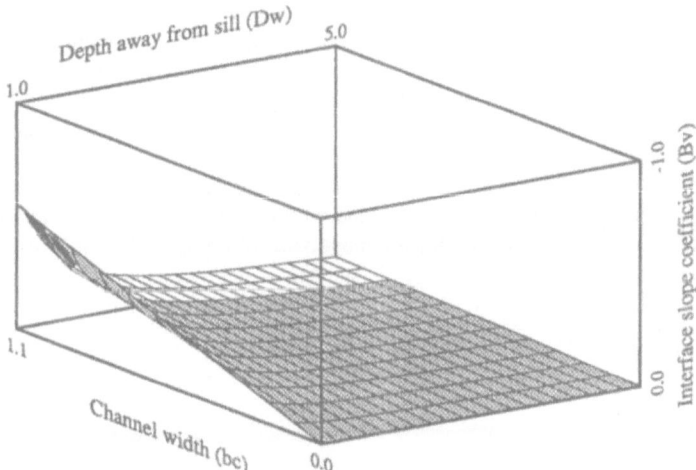

Figure 12: Variations in the interface slope coefficient B_v at the foot control with the channel width b_c and depth D_w away from the sill. Surface shading as for figure 8.

the surface shading indicates separation; note the surface is viewed from a different perspective to that of the preceding figures) decreases for all values of $D_w \geq 1$. The rate of decrease is similar to that for a channel of constant depth. Variations with D_w follow closely the nonrotating form, the magnitude of the variations with D_w decreasing as b_c increases in the

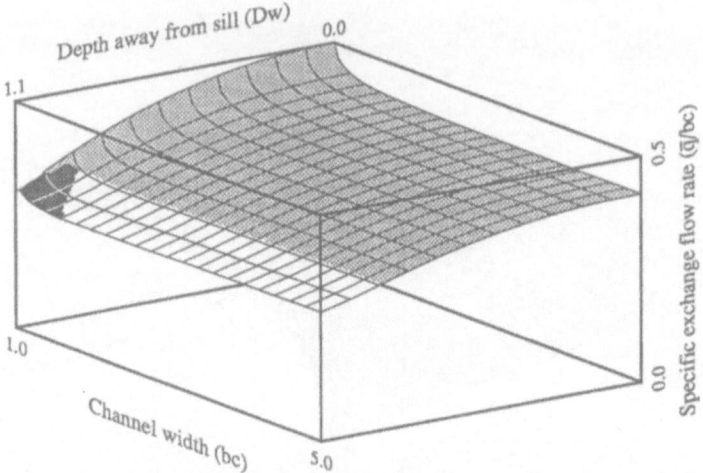

Figure 13: The response of the specific exchange flow rate $\overline{q}/b_c$ to changes in the channel width b_c and depth at the foot control D_w. Surface shading as for figure 8.

manner predicted by the asymptotic limits of equations (33) and (35).

In section 3.2 we required that the denominator of equation (27) remained positive for the definition of either the local or the section composite Froude numbers to be meaningful. A sufficient condition is that $|\partial B/\partial A|<1$ at both controls (and elsewhere in the subcritical region). Figures 14 and 15 plot how $\partial B/\partial A$ varies at the two control sections. Surface shading is for separation. Note that for $b_c \leq 1$ the condition $|\partial B/\partial A|<1$ is satisfied at both controls (and elsewhere in the channel), justifying the introduction of a section composite Froude number F_s. The region in figure 15 with $\partial B/\partial A<-1$ corresponds to a flow in which a velocity reversal occurs in the upper layer at both control sections (the solution is then not valid because the streamlines can not be traced) and so does not further reduce the range of validity for the solution. If $|\partial B/\partial A|>1$ the local denominator of equation (27) may change sign across the channel. This would give an imaginary local composite Froude number for some portion of the channel width. Such a flow is unstable and so would adopt a different configuration (*eg.* the channel-crossing seen by Dalziel, 1988).

6. Submaximal Flow

Fully controlled hydraulic flows are maximal in the sense that the flow adopts a configuration which maximises the exchange flow rate through the channel. In order to understand this concept more fully, we must consider the process by which one of the controls is flooded and the flow becomes only partially controlled. A detailed description of this process is given in Dalziel (1989) for nonrotating channels. In this section we shall summarise the nonrotating results for a simple sill and discuss how they apply to rotating channels. We shall assume the level of the interface in the left-hand (dense) reservoir is higher than in the right-hand (light)

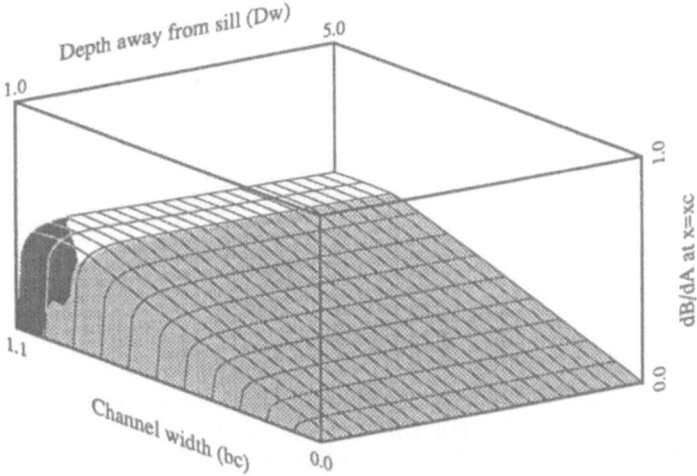

Figure 14: Variations in $\partial B/\partial A$ at the sill crest (x_c) as a function of the channel width and depth away from the sill crest. Note that $|\partial B/\partial A|<1$ over domain plotted. Surface shading as for figure 8.

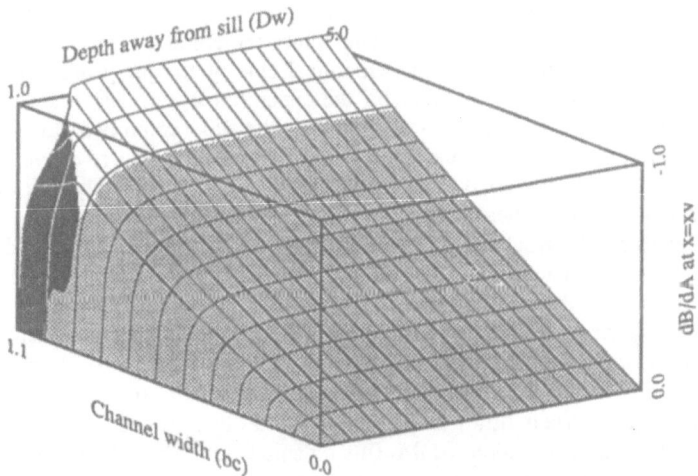

Figure 15: Variations in $\partial B/\partial A$ at the foot of the sill (x_v) as a function of the channel width and depth away from the sill crest. Note that $|\partial B/\partial A|>-1$ over most of the domain; $\partial B/\partial A<0$ only when u_2 changes sign at *both* controls. Surface shading as for figure 8.

reservoir. If the reverse is true we need only interchange the two reservoirs.

Figure 16 shows a range of possible interface profiles for nonrotating flow over a simple sill for a variety of different reservoir conditions. Solid lines represent supercritical solution branches and dashed lines subcritical branches. The heavy lines signify the unique,

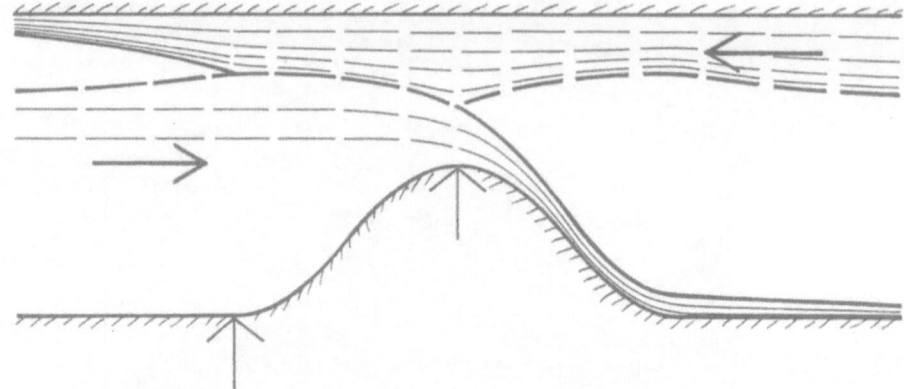

Figure 16: Interface profiles for flow over a simple sill. The heavy lines denote the fully controlled (maximal) solution and its associated subcritical root. Light lines represent sub-maximal flows. Continuous lines indicate supercritical solution branches and dashed lines subcritical branches. See text for more details.

fully controlled solution and its associated subcritical root, and lighter lines indicate submaximal, partially controlled flows. All the profiles have at least one hydraulic transition.

Hydraulic jumps may form only on supercritical regions of the flow. Such jumps are able to adjust the flow to subcritical conditions within the reservoir. As dissipation occurs within the jumps, the subcritical side of the jump must represent a lower energy state than the supercritical side. This means that the subcritical side of the jump must have a position of the interface somewhere between that on the supercritical side and the subcritical solution branch associated with the supercritical flow. For the fully controlled flow in figure 16, the flow may match on to reservoir conditions with the interface anywhere above the heavy dashed line (and below the heavy solid line) in the dense reservoir, or anywhere below the heavy dashed line (and above the heavy solid line) in the light reservoir.

If the interface in the dense reservoir were to fall below the position described by the subcritical root associated with the fully controlled flow (heavy dashed line) then the hydraulic control at the foot of the sill x_v would be flooded. The flow would then be subcritical everywhere to the left of the sill crest. The hydraulic transition at the crest would be maintained, allowing the formation of a subsequent hydraulic jump to adjust the flow to the conditions in the light reservoir. The interface height coefficient at the sill crest A_c is more negative for such flows than for fully controlled flows. From the requirement for critical conditions at this section ($\partial J/\partial A=0$) we demonstrated in section 4 that when $A_c<0$ any decrease in A_c results in a lower exchange flow rate ($\partial \bar{q}_{crit}/\partial A>0$). This partially controlled flow is therefore submaximal.

Similarly, if the interface in the light reservoir were above the heavy dashed line, the hydraulic control at the sill crest x_c would be flooded. The flow would be subcritical everywhere to the right of the foot control (the foot of the sill on the dense reservoir side). While conditions at the foot of the sill on the light reservoir side are also critical, no hydraulic transition can occur at this point as the phase velocities would be towards the transition. A hydraulic transition would be maintained at the foot on the dense reservoir side, producing a supercritical flow into the reservoir and allowing a hydraulic jump to form to match on to the reservoir conditions. Since A_v (>0) is more positive for all flows where the crest control is lost (critical conditions are maintained at x_v), and $\partial \overline{q}_{crit}/\partial A < 0$ for such solutions, the associated exchange flow rate is lower than for the fully controlled flow. We are thus able to say the fully controlled flow produces the *maximal* exchange flow rate.

Rotation does not alter the fundamental aspects of these arguments. We need consider only the relationship between the conditions in the reservoir and the subcritical root associated with the fully controlled flow to determine whether full maximal or submaximal flow will occur. The cross-channel structure of the flow is not important. If the hydraulic jump is large, potential vorticity changes across the jump may be important, and so the cross-channel structure of the subcritical flow after the jump may be very different from that of the subcritical root associated with the supercritical flow. However, large jumps will only occur when the flow is very supercritical and the state after the jump is a long way from the subcritical root of the supercritical flow. Thus the change in potential vorticity is unlikely to present any difficulty. As discussed in section 3.2, the work of Pratt (1983) and Nof (1986) suggest the potential vorticity generation for a small jump will be very small. As the subcritical flow on the *downstream* side of a jump gets closer to flooding the jump the cross-channel structure of the flow on each side will tend towards each other rapidly. As a result, a control will always be flooded across the entire width of the channel and not just at one point across the cross-section.

7. Conclusions

In this paper we have demonstrated the usefulness of a functional approach to two-layer rotating hydraulics. An important feature is the ability to define both a local composite Froude number and a section composite Froude number characterising the flow in terms of the phase speeds of long, small amplitude gravity waves. The simple relationship between the hydraulic functional and such a Froude number greatly simplifies the derivation for a wide class of flows. For attached flows it is possible to write the composite Froude number explicitly in terms of the interface height coefficient A and the exchange flow rate $\overline{q}$; unfortunately this is not the case when two layers are present over only part of the channel width (we can not write $B=B(A,\overline{q})$ explicitly; the Froude number would only apply to the two-layer zone).

The asymmetry introduced to the flow by the presence of a sill in the lower layer is of equal importance in rotating and nonrotating flows. Even relatively small departures from symmetry (about a horizontal plane) in the channel geometry lead to a significant change in the flow rates and the interface profile. While the increased vertical penetration, produced by rotation, of the bottom topography reduces the asymmetry of the fluid response, the interface remains displaced from the mid-depth point by around 10% of the channel depth even for comparatively low sills.

The leading order effect of rotation is the introduction of a cross-channel tilt to the interface. The Coriolis force causes the layers to bank up against the right hand side of the channel when looking downstream with respect to a given layer. If the Coriolis force is sufficiently strong the interface will intersect the channel top or bottom at some point across the channel. The resulting current will consist of a two-layer region, the structure of which is governed by geostrophic balance and conservation of potential vorticity, and a single-layer region in which potential vorticity alone must be conserved. The two regions are matched by the requirement that there are no velocity discontinuities within a layer.

As found by Whitehead *et al.* (1974) for a very simple channel geometry, rotation reduces the exchange flow rate along the channel. For narrow channels this change is a second order effect. In channels with the width at the crest close to the local Rossby radius at that point ($b_c \sim 1$), the reduction in the specific exchange flow rate is up to ⅓ of the nonrotating value.

The situation when $b_c > 1$ has been covered only briefly. Dalziel (1988) found experimentally that under these circumstances the exchange flow rate is *independent of the channel width* (*ie.* the specific exchange flow rate goes down like $1/b_c$). This behaviour is a consequence of the possibility of a change of sign in the velocity of the upper layer at *both* control sections. The apparent paradox of fluid flowing from the *downstream* to the *upstream* reservoir (with respect to the upper layer) may be resolved by the introduction of a zone of stagnant fluid.

Net barotropic flow may be expected to affect rotating flows in a similar manner to the nonrotating limit. Depending on D_w and the strength of the specific net flow, Q/b_c, the foot control may remain at the exit (foot control), move out towards the dense reservoir (virtual control), jump to the widening region leading towards the light reservoir (virtual control), or jump to coincide with the crest of the sill (coincident virtual and crest controls). The interesting structure introduced by these movements of the foot (virtual) control are beyond the scope of this paper. Discussion of this feature may be found in Dalziel (1988).

Conditions for maximal exchange to occur are found to be the same as for nonrotating channels. The important feature is whether or not the subcritical flow in a reservoir may be matched on to by a hydraulic jump from the fully controlled flow. If not, the flow will be partially controlled, with a single hydraulic transition (uncontrolled flows with no hydraulic transitions are at rest). The concept of hydraulic control remains applicable to rotating channels provided the width at the sill crest is less than the local internal Rossby radius at that point. If the width is greater than this the two-layer region will cross the channel over a length scale of the order of the Rossby radius and this process will control the flow.

Acknowledgements

This work was undertaken while I was in receipt of a Commonwealth Scholarship as a postgraduate student at the University of Cambridge. I am grateful to the Association of Commonwealth Universities for this opportunity, and to my supervisor over this time, Dr Paul Linden, for his continued help and guidance.

References

Armi, L (1986): The hydraulics of two flowing layers of different densities; *J. Fluid Mech.* **163**, 27-58.

Armi, L & Farmer, D M (1986): Maximal two-layer exchange through a contraction with barotropic net flow; *J. Fluid Mech.* **164**, 27-51.

Borenas, K & Lundberg, P (1986): Rotating hydraulics of flow in a parabolic channel; *J. Fluid Mech.* **167**, 309-326.

Bormans, M & Garrett, C (1989): The effect of rotation on the surface inflow through the Strait of Gibraltar; *J. Phys. Oceanog.* (in press).

Dalziel, S B (1988): *Two-layer hydraulics: maximal exchange flows*, PhD thesis, Department of Applied Mathematics and Theoretical Physics, University of Cambridge.

Dalziel, S B (1989): Two-layer hydraulics: a functional approach; *submitted to J. Fluid Mech.*

Deacon, M (1971): *Scientists and the sea 1650-1900, a study of marine science*; Academic Press, London.

Deacon, M (1985): An early theory of ocean circulation: J S von Waitz and his explanation of the currents in the Strait of Gibraltar; *Prog. in Oceanog.* **14**, 89-101.

Farmer, D M & Armi, L (1986): Maximal two-layer exchange over a sill and through the combination of a sill and contraction with barotropic flow; *J. Fluid Mech.* **164**, 53-76.

Gill, A E (1976): Adjustment under gravity in a rotating channel; *J. Fluid Mech.* **77**, 603-621.

Gill, A E (1977): The hydraulics of rotating-channel flow; *J. Fluid Mech.* **80**, 641-671.

Hermann, A J, Rhines, P B & Johnson, E R (1989): Nonlinear Rossby adjustment in a channel: beyond Kelvin waves; *J. Fluid Mech.* **205**, 469-502.

Lawrence, G A (1985): *The hydraulics and mixing of two-layer flow over an obstacle*; PhD Thesis, Hydraulic Engineering Laboratory, Department of Civil Engineering, University of California, Berkeley.

Long, R R (1956): Long waves in a two-fluid system; *J. Meteorology* **13**, 70-74.

Nof, D (1986): Geostrophic shock waves; *J. Phys. Oceanog.* **16**, 886-901.

Pedlosky, J (1979): *Geophysical fluid dynamics*, Springer-Verlag, New York.

Pratt, L J (1983): On inertial flow over topography. Part 1. Semigeostrophic adjustment to an obstacle; *J. Fluid Mech.* **131**, 195-218.

Pratt, L J (1984): On nonlinear flow with multiple obstructions; *J. Atmos. Sci.* **41**, 1214-1225.

Pratt, L J & Armi, L (1987): Hydraulic control of flows with nonuniform potential vorticity; *J. Phys. Oceanog.* **17**, 2016-2029.

Rossby, C (1938): On the mutual adjustment of pressure and velocity distributions in certain simple current systems; *J. Marine Res.* **1**, 15-28.

Shen, C (1981): The rotating hydraulics of the open channel flow between two basins; *J. Fluid Mech.* **112**, 161-188.

Stommel, H & Farmer, H G (1953): Control of salinity in an estuary by a transition; *J. Marine Res.* **12**, 13-20.

Whitehead, J, Leetmaa, A & Knox, R (1974): Rotating hydraulics of strait and sill flows; *Geophys. Fluid Dynamics* **6**, 101-125.

Wood, I R (1968): Selective withdrawal from a stably stratified fluid; *J. Fluid Mech.* **32**, 209-223.

Wood, I R (1970): A lock exchange flow; *J. Fluid Mech.* **42**, 671-687.

Wood, I R & Simpson, J E (1984): Jumps in layered miscible fluids; *J. Fluid Mech.* **140**, 329-342.

ROLE OF LABORATORY EXPERIMENTS AND MODELS
IN THE STUDY OF SEA STRAIT PROCESSES

T.A. McClimans
Norwegian Hydrotechnical Laboratory
Norwegian Institute of Technology
N-7034 Trondheim, Norway

ABSTRACT

A brief review of fluid mechanical laboratory studies shows that many
physical processes of flows in sea straits can be studied quantitatively, and that physical insight in the interpretation of field data
has often been gained by means of very simple, but careful laboratory
experiments. Multi-scale energetics have, for example, provided algorithms for computing mass and momentum exchanges in straits.
    Laboratory models can provide fine spatial resolution and coupled
physics for testing advanced numerical schemes. Studies of large scale
geophysical flows must conform to Froude-Rossby similitude. Effects
like surface tension, molecular friction, improper boundary conditions
and noisy measurements must be evaluated before making comparisons
with theory or field data.

1. INTRODUCTION

A sea strait is a narrow passage between two large bodies of water
which have different density stratification and sea level elevations
due to the independent, natural processes in each basin. Sea level
slopes along a sea strait induce so-called barotropic currents while
the density fields induce baroclinic currents. Fluid mechanical laboratory studies provide physical understanding of the dynamics and
energetics of these currents for testing theory and constructing algorithms for numerical models and give a framework for planning and
interpreting field measurements.
    Several laboratory studies are presented to show the state-of-
the-art and indicate future activities. Emphasis is placed on the
theoretical framework for laboratory simulations of natural phenomena
and some important limiting factors.

2. PHYSICAL PROCESSES FOR LABORATORY STUDIES

Tides, winds and thermohaline processes are the most common agents

373

L. J. Pratt (ed.), The Physical Oceanography of Sea Straits, 373–388.
© 1990 Kluwer Academic Publishers.

which drive ocean currents through sea straits. They act as *remote*
boundary conditions to the local flow characteristics. The strait of
Gibraltar is a good example where excessive evaporation in the Mediterranean Sea produces both barotropic and baroclinic flows in
addition to the (mainly barotropic) flow driven by the external Atlantic tides (Almagan, et al., 1989). Local wind is seldom an important
driving agent although it is often correlated with the remote forcing
that sets up the pressure-driven strait flow. Local strait flows can
therefore be modelled in the laboratory without having to model the
surface shear stress at the narrows.

Due to the stratified nature of the world's oceans, most straits
support baroclinic flows. In addition to their obvious barotropic
effects, tides and winds cause internal waves that also have to be
modelled to account for the (time-varying) remote forcing. There are,
however, several important regions like shallow tidal sills where the
flow is essentially barotropic, i.e. homogeneous.

Factors of importance which control the kinematics, dynamics and
energetics are:

- topography including roughness
- hydraulic control
- mixing, and
- rotation

In the laboratory, topography and rotation are imposed while the
hydraulic control and mixing are internal fluid processes which depend
on topography, rotation and the external forcing. Much of the
laboratory work in the literature deals with the determination of the
role of these two processes, since they have a decisive control on the
flow field.

3. EXPERIMENT OR MODEL

Laboratory studies may be broadly divided into two categories which I
will denote experiments and model simulations. An *experiment* is a
highly simplified study to provide basic quantitative data either to
test a theory, produce an unknown empirical constant, or produce an
empirical law which may be needed to account for subgrid processes in
a numerical model. In many cases experiments have filled in large gaps
in our understanding of ocean circulation and have led the way to new
analytical approaches to solving the physics.

Laboratory *models*, on the other hand, are intended to simulate
particular flows in nature using actual topography and boundary
conditions. Laboratory models seldom produce the same level of
precision as simple, well-controlled experiments, due in part to the
fact that implementing complicated boundary conditions often leads to
several sources of errors (e.g. time varying velocity profiles), but
are expected to produce a wealth of details of the kinematics of a
scenario and thereby a different kind of insight. Model results are

often compared with corresponding field measurements, which usually contain experimental noise and very crude information on the actual boundary conditions. At present this appears to be a limiting factor for the need to develop model techniques. The need for advanced experimental techniques, however, increases as higher order theories are developed.

The theoretical basis for laboratory experiments and models and the inherent limitations involved will be presented after a brief review of keynote studies which are pertinent to the physical oceanography of sea straits.

4. SOME KEYNOTE LABORATORY STUDIES

Table 1 gives a list of some laboratory works to illustrate various sea strait processes that have been studied. The list is not exhaustive. It is intended to give the flavor of the variety of processes and experimental approaches to treat both fundamental problems and model specific situations. The comments in Table 1 will be clarified in the text. Here, anonymous referees have been quite helpful.

It is perhaps easiest to begin with the homogeneous tidal strait models of which I have included two. Bonnefille & Chabert d'Hieres (1967) studied the ocean currents through the Strait of Dover on a large rotating platform, simulating the well-known Kelvin wave dynamics which give larger tides along the French coast than along the English coast. The primary boundary conditions were the external tidal elevations on both ends of the English Channel. This region is so shallow that bottom friction had to be tailored to simulate the proper dynamics and energetics. After adjustment, using vertical strips mounted on the bottom, the model tides simulated the major observed constituents throughout the channel very well. McClimans & Gjerp (1979) studied the Tromsø Sound with three narrow openings to the ocean. The tidal jets were highly non-linear and rotational effects were small. Advective accelerations dominated over frictional effects. A high frequency seiching of one of the side arms caused a feedback to the controls of the sea boundaries which had to be damped (electronically). In spite of these small overtones the results compared favorably with field observations of tidal currents without frictional adjustments.

Among the early laboratory experiments on stratified flows in straits were those of Stommel and Farmer (1952; 1953) in which the issue of hydraulic control was studied. It was shown that for short straits the control of the flow was essentially inviscid and could be expressed in terms of the ratio of the flow speed to the interfacial gravity wave speed, or a densimetric Froude number F. The double control on both inflow and outflow in the mouth of an estuary was shown experimentally to limit the amount of mixing within the estuary. They called this condition overmixing. Stommel and Farmer also studied the combined effects of overmixing and tides but did not attempt to

Table 1. Some laboratory works on strait processes.

AUTHORS	YEAR	PROCESS	MODEL OR EXPERIMENT	COMMENTS
Stommel & Farmer	(1952)	Hydraulic control	E	Fjord narrows
Stommel & Farmer	(1953)	Double hydraulic control	E	Overmixing and tides
Keulegan	(1955)	Mixing, salt wedge	E	Law of sea, Keulegan No.
Thorpe	(1973)	Mixing	E	Ri ≈ 1/3
Anati, et al	(1977)	Mixing	E	End controls
Welander	(1974)	Overmixing	E	Minimum thickness
Stigebrandt	(1977)	Hydraulic control, time dependent	E	Barotropic/ baroclinic
Long	(1954)	Sill control	E	Bores/ blocking
Baines	(1984)	Sill control	E	Blocking
Wood	(1970)	Hydraulic control	E	Narrows
Armi	(1986)	Hydraulic control	E	Narrows
Lawrence	(1985)	Hydraulic control	E	Mixing
Wilkinson & Wood	(1987)	Transient control	E	Blocking
Stigebrandt	(1976)	Tides over sill	E	Internal waves
Lee & Beardsley	(1974)	Tides over sill	E	Solitons
Lansing & Maxworthy	(1984)	Tides over sill	E	Solitons
Long	(1977)	Overmixing	E	Three layer control
Denton	(1987)	Sill control	E	Three layer
Whitehead, et al	(1974)	Hydraulic control	E	Rotation
Shen	(1981)	Hydraulic control	E	Rotation
Whitehead	(1986)	Separation in strait	E	Frontal jet
Pratt	(1987)	Supercritical outflow	E	3-D bores
Dalziel	(1988)	Rotating exchange flow	E	Viscous effects
Bonnefille & Chabert d'Hieres	(1967)	Tidal strait	M	English channel
McClimans & Gjerp	(1979)	Tidal straits	M	Archipelago
Rattray & Lincoln	(1955)	Tidal sound	M	Stratified
Monismith & Maxworthy	(1989)	Selective withdrawal	E	Spinup of circulation
Whitehead	(1985)	Uplift/withdrawal	E/M	Gibraltar
Nof	(1978)	Outflow circulation	E	Effect of PV
Whitehead & Miller	(1979)	Outflow circulation	E/M	Alboran sea
Bormans & Garrett	(1989)	Gyre formation	E/M	Effect of F

formulate the system mathematically. Rattray and Lincoln (1955) ran a topographically realistic model of the Puget Sound with both stratification and tides. They showed qualitatively good agreement with deep tidal mixing but did not address the issue of control.

The stationary salt wedge was studied by Keulegan (1955) to establish both the dynamics (shape) and mixing characteristics. He obtained a criterion for turbulent mixing along the pycnocline (Keulegan number). A law of mixing in the sea in which the outflowing surface layer entrains an additional 60 % of bottom water was later shown to be tied to the aspect ratio of his flume (McClimans, 1979).

Thorpe (1973) found that the local gradient Richardson number along the pycnocline $Ri = g'_z/u_z^2 \approx 1/3$ during fully developed turbulent mixing. Here g'_z is the vertical buoyancy gradient and u_z is the vertical gradient of current velocity along the strait. In many of the experimental works which have been published, this appears to be a very robust result.

Anati, et al. (1977) tested two layer exchange flow in a long strait. The densimetric Froude number was critical ($F \approx 1$) at the downstream end of each layer. The shear induced mixing along the pycnocline adjusted the pycnocline thickness to 1/5 of the total depth.

Welander (1974) demonstrated that the principle of overmixing and double hydraulic control in the narrow, shallow Danish Belt could determine the salinity of the Baltic Sea. He found a minimum brackish layer thickness in the basin at moderate fresh water discharges.

Stigebrandt (1977) investigated the transport capacity of a constriction to a two layer tidal flow which was only intermittently baroclinic. The results support the hypothesis that barotropic flow can be important for exchange through straits.

Long (1954) showed that stationary two layer flow in a channel (with or without sills and narrows) develops only after the initial disturbances propagate to large distances. Baines (1984) investigated blocking in stratified flows and obtained experimental data for hydraulic jumps, drops and rarefactions to support a generalized theory of flow over sills. Hysteresis effects were revealed including multi equilibria dependent on downstream conditions.

The type of (steady) flow over a given topography depends both on the upstream and downstream conditions. For non-rotating flows Wood (1970) investigated 2-layer flow through a 2 cm narrows. There was little evidence that friction was important. Armi (1986) also studied the duality of flow through a narrows. Lawrence (1985) studied the flow and mixing over a sill. Transitions between subcritical (F < 1) and supercritical (F > 1) flows are seen to produce energy loss and mixing and indications of three dimensional flow. Mixing usually occurs along the laterally spreading fronts in three dimensional flows (McClimans, 1979).

Wilkinson and Wood (1987) demonstrated experimentally the quasi-stationary advance and retreat of blocked flow through a contraction. Stagnation pressure seems to be important for the dynamics of advancing fronts.

Stigebrandt (1976) demonstrated internal wave generation by tides over a sharp sill. For subcritical flow there were regular waves propagating from the sill while supercritical flow led to turbulent rotors that greatly reduced the amplitude of the internal tides. Lee and Beardsley (1974) showed that tidal motions of a two-layer fluid over a streamlined sill could generate soliton wave packets with frequencies much higher than tidal. Lansing and Maxworthy (1984) showed that solitons are generated provided that the maximum densimetric Froude number (based on the barotropic current) is sufficiently large, but not so large that separation and mixing dominate.

An early contribution of Hachey (1934) inspired Long (1977) to model a three-layer "overmixing" control at fjord narrows. A more complete theory of 3 layer controls is needed to utilize experiments of this type. Denton (1987) studied experimentally the multiple control aspects of flow intrusions in three-layer sill flow. Here, too, multiple equilibria occur.

Rotational sill overflow was simulated in the laboratory by Whitehead, et al (1974) for zero potential vorticity upstream. When the width of the strait is much greater than the Rossby radius of deformation $r_0 = c_1/f$, where f is the Coriolis parameter (planetary vorticity) and c_1 is the celerity of interfacial waves, fronts are formed along the strait. Shen (1981) studied further aspects of separation and found that the transport decreases when the upstream potential vorticity is increased.

Whitehead (1986) demonstrated the development of a frontal jet along a rotationally controlled sill overflow. In applying these results to the Strait of Gibraltar, he found that the inertial radius $r_1 = u/f$ is an important scaling length for laboratory simulations where $F \approx 1$.

Pratt (1987) studied details of the effects of rotation on supercritical overflows with downstream jumps. Strong rotation caused the flow to separate from the left hand wall of the channel (northern hemisphere) and transformed the jump to a vertical undular motion with a large horizontal excursion (back to the wall). The local energy dissipation appeared to be greatly reduced.

Dalziel (1988) investigated two-layer rotational exchange flow in the laboratory. A similar separation frontal pattern was observed and quantified for a wide range of conditions. The exchange flow rate is only a weak function of the potential vorticity at least when it is the same in both layers. The exchange flow rate is maximal for zero potential vorticity. The flow rate is shown to be independent of channel width for a variety of geometries when the width is greater than the Rossby deformation radius. Effects of barotropic forcing on exchange flow were studied for narrow channels. The viscous effects on the side walls were taken into account to explain the results theoretically.

Upstream influence induced by the selective withdrawal associated with sill overflows has been reported by Whitehead (1985) and Monismith & Maxworthy (1989). Signals propagate along the right hand wall (looking upstream, northern hemisphere) as internal Kelvin waves.

Whitehead (1985) studied the uplift of deeper Mediterranean water due to the selective withdrawal of the exchange flow through the Strait of Gibraltar. Rotation enhanced the depth of withdrawal as expected to comply with the Taylor-Proudman theorem. Since the experiment was transient, the filling of the upper layer compressed the ambient water and imposed a slight anticyclonic vorticity which may have reduced the widthdrawal thickness.

The above studies of stratified and/or rotating flows were mostly experimental in nature and have all given insight (and or quantitative results) for physical processes in sea straits. The use of laboratory models for baroclinic flows over real topography seems to have been focussed on the Strait of Gibraltar although the most detailed topography (Whitehead and Miller, 1979) was two dimensional (in the horizontal). These models are all simplifications of the real system and have focussed on various aspects of exchange flow in straits.

Nof (1978) showed experimentally that the development of the downstream circulation depends on the lateral shear, at least for $F \ll 1$. For shear equal to $-f$ the flow diverted to the right as a coastal current. For zero lateral shear the downstream flow developed a large anticyclonic gyre in the Alboran Sea. Whitehead and Miller (1979) studied small lateral shear for $F \approx 1$ and found the downstream current development to depend on the curvature of the right hand exit wall. When the inertial radius was less than the wall curvature a coastal current developed.

Bormans and Garrett (1989) found that the type of downstream flow (coastal current or gyre) depended on the wall curvature and the densimetric Froude number (super or subcritical). This type of behaviour was also found for a very simple topography by McClimans, et al. (1985).

5. LABORATORY MODEL THEORY AND SCALE EFFECTS

The experiments and models described above were subjected to certain limitations in the laboratory. Often, the small size of a laboratory is the primary cause of limitations. In the following sections model theory and scale effects will be presented to indicate the significance of the earlier work and to inspire newer laboratory studies of the physical processes in sea straits. These rules and experiences are important for planning and evaluating good laboratory work.

5.1. Model theory

The dynamic simulation of stratified flows in the sea must follow the rules of model theory as well as the limitations which are often called scale effects or model effects. Experiments are usually performed with very simple geometric basins with no claim of geometrical similitude except in a parametric manner (e.g. width equal to a deformation radius). Simulating large geophysical flows in the laboratory usually requires the use of distorted hydraulic models in which the vertical scale is exaggerated vis-à-vis the horizontal scale, but with realistic horizontal geometrical similitude. Most straits with their large width (B) to depth (H) aspect ratios are difficult to simulate geometrically within a reasonable (economic) size. The validity of testing the dynamics of stratified flows in distorted laboratory models is thus an important issue for this work.

An essential feature is that buoyant spreading is a process for which the pressure p is approximately hydrostatic, except in the region of frontal convergence where the vertical accelerations are significant and turbulent mixing is quite vigorous. Thus, with surface elevation z, vertical coordinate z positive upwards, density ϱ and acceleration of gravity g, the flow is assumed hydrostatic

$$p(z) = \int_z^z g\varrho dz' \qquad (1)$$

except, e.g., in the narrow region near a front. In this region the energetics of frontal turbulence is often an important factor for momentum exchange. Model distortion cannot be so large that (1) is significantly violated in the laboratory. This sets the stage for the use of shallow water (long wave) considerations inherent in geo-physical hydrodynamic models.

The equation of motion for a nearly horizontal velocity field, to the Boussinesq approximation, is

$$\frac{\partial \mathbf{v}}{\partial t} + \mathbf{v} \cdot \nabla_H \mathbf{v} = \frac{1}{\varrho_0} \nabla_H p - f\mathbf{k}\mathbf{x}\mathbf{v} + \nu\nabla^2\mathbf{v} \qquad (2)$$

where $\mathbf{v}$ is the horizontal velocity vector, ∇_H is the horizontal gradient operator, f is the Coriolis parameter (or local vertical planetary vorticity), $\mathbf{k}$ is the unit vector along the vertical (z) axis, ν is the kinematic viscosity and ϱ_0 is a (constant) reference density.

In nature the molecular viscosity is negligible and the nonlinear Reynolds stresses generated by the stochastic (turbulent) motions encompassed in the second term on the left hand side of (2) account for a multi-scale momentum exchange. However, the introduction of the hydrostatic approximation means that (2) does not strictly apply to three-dimensional turbulence with significant vertical accelerations. In the laboratory the division between Reynolds stresses and viscous stresses may be different, so the terms on the right are retained for the discussion of model laws and scale effects. Other physics which are not included in (2), like surface tension, may also be important in the laboratory. This is discussed in section 5.3.

For densimetric or baroclinic flows the hydrostatic pressure can be conveniently divided into barotropic and baroclinic parts

$$p(z) = \int_z^z \varrho_0 g dz' + \int_z^z (\varrho-\varrho_0) \, g dz' \qquad (3)$$

or

$$p(z) = \varrho_o g(z-z) - \varrho_o \int_z^Z g'dz' \qquad (4)$$

where $g' = g(\varrho_o-\varrho)/\varrho_o$ is the specific buoyancy or reduced gravity. This must be simulated correctly in the laboratory for buoyancy driven flows.

Combining (2) and (4) gives

$$\frac{\partial v}{\partial t} + v \cdot \nabla_H v = -g\nabla_H z + \nabla_H \int_z^Z g'dz' - fkxv + \nu\nabla^2 v \qquad (5)$$

The fluid dynamics in the model are the same as the fluid dynamics in nature when (5) is identical for the two realizations and when identical boundary conditions are imposed. This is dynamic similitude.

Scaling laws for the dynamic simulation and the interpretation of laboratory results are derived from the dimensionless equations of motion. This requires a choice of scaling factors. For distorted models two length scales are necessary. Thus, all variables in (5) are normalized by the following (constant) characteristic scales for a given realization.

H = vertical length
B = horizontal length
U = horizontal velocity $\qquad\qquad$ (6)
T = time scale
G' = buoyancy

For the horizontal advective processes at hand I choose $T = B/U$ for a proper time scale to unify the left hand side of (5) into the nondimensional total derivative $D/dt = \partial/\partial t + v\cdot\nabla_H$. This simplification emphasizes that only the geostrophic (2-D) turbulence is simulated correctly within the framework of the derived model laws. Three dimensional turbulence can by simulated in terms of energetics, but not in dynamic detail with distortion (B ≠ H).

Equation (5) may be expressed in dimensionless form as

$$\frac{Dv}{dt} = - \{\frac{gH}{U^2}\} \nabla_H z + \{\frac{G'H}{U^2}\} \nabla_H \int_z^Z g'dz' - \{\frac{fB}{U}\} kxv + \{\frac{\nu}{UB}\} \nabla_H^2 v$$

$$+ \{\frac{\nu B}{UH^2}\} \frac{\partial^2 v}{\partial z^2} \qquad (7)$$

in which all quantities outside the curly brackets are now dimensionless. A distorted laboratory model simulates the natural flow described in (7) only when the assumptions leading to (5) are fulfilled, when the products in curly brackets are the same in the model as they are in nature, and when the proper boundary conditions are applied.

The dimensionless groups in curly brackets containing all the characteristic sizes are from left to right,

Froude number $Fr = U/(gH)^{1/2}$ (8)

Densimetric Froude number $F = U/(G'H)^{1/2}$ (9)

Rossby number $Ro = U/fB$ (10)

and two Reynolds numbers:

$Re_H = UB/\nu$; $Re_v = UH^2/\nu B$ (11)

The importance of molecular friction increases inversely with Re. Thus, for geophysical flows with $B \gg H$, the vertical shear is most important. In the laboratory, this is to some extent reduced due to distortion for which H/B is increased from 10 to 1000 over natural conditions. In straits, however, the advective acceleration term on the left hand side of (7) is balanced primarily by pressure terms or by the Coriolis term. It is therefore important that the model does not exaggerate the molecular friction to the point that it becomes a dominant term in the balance. The turbulent momentum exchange is reduced somewhat by a limited spectrum of turbulence (when it is present in the laboratory), which counteracts the larger friction expected at the lower laboratory Reynolds numbers. However, when turbulent friction dominates the dynamics of the natural strait flow, it must be simulated in the model, rendering a combined Froude-Reynolds model law. This limits the possibility of scaling model results to natural flows over a wide range of boundary conditions. Fortunately, the momentum exchange along fronts and pycnoclines occurs primarily in large wave instabilities and vortices which derive their energy from the spreading, buoyant flow (Thorpe, 1973; McClimans, 1979). Although the details of these three-dimensional processes are not simulated in a distorted model, the energetics of the process has been argued to be independent of distortion provided $B/H \gg 1$ in the laboratory (McClimans, 1979). The importance of horizontal momentum exchange in distorted models was treated by McClimans & Gjerp (1979).

The exact simulation of horizontal friction is not usually necessary since it represents a smaller contribution to the dynamics. It is essential, however, that it is not greatly amplified. Here wall roughness is an important factor. Fortunately, the boundary layers along smooth model walls are often of the same (scaled) size as the boundary layers along a natural coastline. With these reservations the last two terms of (5) can usually be ignored.

By ignoring molecular processes, choosing a time scale f^{-1} and simulating the density, the set (8)-(10) makes (7) a complete model of (2). This model gives Froude-Rossby similitude.

Long shallow water waves like tides and seiches have negligible vertical accelerations in nature. In most cases these waves will be reflected along the sloping bottom. Vertical distortion, however, moves the limit of total reflection for internal waves to shorter wavelengths (Nagashima, 1971). Thus, for gently sloping bottom topography, distortion may pose an incorrect reflection condition.

5.2. Scaling laws

Denoting the ratio of natural quantities to laboratory quantities by the index r, the requirement for Froude-Rossby similitude can be expressed as

$$(Fr)_r = (F)_r = (Ro)_r = 1 \tag{12}$$

Choosing the primary similarity requirement $(F)_r = 1$ leads to

$$U_r = (G'_r H_r)^{1/2} \tag{13}$$

The kinematic relation for horizontal currents yields a time scale

$$T_r = B_r/U_r = B_r/(G'_r H_r)^{1/2} \tag{14}$$

This constitutes Froude's scaling law for densimetric flows. If both Fr and F are constraints, then $G'_r = 1$. If Ro is also a constraint (for rotational flows) then $f_r = T_r^{-1}$.

Scaling laws for all hydrodynamic quantities can be derived from the above. Take, for example, discharge Q:

$$Q_r = U_r B_r H_r = B_r^4 f_r^3/G'_r \tag{15}$$

showing the strong influence rotation has on the interpretation of model results of large scale flows. For non-rotating flows, f_r has to be replaced by $(G'_r H_r)^{1/2}/B_r$.

Length scales are often helpful in judging parameter space. The characteristic inertial radius is

$$r_i = U/f \tag{16}$$

and the characteristic baroclinic Rossby deformation radius is

$$r_o = (G'H)^{1/2}/f \tag{17}$$

Their ratio is the densimetric Froude number

$$F = r_i/r_o \tag{18}$$

Further

$$Ro = r_i/B \qquad\qquad (19)$$

$$F/Ro = B/r_o \qquad\qquad (20)$$

These ratios are important for interpreting the large scale flow simulations for which fluid rotation is important.

5.3. Scale effects

Effects not modelled by Froude-Rossby similitude give spurious results which are interpreted as scale effects. They are different in the laboratory than in nature. The simulation of friction has already been discussed.

It may be important for momentum exchange as well as mass exchange that the flow is naturally unstable and three-dimensionally turbulent. This is one of the limiting conditions beyond which the simulation is not strictly valid. For momentum exchange in jets at large F, it is well accepted that fully turbulent flow occurs when Re > 1500. For the case of F ≈ 1 there is a large reserve of potential energy to create turbulence, and the critical Reynolds number appears to be between 500 and 1000. For the frontal regions, there appears to be a highly energetic horizontal exchange. For a sharp pycnocline the gravitational stability and or viscosity dampens out the turbulence above the critical value (Keulegan, 1955)

$$\theta = \frac{(vg')^{1/3}}{\Delta u} \approx 0.178 \qquad\qquad (21)$$

where Δu is the velocity difference across the pycnocline. θ is sometimes referred to as the Keulegan number. The exchange of mass is also highly dependent on turbulence, so (21) is an important criterion.

Surface tension gradients in small laboratory experiments pose an important limiting factor for frontal studies. After some bad experience with spurious surface currents induced by surface tension reducing agents, I avoid surface tension effects by dimensioning the experiments so that surface films have no dynamic significance. This of course leads to a larger apparatus. It can be shown (McClimans & Sægrov, 1982) that a reasonable lower limit of source thickness for the simulation of river plumes is 1 cm for saline and 4 cm for thermal stratification. Thermal plume studies require an apparatus which is 4 times larger than that for fresh water plumes in seawater.

A stringent requirement $G'_r = 1$ for combined Fr and F similitude would limit the applicability of the laboratory results. In the absence of surface forcing (atmosphere and tide) and with $g' \ll g$, there is very weak coupling between the surface slope and the internal density field. With entrainment, however, g' is not a constant and there may be an argument to maintain $G'_r = 1$. It should therefore be mentioned that laboratory studies of baroclinic coastal currents over a wide range of g' show little reason for concern (Griffiths & Linden, 1981).

5.4. Laboratory conditions and control of experiments

In addition to physics not accounted in the model laws (scale effects) another cause for concern are spurious results caused by noisy, faulty or incomplete boundary conditions. The wave reflection condition was mentioned earlier. An example of a noisy boundary condition was observed by McClimans & Gjerp (1979) where a control feedback due to seiching was not fully dampened (reflecting open boundary).

The deviation of the axis of rotation from the vertical in the Barents Sea model (McClimans & Myhr, 1989) caused barotropic waves that were amplified over shallow sea mounts. For simulating flows in straits between large ocean basins in which the strait is in the middle, the deviation of the axis of rotation from the vertical should be less than 10^{-6} rad (0.2 sec arc) to limit spurious barotropic flows. (It is impossible to level a rotating platform to better than 10^{-7} rad which is the deviation of the vertical in the earth-moon gravitational system. This, then, defines the limits of the type of motion which can be simulated in the laboratory without taking into account the phase of the moon and the time of day.)

Incomplete boundary conditions are often used by replacing the desired condition by an ad-hoc simulation of e.g. wave radiation, wind shear, etc. Each case should be tested against the desired performance to determine to which degree the boundary condition is fulfilled.

5.5. Measurements

The value of laboratory work depends on how well it is observed and documented. In general, the accuracy of modern electronic devices is not a limiting factor. A well defined experiment with a limited requirement on information to test a theory will usually be documented satisfactorily using in situ or remote acoustic or laser profilers for high precision data. In situ temperature and conductivity probes together with surface followers often provide satisfactory experimental documentation. Photographs are very often used for quantitative analysis.

The big problem with time-varying flow through model topography is to register the enormous amount of action going on over such a short period of time without disturbing the flow. Thus photographic methods are generally used to establish a kinematic field. The present development of three dimensional photogrammetric image analyses will soon provide flow fields that can be compared with and used by three dimensional numerical models. The dynamics and energetics are thus derived from the flow fields with only a few necessary instrumental measurements for reference data.

6. STATE-OF-THE-ART

Many important physical processes associated with flows in sea straits have been tested or simulated in laboratory studies. The details and accuracy of laboratory experiments provide a necessary input for analyses and numerical modelling of strait processes. Most studies have been focused on the questions of hydraulic control and mixing. Both three dimensionality and rotational effects have been studied.

The role of the laboratory for experimental verification of analytical solutions of simplified physical situations is well known. Laboratory experiments of nonlinear and coupled physics for which no analytical solutions are yet available provide a good test for nonlinear numerical solutions. In this respect, field data are often of less scientific value since a truly synoptic coverage of a scenario is seldom found and average conditions of a nonlinear system are in general not obtained by applying average boundary conditions.

Laboratory models provide the basis for simulating coupled processes which are consistent with the combined Froude-Rossby model law and lie within acceptable limits of model scale effects. Few detailed topographic models of baroclinic strait flows with real world boundary conditions have been simulated to date, and there is a great potential for them. On several occasions laboratory models have proven to produce an economical solution to a difficult problem. Particular issues on physical processes that need further study are mixing, internal waves and arbitrary potential vorticity.

Although the super computer is modelling progressively more complicated systems, the role of laboratory facilities for modelling flow details is still indisputable, especially for hydrodynamic instability and turbulence. Consider the 13 m diameter rotating platform in Grenoble, which provides a 1 cm resolution over a depth of 1.2 m. This amounts to more than 10^8 computational points at uniform spatial resolution. If a strait model were to have a horizontal extent of 1300 km ($B_r = 10^5$) and a vertical extent of 1200 m ($H_r = 10^3$), the time scale from (14) is $T_r = 3162$. In other words, a day is simulated in about 27 s and a year in less than 3 hours. How long will it be before a super computer will be able to compute so much, so fast and within the same budget? And how much testing of algorithms is necessary to have confidence in the physical relevance of the results? Here, the laboratory has an important role.

These concluding remarks apply primarily for advective processes which seem to dominate the fluid dynamics of many sea straits. For regions where friction, wind energy and thermodynamic processes (cooling, freezing, evaporation) dominate, as in the large basins adjacent to sea straits, the shortcomings of the limitations of Froude-Rossby similitude concede to the versatility of numerical modelling. Furthermore, numerical models are used to generate forced boundary conditions for laboratory models of limited, topographically complicated regions.

7. REFERENCES

Almagan, J.L., Bryden, H., Kinder, T. and Parrilla, G. (Editors) (1989) Proceedings, Seminario sobre la oceanografia fisica del Estrecho de Gibraltar. Madrid, 24-28 Oct 1988. SECEG/ONR.

Anati, D.A., Assaf, G. and Thompson, R.O.R.Y. (1977) Laboratory models of sea straits. J. Fluid Mech. 81:341-351.

Armi, L. (1986) The hydraulics of two flowing layers with different densities. J. Fluid Mech. 163:27-58.

Baines, P.G. (1984) A unified description of two-layer flow over topography. J. Fluid Mech. 146:127-167.

Bonnefille, R. and Chabert D'Hières, G. (1967) Étude d'un modèle tournant de mer littorale. Application au problème de l'usine marémotrice des iles chausey. Houille Blanche. 1967(6):651-658.

Bormans, M. and Garrett, C. (1989) A simple criterion for gyre formation by the surface outflow from a strait, with application to the Alboran Sea. J. Geophys. Res. 94:12637-12644.

Dalziel, S.B. (1988) Two-layer hydraulics: Maximal exchange flows. Ph.D. thesis, DAMPT, The Univ. of Cambridge. Cambridge, England.

Denton, R.A. (1987) Hydraulic control of multilayered exchange flow through obstructions. Paper presented at Third International Symposium on Stratified Flows, Pasadena, CA. Vol 1, Session B3.

Griffiths, R.A. and Linden, P. F. (1981) The stability of buoyancy driven coastal currents. Dyn. of Atmos. and Oceans 5:281-306.

Hachey, H.B. (1934) Movements resulting from mixing of stratified waters. J. Biol. Board Can. 1:133-143.

Keulegan, G. (1955) Seventh progress report on model laws for density currents. Interfacial mixing in arrested saline wedges. NBS Report 4142.

Lansing, F.S. and Maxworthy, T. (1984) On the generation and evolution of internal gravity waves. J. Fluid Mech. 145:127-149.

Lawrence, G.A. (1985) The hydraulics of mixing of two-layer flow over an obstacle. Hyd. Eng. Lab., U. of Cal., Berkeley. Report UCB/HEL-85/02.

Lee, C-Y. and Beardsley, R.C. (1974) The generation of long nonlinear internal waves in a weakly stratified shear flow. J. Geophys. Res. 79:453-462.

Long, R.R. (1954) Some aspects of the flow of stratified fluids. II. Experiments with a two-fluid system. Tellus 6:97-115.

Long, R.R. (1977) Three-layer circulations in estuaries and harbors. J. Phys. Ocean. 7:415-421.

McClimans, T.A. (1979) On the energetics of river plume entrainment. Geophys. Astrophys. Fluid Dyn. 13:67-81.

McClimans, T.A. and Gjerp, S.A. (1979) Numerical study of distortion in a Froude model. Proceedings, 16th International conference on Coastal Engineering, Hamburg, 29 Aug - 1 Sep. III: 2887-2904, ASCE.

McClimans, T.A. and Myhr, B. (1989) Laboratory model of the Barents Sea. NHL Video.

McClimans, T.A. and Sægrov, S. (1982) River plume studies in distorted Froude models. J. Hyd. Res. 20:15-27.

McClimans, T.A., Vinger, Å. and Mork, M. (1985) The role of Froude number in models of baroclinic coastal currents. Ocean Modelling 62:14-17.

Monismith, S.G. and Maxworthy, T. (1989) Selective withdrawal and spin-up of a rotating stratified fluid. J. Fluid Mech. 199: 377-401.

Nagashima, H. (1971) Reflection and breaking of internal waves on a sloping beach. J. Ocean. Soc. Jap. 27:1-6.

Nof, D. (1978) On geostrophic adjustments in sea straits and wide estuaries: Theory and laboratory experiments. Part II - Two-layer system. J. Phys. Ocean. 8:861-872.

Pratt, L.J. (1987) Rotating shocks in a separated laboratory channel flow. J. Phys. Ocean. 17:483-491.

Rattray, M., Jr. and Lincoln, J.H. (1955) Operating characteristics of an oceanographic model of Puget Sound. Trans. Amer. Geophys. Union 36:251-261.

Shen, C. (1981) The rotating hydraulics of the open-channel flow between two basins. J. Fluid Mech. 112:161-188.

Stigebrandt, A. (1976) Vertical diffusion driven by internal waves in a sill fjord. J. Phys. Ocean 6:486-495.

Stigebrandt, A. (1977) On the effect of barotropic current fluctuations on the two-layer transport capacity of a constriction. J. Phys. Ocean. 7:118-122.

Stommel, H. and Farmer, H.G. (1952) Abrupt change in width in two-layer open channel flow. J. Mar. Res. 11:205-214.

Stommel, H. and Farmer, H.G. (1953) Control of salinity in an estuary by a transition. J. Mar. Res. 12:13-20.

Thorpe, S.A. (1973) Experiments on instability and turbulence in a stratified shear flow. J. Fluid Mech. 61:731-751.

Welander, P. (1974) Two-layer exchange in an estuary basin, with special reference to the Baltic Sea. J. Phys. Ocean. 4:542-556.

Whitehead, J.A. (1985) A laboratory study of gyres and uplift near the Strait of Gibraltar. J. Geophys. Res. 90:7045-7060.

Whitehead, J.A. (1986) Flow of a homogeneous rotating fluid through straits. Geophys. Astrophys. Fluid Dynamics. 36:187-205.

Whitehead, J.A., Leetmaa, A. and Knox, R.A. (1974) Rotating hydraulics of strait and sill flows. Geophys. Fluid Dyn. 6:101-125.

Whitehead, J.A. and Miller, A.R. (1979) Laboratory simulation of the gyre in the Alboran Sea. J. Geophys. Res. 84:3733-3742.

Wilkinson, D.L. and Wood, I.R. (1987) Blocking of layered flows in channels of gradually varying geometry. Paper presented at Third International Symposium on Stratified Flows, Pasadena, CA Vol 1, Session B3.

Wood, I.R. (1970) A lock exchange flow. J. Fluid Mech. 42:671-687.

III. Waves, Tides and Time-Dependence

REVIEW OF DISPERSIVE AND RESONANT EFFECTS IN INTERNAL WAVE PROPAGATION

K. R. HELFRICH
Woods Hole Oceanographic Institution
Woods Hole, Massachusetts 02543

and

W. K. MELVILLE
R. M. Parsons Laboratory
Massachusetts Institute of Technology
Cambridge, Massachusetts 02139

ABSTRACT. Theories and models for the generation, propagation and dissipation of long, nonlinear internal waves are reviewed. The roles of dispersive and resonant effects are then discussed in the context of transcritical flow through channels. Recent work on the resonant generation of upstream advancing solitary waves is reviewed. It is shown that these waves (which include non-hydrostatic effects) may be important in the time-dependent hydraulic control problem. The instability of non-linear Kelvin waves in a rotating channel is also discussed and it is shown that this instability may be responsible for observations of wave-front curvature. These processes may be significant in resolving the dynamics of sea straits.

1. Introduction

Long internal waves are ubiquitous features in many of the world's marginal seas, coastal waters and straits. Examples of measurements of these waves include those from Massachusetts Bay (Haury, Briscoe and Orr, 1979), the Andaman Sea (Osborne and Burch, 1980) and the Strait of Gibraltar (Kinder, 1984; La Violette and Arnone, 1988). In addition to *in situ* measurements, remote sensing techniques reveal the extent of internal wave activity. Fu and Holt (1982) state "internal waves constitute a major element in the wealth of information contained in SEASAT SAR imagery."

The dynamics and evolution of long internal waves are of interest not only because the waves are common, but also because the waves affect other processes. In coastal waters, these waves are observed to propagate primarily shoreward with little evidence of reflection (Fu and Holt, 1982). Some of this incident energy may cause vertical mixing. Haury *et al.* (1979) observed overturning instabilities. Internal wave breaking is an effective mechanism for mixing nutrient rich water from the bottom to the biologically active upper

L. J. Pratt (ed.), The Physical Oceanography of Sea Straits, 391–420.

layer (Sandstrom and Elliot, 1984). Waves incident on the coast have also been suggested as a mechanism for coastal seiche excitement (Giese *et al.*, 1982). In straits, internal waves contribute to vertical mixing (La Violette and Arnone, 1988). Also, Kinder (1984) estimates that the large internal waves in the Strait of Gibraltar are responsible for up to 40 percent of the net transport through the strait. Internal wave processes may also be important in the hydraulic control problem (Melville and Helfrich, 1987).

In this paper we review theories and models of long internal wave evolution and then focus on several aspects relevant to strait dynamics. However, before considering the details of any theories it is useful to examine typical wave characteristics. Most observations show long, large amplitude waves traveling on the main pycnocline. The waves usually have a vertical structure corresponding to the first internal mode (i.e., one zero-crossing in the velocity profile). Two important parameters which characterize dynamics are

$$\alpha = a/D , \ \beta = (kD)^2.$$

The parameter α represents nonlinearity and β dispersion. Here a is a typical wave amplitude, D is a depth scale and k^{-1} is a longitudinal wavelength. Measurements of waves in the Massachusetts Bay give $\alpha \approx (10m/80m) = 0.13$ and $\beta \approx (80m/300m)^2 = 0.07$. From the Strait of Gibraltar $\alpha \approx (40m/500m) = 0.08$ and $\beta \approx (500m/1500m)^2 = 0.11$. Both nonlinear and dispersive effects are small but significant.

The presence of dispersion (i.e., non-hydrostatic effects) is fundamentally important. Without dispersion, nonlinearity would lead to wave steepening and breaking over short propagation times or distances. This behavior is a feature of hydraulic models which are hydrostatic by definition. Dispersion counteracts the nonlinear tendency for steepening and the formation of discontinuities. The result is nonlinear dispersive waves which may propagate over large distances without forming discontinuities. Stability over long space and timescales is an observed feature in both coastal waters and straits.

The importance of non-hydrostatic effects in straits is discussed in Section 3 in the context of transcritical flows. Transcritical flows are those in which the Froude number $F \equiv U/c_0 \approx 1$. Here U is the mean flow speed and c_0 is the linear long internal wave phase speed. The resonant generation of upstream advancing waves by flow over topography is reviewed. The necessity of considering these waves in any description of the internal hydraulics of straits is demonstrated. Recent results on the stability of Kelvin waves in channels are then reviewed and the results are offered as an explanation of laboratory observations of long internal waves in rotating channels.

2. Theories and Models of Generation, and Dissipation and Propagation

2.1. GOVERNING EQUATION

In this section a general evolution equation for long nonlinear internal waves in a rotating channel is presented. Weak nonlinearity, $\alpha \ll 1$, and weak dispersion, $\beta \ll 1$, have both been shown to be important in describing wave evolution. When $\beta = O(\alpha) \ll 1$, one-dimensional wave propagation is governed by the Korteweg-de Vries (KdV) equation (Whitham, 1974). KdV models have been shown to give reasonably good agreement with oceanic measurements of wave characteristics (Osborne and Burch, 1980; Sandstrom and Elliot, 1984) and laboratory experiments (Koop and Butler, 1981; Segur and Hammack,

1982; Helfrich and Melville, 1986). However, we present a more general evolution equation which incorporates the effects of rotation, lateral variation (two-dimensional propagation), finite channel width, boundary layer damping and near-resonant forcing. Specific cases will be examined from simplifications of this general evolution equation. Cubic nonlinearity, which may dominate quadratic nonlinearity when the coefficient of the quadratic term in the KdV equation is small (Kakutani and Yamasaki, 1978; Miles, 1979) will also be included. In the context of a two layer system the appropriate balance becomes

$$\beta = O\left(\frac{|d_+^* - d_-^*|}{D}\alpha\right),$$

where $d_\pm^*$ is the depth of the upper/lower layer. If $|d_+^* - d_-^*|/D = O(\alpha)$ then $\beta = O(\alpha^2)$ and cubic nonlinearity should be incorporated into the theory.

In order to demonstrate the essential physics, yet retain simplicity, consider the case of a two-layer system in a channel of width W^* rotating about the vertical axis with angular velocity $\frac{1}{2}f$. The density of the upper/lower layer is given by $\rho_\pm^*$. The side walls at $y^* = 0, W^*$ are vertical and a uniform flow of magnitude U in the negative x-direction has been imposed in both layers. The channel contains a transverse ridge $H^*(x^*)$ centered at $x^* = 0$. The upper surface is treated as a rigid lid.

The important nondimensional parameters representing the magnitudes of the various effects, in addition to nonlinearity α and dispersion β, are:

$$\text{lateral variation } \epsilon = (\ell/k)^2, \tag{1a}$$

$$\text{rotation } \Gamma = (f/kc_0)^2 \tag{1b}$$

$$\text{topographic height } \gamma = H_0/D, \tag{1c}$$

$$\text{dissipation } \delta = (\nu/kc_0)^{1/2}D^{-1}, \tag{1d}$$

and

$$\text{near–resonant flow } \Delta = U/c_0 - 1 = F - 1. \tag{1e}$$

Here

$$c_0 = (\sigma g D)^{1/2} \tag{2}$$

is the linear long wave phase speed,

$$D = \frac{d_+^* d_-^*}{d_+^* + d_-^*}, \tag{3}$$

and

$$\sigma = (\rho_-^* - \rho_+^*)/\rho_0^*.$$

Also, ℓ^{-1} is a length scale in the cross-channel direction, H_0 is a topographic height scale and ν is the kinematic viscosity.

In order to bring all the effects into the evolution equation at the highest order we require ϵ, Γ, $\delta = O(\beta)$ and $\beta = O(\alpha^2) \ll 1$ so that cubic nonlinearity is included. The case $\Gamma \ll 1$ corresponds to the weak rotation limit considered by Grimshaw (1985). Furthermore, for inhomogeneous forcing due to near-resonant flow over isolated topography to enter at the same order we require $\Delta \ll 1$ and $\gamma = O(\alpha\beta)$ (Grimshaw and Smyth, 1986; Melville and Helfrich, 1987; Wu, 1987).

The dimensional (starred) variables are normalized according to

$$(x^*, y^*, z^*) = (k^{-1}x, \ell^{-1}y, Dz)$$

$$(u^*, v^*, w^*) = \alpha c_0(u, \epsilon^{1/2}v, \alpha^{1/2}w) \tag{4}$$

$$\eta^* = \alpha D\eta, \quad t^* = t/kc_0, \quad H^* = \gamma DH(x)$$

$$W^* = \ell^{-1}W, \quad d_\pm^* = Dd_\pm, \quad F = U/c_0.$$

The momentum equations for each layer, the interfacial kinematic and dynamic conditions, the top, bottom and side wall ($v^* = 0$ at $y^* = 0, W^*$) boundary conditions are normalized with (4). Following the standard expansion procedure employed by Grimshaw (1985) and Grimshaw and Melville (1989), the governing equation for wave amplitude η is found to be

$$\frac{\partial}{\partial x}\left[\eta_t - \Delta\eta_x + \frac{3}{2}\alpha d_{-2}\eta\eta_x - 3\alpha^2 d_{-3}\eta^2\eta_x\right.$$

$$\left. +\frac{1}{6}\beta d_1\eta_{xxx} + (\gamma/\alpha)\frac{F}{2d_-}H_x + \mathcal{F}\right] \tag{5}$$

$$+\frac{1}{2}(\epsilon\eta_{yy} - \Gamma\eta) = 0(\alpha\beta).$$

where

$$d_n = \{d_-^n + (-1)^{n-1}d_+^n\}, (n = \pm1, \pm2, \ldots). \tag{6}$$

Equation (5) is valid for waves moving in the positive x-direction and is subject to the boundary conditions

$$(\Gamma/\epsilon)^{1/2}\eta + \eta_y = 0 \quad \text{on} \quad y = 0, W \tag{7}$$

(i.e. $v = 0$ at $y = 0$, W).

The boundary layer dissipation term,

$$\mathcal{F} = \frac{\delta}{4\pi^{1/2}}\left[d_+^2 + \frac{1}{2}d_1^2\right]\int_{-\infty}^{+\infty}\frac{\partial\eta}{\partial x}\frac{1 - \text{sgn}(x - x')}{|x - x'|^{1/2}}dx', \tag{8}$$

includes the effects of a viscous boundary layer at $z = -d_-$ (first term in brackets) and an interfacial boundary layer (second term). The bottom dissipation term is found from a boundary layer analysis (Kakutani and Matsuuchi, 1975; Miles, 1976; Grimshaw, 1981). The interfacial term comes from a similar analysis which matches velocity and shear stresses at the interface (Leone, Segur and Hammack, 1982; Helfrich, 1985). The boundary layers are uneffected by rotation at the order of (5).

Before considering simplifications, we note that (5) is a generalization of the Kadomtsev-Petviashvili (KP) equation (Kadomtsev and Petviashvili, 1970). The standard KP equation for internal waves is obtained when $\gamma = \Gamma = \delta = 0$, $\Delta = -1$ and cubic nonlinearity is discarded (i. e., $| d_+ - d_- | / D = O(1)$). When $\Gamma \neq 0$ we have the rotationally modified KP equation (Grimshaw, 1985; Katsis and Akylas, 1987; Grimshaw and Melville, 1989). When two-dimensional ($\epsilon = \Gamma = 0$) and inhomogeneous ($\gamma = \delta = 0$) effects are removed the extended KdV (EKdV) equation (i.e., with cubic nonlinearity) is recovered.

For continuous, rather than two-layer, stratification the evolution of one internal mode is given by a version of (5) in which the coefficients become integrals of the vertical structure function $\phi_n(z)$ found from the eigenvalue problem

$$\frac{d^2\phi_n(z)}{dz^2} - \frac{\overline{p}_z(z)}{\overline{p}(z)c_n^2}\,\phi_n(z) = 0$$

subject to

$$\phi_n(0) = \phi_n(-D) = 0.$$

D is now the total depth, c_n is the linear phase speed for mode n and $\overline{p}(z)$ is the basic state stratification. Benney (1966) and Miles (1979) give derivations and coefficients for the KdV and EKdV equations, respectively. Grimshaw (1985) discusses the continuously stratified version of the rotationally modified KP equation.

2.2. PROPAGATION

It is useful to examine the free linear modes of the rotationally modified KP equation in a channel. Setting $\gamma = \delta = \alpha = 0$, $\Delta = -1$ in (5) and substituting $\eta \sim e^{i(\tilde{k}x + \tilde{l}y - \omega t)}$ gives the dispersion relation

$$\omega = \tilde{k} - \frac{1}{6}\beta d_1\tilde{k}^2 + \frac{1}{2}\tilde{k}^{-1}(\epsilon\tilde{l}^2 + \Gamma). \tag{9}$$

For $\tilde{k} \ll 1$ (9) is singular and the group velocity is in the negative x-direction (Katsis and Akylas, 1987). To avoid the singularity Melville et $al.$ (1989) proposed a regularized version of the KP equation that is asymtotically equivalent to (5). The linear dispersion relation then becomes

$$\omega^2(1 + \frac{1}{6}\beta d_1\tilde{k}^2) - \omega\tilde{k} - \frac{1}{2}(\epsilon\tilde{l}^2 + \Gamma) = 0. \tag{10}$$

The primary root of (10) has $\omega = O(1)$ and includes a linear Kelvin wave for $\tilde{l}^2 + \Gamma/\epsilon = O$ and Poincaré modes for $\tilde{l} = n\pi/W$ ($n = 1, 2, \ldots$). For $\tilde{k} \geq O(1)$ and ϵ, $\Gamma \ll 1$ (9) and (10) are equivalent. Figure 1 shows an example of the dispersion curves from (10). The weak dispersion causes the concave behavior of the dispersion curves at large k. Graphical construction using Figure 1 shows that a triad resonance between two Kelvin waves and one Poincaré mode is possible. When the hydrostatic approximation is made (i.e., $\beta = 0$)

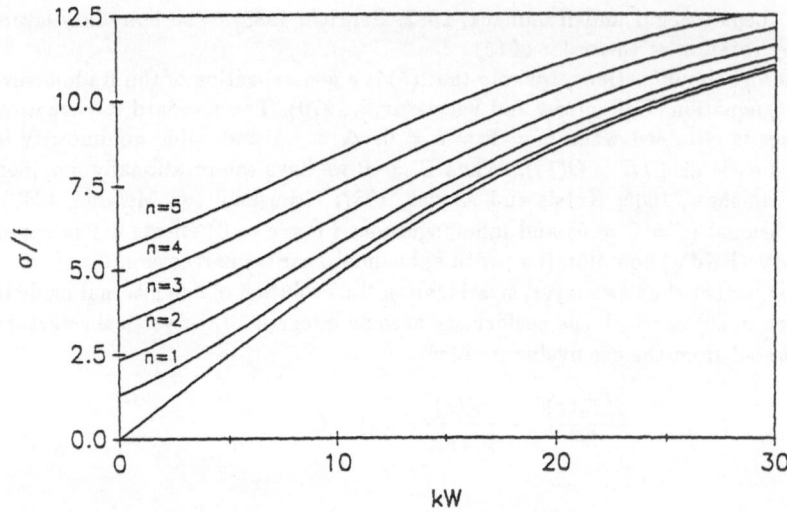

Figure 1. Dispersion curves (10) showing the Kelvin wave and the first five Poincaré modes (Tomasson and Melville, 1989).

the concave behavior is lost (c.f. Figure 11) and this resonance is lost. Details of the triad interactions are discussed by Tomasson and Melville (1989). More importantly, the system permits a direct resonance between a nonlinear Kelvin wave and Poincaré modes. This interaction is discussed in Section 3.

2.2.1. *One-Dimensional Propagation.* When lateral variation, rotation, dissipation, topography and mean flow are ignored ($\epsilon = \Gamma = \delta = \gamma = 0$ and $\Delta = -1$) the EKdV equation,

$$\eta_t + \eta_x + (\tfrac{3}{2} \alpha d_{-2} - 3\alpha^2 d_{-3}\eta) \eta_x + \tfrac{1}{6}\beta d_1 \eta_{xxx} = 0, \tag{11}$$

is recovered from (5). Kakutani and Yamasaki (1978) and Miles (1979) have shown that (11) has a family of solitary wave solutions given by

$$\eta(x, t; \mu) = \eta_0 (\cosh^2\theta - \mu \sinh^2\theta)^{-1} \tag{12a}$$

where

$$\theta = \left[\frac{3}{4} \frac{\alpha}{\beta} \frac{d_{-2}}{d_1} \eta_0 \left(1 - \frac{d_{-3}}{d_{-2}} \alpha\eta_0\right)\right]^{1/2} (x - ct) \tag{12b}$$

$$c = 1 + \frac{1}{2} d_{-2} \alpha\eta_0 \left(1 - \frac{d_{-3}}{d_{-2}}\alpha\eta_0\right) \tag{12c}$$

and

$$\mu = \frac{d_{-3}}{d_{-2}} \alpha \eta_0 \left(1 - \frac{d_{-3}}{d_{-2}} \alpha \eta_0\right)^{-1}. \tag{12d}$$

When $0 < \mu < 1, \eta$ is real and bounded for $-\infty < \theta < \infty$. Equation (11) also admits periodic cnoidal wave solutions.

A non-dissipative monotonic bore solution is obtained for $\mu = 1$:

$$\eta_1(x, t; 1) = \eta_{max} \frac{1}{2}(1 - \tanh\theta_1) \tag{13a}$$

where

$$\theta_1 = \left[\frac{3}{16} \frac{d_{-2}^2}{\beta d_1 d_{-3}}\right]^{1/2} (x - c_{max}t) \tag{13b}$$

$$\alpha \eta_{max} = \frac{1}{2} \frac{(1 - 2R)}{(1 - 3R + 3R^2)} \tag{13c}$$

$$c_{max} = 1 + \frac{(1 - 2R)^2}{8(1 - 3R + 3R^2)} \tag{13d}$$

and

$$R = d_-(d_+ + d_-)^{-1}. \tag{13e}$$

This solution corresponds to a forward facing transition from $\eta = 0$ to $\eta = 1$ propagating into the $\eta = 0$ region. The presence of cubic nonlinearity gives rise to this solution and also limits the maximum amplitude of a solitary wave to (13c) and the maximum speed to (13d). Solitary waves (12) and bores (13) are waves of elevation (depression) for $R < 1/2 (> 1/2)$. The EKdV equation is invariant under the transformation $\tilde{\eta} = 1 - \eta$, leading to a rearward facing bore (transition from $\eta = 1$ to $\eta = 0$). It should be noted that the bore solutions contain infinite mass and energy and therefore cannot be uniformly valid solutions to (11). They may represent inner solutions which must be matched to an appropriate outer solution (Miles, 1979; Melville and Helfrich, 1987).

When $| d_+ - d_- | = O(1)$, $\beta = O(\alpha)$, and the cubic nonlinear term in (11) should be discarded. The resulting KdV equation has the usual solitary wave solutions which can be obtained from (12) by setting $\mu = 0$ and dropping all terms of $O(\alpha^2)$ (or by setting $d_{-3} = 0$). Again the solution is a wave of elevation (depression) for $R < 1/2$ $(> 1/2)$. However, there is no maximum wave height and no non-dissipative monotonic bore solution.

For an arbitrary initial condition the KdV equation can be solved using the inverse scattering transform method (Whitham, 1974). The EKdV equation is also amenable to this technique (Miles, 1979). The result is that for initial disturbances of elevation (depression) for $R < 1/2$ $(> 1/2)$ the solution asymptotically evolves into a rank-ordered sequence of solitary waves followed by a dispersive wave train. If the polarity of the intitial disturbance is opposite that necessary for the existence of solitary waves, a dispersive wave train (but no solitary waves) develops (Hammack and Segur, 1978). Note that generation of individual solitary waves from an arbitrary initial condition is an *asymptotic* result (in propagation time or distance). Other factors such as dissipation or variable topography should be expected to interfere with this process.

When variable topography is present an EKdV (or KdV) equation incorporating this effect can be derived provided that $(kL)^{-1} \ll 1$, where L is a length scale of the topography (Djordjevic and Redekopp, 1978; Grimshaw, 1981; Helfrich *et al.*, 1984). The total depth change can be O(1). Internal solitary waves may fission upon passing from regions of one depth to a region with another depth. Helfrich *et al.* (1984) considered the "turning point" problem when a solitary wave of depression $(R > 1/2)$ propagates up a slope onto a shelf where $R < 1/2$ and only solitary waves of elevation can exist. They showed that the incident solitary wave will scatter into an oscillatory wave train from which one or more solitary waves of reversed polarity may emerge asymptotically. However, the presence of weak dissipation may act to inhibit the transmission of solitary waves (Helfrich, 1985).

The effect of a vertically sheared mean flow $U(z)$ on the one-dimensional propagation of long nonlinear internal waves (KdV and EKdV equations) is primarily kinematic (Miles, 1979; Maslowe and Redekopp, 1980; Tung *et al.*, 1981; Rockliff, 1984). The coefficients of (11) now depend on the mean flow structure. Provided that no critical layers exist $(U(z) - c \neq 0$ for all $z)$ solitary wave characteristics such as vertical structure, length and phase speed are altered, but the waves are stable. If a critical layer does exist then the effect is dynamic and the wave may exchange energy and momentum with the mean flow. Details of the energy exchange depend on the structure of the critical layers (Tung *et al.*, 1981).

2.2.2. *Two-Dimensional Propagation.*

The KP equation [(5) with $\gamma = \Gamma = \delta = 0$ and $\Delta = -1$] has been proposed as a model of two-dimensional wave propagation when lateral variations are small. Short of solving this two-dimensional equation, the EKdV equation can be modified to account for geometric (radial) spreading. In this case the waves are governed by a cylindrical EKdV equation (Miles, 1978, 1980). This new equation is (11) with the term $(2x)^{-1}\eta$ added to the left hand side. Here x is now the radial distance from the source. The cylindrical EKdV equation is valid provided $(kx_0)^{-1} = O(\beta) \ll 1$, where x_0 is the distance scale from the wave source. Liu *et al.* (1985) used this geometric spreading correction with good results for their comparison of theory and observations from the Sulu Sea.

Miles (1978, 1980) has studied the cylindrical KdV equation and found that the amplitude of a solitary wave varies according to $\eta \sim \eta_0(x/x_0)^{-2/3}$, where η_0 and x_0 are the initial amplitude and radial location, respectively. This result assumes adiabatic changes in solitary wave characteristics and conservation of wave energy. However, Miles also found that this adiabatic approximation, while conserving energy, does not conserve mass. For a consistent solution the slowly varying solitary wave must be followed by a secondary, linear wave.

Further discussion of two-dimensional propagation in a channel is taken up in Section 3 on transcritical flows.

2.3 DISSIPATION

Wave dissipation may occur through several processes. The simplest is viscous boundary damping, represented in (5) by the term $\mathcal{F}$ [defined in (8)]. Assuming slow adiabatic changes in wave properties the amplitude of a KdV solitary wave will decay according to

$$
\frac{\eta}{\eta_0} = \left[1 + 0.06\delta \left(\frac{d_+^2}{d_1^2} + \frac{1}{2} \right) x \right]^{-4},
$$

where η_0 is the initial amplitude. Good agreement between this relationship and laboratory measurements of internal solitary wave decay has been found (Leone, et al., 1982; Helfrich, 1985).

Solitary waves may also decay through the inviscid process of radiation damping (Maslowe and Redekopp, 1980). If one layer is continuously stratified, rather than uniform, first-mode solitary wave energy is lost to vertically propagating internal waves generated in the continuously stratified layer by the passage of the solitary wave. In cases where the stratification is such that two vertical modes have phase speeds that differ by $O(\alpha)$ mode coupling leads to energy transfer between modes (Gear and Grimshaw, 1984; Gear, 1985). Evolution of the modes is governed by a pair of coupled KdV equations.

Observations (Haury et al., 1979) suggest that interfacial shear instabilities contribute significantly to wave dissipation. When gradient Richardson numbers in the wave-induced flow field drop below $\frac{1}{4}$, local instability may occur if the growth rate of any unstable disturbances is fast compared to the time in which the flow field is changing. Sandstrom and Elliot (1984) estimate that this mechanism is the dominant source of wave dissipation on the Scotian Shelf. If the waves are imbedded in a shear flow critical layers may develop and energy and momentum may be transferred from the wave to the mean flow. The critical layers may become unstable leading to local mixing (Tung et al., 1981).

Wave interaction with topographic features will also result in instabilities and dissipation. Kao, Pan and Renouard (1985) conducted laboratory experiments on the propagation of an internal solitary wave up a slope on to a shelf. They found that for incident waves of depression the rear face would steepen over the slope and interfacial instability occurred when the local gradient Richardson number fell below $\frac{1}{4}$. Helfrich and Melville (1986) conducted similar slope–shelf topography experiments but considered the case when a turning point ($d_+/d_- = 1$) was encountered on the slope. For certain parameter ranges incident solitary waves of depression became unstable and broke in the neighborhood of the turning point. They attributed the breaking to a kinematic (gravitational), rather than a shear, instability. Measurements of local fluid particle velocities were equal to the wave phase speed. Breaking caused significant vertical mixing and in some cases a second-mode solitary wave was generated from the collapse of the mixed region.

2.4. GENERATION

The most commonly referenced model for the generation of long internal waves involves the interaction of a barotropic tide with an isolated topographic feature. Using laboratory

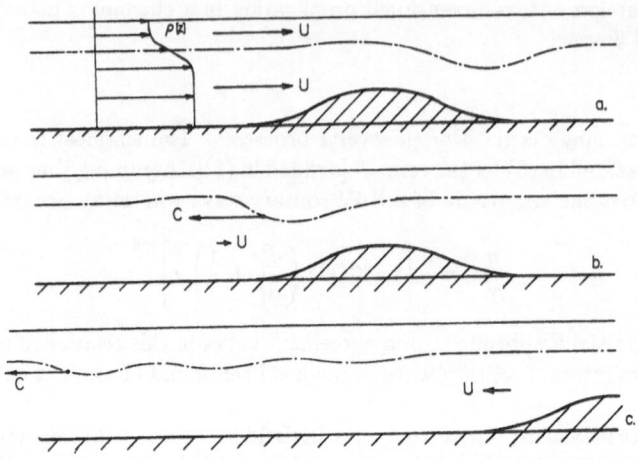

Figure 2. Cartoon illustrating long internal wave generation by stratified tidal flow over topography (from Maxworthy, 1979). (a) Flow generates and interfacial disturbance downstream of the topography. (b) As the tide slackens the disturbance moves upstream and (c) eventually evolves into a train of nonlinear waves.

experiments with oscillating topography in a two-layer system Maxworthy (1979) arrived at the generation scenario shown in Figure 2. As the tidal flow increases a quasi-stationary lee-wave is generated over and downstream of the obstacle. When the tide slackens the disturbance is able to propagate upstream and evolve into a packet of nonlinear internal waves. Asymptotically these waves would become a rank-ordered sequence of solitary waves (c.f. description of the inverse scattering solution to the EKdV equation). For certain conditions intense mixing was observed to occur in the lee-wave resulting in the generation of a front of mixed fluid in addition to a wave train. Lansing and Maxworthy (1984) and Hibiya (1986, 1988) have refined this model with theoretical and numerical studies.

Farmer and Smith (1980) reported field observations from a sill in Knight Inlet, British Columbia, which support this generation mechanism. The study pointed out that details of the stratification, sill geometry and tidal flow strength, as measured by the internal Froude number, are important in determining the flow response. Chereskin's (1983) study of the tidal flow at Stellwagen Bank in Massachusetts Bay also support these conclusions.

In addition to Maxworthy's (1979) model, a steady transcritical stratified flow over topography can generate long internal waves which propagate upstream. This generation mechanism, along with other aspects of transcritical flows, will be taken up in the next section.

3. Transcritical Flows

Steady hydraulic modelling of stratified flows in straits and fjords has become increasing sophisticated in the last decade due in great measure to the efforts of Armi and Farmer (1985) and Farmer and Armi (1986). At the same time it has been recognized that transcritical flows over localized topography may be intrinsically unsteady, with the unsteadiness manifested by the continuous generation of solitary waves upstream. The phenomenon appears to have been discovered in a numerical solution of Wu and Wu (1982) and the ship towing tank experiments of Huang et al. (1982). It is worth noting that the first observation of a solitary water wave was associated with the motion of a barge in a canal (Russell, 1838).

According to linear theory a disturbance moving at the speed of a free mode of the system will resonate with the flow causing a linear growth in time of the waves in the neighborhood of the disturbance. In hydraulic theory the forcing is balanced by the nonlinearity of the response and a steady solution is attainable in the neighborhood of the forcing. This locally steady solution may be terminated by shocks upstream and downstream of the forcing. The introduction of weak dispersion leads to the result that steady forcing (whether by a moving disturbance or by a steady flow over stationary topography) can lead to an unsteady response in a transcritical regime with an almost periodic generation of solitary waves upstream. The essential features of this phenomenon have been reviewed by Wu (1987) for the single layer flow, which is comparable to the stratified flow in the absence of cubic nonlinearity.

The governing forced EKdV equation for transcritical two-layer flow over topography can be found from (5) by taking $\delta = \epsilon = \Gamma = 0$:

$$\eta_t - \Delta\eta_x + \frac{3}{2}\alpha d_{-2}\,\eta\eta_x - 3\alpha^2 d_{-3}\eta^2\eta_x + \frac{1}{6}\beta d_1\eta_{xxx} = \frac{-\gamma}{\alpha}\frac{F}{2d_-}\,H_x. \quad (14)$$

The linear resonance at $\Delta = 0$ can be clearly seen in (14) when the nonlinear and dispersive terms are ignored.

When the cubic nonlinear term is ignored, numerical solutions of the forced KdV equation have been found by a number of authors to give Boussinesq solitary waves upstream for a range of transcritical Froude numbers ($F = U/c_0$). Figures 3 and 4, from Melville and Helfrich (1987), show examples of these solutions and their occurrence for various Froude numbers and forcing strengths (topographic heights γ). In Figures 3 and 7 the governing equation has been solved for flow from the left. Miles (1986) found that steady solutions were not possible in a transcritical range of Froude numbers; a result which was extended to the two-layer flow by Melville and Helfrich to give the unsteady regime as

$$1 - \left(\frac{9\mathcal{A}}{2D^2}\right)^{\frac{2}{3}} < F^2 < 1 + \left(\frac{9\mathcal{A}}{4D^2}\right)^{\frac{2}{3}}$$

where

$$\mathcal{A} = \frac{d_{-2}}{d_- d_1^{1/2}}\gamma.$$

The upper bound is shown in Figure 4 by the dashed line. Baines (1984) conducted laboratory experiments on transcritical forcing and some of his photographs are reproduced in Figure 5. Figures 5a and 5b are from runs in the KdV regime and show good qualitative agreement with the numerical results in Figure 3 for $F = 1.1$.

Hydraulic models, which contain no dispersive effects, cannot capture this phenomenon. Therefore the use of the quasi-steady approximation in hydraulic modeling of straits and fjords needs to be examined. This may be done by comparing the timescale for solitary wave production in the transcritical problem to the tidal timescale.

Following the work of Grimshaw and Smyth (1986), who undertook an extensive study of the forced KdV equation, (14) without the cubic term can be renormalized to give

$$A_t - \Delta A_x + 6AA_x + A_{xxx} = -G_0 G_x(x) \tag{15}$$

Here G_0 is the renormalized amplitude of the forcing and $G(x)$ is the forcing shape with a maximum of 1. Grimshaw and Smyth used the hydraulic approximation to (15) (i.e., dropped the A_{xxx} term) to derive a solution which was steady in the neighborhood of the forcing and terminated by upstream and downstream moving shocks. This solution is sketched in Figure 6.

$$A_\pm = \frac{\Delta}{6} \pm \frac{(12G_0)^{1/2}}{6} \tag{16a}$$

$$V_\pm = \frac{1}{2} \left[(12G_0)^{1/2} \mp \Delta \right]. \tag{16b}$$

The subscripts refer to upstream (+) and downstream (−) propagation.

The integral flows of mass and energy conservation laws in the advancing bore along with the mass and energy of a Boussinesq solitary wave can then be used to infer the amplitude a and separation h (or equivalent period of generation $\tau_s = c/h$, where c is the nonlinear phase speed of the solitary waves) of the upstream solitary waves. Wu (1987) also undertook a similar analysis. Taking the case $\Delta = 0$ (resonant flow) for simplicity, the hydraulic approximation to (15) and its corresponding energy equation [(15) multiplied by A] are integrated from the neighborhood of the topography to upstream of the bore (or solitary waves) to give (see Figure 6 for the location of the integration limits),

$$\frac{\partial}{\partial t} \int_{x_0}^{x_2} A\,dx = 3A_-^2 \tag{17a}$$

$$\frac{\partial}{\partial t} \int_{x_0}^{x_2} \frac{1}{2} A^2\,dx = \overline{C}_D + 2A_-^3 \tag{17b}$$

$$\frac{\partial}{\partial t} \int_{x_1}^{x_2} \frac{1}{2} A^2\,dx = 2A_+^3 = -2A_-^3. \tag{17c}$$

The topographic (or wave) drag is

$$C_D = \frac{\mathcal{D}}{\rho g' D L},$$

403

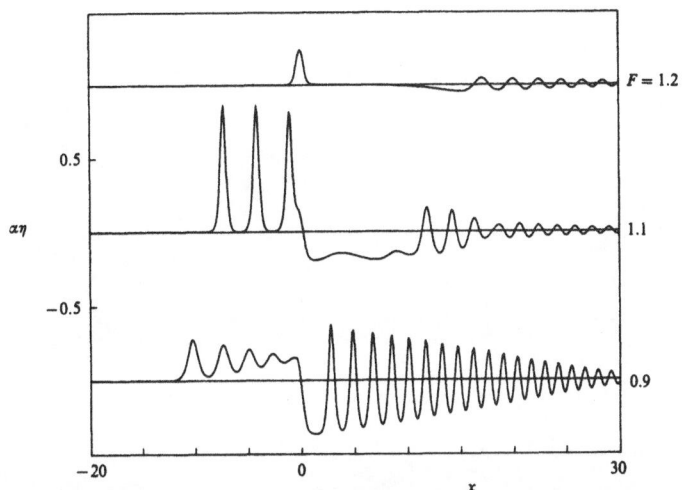

Figure 3. Examples of three classes of solutions to the inhomogeneous KdV equation [(14) without the cubic term] at $t = 63.1$: Steady supercritical ($F = 1.2$); sequence of solitary waves (1.1); undular upstream bore (0.9). $(R, \alpha, \beta, \gamma) = (0.2, 0.577, 0.0076, 0.0625)$. In this figure flow is from left to right. From Melville and Helfrich (1987).

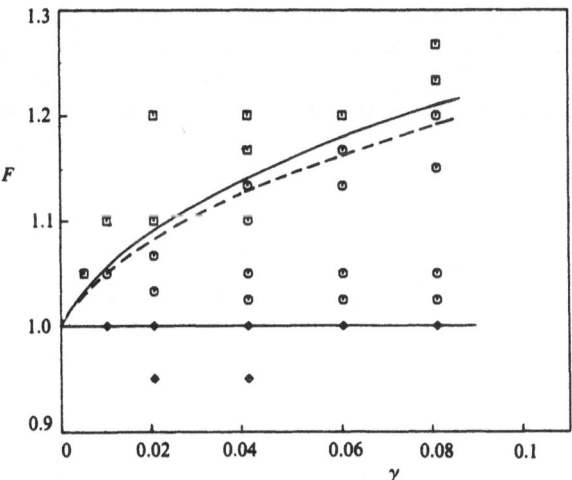

Figure 4. Regions in the (F, γ)-plane occupied by solutions to the inhomogeneous KdV equation: □, locally steady supercritical solution; ○, sequence of solitary waves upstream; ◇, undular bore upstream. $(R, \alpha, \beta) = (0.2, 0.577, 0.0076)$. The dashed line corresponds to the theoretical upper bound for the unsteady regime. The solid lines denote empirical regime boundaries. From Melville and Helfrich (1987).

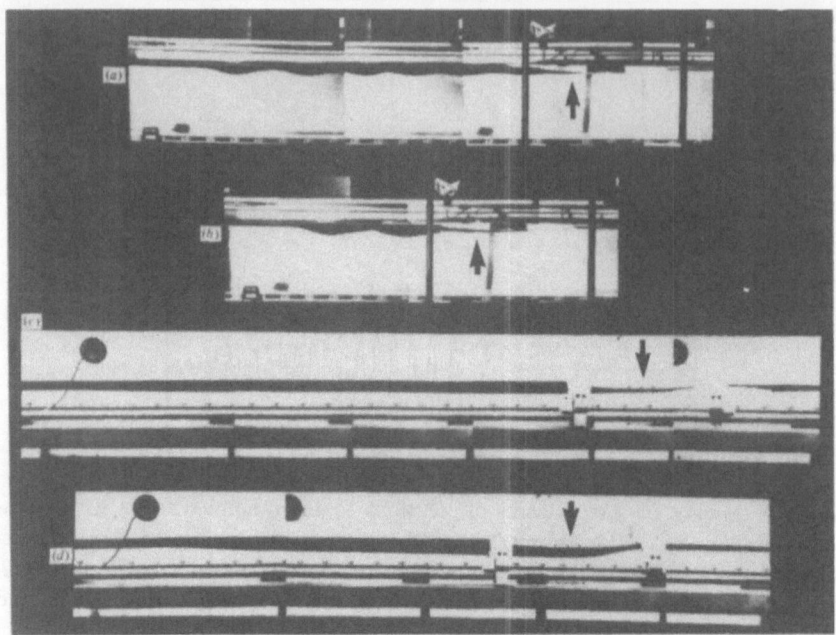

Figure 5. Photographs from Baines's (1984) experiments showing disturbances upstream of towed topography in a two-layered system. The topography is moving from right to left. The arrow shows the location of the topography and the upper layer is dyed. (a) and (b) are in the KdV regime and show upstream undular bores. (c) and (d) are in the extended KdV regime ($d_+/d_- \approx 1$) and show a monotonic upstream bore.

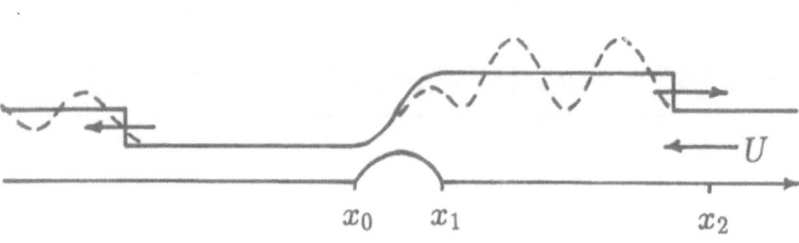

Figure 6. Sketch illustrating the hydraulic solution (——). The corresponding solution with dispersion (– – – –) is also shown.

where $\mathcal{D}$ is the unsteady drag and L is the length of the topography. $\overline{C}_D$ is the average value.

The mass M_s and energy E_s of a solitary wave solution to (15) of amplitude a are,

$$M_s = 2(2a)^{1/2},$$

and

$$E_s = \frac{1}{3}(2a)^{3/2}.$$

Equating (17a), the time rate of change of mass, with M_s/τ_s and (17 b, c) with E_s/τ_s, the amplitude, period and average drag coefficient are found to be

$$a = 2\,|\,A_+\,| = \frac{1}{3}(12G_0)^{1/2} \tag{18a}$$

$$\tau_s = 24(3a)^{-3/2} \tag{18b}$$

$$\overline{C}_D = \frac{1}{2}a^3. \tag{18c}$$

Grimshaw and Smyth (1986) and Wu (1987) found excellent agreement between this approximate theory and numerical solutions to (15) for Froude numbers close to unity ($\Delta << 1$). Returning to the original variables [i.e., as in (14)] (18 a–c) become

$$\frac{\eta_s^*}{D} = \left(\frac{8}{3}\frac{\gamma}{d_{-2}d_-}\right)^{1/2} \tag{19a}$$

$$\frac{c_0}{D}t_s^* = \frac{32}{5}(3d_1)^{1/2}\left(d_{-2}\frac{\eta_s^*}{D}\right)^{-3/2} \tag{19b}$$

$$\overline{C}_D = \frac{d_{-2}^3}{128}\left(\frac{\eta_s^*}{D}\right)^3. \tag{19c}$$

Here η_s^* and t_s^* are the dimensional amplitude and generation period. For a two-layer flow with $d_+^* = 100$ m, $d_-^* = 400$ m, $H_0 = 30$ m and $c_0 \approx 1$ m/s (19 a-c) give $\eta_s^* \approx 46$ m, $t_s^* \approx 3$ hr and $\overline{C}_D \approx 3 \times 10^{-4}$. The period of generation is the same order or less than the tidal period $O(10$ hr$)$, therefore additional unsteadiness due to transcritical phenomena may result. The use of quasi-steady hydraulics may be inappropriate. Furthermore, the average wave drag is small but not insignificant.

When the cubic nonlinear term is important the flow response in the transcritical regime is quite different than the KdV limit. Melville and Helfrich (1987) solved the forced EKdV equation (14) and found that instead of periodic solitary wave generation the transcritical regime was marked by the appearance of the monotonic non-dissipative bore solution (13). Figures 7 and 8 (from Melville and Helfrich) show examples of the solutions and their occurrence for various Froude numbers and forcing strengths. Their numerical solutions showed qualitative and quantitive agreement with their own and Baines's (1984) experiments and appeared to explain Baines's observations (see Figure 5c, d) of a laminar upstream bore that "mainly consisted of a forward face that propagated without change in shape." Melville and Helfrich (1987) also showed that for typical parameters ($D \simeq 100$ m, $F = 1, c_0 \simeq 1$ m/s) the time to establish the steady flow upstream of the topography was O(10 hr). This again brings into question the quasi-steady hydraulics approximation. Their numerical solutions also demonstrated that monotonic forward and rearward facing bores could be generated by unsteady transcritical forcing. Thus solitary waves of arbitrary length may be generated by complementary pairs of monotonic bores.

The theoretical extension of resonant phenomena to three dimensions (two propagation dimensions) was undertaken for single layer flows by Mei (1986) and Katsis and Akylas (1987). The latter authors used the KP equation to study the evolution of a three-dimensional disturbance (two propagation dimensions), and the resonant forcing by a moving pressure distribution in a channel. In both cases they found that two-dimensional solitary waves were generated which spanned the channel. By using an empirical constant Katsis and Akylas were able to relate their surface pressure distribution to the blockage coefficient in the ship towing tank experiments of Ertekin *et al.* (1984) and then found good agreement with the measured amplitude and period of generation of the solitary waves. The amplitude increased and period of generation decreased as the blockage coefficient increased.

Subsequently, Macomb and Melville (1987), concerned by the singularities in the KP equation, used coupled evolution equations to model the ship generated solitons. They found qualitative agreement with the results of Katsis and Akylas (1987) after deriving from first principles the empirical constant used by Katsis and Akylas. They also looked in more detail at the establishment of the two-dimensional solitary waves by reflection at the side wall of the channel. They suggested that the process was related to the Mach reflection of solitary waves first established by Miles (1977). Macomb and Melville (1987) showed how the blockage coefficient in the single layer case corresponded to the modal decompostion of the topography in the two-layer case.

Very recently, Pedersen (1988) has undertaken a study of the transcritical forcing of waves in a channel. In contrast to the 3-D experiments of Ertekin *et al.* (1984) and the numerical results of Katsis and Akylas (1987) and Macomb and Melville (1987), he finds that the amplitude of the solitary waves is determined by the blockage coefficient but that the period of generation and the limiting Froude number for which upstream radiation ceases are not clearly related to the blockage coefficient. This would also appear to be in constrast to the two-dimensional case in which the period of generation is explicitly related to the amplitude (Wu, 1987) and the bounds of the transcritical regime depend on the blockage coefficient (Miles, 1986; Melville and Helfrich, 1987). Pedersen also concluded that upstream radiation in wide channels may be qualitatively analyzed through the properties of Mach reflection. While the reflection process is quantitatively similar to Mach reflection, the quantitative application of Mach reflection appears to be heuristic.

The introduction of rotation to this class of problems leads to the introduction of additional modes and concomitant complexity. Nevertheless, the paradigm of transcritical forcing appears to carry over in resolving some of the observations made in recent physical and numerical experiments on the evolution of solitary waves in a rotating channel.

Maxworthy (1983) conducted experiments on the second-mode waves evolving from the collapse of mixed fluid in a rotating stratified channel. He found that the leading wave curved backwards (unlike the straight-crested linear Kelvin waves) and that the amplitude decayed significantly along the channel (Figure 9).

A careful theoretical investigation by Grimshaw (1985) led to the development of evolution equations for weakly nonlinear long waves in rotating channels. He found that in the case of strong rotation, for which the Rossby radius was at most comparable to the wavelength, the effects of rotation were separable from those of nonlinearity and dispersion. The exponential decay scale across the channel was the Rossby radius and the evolution along the channel was described by the KdV equation. For weak rotation separation is not possible and the evolution equation is a rotation-modified KP equation (c.f. (5)). In neither case could the wave-front curvature in a channel be explained.

Renouard, Chabert d'Hieres and Zhang (1987) generated first mode internal waves in a controlled fashion in a rotating channel on the rotating platform at the Coriolis Laboratory, Grenoble, France. They too found wave-front curvature (Figure 10), but attributed the decay along the channel to viscous dissipation.

Katsis and Akylas (1987) numerically solved the rotation-modified KP equation and found solutions qualitatively similar to those of Renouard et al. (1987); however, the KP equations contained no dissipation. Katsis and Akylas noted that the leading disturbance was followed by smaller amplitude waves. They did not relate these to the decay of the leading wave, but did conclude that wave-front curvature was consistent with decay along the channel (c.f. Grimshaw, 1985). Subsequently, Grimshaw and Melville (1989) reviewed the derivation of the rotation-modified KP equation and its integral constraints and concluded that the radiation of Poincaré waves behind the leading wave is not consistent with the assumption that the solutions of the equations are locally confined.

Recently Melville, Tomasson and Renouard (1989) have shown that due to the speed correction, nonlinear Kelvin waves may directly resonate with the linear Poincaré modes of a channel. This is most easily seen in Figure 11 where we show the dispersion curves for a channel [equation (10) with $\beta = 0$] and the dispersion curve for a weakly nonlinear Kelvin wave whose speed is a constant fraction above the speed of the linear Kelvin wave. Melville et al. (1989) anticipated that at the intersection points direct resonance between the nonlinear Kelvin wave and the linear Poincaré modes would occur.

Melville et al. used the regularized KP equations

$$u_t \;+\; u_x + \frac{3}{2}\,\alpha u u_x \;-\; \frac{1}{6}\beta u_{xxt} + \frac{1}{2}\Gamma(v_y - v) = O(\alpha\beta) \tag{20a}$$

$$v_t \;+\; u_y + u = O(\alpha\beta) \tag{20b}$$

$$v \;=\; 0 \text{ at } y = 0, W \tag{20c}$$

where α and β are the small nonlinear and dispersive parameters, $\Gamma(= \epsilon)$ is a small parameter measuring 3-D and rotational effects, and u and v are the longitudinal and transverse velocities, respectively.

These equations may be rewritten to give

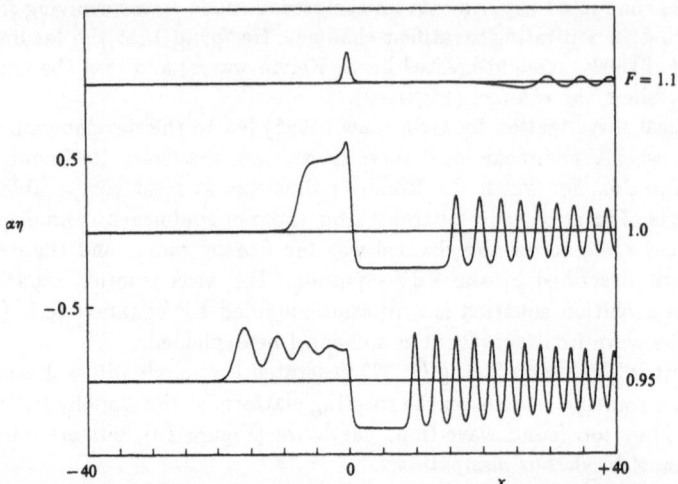

Figure 7. Examples of three classes of solutions to the inhomogeneous EKdV equation (14) at $t = 221.6$: Steady supercritical ($F = 1.1$); monotonic bore upstream (1.1); undular upstream bore (0.95). $(R, \alpha, \beta, \gamma) = (0.35, 0.472, 0.0154, 0.055)$. From Melville and Helfrich (1987).

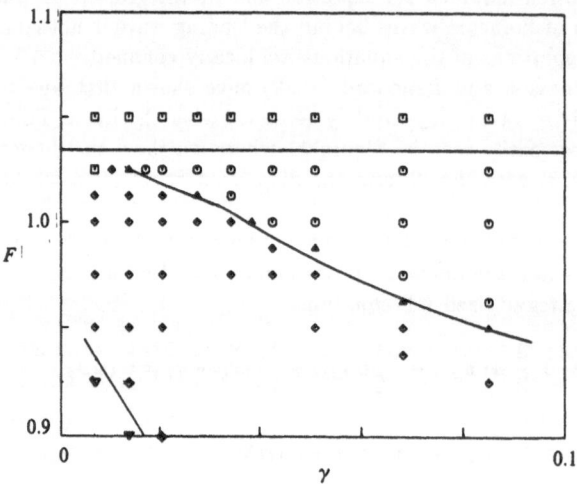

Figure 8. Regions in the (F, γ)-plane occupied by solutions to the inhomogeneous EKdV equation: □, locally steady supercritical solution; ○, monotonic bore upstream; ◇, undular bore upstream; and ▽, no upstream influence. Symbol, △, marks a transition solution. $(R, \alpha, \beta) = (0.35, 0.472, 0.0154)$. The solid lines denote empirically determined regime boundaries. From Melville and Helfrich (1987).

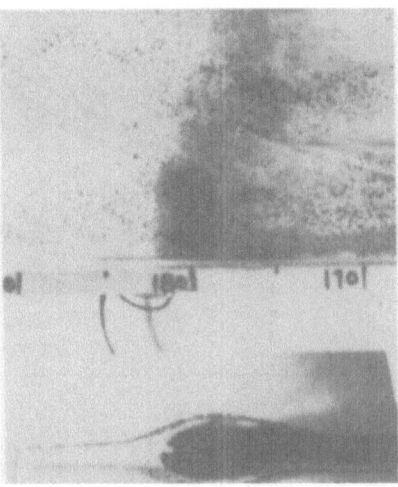

Figure 9. Photographs of the plan and side view of an internal wave in a rotating channel showing the wave-front curvature. The dye shows the wave which is propagating from right to left. From Maxworthy (1983).

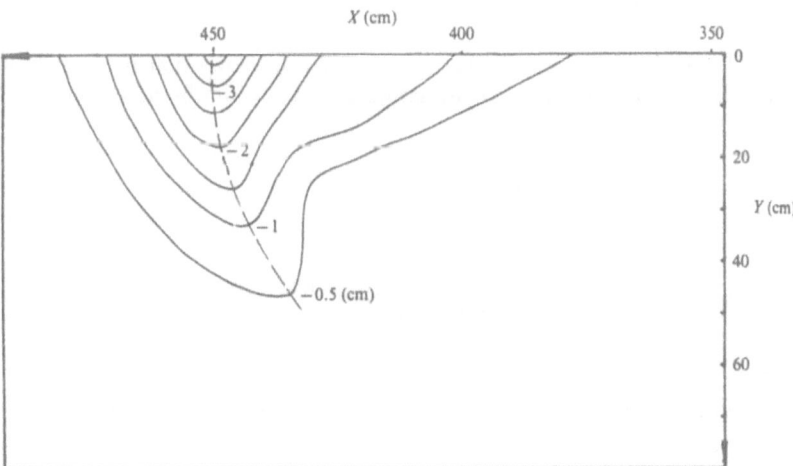

Figure 10. Contours of internal wave amplitude in a rotating channel showing wave-front curvature. The wave is propagating front right to left. From Renouard et al. (1987).

410

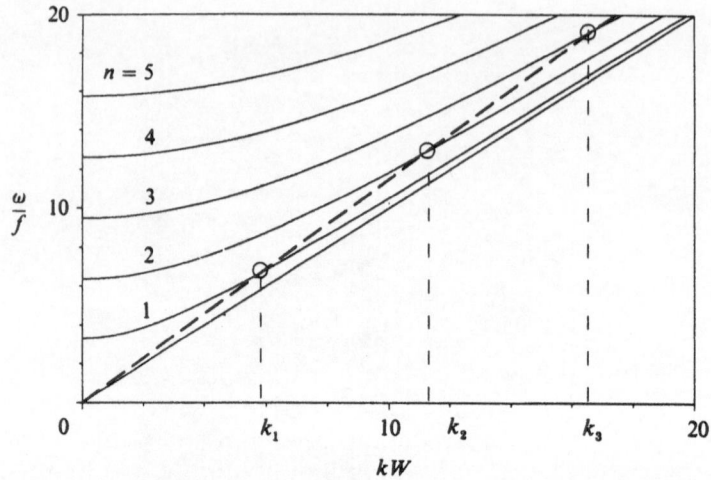

Figure 11. The dispersion curves for linear hydrostatic modes in a channel [(10) with $\beta = 0$] (———), showing the possibility of direct resonant forcing of Poincaré waves by weakly nonlinear Kelvin waves (– – – – –) having a small positive speed correction. k_n denotes the wavenumber of the nth Poincaré mode. From Melville *et al.* (1989).

$$v_{tt} + v_{xt} - \frac{1}{6}\,\beta v_{xxtt} + \frac{1}{2}\Gamma(v - v_{yy}) = \frac{3}{2}\,\alpha(1 + \frac{\partial}{\partial y})\,[u u_x]. \tag{21}$$

With initial data for which $v = 0$, this equation clearly shows that it is the nonlinearity in u which forces the response in v. They assumed that u could be decomposed into

$$\begin{aligned}
u &= \overline{u}(x,t)e^{-y} + \epsilon \tilde{u}(x,y,t) \\
(\overline{u} &= O(1),\ \epsilon << 1),
\end{aligned} \tag{22}$$

where

$$\overline{u} = a\,\mathrm{sech}^2\left[\left(\frac{\alpha s a}{2\beta}\right)^{1/2}(x - ct)\right], \tag{23a}$$

$$c = 1 + \frac{1}{3}\alpha s a \tag{23b}$$

and

$$s = \frac{e^W - e^{-2W}}{e^W - e^{-W}}. \tag{23c}$$

Here $\bar{u}$ is the leading order solitary wave solution, and the residual, $\tilde{u}$, is dominated by Poincaré waves travelling at a speed c along the channel. To leading order at small t,

$$\frac{3}{2}(1 + \frac{2}{\alpha y})[uu_x] \simeq -\frac{3}{2}\,\alpha\bar{u}\,\bar{u}_x e^{-2y}. \tag{24}$$

Thus, the response in v will be resonant when the free modes of the homogeneous form of (21) resonate with the forcing on the right-hand side. In this context the nonlinear Kelvin wave acts as a transcritical forcing for the Poincaré waves.

With the approximations (22) and (24) the problem is linear and (21) may be solved for v. This solution is substituted into

$$v_t + \epsilon(\tilde{u} + \tilde{u}y) = 0 \tag{25}$$

to solve for $\tilde{u}$. Figure 12 shows contour plots of $\bar{u}$, $\tilde{u}$ and $\bar{u} + \tilde{u}$ derived in such a manner. Figure 13 shows contour u and v obtained by direct numerical solution of (20) for initial data corresponding to (23). The plots of $(\bar{u}+\tilde{u})$ from the approximate analytical solution in Figure 12 and u at $t = 60$ from the numerical solution in Figure 13 are directly comparable (except for an arbitrary shift in origin of x). These figures show that direct resonance of the Poincaré modes by a nonlinear Kelvin wave may lead to curvature of the wave front.

Figure 14 shows longitudinal spectra of the first three Poincaré modes at time $t = 360$. Table 1 shows the predicted resonant wavenumbers for these three modes and the peaks of the computed spectra for times up to $t = 360$. Tuning to the resonant wave numbers as t increases is clear and by $t = 360$ the differences between predicted and computed spectral peaks are not more than 1%. Further support for the direct resonance mechanism is shown in Figure 15, where approximate analytical spectra and computed spectra for the first two modes are compared at $t = 120$.

Given the fact that the dispersion curves for the first few Poincaré modes are proximate to the Kelvin mode for stratified channels at mid-latitudes we anticipate that the effects described here may be significant in resolving the dynamics of a number of sea straits.

4. Summary

Models and theories for the generation, propagation and dissipation of long internal waves in coastal oceans and sea straits have been reviewed. The importance of nonlinear and dispersive (i.e., non-hydrostatic) effects on wave evolution was emphasized. In particular, it was noted that hydraulic models cannot account for observations of waves that propagate for long spatial and temporal scales without breaking. A general KP-type equation (5) governing weakly nonlinear, weakly dispersive long internal waves in a two-layer fluid in a rotating channel was developed. The effects of weak transverse variations, boundary layer damping and forcing by transcritical flow over topography were included. Cubic nonlinearity, which may dominate quadratic nonlinearity when the layer depths are nearly equal, was also considered. Simplifications of this equation were discussed as models for wave evolution in several situations.

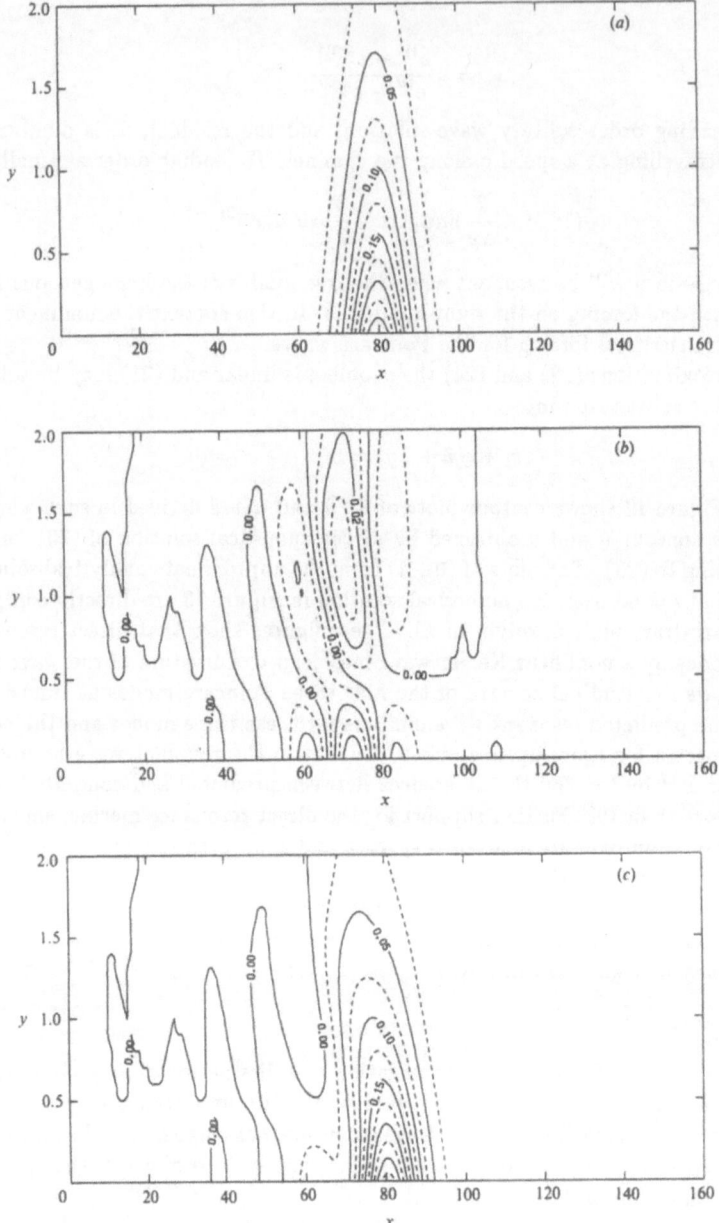

Figure 12. Contour maps of a typical analytic solution for u showing the apparent curvature of the wave as Poincare modes are superposed on the straight-crested Kelvin wave at $t = 60$. (a) $\overline{u}(x,t)e^{-y}$ with $\overline{u}$ given by (23): (b) $\tilde{u}(x,y,t)$ from (26): (c) $u = \overline{u}(x,t)e^{-y} + \tilde{u}(x,y,t)$. From Melville et $al.$ (1989).

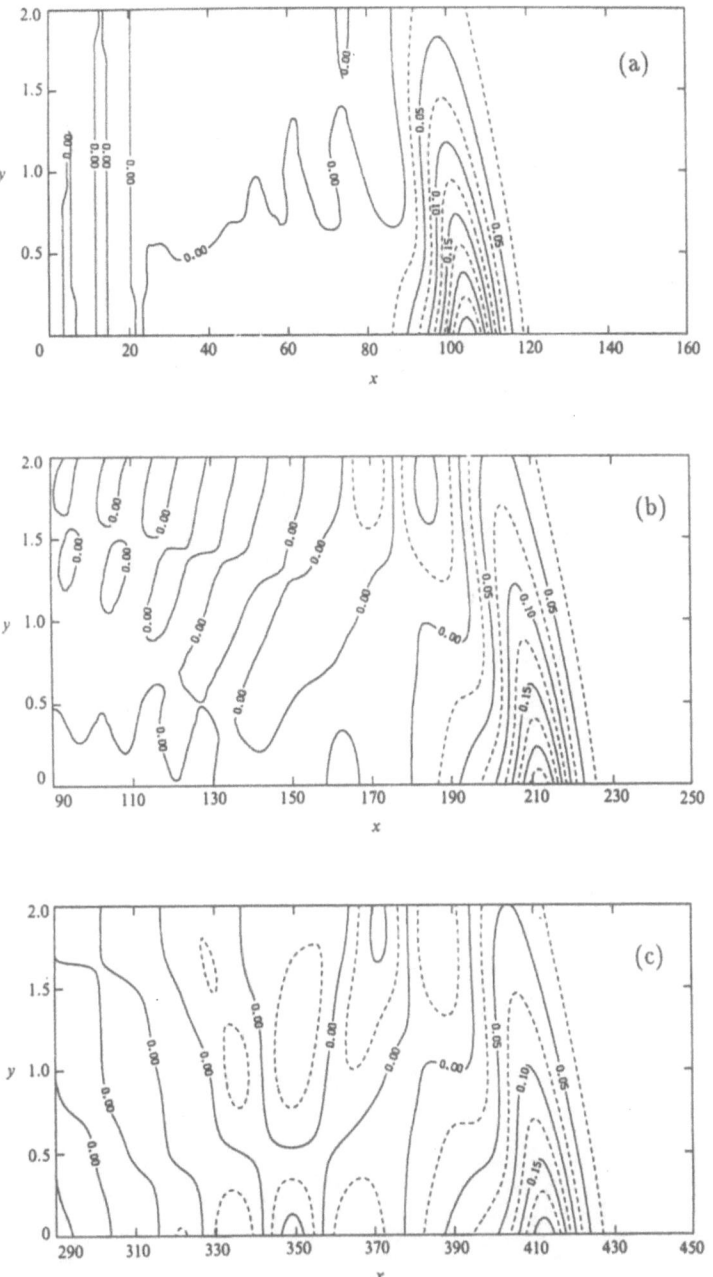

Figure 13. Contour maps of u and v from direct numerical solution of (20) for initial conditions given by (23). (a) u at $t = 60$, (b) 160, (c) 350, (d) v at $t = 60$, (e) 160, (f) 350. From Melville *et al.* (1989).

414

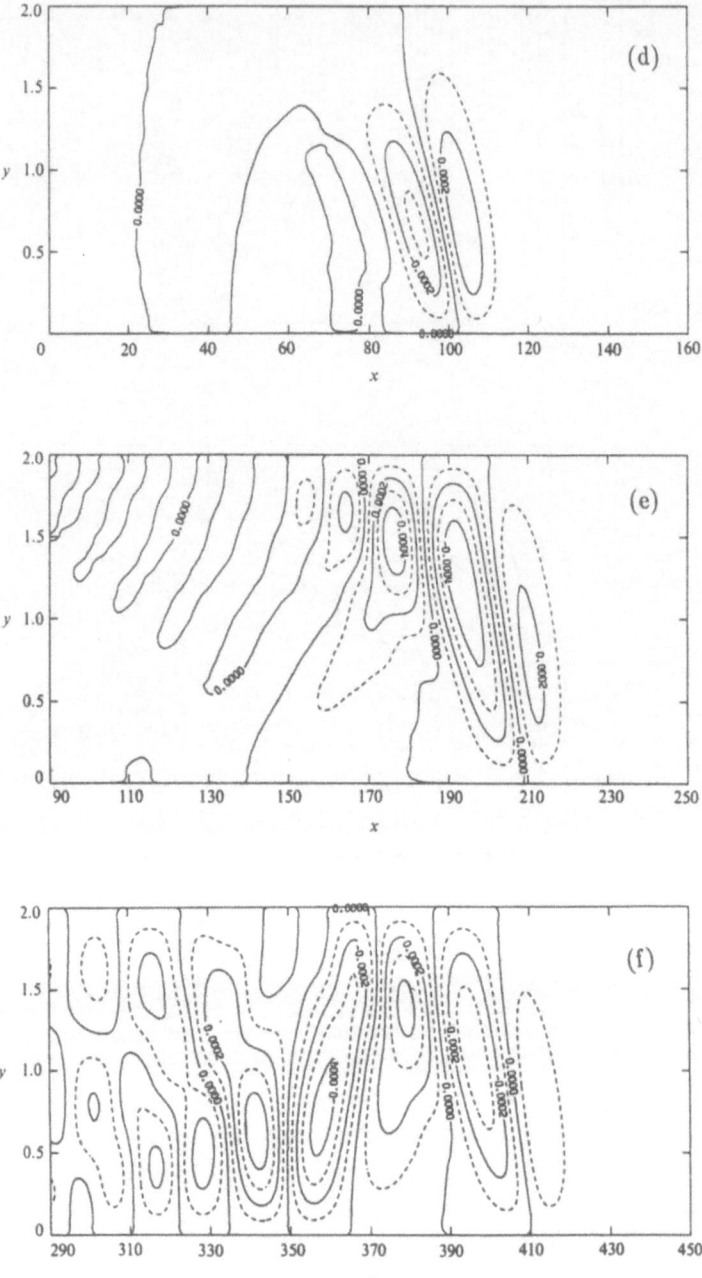

Figure 13. (continued)

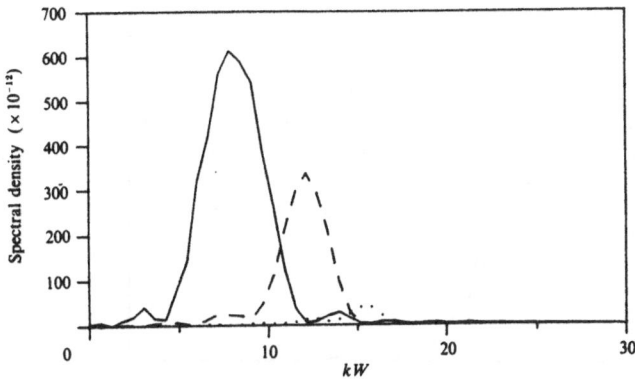

Figure 14. Longitudinal wavenumber spectra of the first three transverse modes of v from Figure 13 at $t = 350$: ———, $n = 1$; – – – –, $n = 2$; $\cdots\cdots$, $n = 3$. From Melville *et al.* (1989).

Table 1. Resonant wavenumbers predicted from initial data and the wavenumbers of the peaks of the computed spectra of v (Melville *et al.*, 1989).

		Computed kW			
n	Predicted kW	t = 60	t=160	t=250	t=350
1	8.2	7.1	7.6	8.0	8.1
2	12.3	10.6	11.7	12.0	12.2
3	15.7	13.8	14.6	15.4	15.6

416

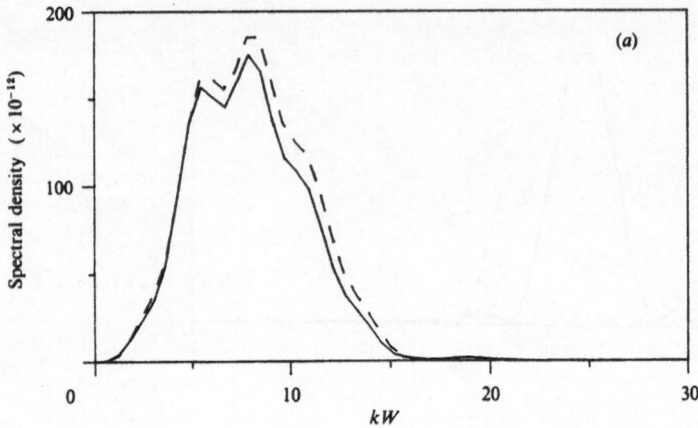

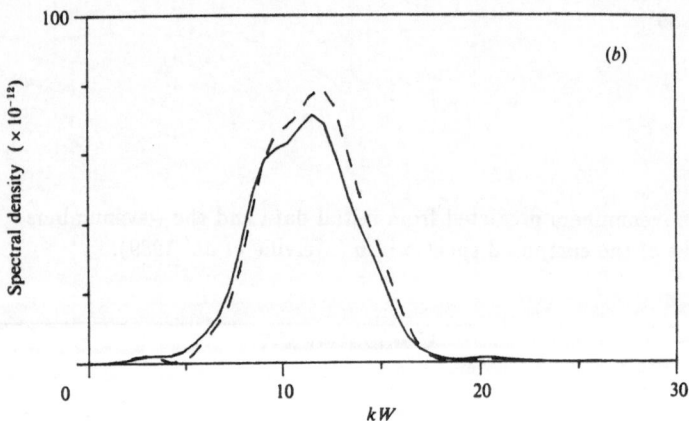

Figure 15. Longitudinal wavenumber spectra of individual transverse modes of v at $t = 120$ from the analytical theory (– – – – –) and direct numerical solution (————). (a) $n = 1$, (b) $n = 2$. From Melville *et al.* (1989).

Resonant forcing of long waves was then reviewed and the importance of this process in the dynamics of straits was highlighted. For steady transcritical flow ($F \approx 1$) over topography a forced KdV (or EKdV) equation was shown to govern wave evolution. The steady flow results in the continuous production of upstream advancing solitary waves and unsteady flow in the neighborhood of the topography. The theory agrees quite well with laboratory studies. This constrasts with the hydraulic solution in which shocks propagate upstream and downstream, but the flow over the topography is steady. Because of the additional unsteadiness introduced by dispersion, the use of the quasi-steady approximation in internal hydraulic modeling of time-dependent strait flow may lead to significant errors. Estimates with parameters typical of the Strait of Gibraltar showed that the timescale for generation of a solitary wave (i.e. the unsteadiness due to dispersion) was less than a tidal period and that the wave drag coefficient was $O(10^{-4} - 10^{-3})$. Thus time-dependent strait flows which pass through the transcritical regime may be inadequately described by hydraulic theory.

Finally, the concept of transcritical forcing was shown to help resolve some laboratory and numerical experiments on the evolution of solitary nonlinear Kelvin waves in a rotating channel. The experiments showed that an initially straight crested Kelvin wave quickly developed backward curvature and the amplitude decayed as the wave propagated down the channel. The explanation was shown to be due to the nonlinear correction to the phase speed of the Kelvin wave. The Kelvin wave is unstable to a direct resonance with Poincaré modes of the channel in which the Kelvin wave acts as a transcritical forcing of the Poincaré modes. The sum of the decaying Kelvin wave and growing Poincaré modes results in a curved wave front. For parameters typical of the Strait of Gibraltar ($c_0 = 1$ m/s, $f \simeq 10^{-4}$ s^{-1}, $k^{-1} \simeq 1000$ m) $\Gamma \simeq 10^{-2}$ and the weak rotation KP model (20a,b,c) is valid. From the numerical solutions wave-front curvature occurs on a timescale $t = t^* k c_0 \approx 50$, or a propagation distance $x^* \approx c_0 t^* = 50$ km using the Gibraltar parameters. This distance is the same order as the length of the Strait of Gibraltar (80 km). Thus, this resonance mechanism may occur in the Strait.

Acknowledgements

The preparation of this review was supported by National Science Foundation Grant OCE-89-02671 to KRH and by a contract from The Office of Naval Research, Coastal Sciences, with WKM.

References

Armi, L. and Farmer, D. M., 1985. The internal hydraulics of the strait of Gibraltar and associated sills and narrows. *Oceanologica Acta*, 8, 37–46.

Baines, P. G., 1984. A unified description of two-layer flow over topography. *J. Fluid Mech.*, 146, 127–167.

Benney, D. J., 1966. Long nonlinear waves in fluid flows. *J. Math. Phys.*, 45, 52–63.

Chereskin, T., 1983. Generation of internal waves in Massachusetts Bay. *J. Geophys. Res.*, 88, 2649–2661.

Djordjevic, V. D. and Redekopp, L. G., 1978. The fission and disintegration of internal solitary waves moving over two-dimensional topography. *J. Phys. Ocean.*, 8, 1016–1024.

Ertekin, R. C., Webster, W. C. and Wehausen, J. V., 1984. Ship generated solitons. *Proc. 15th Symp. Naval Hydrodynamics*, pp. 347–364.

Farmer, D. and Armi, L., 1986. Maximal two-layer exchange over a sill and through the combination of a sill and contraction with barotropic flow. *J. Fluid Mech.*, **164**, 53–76.

Farmer, D. M. and Smith, J. D., 1980. Tidal interaction of stratified flow with a sill in Knight Inlet. *Deep-Sea Res.*, **27 A**, 239–254.

Fu, L. L. and Holt, B., 1982. Seasat views oceans and sea ice with synthetic aperature radar. *JPL Publications*, 81–120, Feb. 15.

Gear, J. A., 1985. Strong interactions between solitary waves belonging to different wave modes. *Stud. Appl. Maths.*, **72**, 95–124.

Gear, J. A. and Grimshaw, R., 1984. Weak and strong interactions between internal solitary waves. *Stud. Appl. Maths.*, **71**, 235–258.

Giese, G. S., Hollander, R. B., Fancher, J. E. and Giese, B. S. 1982. Evidence of coastal seiche excitation by the tide generated internal solitary waves. *Geophys. Res. Lett.*, **9**, 1305–1308.

Grimshaw, R., 1981. Evolution equations for long, nonlinear internal waves in stratified shear flows. *Stud. Appl. Maths.*, **65**, 159–188.

Grimshaw, R., 1985. Evolution equations for weakly nonlinear, long internal waves in a rotating fluid. *Stud. Appl. Maths.*, **73**, 1–33.

Grimshaw, R. and Melville, W. K., 1989. On the derivation of the modified Kadomtsev-Petviashvili equation. *Stud. Appl. Maths.*, **80**, 183–202.

Grimshaw, R. and Smyth, N., 1986. Resonant flow of a stratified fluid over topography. *J. Fluid Mech.*, **169**, 429–464.

Hammack, J. L. and Segur, H., 1978. Modelling criteria for long water waves. *J. Fluid Mech.*, **84**, 359–373.

Haury, L. R., Briscoe, M. G. and Orr, M. H., 1979. Tidally generated internal wave packets in Massachusetts Bay. *Nature*, **278**, 312–317.

Helfrich, K. R., 1985. On long nonlinear internal waves over topography. Ph.D. thesis, Dept. of Civil Engineering, MIT.

Helfrich, K. R. and Melville, W. K., 1986. On long nonlinear internal waves over slope-shelf topography. *J. Fluid Mech.*, **167**, 285–308.

Helfrich, K. R. Melville, W. K. and Miles, J. W., 1984. On interfacial solitary waves over slowly varying topography. *J. Fluid Mech.*, **149**, 305–317.

Hibiya, T., 1986. Generation mechanism of internal waves by tidal flow over a sill. *J. Geophys. Res.*, **91**, 7697–7708.

Hibiya, T., 1988. The generation of internal waves by tidal flow over Stellwagen Bank. *J. Geophys. Res.*, **93**, 533–542.

Huang, D. D., Sibul, O. J., Webster, W. C., Wehausen, J. V., Wu, D. M. and Wu, T. Y., 1982. *Proc. Conf. on Behavior of Ships in Restricted Waters*, Vol. II, pp. 26-1 – 26-10. Varna, Bulgaria.

Kadomtsev, B. B. and Petviashvili, V. I., 1970. On the stability of solitary waves in weakly dispersing media. *Soviet Phys. Dokl.*, **15**, 539–541.

Kakutani, T. and Matsuuchi, K., 1975. Effect of viscosity on long gravity waves. *J. Phys. Soc. Japan*, **34**, 237–246.

Kakutani, T. and Yamasaki, N., 1978. Solitary waves on a two-layer fluid. *J. Phys. Soc. Japan*, **45**, 674–679.

Kao, T. W., Pan, F.-S. and Renouard, D., 1985, Internal solitons on the pycnocline: generation, propagation, and shoaling and breaking over a slope. *J. Fluid Mech.*, **159**, 19–53.

Katsis, C. and Akylas, T. R., 1987. Solitary internal waves in a rotating channel. A numerical study. *Phys. Fluids*, **30**, 297–301.

Kinder, T. H., 1984. Net mass transport by internal waves near the Strait of Gibraltar. *Geophys. Res. Lett.*, **11**, 987–990.

Koop, C. G. and Butler G., 1981. An investigation of internal solitary waves in a two-fluid system. *J. Fluid Mech.*, **112**, 225–251.

Lansing, F. S. and Maxworthy, T., 1984. On the generation and evolution of internal waves. *J. Fluid Mech.*, **145**, 127–149.

La Violette, P. E. and Arnone, R. A., 1988. A tide-generated internal waveform in the western approaches to the Strait of Gibraltar. *J. Geophys. Res.*, **93**, 15653–15667.

Leone, C., Segur, H. and Hammack, J. L., 1982. The viscous decay of long internal solitary waves. *Phys. Fluids*, **25**, 942–944.

Liu, A. K., Holbrook, J. R. and Apel, J. R., 1985. Nonlinear internal wave evolution in the Sulu Sea. *J. Phys. Ocean.*, **15**, 1613–1624.

Macomb, E. S. and Melville, W. K., 1987. On the generation of long nonlinear waves in a channel. (unpublished manuscript).

Maslowe, S. A. and Redekopp, L. G., 1980. Long nonlinear waves in stratified shear flows. *J. Fluid Mech.*, **101**, 321–348.

Maxworthy, T., 1979. A note on internal solitary waves produced by tidal flow over a three-dimensional ridge. *J. Geophys. Res.*, **84**, 338–346.

Maxworthy, T., 1983. Experiments on solitary internal Kelvin waves. *J. Fluid Mech.*, **129**, 365–383.

Mei, C. C., 1986. Radiation of solitons by slender bodies advancing in a shallow channel. *J. Fluid Mech.*, **162**, 53–67.

Melville, W. K., and Helfrich, K. R., 1987. Transcritical two-layer flow over topography. *J. Fluid Mech.*, **178**, 31–52.

Melville, W. K., Tomasson, G. G. and Renouard, D. P., 1989. On the stability of Kelvin waves. *J. Fluid Mech.*, **206**, 1–24.

Miles, J. W., 1976. Korteweg de Vries equation modified by viscosity. *Phys. Fluids*, **19**, 1063.

Miles, J. W., 1977. Resonantly interacting solitary waves. *J. Fluid Mech.*, **79**, 171–179.

Miles, J. W., 1978. An axisymmetric Boussinesq solitary wave. *J. Fluid Mech.*, **84**, 181–191.

Miles, J. W., 1979. On internal solitary waves. *Tellus*, **31**, 456–462.

Miles, J. W., 1980. Solitary waves. *Ann. Rev. Fluid Mech*, **12**, 11–43.

Miles, J. W., 1986. Stationary, transcritical channel flow. *J. Fluid Mech.*, **162**, 489–499.

Osborne, A. R. and Burch, T. L., 1980. Internal solitons in the Andaman Sea. *Science*, **208**, 451–459.

Pedersen, G., 1988. Three-dimensional wave patterns generated by moving disturbances at transcritical speeds. *J. Fluid Mech.*, **196**, 39–69.

Renouard, D. P., Chabert d'Hieres, G. and Zhang, X., 1987. An experimental study of strongly nonlinear waves in a rotating system. *J. Fluid Mech.*, **177**, 381–394.

Rockliff, N., 1984. Long nonlinear waves in stratified shear flows. *Geophys. Astrophys. Fluid Dyn.*, **28**, 55–74.

Russell, Scott, 1838. Report of the Committee on Waves. Rep. Meet. Brit. Assoc. Advc. Sci., 7th, Liverpool/1837, pp. 417–496.

Sandstrom, H. and Elliott, J. A., 1984. Internal tide and solitons on the Scotian Shelf: a nutrient pump at work. *J. Geophys. Res.*, **89**, 6415–6426.

Segur, H. and Hammack, J. L., 1982. Soliton models of long internal waves. *J. Fluid Mech.*, **118**, 285–304.

Tomasson, G. G. and Melville, W. K., 1989. Nonlinear and dispersive effects in Kelvin waves. *Phys. Fluids*, in press.

Tung, K.-K., Ko, D. R. S. and Chang, J. J., 1981. Weakly nonlinear internal waves in shear. *Stud. Appl. Maths.*, **65**, 189–221.

Whitham, G. B., 1974. *Linear and nonlinear waves.* Wiley, New York.

Wu, T. Y., 1987. Generation of upstream advancing solitons by moving disturbances. *J. Fluid Mech.*, **184**, 75–99.

Wu, D. M. and Wu, T. Y., 1982. Three-dimensional nonlinear long waves due to moving surface pressure. *Proc. 14th Symp. Naval Hydrodyn.*, pp. 103–129.

TIME-DEPENDENT, TWO-LAYER FLOW OVER A SILL

W. ROCKWELL GEYER
Woods Hole Oceanographic Institution
Woods Hole, Massachusetts 02543
U.S.A.

ABSTRACT. A one-dimensional, two-layer model is used to illustrate the time-dependent nature of two-layer flow over a sill forced by barotropic tides. The quasi-steady approximation is valid in the vicinity of the sill for the portion of the tidal cycle when the flow is transcritical, but the time-dependent term in the momentum equation becomes significant during the subcritical portion of the tidal cycle. The results suggest that to understand the influence of tidal forcing on sill flows requires the application of fully time-dependent models.

1 Introduction

Steady, two-layer hydraulic theory provides considerable insight into the dynamics of stratified flows through straits and over sills, since some of the dominant features of the flow, such as the transition to supercritical flow and hydraulic jumps, are well represented by the steady approximation. Farmer and Smith (1980) and Armi and Farmer (1986) have shown the applicability of the steady state approach to hydraulic transitions in Knight Inlet, British Columbia and the Strait of Gibraltar. This approach is very attractive in that it focuses on the advective processes that dominate the dynamics near the sill crest and doesn't get obscured by minor influences of other terms in the momentum equation.

However, there are important aspects of the flow in Knight Inlet, Gibraltar and every other strait with appreciable tidal currents that are fundamentally time-dependent. The most obvious example is the propagation of bores away from the sill crest, observed in many stratified straits (Kinder 1984; Osborne and Burch 1980). There are less dramatic effects that may be equally important with respect to the exchange flow, such as the excitation of linear waves upstream of the obstacle, demonstrated by Hibiya (1986), in an analytic, linear model of two-layer sill flow in transcritical flow. Hibiya's results indicate that disturbances form both upstream and downstream of the obstacle during the subcritical portion of the tidal cycle, so the upstream conditions for the supercritical portion of the cycle are strongly influenced by the upstream wave, even though the wave can no longer propagate upstream. While a quasi-steady regime may obtain for the period of maximum flow, the flow will depend on the upstream conditions that are determined by the transient response of the flow over the sill.

So, it seems that while steady hydraulic theory is useful locally to describe the dynamic balance at the sill crest, one must consider time dependence to solve the entire problem,

L. J. Pratt (ed.), The Physical Oceanography of Sea Straits, 421–432.

including upstream influence as well as the formation and release of bores. At the same time, it is advantageous to maintain as simple a depiction of the flow as possible, in order not to obscure the important processes going on within the strait.

This paper presents the application of a simple, numerical model of the two layer flow in attempt to maintain as much of the simplicity of the steady hydraulic approach but expand the problem to the consideration of time dependence. Unlike Hibiya's analytical approach, this model is fully nonlinear, thus it includes the contributions of momentum advection in the two layers as well as time-dependence.

2   The Numerical Model

The model is a finite-difference, time-dependent, nonlinear, one-dimensional, two-layer representation of the flow in stratified channels in which rotational effects are unimportant, (i.e., in which the width is less than the internal Rossby radius). The model is designed to handle transitions between subcritical and supercritical flow and hydraulic jumps while conserving momentum and mass, using conservation forms of the equations. The model has a staggered spatial grid and a leapfrog scheme in time, providing second-order accuracy in space and time. To maintain the proper direction of information propagation, the differencing scheme shifts in supercritical conditions to upwind, which introduces numerical diffusion but maintains stability in transcritical, bidirectional flow. Momentum conservation at jumps is handled automatically by the scheme, and to satisfy mass conservation a jump tracking algorithm is implemented.

2.1   GOVERNING EQUATIONS

The equations of motion in each layer are represented as follows:

$$\frac{\partial Q_1}{\partial t} + \frac{\partial}{\partial x}\left(\frac{Q_1^2}{h_1}\right) + gh_1\frac{\partial \eta}{\partial x} = 0 \tag{1}$$

$$\frac{\partial Q_2}{\partial t} + \frac{\partial}{\partial x}\left(\frac{Q_2^2}{h_2}\right) + gh_2\frac{\partial \eta}{\partial x} + g'h_2\frac{\partial h_i}{\partial x} = 0 \tag{2}$$

$$\frac{\partial h_1}{\partial t} + \frac{\partial Q_1}{\partial x} = 0 \tag{3}$$

$$\frac{\partial h_2}{\partial t} + \frac{\partial Q_2}{\partial x} = 0 \tag{4}$$

where Q_1 and Q_2 are transport per unit breadth in each layer, h_i is the interface height above datum, h_1 and h_2 are cross-sectional areas of each layer, g is gravitational acceleration, g' is reduced gravity, and η is the elevation of the free surface above datum. Using the specified total transport, the two momentum equations can be combined to generate a single equation for the transport in the lower layer

$$\left(\frac{1}{h_1} + \frac{1}{h_2}\right)\frac{\partial Q_2}{\partial t} = \frac{1}{h_1}\frac{\partial Q_0}{\partial t} + \frac{1}{h_1}\frac{\partial}{\partial x}\left(\frac{Q_1^2}{h_1}\right) - \frac{1}{h_2}\frac{\partial}{\partial x}\left(\frac{Q_2^2}{h_2}\right) - g'\frac{\partial h_i}{\partial x}. \tag{5}$$

The first term on the right side is the forcing by the time-dependent barotropic flow; the second and third terms are the advection or momentum flux terms, and the fourth term is the baroclinic pressure gradient. Either bottom or interfacial friction can be added with no difficulty; in fact they improve the stability characteristics of the model. However for the purpose of simplicity they were not included in the present application.

The flow is assumed to be uniform in the upper and lower layers. The elevation of the interface is held fixed at the left boundary, and there is a radiation condition at the right boundary. The "sill" is represented by $sech^2(x)$.

2.2 Differencing Scheme

Two differencing schemes are used in the model, one for subcritical flow and the other for supercritical flow, in order to maintain the proper direction of flow of information. In subcritical conditions, the grid is staggered, so that both continuity and momentum have centered differences. Information is carried upstream and downstream. When the flow is supercritical, the differencing scheme shifts, and both momentum and continuity only depend on variables at the same longitudinal position or upstream. Thus disturbances cannot be transmitted upstream. This is consistent with the direction of the characteristics during supercritical conditions. This method thus has some of the attributes of the method of characteristics while maintaining the simpler finite difference representation in space.

Another attribute of this model is its explicit representation of the propagation of jumps. The advance or retreat of jumps is determined by the gradient of transport between the supercritical flow just upstream of the jump and the jump. Once there is a large enough accumulation or deficit of mass, the jump either advances or retreats one grid point. By this procedure, mass and momentum are conserved across the jump, and the sharp gradient is preserved.

3 Results

3.1 Steady Barotropic Forcing

An example of the model representation of a flow with steady forcing conditions is shown in fig. 1, with the bottom and interface elevation plotted against horizontal position. The flow in the lower layer is from right to left, and flow in the upper layer is in the opposite direction, with zero net transport. The flow is critical at the crest, is supercritical for some distance downstream, then it jumps back to subcritical. Because the lower layer is thinner than the upper layer, the upper layer is nearly passive in this case. The model was started with a flat interface, and run for 72 time units (the time it would take for a linear internal wave to go 72 units in space). The flow is nearly steady, but not quite, since there are still transient disturbances bouncing back and forth between the left boundary and the jump. This is a consequence of the lack of dissipation.

Another, more interesting example with steady forcing has a thinner upper layer (fig. 2). The left panel shows the interface structure at t=50, and the right case shows the flow at t=90. Because the upper layer is initially thinner, the flow first becomes supercritical in the upper layer. As the flow develops, it becomes supercritical in the lower layer. Eventually the two supercritical regions coalesce, resulting in a "hydraulic drop" (Long 1954), in which a discontinuity in interface elevation connects two conjugate, supercritical states.

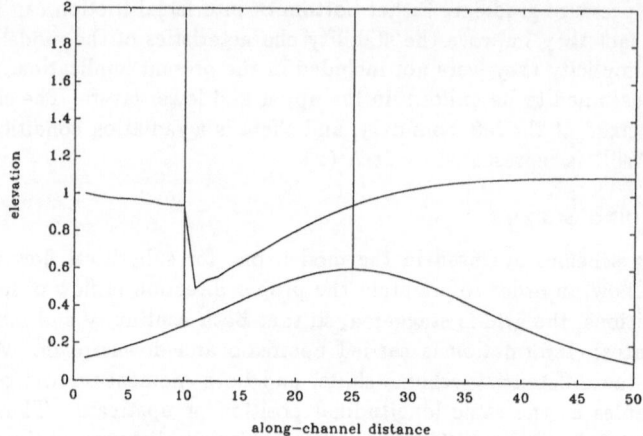

Figure 1. Model simulation of unforced, frictionless exchange flow. Lower layer is moving from right to left, with an upstream Froude number of 0.2 and a maximum Froude number of 2.5. The two vertical lines indicate hydraulic transitions; at the crest there is a control section and to the left there is a strong jump. There is a reservoir condition at $x = 0$ and a radiation condition at $x = 50$.

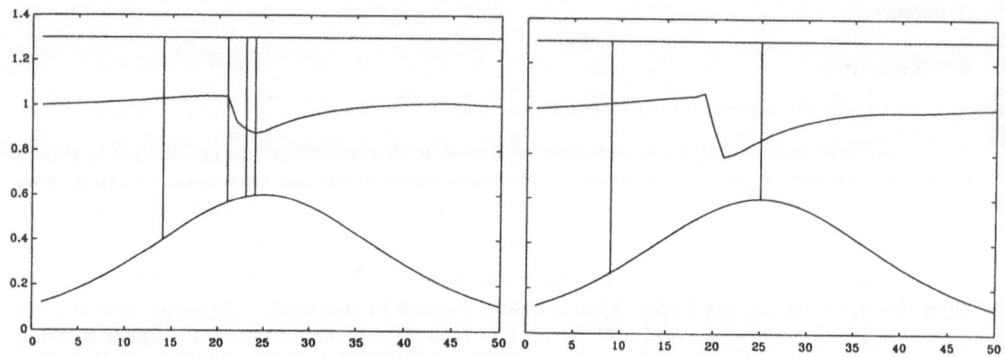

Figure 2. Model simulation of exchange flow with a thin upper layer, showing the development of a hydraulic drop. At $t = 50$ (upper panel), there are two zones of supercritical flow, a narrow one at the crest of the obstacle and a broader one to the left, where the upper layer has a local maximum in velocity. At $t = 90$ the two supercritical regions have coalesced, and a hydraulic drop is evident.

The last case with steady forcing is one which illustrates the "approach control" condition (Lawrence, 1985), where the control section occurs at the leading edge of an obstacle, rather than at its crest. The upstream conditions for this case were taken from laboratory data reported by Lawrence (1985), and the interface was initially flat. Fig. 3 indicates the evolution of the interface in time, with each plot of the interface profile displaced down-ward from the one preceding it. The control section (i.e., the transition from subcritical to supercritical conditions) is seen to progress upstream of the obstacle through the course of the simulation, and by t=80 the "approach control" condition has nearly been reached.

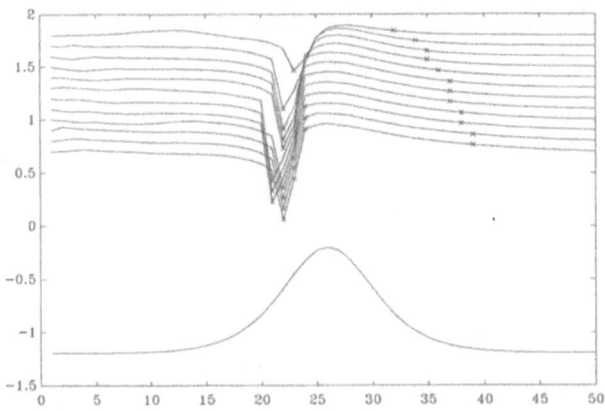

Figure 3. Model simulation of the development of approach control (Lawrence 1985) from an initially level interface. Each trace represents a snapshot of the interface elevation, displaced downward from the one preceding it by 0.05 units. The transitions between subcritical and supercritical conditions are noted by ×. In this case, the control does not occur at the crest, but rather it occurs upstream, and it moves further upstream as the flow develops. A complex combination of hydraulic transitions occurs downstream of the obstacle.

Just after the crest of the obstacle, the interface drops sharply, and following this drop is a short region of supercritical flow followed by a hydraulic jump. The structure is very similar to that reported in Lawrence's experiment, as indicated by the variation in layer Froude numbers across the obstacle

$$F_1^2 = \frac{u_1^2}{g'h_1} \tag{6}$$

and

$$F_2^2 = \frac{u_2^2}{g'h_2} \tag{7}$$

and composite Froude number

$$G^2 = F_1^2 + F_2^2. \tag{8}$$

Lawrence's estimate of G (fig. 4) shows a drop to subcritical conditions just past the crest of the sill, indicating a weak jump. The model shows a similar decrease in G, just reaching $G = 1$ behind the obstacle and then increasing in the downstream direction before the jump. The model-generated jump occurs over just one grid point, a consequence of the absence of explicit dissipation in the model that would smear out the transition to subcritical flow, as indicated in the laboratory results.

426

Lawrence (1985) attributed the drop in the composite Froude number behind the jump to non-hydrostatic effects, but since the hydrostatic model also indicates the drop, it appears that there is actually a complex hydraulic transition behind the obstacle. The feature may be a hydraulic jump, although it is not clear from the laboratory results that there is dissipation associated with the feature. In any case the similarity between the laboratory results and the model provide some confidence in the model's applicability, at least in unforced flows.

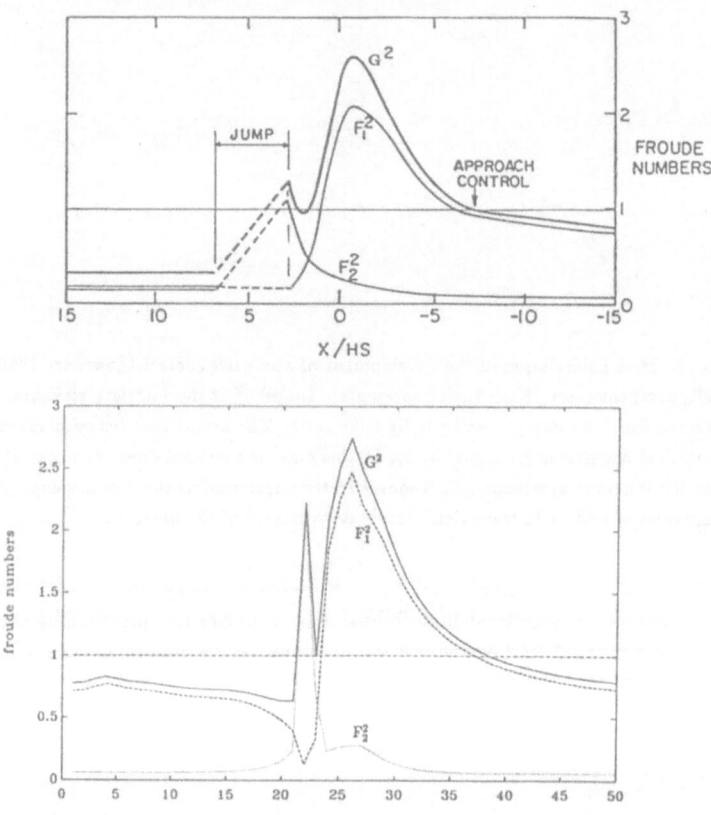

Figure 4. Layer Froude numbers F_1^2 and F_2^2 and composite Froude number G^2 from Lawrence's (1985) experiment (upper panel) and from the model simulation (lower panel). Many of the features of the experimental results are reproduced in the model, including the approach control, the maximum value of G^2, the dip in G^2 behind the obstacle, and the jump. The major difference between the experiment and the model is the longitudinal extent of the jump, which confined to one grid point in the model due to lack of explicit dissipative mechanisms.

3.2 UNSTEADY BAROTROPIC FORCING

The following cases illustrate the time evolution of the hydraulics with a time-varying barotropic flow, such as tidal forcing. The tidal frequency has been selected such that the internal tidal wavelength is approximately twice the length of the domain. In the first case, the barotropic transport is just great enough to cancel out the transport in the lower layer during the flood (left to right), and it cancels out the flow in the upper layer during the ebb (right to left). In the second case, the barotropic flow is doubled, so that the flow in both layers is reversed by the tide. The first case is designated the "neap tide" and the second the "spring tide" case, since this range of variation in the barotropic forcing is comparable to the fortnightly modulation of the semi-diurnal tide in many locations.

The results of the simulation for the neap case are indicated in fig. 5. As in the previous example, the time-evolution of the interface is indicated by multiple lines, offset by 0.05, representing the interface position at half hour intervals. At the beginning of the ebb (upper panel), the most pronounced feature is a bore that is advancing against the barotropic flow.

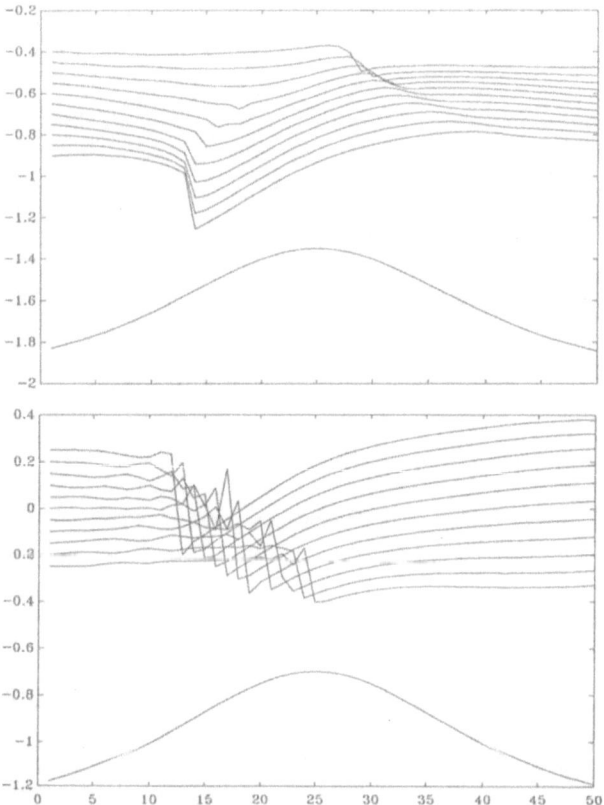

Figure 5. Model simulation of tidally forced exchange flow, "neap tide". The tidally varying barotropic transport is equal in magnitude to the mean exchange transport. The upper panel indicates ebb conditions (flow from right to left), and the lower panel indicates the flood. Each trace is a snapshot of interface elevation, plotted at half-hour intervals with a vertical offset of 0.05 units. During the ebb (upper panel), a strong jump forms downstream of the obstacle, while during the flood (lower panel), the jump turns into a bore and propagates over the sill.

This bore weakens as the ebb progresses, and it virtually disappears by the end of the ebb. On the downstream side of the sill, however, a transition to supercritical flow develops, and a moderate jump forms approximately halfway down the back of the obstacle. The position of the jump remains stationary for approximately 2 hours in the latter portion of the ebb. During the flood (lower panel) the hydraulic jump evolves into a bore and propagates across the sill. There is considerable steepening of the front as it approaches the sill crest; in fact it nearly becomes unstable before reaching the crest. The flood does not last long enough for the bore to propagate very far; it only reaches the sill crest before the flow starts to ebb again. As the upper panel indicates, the bore continues landward during the ebb, but it is completely attenuated before it reaches the boundary of the domain.

The spring tide case (fig. 6) has the same shear as the previous one, but the maximum

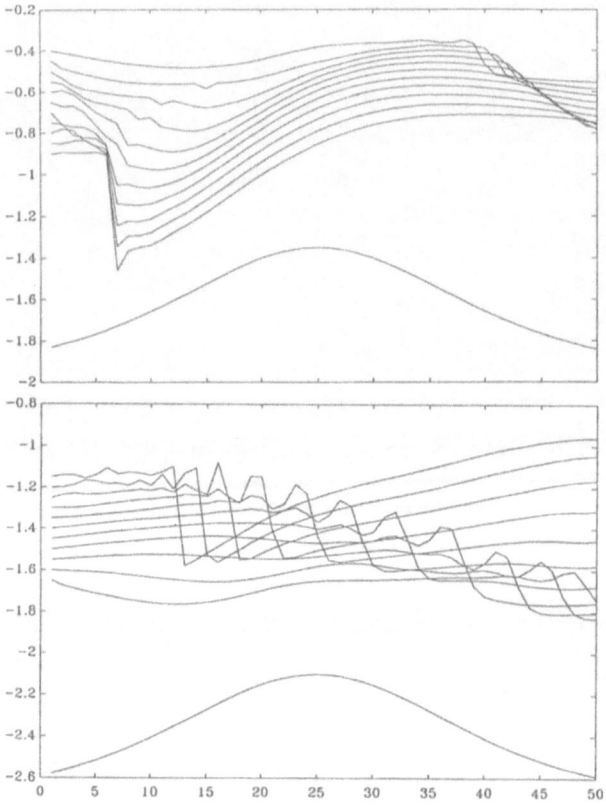

Figure 6. Model simulation of the tidally forced exchange flow, "spring tide". The tidal transport is twice the magnitude of that in the "neap tide" case. The jump is stronger and is farther from the sill crest during the ebb (upper panel). During the flood, the bore is larger and travels more rapidly, crossing the sill and reaching the landward boundary before the end of the flood (lower panel).

429

barotropic flow is twice the mean flow in the lower layer (at the boundary). Since the
flow is stronger both during the flood and the ebb, the response is more energetic in this
case. During the ebb (upper panel), the bore from the previous ebb is evident, and it is
observed to propagate out of the domain. A control section develops at the sill crest early
in the ebb, and a strong jump is formed near the base of the obstacle. There is swelling of
the interface upstream of the sill, similar to the linear response found by Hibiya (1986) to
supercritical barotropic flow. As in the neap tide case, the hydraulic jump evolves into a
bore during the flood (lower panel), but in this case the bore moves faster, due to its larger
amplitude and the stronger tidal current, and it propagates across the sill to the landward
boundary in the course of the flood. Again there is considerable steepening of the front as
it approaches the sill crest, after which it attains nearly constant form. (Due to the neglect
of non-hydrostatic effects, the model does not represent the development of solitary waves,
however.)

Further analysis of the time-dependent case reveals some interesting features. First,
taking the mean of the interface elevation over three tidal cycles (fig. 7), one finds a distinct
asymmetry in the mean baroclinic pressure field. The dip in the mean height to the left
of the sill results from the hydraulic transition that occurs only during the ebb. This
asymmetry results from the presence of a mean shear, which causes considerably stronger
flows in the lower layer during the ebb than the flood. The mean interface elevation indicates
that there is a net pressure difference between the upstream and downstream sides of the
sill, i.e., there is a form drag associated with the internal hydraulic response. The net
force by the mean pressure difference for the spring tide case was calculated, and a drag
coefficient for the obstacle was calculated based on the relation

$$Force = \frac{1}{2}\rho C_{FD} A_{sill} \overline{u_2^2}$$
(9)

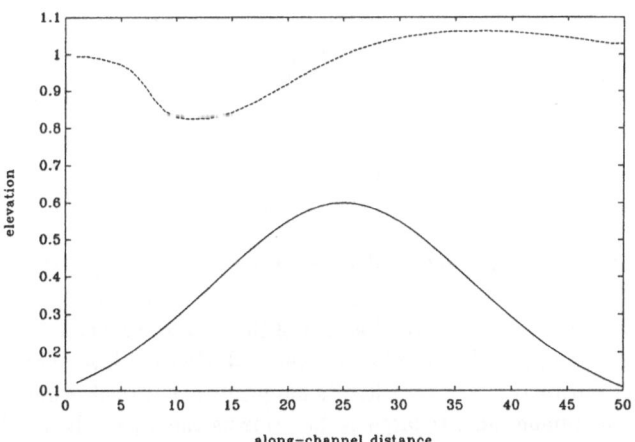

Figure 7. Mean elevation of the interface, based on three tidal cycles of the "spring tide" case.
The mean depression of the interface indicates that there is a net form drag associated with
the internal hydraulic response, which retards the exchange flow.

where C_{FD} is the drag coefficient due to form drag, A_{sill} is the projected area of the sill, and the overbar indicates averaging over the tidal cycle. For the spring tide case, the value of C_{FD} was found to be 1.3, comparable to the form drag associated with bluff bodies.

Fig. 8 indicates the time-variation of the elevation for the spring tide case for two points in the domain, one on either side of the sill. Both time series show a strong internal tidal signal, and both indicate the passage of the bore, which is characterized by a sudden change in interface height. Comparing subsequent tidal cycles, it is evident that the bore passage is earlier each tidal cycle. Because there is no frictional dissipation in the model, the spin-up time is longer in the model than it would be in real flows, but the results suggest that the timing of the bore passage is very sensitive to initial conditions, so it may vary considerably in real straits from one tidal cycle to another.

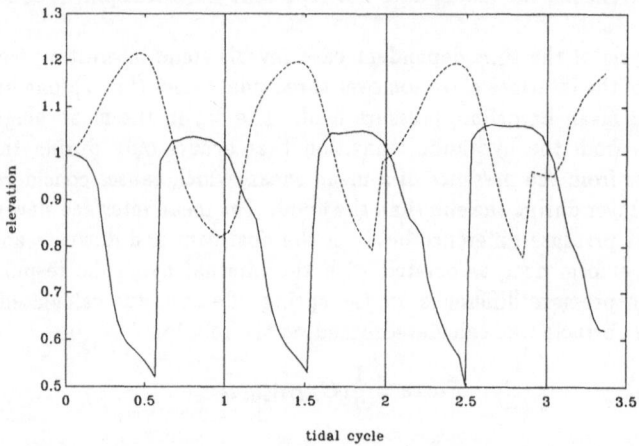

Figure 8. Time series of interface elevation at $x = 10$ (solid trace) and $x = 40$ (dashed trace), from the "spring tide" case. The internal tidal signal is evident in both cases, as is the passage of the bore. The amplitude of the bore is much larger at $x = 10$, since this position is just to the right of the position where the jump forms (cf., fig. 6).

4 Discussion

The numerical simulations of barotropically forced exchange flow clearly do not indicate quasi-steady dynamics, except perhaps during a portion of the ebb when the flow is controlled at the sill crest. In these examples the upper layer Froude number F_1 is always significantly less than 1 (due to the greater thickness of the upper layer), so the hydraulic state of the flow is determined by F_2. Because in these cases the exchange flow and barotropic velocities are comparable, there is a large fraction of the tidal cycle when the baroclinic and barotropic velocities are opposed in the lower layer, reducing the layer Froude number F_2 below critical. Thus during the flood, the flow is uniformly subcritical, while during the ebb the composite Froude number gets as large as 2.5. During these periods of subcritical flow, the flow is inherently unsteady, since the time-dependent term in the momentum equation becomes larger than the advective term.

The unsteady flow noted in these simulations will occur for a large range of the ratio $u_{tide}/\Delta u_{mean}$ because of the nonlinear nature of the hydraulic response. While Pratt (1982) found that weak periodic forcing would not significantly alter the nature of a controlled flow, as soon as $u_{tide}/\Delta u_{mean}$ becomes order 1, the control is likely to be flooded for a fraction of the tidal cycle, and bores will propagate across the sill, potentially altering the mean exchange flow. Taking the other extreme, if $u_{tide}/\Delta u_{mean}$ is much greater than 1, then hydraulic control will be established in both directions during the periods of maximum flow, but the flow cannot be quasi-steady in the far field, since the Bernoulli function must drop across a jump, and it cannot drop in both directions.

While these simulations clearly indicate the inherent transience of tidally forced exchange flow, the problem of the actual influence of tides on the exchange flow was not resolved. In these model runs, the exchange flow was imposed as an initial condition, and the simulations did not continue for long enough to allow the exchange to achieve a state independent of the initial conditions. Nevertheless, some inferences can be made about the influence of tides on the exchange flow based on these results. There is appreciable net form drag that acts to retard the mean exchange flow; in fact this may be comparable or larger than the influence of bottom friction in straits such as Gibraltar. Because of the quadratic dependence of form drag, higher tidal velocities will contribute to higher values of the net form drag.

Another influence of the tides on the exchange flow is the formation of bores, which has been found to augment the exchange through the Strait of Gibraltar (Kinder 1984; Pettigrew, this volume). The model results indicate a strong influence of tidal amplitude on the strength and propagation speed of the bore, so there should be a corresponding variability in the transport associated with it. In addition to the bore, there may be a significant Stokes drift associated with the longer wavelength internal tidal oscillations, which are quite apparent in fig. 8. Further use of this sort of simple numerical model should provide valuable insight into the influence of tides on the exchange flow in straits, including the influences of form drag, and Stokes transport and bottom frictional influences.

5   References

Armi, L. and D.M. Farmer, 1986. Maximal two-layer exchange through a contraction with barotropic net flow. *J. Fluid Mech.*, **164**, 27-51.

Farmer, D.M. and J.D. Smith, 1980. Tidal interaction of stratified flow with a sill in Knight Inlet. *Deep Sea Research*, **27A**, 239-254.

Hibiya, T., 1986. Generation mechanism of internal waves by tidal flow over a sill. *J. Geophys. Res.*, **91**, 7697-7708.

Kinder, T.H., 1984. Net mass transport by internal waves near the Strait of Gibraltar. *Geophys. Res. Letters*, **11**, 987-990.

Lawrence, G.A., 1985. The hydraulics and mixing of two-layer flow over an obstacle. PhD thesis, University of California, Berkeley, 122 pp.

Long, R.R., 1954. Some aspects of the flow of stratified fluids II. Experiments with a two-fluid system. Tech. Rept. 4, Johns Hopkins Univ., Baltimore, 54 pp.

Osborne, A.R. and T.L. Burch, 1980. Internal solitons in the Andaman Sea. *Science*, **208**, 451-460.

Pettigrew, N., *this volume*. Doppler measurements of bores in the Strait of Gibraltar.

432

Pratt, L., 1982. The dynamics of unsteady strait and sill flow. Doctoral dissertation, WHOI-MIT Joint Program in Oceanography and Ocean Engineering, 140 pp.

LONG PROGRESSIVE WAVES IN ROTATING FLUID

J.-P. GERMAIN and D. P. RENOUARD
Institut de Mecanique de Grenoble
B.P. 53 X
38041 Grenoble-Cedex
France

ABSTRACT. We propose a general theoretical framework which emcompasses all possible long progressive waves in rotating fluid, whether in a channel or in an infinite or semi-infinite ocean. There are different governing equations depending upon the value of the Coriolis parameter (f). When f is large, we find either nonlinear solitary Kelvin waves or nonlinear Poincare waves in a channel, or nonlinear Sverdrup waves in an infinite ocean. But in the latter, there are no solitary waves in an infinite ocean. When f is small, such waves may exist. For intermediate values of f, there are either solitary waves with a horizontal crest in an infinite ocean, or Poincare type of waves in a channel. In that last case, we recover the equation first established by Grimshaw and Melville (1989).

1. Formulation

Let us consider long permanent progressive waves in an inviscid and incompressible fluid, and choose the z-axis as the rotation axis, oriented upward, the x-axis being oriented in the direction of the wave propagation. The depth is uniform, $z = -h$, and the fluid is bounded by a free surface, $z = 0$ at rest. We denote g the acceleration due to gravity and f the Coriolis parameter, which is assumed to be constant over the entire basin (f-plane approximation). At the point (x,y,z) and time t the velocity components will be denoted by $(u,v,w)(x-ct,y,z)$, the pressure by $p(x-ct,y,z)$ and the free surface displacement by $\eta(x-ct,y)$, c being the wave's celerity.

Since we are interested in progressive long waves, let us introduce the stretched variables:

$$\bar{x} = \varepsilon(x-ct), \quad \bar{y} = \varepsilon y, \quad \bar{z} = z$$

and assume that $f = \varepsilon^{(1+p)}\bar{f}$.

433

L. J. Pratt (ed.), The Physical Oceanography of Sea Straits, 433–440.
© 1990 *Kluwer Academic Publishers.*

2. Case $f = \varepsilon \bar{f}$ ((p= 0, Strong Rotation)

The governing dynamics and kinematic equations then become:

$$\varepsilon\left[(u-c)\frac{\partial u}{\partial x} + v\frac{\partial u}{\partial y} - fv + \frac{1}{\rho}\frac{\partial p}{\partial x}\right] + w\frac{\partial u}{\partial z} = 0$$

$$\varepsilon\left[(u-c)\frac{\partial v}{\partial x} + v\frac{\partial v}{\partial y} + fu + \frac{1}{\rho}\frac{\partial p}{\partial y}\right] + w\frac{\partial v}{\partial z} = 0$$

$$\varepsilon\left[(u-c)\frac{\partial w}{\partial x} + v\frac{\partial w}{\partial y}\right] + w\frac{\partial w}{\partial z} + \frac{1}{\rho}\frac{\partial p}{\partial z} + g = 0$$

$$\varepsilon\left[\frac{\partial u}{\partial x} + \frac{\partial v}{\partial y}\right] + \frac{\partial w}{\partial z} = 0$$

with the boundary conditions:
at the bottom, z= -h:

$$w(z=-h) = 0$$

at the free surface, $z = \eta(x,y)$:

$$w(z=\eta) = \varepsilon\left[(u-c)\frac{\partial \eta}{\partial x} + v\frac{\partial \eta}{\partial y}\right] (z= \eta)$$

$$p= 0$$

As yet we make no assumption about the vorticity components:

$$\frac{\partial v}{\partial z} - \varepsilon\frac{\partial w}{\partial y}, \quad \varepsilon\frac{\partial w}{\partial x} - \frac{\partial u}{\partial z}, \quad \varepsilon\left[\frac{\partial u}{\partial y} - \frac{\partial v}{\partial x}\right]$$

Now let us assume that the velocity components, the pressure, the free surface displacement, as well as the celerity, are power series of the distorsion parameter ε:

$$u= \sum_{n=1}^{\infty} \varepsilon^{2n} u_{2n}(x,y,z)$$

$$v= \sum_{n=1}^{\infty} \varepsilon^{2n} v_{2n}(x,y,z)$$

$$w= \sum_{n=1}^{\infty} \varepsilon^{2n+1} w_{2n+1}(x,y,z)$$

$$p= -\rho gz + \sum_{n=1}^{\infty} \varepsilon^{2n} p_{2n}(x,y,z)$$

$$\eta= \sum_{n=1}^{\infty} \varepsilon^{2n} h\eta_{2n}(x,y)$$

$$c= c_0 + \sum_{n=1}^{\infty} \varepsilon^{2n} c_{2n}$$

and let us consider small free surface displacements from a state of rest. As for a non-rotating fluid, we impose that u_{2n} and v_{2n} are polynomials in $(z+h)$ of degree $(2n-2)$. Therefore, u_2 and v_2 are z-independent, and thus the vorticity is equal to zero at the second order, $n=1$, and it is straightforward to see that:

$$P_2 = \rho g \eta_2(x,y)$$

$$w_3 = -(z+h)\left(\frac{\partial u_2}{\partial x} + \frac{\partial v_2}{\partial y}\right)$$

It is then easy to see that u_2, v_2 or η_2 must satisfy an equation of the form:

$$(c_0^2 - gh)\frac{\partial^2 x_2}{\partial x^2} - gh\frac{\partial^2 x_2}{\partial y^2} + f^2 x_2 = 0$$

which implies that $c_0^2 \geq gh$ in order to obtain wave type solutions.

2.1. $c_0^2 = gh$

At the second order, either in a semi-infinite ocean or in a channel of width L, we get:

$$\eta_2 = N_2(x)\, e^{-\frac{f}{c_0}y}$$

$$u_2 = \frac{g}{c_0} N_2(x)\, e^{-\frac{f}{c_0}y}$$

$$v_2 = 0$$

$$w_3 = -\frac{g}{c_0}(z+h)\frac{dN_2}{dz}\, e^{-\frac{f}{c_0}y}$$

$$P_2 = \rho g\, N_2(x)\, e^{-\frac{f}{c_0}y}$$

and the vorticity vertical component is then:

$$\frac{\partial u_2}{\partial y} - \frac{\partial v_2}{\partial x} = -\frac{f}{h} N_2(x)\, e^{-\frac{f}{c_0}y}$$

where $N_2(x)$ is an unknown function. In order to get the solution at the following order, we assume that:

$$\frac{\partial w_3}{\partial x} - \frac{\partial u_4}{\partial z} = \lambda(x,y)(z+h)$$

$$\frac{\partial v_4}{\partial z} - \frac{\partial w_3}{\partial y} = \mu(x,y)(z + h)$$

and the compatibility conditions require that:

$$\lambda(x,y) = \left[A(y) \, Cos \, \frac{f}{c_0} x + B(y) \, Sin \, \frac{f}{c_0} x \right] e^{-\frac{f}{c_0} y}$$

$$\mu(x,y) = \left[A(y) \, Sin \, \frac{f}{c_0} x - B(y) \, Cos \, \frac{f}{c_0} x \right] e^{-\frac{f}{c_0} y} - \frac{f}{c_0} \frac{dN_2}{dx} e^{-\frac{f}{c_0} y}$$

i.e. there appear inertial waves which are closely related to the vorticity. At this point it is worth noting that we could make a weaker assumption and set:

$$u_{2n} = \sum_q \bar{u}_{2n,q}(x,y) \, (z + h)^{2q}$$

$$v_{2n} = \sum_q \bar{v}_{2n,q}(x,y) \, (z + h)^{2q}$$

with p<n, and so obtain the flow with an arbitrary vorticity.

Thus, at the following order, we get u_4, v_4, w_5, P_4, η_4 and c_2, and, in a channel, the boundary conditions along the side-walls, y= 0 and y= L, give us an equation for $N_2(x)$:

$$\left[e^{\frac{f}{c_0} L} - e^{-2 \frac{f}{c_0} L} \right] N_2 \frac{dN_2}{dx} +$$

$$\left[e^{\frac{f}{c_0} L} - e^{-\frac{f}{c_0} L} \right] \left[\frac{c_0^2 h^2}{6g} \frac{d^3 N_2}{dx^3} - \frac{c_2 c_0}{g} \frac{dN_2}{dx} \right] = 0$$

which is of the KdV type, and thus we obtain either cnoidal or solitary Kelvin waves, since $\eta_2 = N_2(x) \exp(-fy/c_0)$.

2.2. $c_0^2 > gh$

2.2.1. If we are considering an infinite ocean, hence looking for y-independant progressive waves, we obtain, at the second order:

$$u_2 = A \, \frac{\lambda c_0}{f} \, Cos \, \lambda x$$

$$v_2 = A \, Sin \, \lambda x$$

$$w_3 = A \, \frac{\lambda^2 c_0}{f} \, \text{Sin } \lambda x \, (z + h)$$

$$\eta_2 = A \, \frac{h\lambda}{f} \, \text{Cos } \lambda x$$

with:

$$\lambda^2 = \frac{f^2}{c_0^2 - gh}$$

that is, linear Sverdrup waves. At the following orders, the nonlinearity appears as higher order harmonics and corrective terms for the celerity.

2.2.2. If we are looking for a y-dependent phenomenon, as in a channel for instance, then we get, using the separation of variables method:

$$u_2 = \frac{A}{\lambda(c_0^2 - gh)} \, (gh\Lambda \text{ Cos } \Lambda y - c_0 f \text{ Sin } \Lambda y) \text{ Sin } \lambda x$$

$$v_2 = A \text{ Sin } \Lambda y \cdot \text{Sin } \lambda x$$

$$w_3 = \frac{-A}{c_0^2 - gh} \, (c_0^2\Lambda \text{ Cos } \Lambda y - c_0 f \text{ Sin } \Lambda y) \text{ Cos } \lambda x \, (z + h)$$

$$\eta_2 = \frac{Ah}{\lambda(c_0^2 - gh)} \, (c_0\Lambda \text{ Cos } \Lambda y - f \text{ Sin } \Lambda y) \text{ Sin}\lambda x$$

where:

$$\Lambda^2 = \lambda^2 (c_0^2 - gh) - f^2$$

that is linear Poincare waves, and, as previously, at higher orders, the nonlinearity appears as higher order harmonics, and corrective terms for the celerity.

3. Case $f = \varepsilon^2 \bar{f}$ (p= 1, Weak Rotation)

3.1. SVERDRUP TYPE WAVES (Y-INDEPENDENT)

Let us now expand the velocity components, the free surface, displacements and the pressure as:

$$u = \sum_{n=1}^{\infty} \varepsilon^{2n} u_{2n}$$

$$v = \sum_{n=1}^{\infty} \varepsilon^{2n+1} v_{2n+1} \quad \text{(instead of } v = \sum_{n=1}^{\infty} \varepsilon^{2n} v_{2n}\text{)}$$

$$w = \sum_{n=1}^{\infty} \varepsilon^{2n+1} w_{2n+1}$$

$$\eta = \sum_{n=1}^{\infty} \varepsilon^{2n} \eta_{2n}$$

$$p = -\rho g z + \sum_{n=1}^{\infty} \varepsilon^{2n} p_{2n}$$

$$c = c_0 + \sum_{n=1}^{\infty} \varepsilon^{2n} c_{2n}$$

then the same kind of computation leads to:

$$u_2 = u_2(x)$$

$$v_3 = \frac{f}{c_0} \int_{x_0}^{x} u(\sigma) \, d\sigma$$

$$w_3 = -\frac{du_2}{dx} (z + h)$$

$$\eta_2 = \frac{c_0}{g} u_2 \quad \text{and,} \quad p_2 = \rho g \eta_2$$

with $c_0^2 = gh$, and the compatibility condition that we obtain at higher orders, is written:

$$\frac{2hc_2}{c_0} \frac{du_2}{dx} - \frac{3c_0}{g} u_2 \frac{du_2}{dx} - \frac{h^3}{3} \frac{d^3 u_2}{dx^3} + f^2 \frac{h}{c_0^2} \int_{x_0}^{x} u_2(\sigma) \, d\sigma = 0$$

or, setting $U_2(x) = \int_{x_0}^{x} u_2(\sigma) \, d\sigma$:

$$2h \frac{c_2}{c_0} \frac{d^2 U}{dx^2} - 3 \frac{c_0}{g} \frac{dU_2}{dx} \frac{d^2 U_2}{dx^2} - \frac{h^3}{3} \frac{d^4 U_2}{dx^4} + f^2 \frac{h}{c_0^2} U_2 = 0$$

which is the Ostrovskiy equation for progressive waves (cf Ostrovskiy, 1978).

3.2. GENERAL TYPE (Y-DEPENDENT)

Setting $y = \varepsilon^2 \bar{y}$, and using the same procedure, we get the same expressions as above for v_1, η_2, p_2 and w_3, although with u_2 now a function of x and y, and:

$$v_3 = \int_{x_0}^{x} \left[\frac{f}{c_0} u_2(\sigma, y) + \frac{\partial u_2}{\partial y}(\sigma, y) \right] d\sigma$$

Then, at the higher order, the compatibility equation is:

$$\frac{\partial}{\partial x} \left[-3 u_2 \frac{\partial u_2}{\partial x} + 2c_2 \frac{\partial u_2}{\partial x} - \frac{h^3 c_0}{3} \frac{\partial^2 u_2}{\partial x^3} \right] + c_0 \left[\frac{f^2}{c_0^2} u_2 - \frac{\partial^2 u}{\partial y^2} \right] = 0$$

which is the Grimshaw-Melville equation (cf Grimshaw and Melville, 1989). And if there are side-wall conditions at y= 0 and y= L, we must have:

$$\frac{f}{c_0} u_2(x, {}_L^0) + \frac{\partial u_2}{\partial y}(x, {}_L^0) = 0$$

It should be noted that if u_2 is y-independent, then we get back the Ostrovskiy equation. If $u_2(x,y)$ is in the form $L(x)M(y)$ then we a find Kelvin type solution.

4. Case $f= \varepsilon^3 \bar{f}$ (p= 2, Very Weak Rotation)

We use the same expansion of the unknown function as in 2. Up to the order n= 4 the equations are identical to the irrotational case ,i.e. non-rotating case, except that the dynamics equation projected on the y-axis leads to:

$$v_2 = 0 \text{ ,and: } v_4 = \frac{f}{c_0} \int_{x_0}^{x} u_2(\sigma) \, d\sigma$$

Thus we have Sverdrup type waves, and the compatibility equation for η_2 shows that the waves can then be either cnoidal or solitary waves.

More generally, we can superimpose two Sverdrup waves propagating in different directions in order to get Poincare waves, and, at higher order, the interacton terms produce Mach type waves.

5. Conclusion

For very weak interaction, we have a reasonable chance to apply inverse scattering methods, as in the case of the Andaman Sea (cf Osborne & Burch, 1980; Kabbaj, 1985).

For weak rotation, even if we are able to compute solitons, no direct extension of the inverse scattering method seems possible, for the "mass" of the solitary wave must then be equal to zero, and that the mass of the solitons is not.

For strong rotation, there are no Sverdrup or Poincare solitons, so that no direct inverse scattering is possible.

Hopefully, there will be a way to solve these problems by direct methods.

This study was suggested by the experimental results obtained on the large I.M.G. rotating platform. Some of the experiments were financially supported by the D.R.E.T., under contract 88/089.

6. References

Grimshaw, R. and Melville, W.K. (1989) 'On the derivation of the modified Kadomstsev-Petviashvili equation', Stud. Appl. Maths (in press)

Kabbaj, A. (1985) 'Contribution   l'etude du passage des ondes de gravite sur le talus continental et   la generation ondes internes', These Universite Grenoble

Osborne, A.R. and Burch, T.L. (1980) 'Internal solitons in the Andaman sea', Science, 208, 451-460

Ostrovskiy, L.A. (1978) 'Nonlinear internal waves in a rotating ocean', Oceanology, 18, 2, 119-124

CHARACTERISTIC FEATURES OCCURRING IN THE STRAIT OF GIBRALTAR AS SEEN THROUGH REMOTE SENSING DATA

Claude RICHEZ (*) and Claude KERGOMARD (**)
(*) *Laboratoire d'Océanographie Dynamique et de Climatologie*
LODYC- University of PARIS 6,
4 Place Jussieu, 75252 PARIS CEDEX 05, France
(**) *Laboratoire d'Optique Atmosphérique*
LOA- University of Sciences and Techniques, LILLE, France

ABSTRACT. Remote sensing data contribute efficiently to a better knowledge of oceanographic processes. Surface features, but also the internal dynamics which induce surface signatures, may be revealed by use of some remote sensing techniques. In 1986, we proposed to our Gibraltar Experiment colleagues to undertake a Synthetic Aperture Radar (SAR) survey of the Strait of Gibraltar during two successive tidal cycles at Springs (June 22 and 24, 1986). Our aim was to obtain a synoptic view of the Strait to emphasize characteristic surface features related to the well known generation of internal waves at the Camarinal Sill at each tidal cycle (Lacombe and Richez, 1982) and their eastward propagation during the ebb phase of the tide. Besides this SAR experiment, we used a series of NOAA-AVHRR data, around the same period as our SAR flights (June 21 to July 4, 1986), and we completed our data set with some digital images obtained by the French Satellite SPOT-1, launched in February 1986.

In this paper, through some selected examples, we will show the possibilities and limitations of these different techniques in studying the complex dynamical processes occurring in the Strait of Gibraltar. Complete results of our SAR Experiment are under preparation and will be published separately in a future paper.

1. NOAA-AVHRR Data Set:

A series of 8 NOAA9-AVHRR images, around the end of June-beginning of July 1986, were processed in order to study the surface thermal evolution, during a tidal cycle, in the Strait of Gibraltar and its surroundings (Table I). La Violette and Lacombe (1988) used similar techniques to describe the thermal pattern evolution of the Strait of Gibraltar in October 1982. We were interested in verifying if some of the features they reported in October 1982 could be considered as "characteristic features" occurring regularly, in relation to the tidal phase, at other seasons and other tidal coefficients.

NOAA-images were first totally corrected: (1) Geometrical correction by using an orbital model and warping polynomials established from coastal control points, (2) data resampling on a Mercator grid (precision: 1 pixel), (3) temperature calibration and atmospheric correction of infrared data by applying a "split-window" algorithm, (4) smoothing of the obtained SST data by a 3 x 3 pixels convolution in order to reduce noise.

441

L. J. Pratt (ed.), The Physical Oceanography of Sea Straits, 441–455.
© 1990 *Kluwer Academic Publishers.*

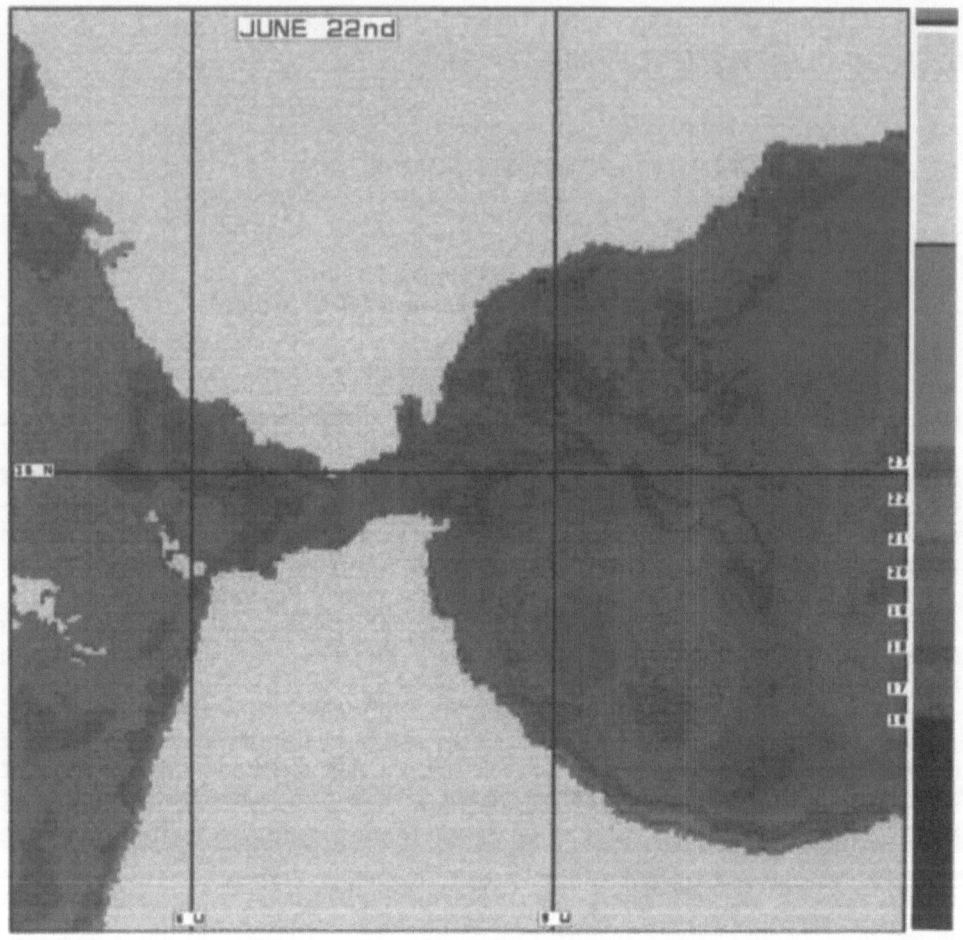

Figure 1.   NOAA-AVHRR Image of the Strait of Gibraltar on June 22, 1986 at 15 H 14 UT. Color
scale is from black (< 16 °C) to red (> 23.5 °C) by 0.5 °C (16 colors).

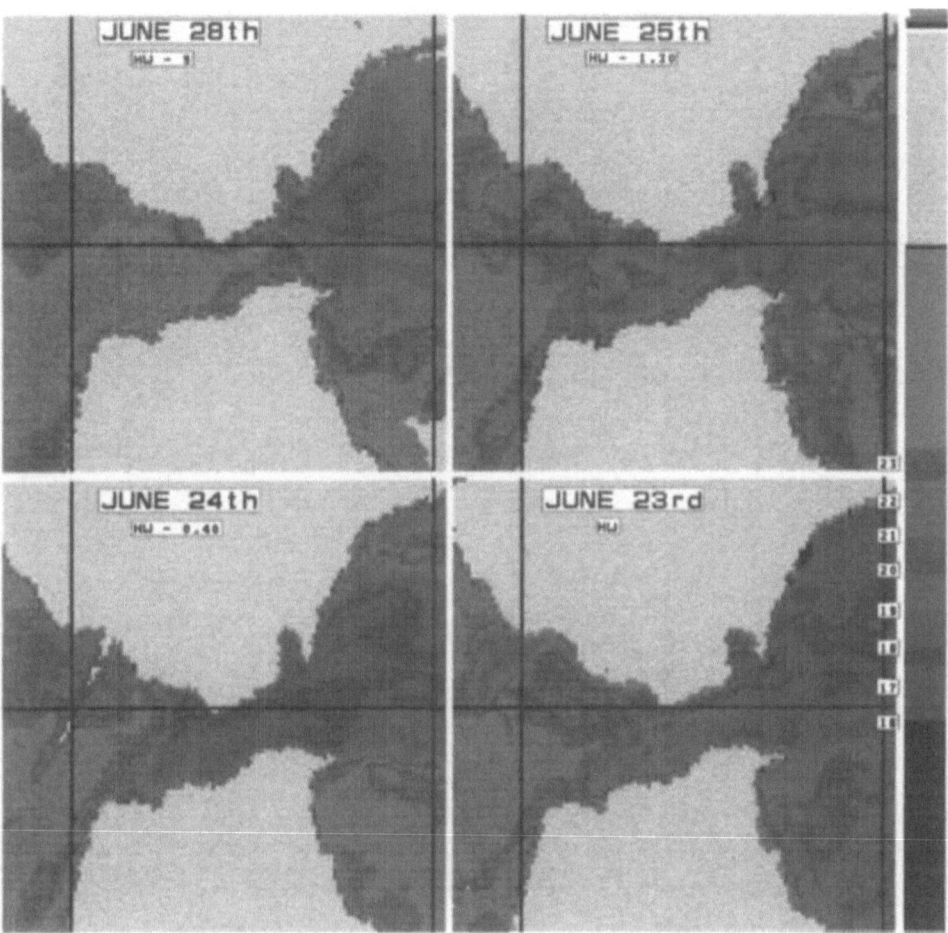

Figure 2. Series of 4 NOAA-AVHRR images of the Strait of Gibraltar between Low Water and High Water. The color scale is the same as in Figure 1. See Table I for the exact references of the images.

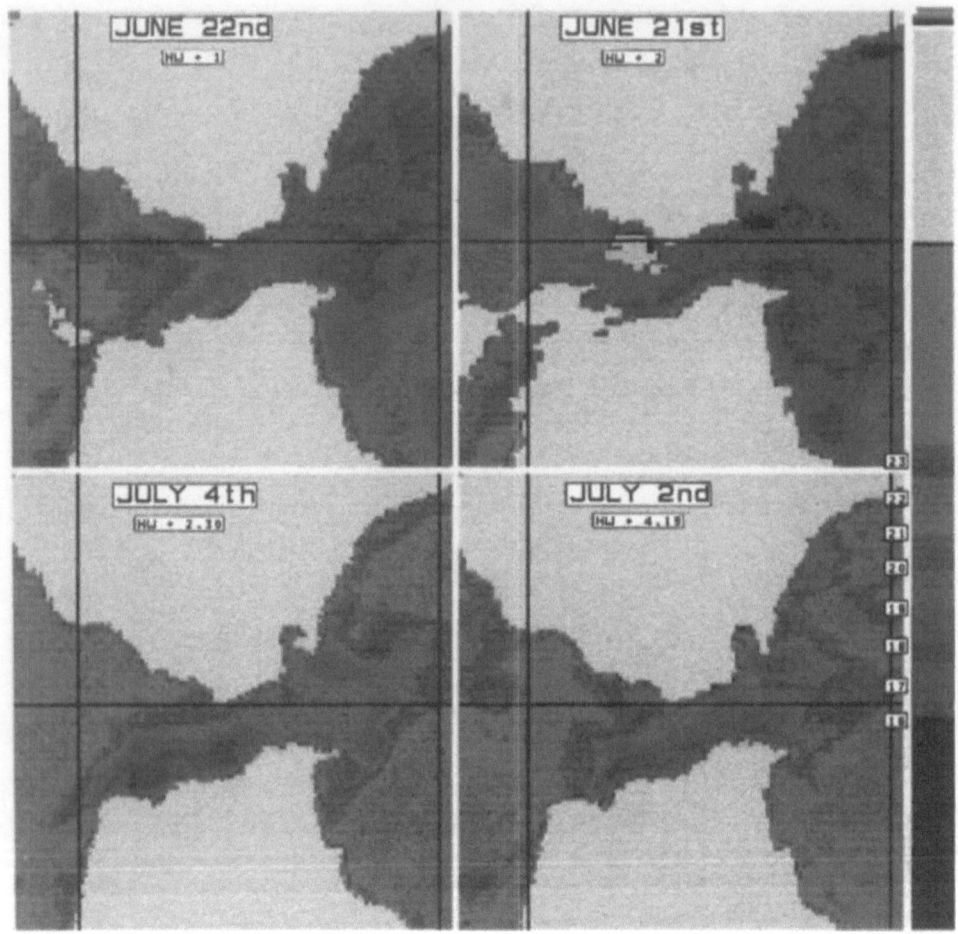

Figure 3.   Series of 4 NOAA-AVHRR images of the Strait of Gibraltar after High Water. The color scale is the same as in Figure 1. See Table I for the exact references of the images.

	Characteristics of NOAA-9 Images						
Orbit	Date	Hour (GMT)	ANL*	ANT*	H/Tide	TidalCoeff	
7846	21/06/86	15.2505	-7.36	15.1635	HW+2.00	88	
7860	22/06/86	15.1413	-4.63	15.0542	HW+1.00	92	
7874	23/06/86	15.0321	-1.89	14.5450	HW	92	
7888	24/06/86	14.5228	0.83	14.4358	HW-0.40	88	
7902	25/06/86	14.4136	3.57	14.3306	HW-1.30	82	
7916	26/06/86	14.3044	6.30	14.2213	HW-2.45	74	
7930	27/06/86	14.2024	9.04	14.1121	HW-3.45	65	
7944	28/06/86	14.0943	11.77	14.0029	HW-5.00	57	
8001	02/07/86	15.0737	-2.80	14.5905	HW-4.15		
8029	04/07/86	14.4604	2.66	14.3720	HW+2.30		

* ANL: Ascending Node Longitude   * ANT: Ascending Node Time

Table I

Figure 1 (June 22, 1986/15H 14) enhances the principal known features encountered in the region of the Strait of Gibraltar. This image shows relatively cool temperatures in the Alboran Sea, with the lowest ones along the Spanish coast and around and south of the Rock of Gibraltar. The anticyclonic gyre of the Alboran Sea is clearly visible, while, west of Gibraltar, relatively higher temperatures are found in the Atlantic Ocean. A core of Atlantic water is flowing through the Strait. This image is a typical situation at the beginning of the eastward phase of the tidal current. Referring to Figure 5, one hour after HW, on June 22, at 15H 14, the current has an eastward component at all depths. Fig. 2 shows a sequence of 4 NOAA images arranged relative to tide from HW - 5 to HW. On June 28, at HW - 5, it is the end of the eastward phase of the tide. This image depicts a classical situation at low water (LW). In the western part of the Strait, a core of warm water denotes a thick upper layer of Atlantic water (the interface between Mediterranean and Atlantic Water is at its lowest position), while in the eastern part of the Strait, gradually decreasing temperatures show the transition between entering Atlantic Water and surface Alboran Sea water.

June 25 image, at HW - 1.30, shows up what could be named a "characteristic feature" as it was already observed by La Violette and Lacombe (1988): Two patches of cold water appear North and South of the Strait, west of the Tarifa meridian. They stand over the continental shelf, north and south of the Camarinal sill region, and in the June 24 image (HW - 0.40), like in the La Violette and Lacombe (1988) October 7 (HW - 2) and October 6 (HW - 1) images, they join to form a cold tongue just over the Camarinal sill. At that time, the interface is at its highest position, with a thick layer of Mediterranean water over the sill, while the tidal current is westward.

On June 24 (HW - 0.40/ 14H 52) a westward current is still strong (see Figure 5) at middle depth, but has turned east at 106 m depth and at 306 m depth. On June 23, at HW/ 15H 03 (see Figure 5), current is eastward at all depths, and Atlantic Water begins to penetrate into the Strait.

Figure 3 shows up a sequence of 4 images during the ebb tidal phase. The first one, at HW + 1 (June 22), shows, as above-mentioned, the beginning of the eastward tidal phase of the current, with the penetration of the Atlantic Water in the Strait, enhanced by higher temperatures in the western part of the Strait.

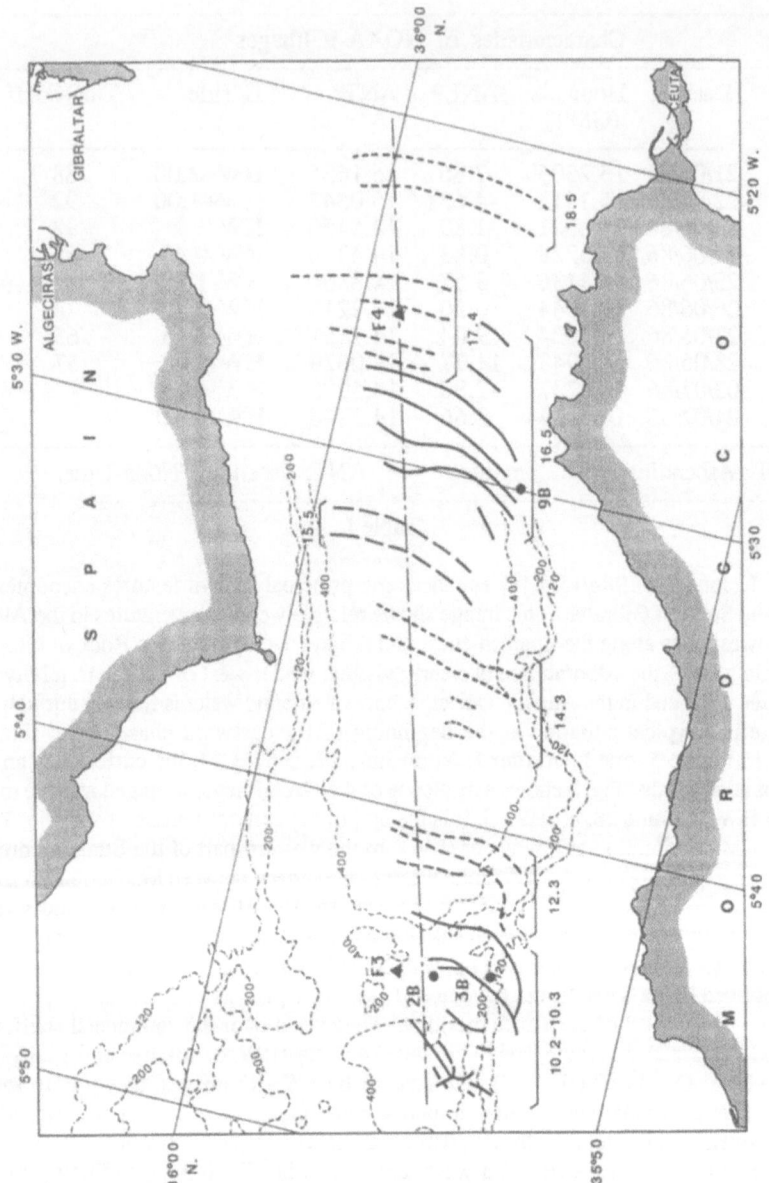

Figure 4. Location of the surface signatures of the internal bore propagating eastwards in the strait of Gibraltar, on June 22, 1986, Images were recorded with a X-band Synthetic Aperture Radar, flying on board an aircraft.
10.2 - 10.3 : 12 H 55 UT; 12.3: 14 H; 14.3: 15 H 07; 15.5: 15 H 37; 16.5: 16 H 21;
17.4: 16 H 47; 18.5: 17 H 23.
Dots: location of moorings (Oregon State University)operating at the same time as the flights
Triangles: location of Armi and Farmer's moorings operating in April 1986

Unfortunately, the three last images cannot be used to characterize permanent or periodic processes usually occurring in the Strait during this part of a tidal cycle, as one (June 21) is cloudy, that affects the surface temperatures, and the two last ones show an anomalous situation over the Alboran Sea with warm temperatures appearing where we usually see an upwelling zone. At the beginning of July, a period of strong levanters (C. Dorman, personal communication, and C. Dorman et al., 1988) could be responsible for this situation, and for the large cold patches appearing in the Strait along the Moroccan coast, on July 2 and 4, denoting upwellings along these coasts. These situations will be investigated later on, in another paper.

As a conclusion, NOAA-AVHRR images allow enhancement of sharp variations of temperatures corresponding to regions of thermal fronts, eddies, gyres, upwellings and their low frequency evolution with time. The resolution is too poor to study quickly evolving phenomena, such as the propagation of internal waves during a tidal cycle, but it gives an interesting synoptic view of the occurrence of oceanic features with a surface thermal signature.

2. Airborne SAR Images :

In 1986, we realized, during the Gibraltar Experiment, an airborne SAR survey of the Strait of Gibraltar, on June 22 and 24, 24 hours apart. We used an X-band synthetic aperture radar, at Springs, overflying the Strait 20 times on June 22 (HW = High Water at Tarifa = 14 H 02 UT, C= 92; 110 to 120 Km long and 10 Km width axis), between HW - 6 to HW + 4.30, and 18 times on June 24 (HW= 3H 12 UT, C= 90; 70 km long axis) between HW and HW + 6 (Richez et al., 1987; Richez, 1988). The results of this experiment will be completely published in a forthcoming paper.

On June 22, the generation of a hydraulic jump (Armi and Farmer, 1988), west of the Camarinal Sill, before HW and during the westward tidal current, is quite well documented through the surface signature of this internal process which evolves with time, closely related to the topography, before passing over the sill, when the westward current relaxes, and propagates eastward as an internal bore, with a speed (relative to bottom) of 2.2 to 2.6 m.s^{-1}. At the surface, its signature is an oblique (NE-SW) double line, about 1300 m distant, which propagates, almost without deformation, along the Strait (Fig. 4). The location of moorings (Pillsbury et al., 1987), working during the SAR flights, is noted in Figure 4. Figures 5 and 6 show data from mooring 2B and Figure 7 shows data from mooring 9B, for June 22, 23 and 24, 1986. In these figures, we delimit the intervals of time during which we were operating with the SAR over the Strait of Gibraltar, on June 22 (SAR I) and June 24 (SAR II). Arrows indicate the times of particular features described in Figures 4 and 8 .

Variation of temperature and pressure at moorings 2B and 9B is reported in Figures 6 and 7. On top of each graph is delta p = p - pmean = departure from the mean pressure, computed for 3 days (22 to 24 June) at the uppermost currentmeter. Temperatures recorded at the moorings show alternating phases of high and low temperatures. High temperatures are the signature of Atlantic Water, while low temperature corresponds to Mediterranean Water, and their alternate occurrence is strongly correlated to the tidal cycle and to the consequent oscillation of the interface between the two principal water masses. It is quite visible, in these data and during these three days, that at one tide out of two, a double oscillation of the interface occurs at the sill of Camarinal, which is signaled by a double increase and decrease of temperature, the second one being sensed by all currentmeters, even the lowest (306 m). In the current data, sharp changes of direction can be noted. On June 22, it is obvious that these changes of direction, at mooring 9B, are related to the passage of the internal

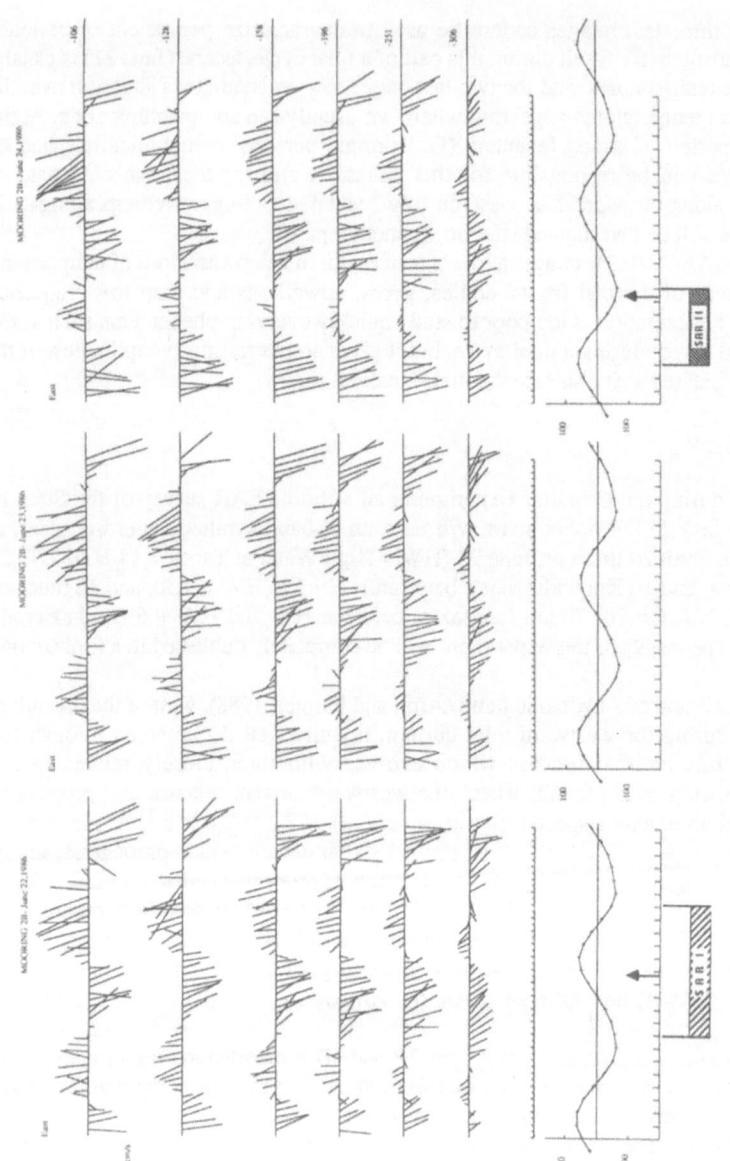

Figure 5.   Stick diagrams of currents recorded at mooring 2B , at the Camarinal Sill, on June 22, 23 and 24, 1986. At the bottom of the figure, bottom pressure at Tarifa. SAR I and SAR II indicate the period during which a SAR was flying over the strait of Gibraltar. The arrow in SAR I indicates the time when images 10.2 and 10.3 (see figure 4) were recorded. The arrow in SAR II indicates the time around which axis 15 and 16 were sampled (see figure 8).

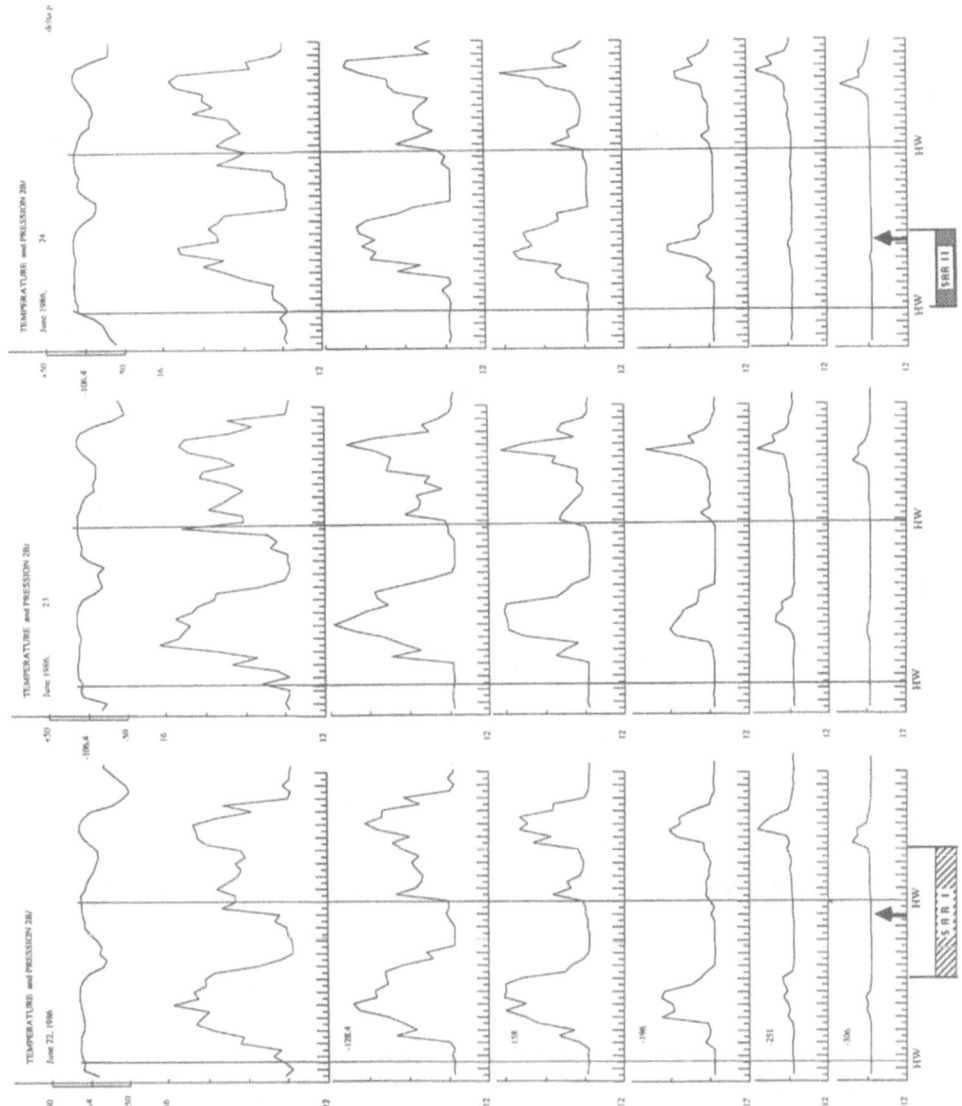

Figure 6. Temperatures recorded at mooring 2B on June 22, 23 and 24, 1986. On the top of the figure is delta p = departure from the mean pressure recorded at the shallowest currentmeter. The arrow in SAR I indicates the time when images 10.2 and 10.3 (see figure 4) were recorded. The arrow in SAR II indicates the time around which axis 15 and 16 were sampled (see figure 8).

450

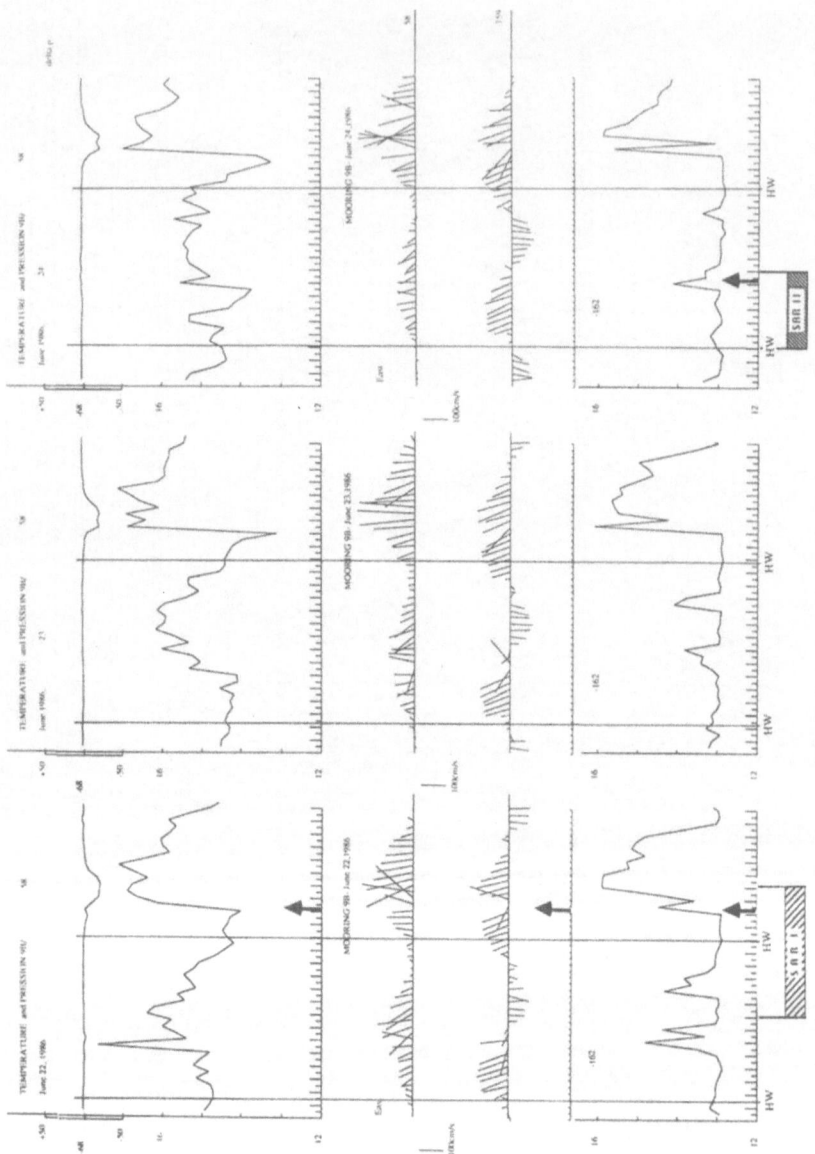

Figure 7. Stick Diagrams and temperature records at mooring 9B, east of Gibraltar. On top of the figure, delta p= departure from the mean pressure recorded at the shallowest currentmeter. The arrow in SAR I indicates the time when image 16.5 (see figure 4) was recorded. The arrow in SAR II indicates the time around which axis 15 and 16 were sampled (see figure 8).

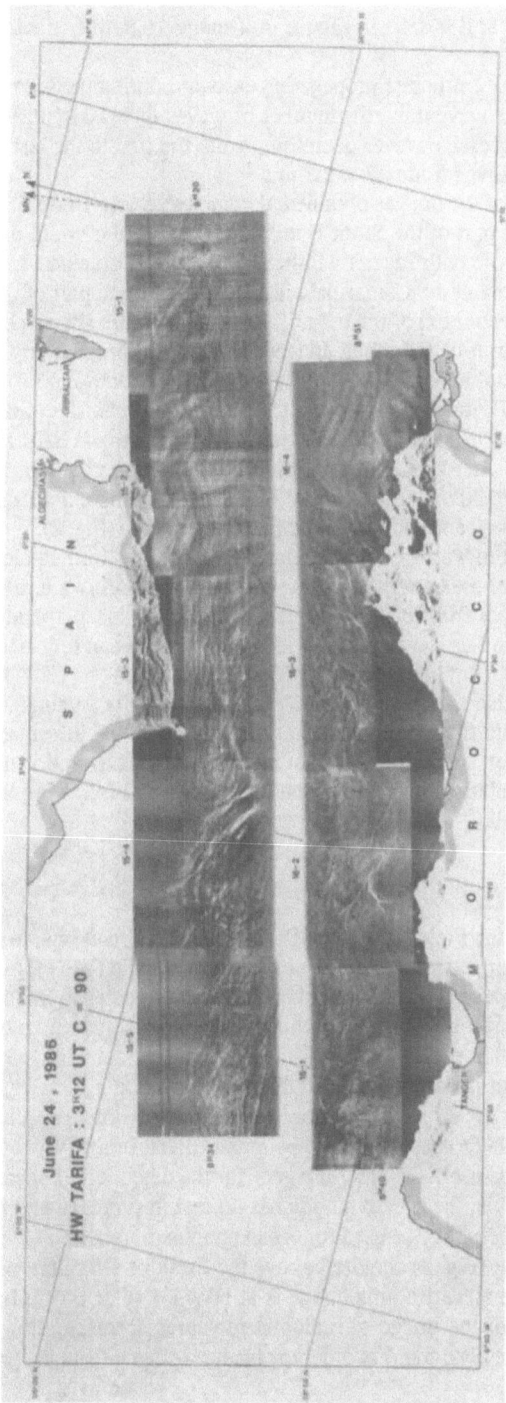

Figure 8: Images of the surface signature of internal features occurring in the Strait of Gibraltar, as recorded by an airborne X-band Synthetic Aperture Radar, on June 24, 1986, between 8 H 20 and 8 H 51.UT, around HW + 5.30

passage of the internal bore, at it is seen in Figure 4 (image 16.5, over the location of mooring 9B).

Quite different is the internal structure propagating eastward during the second flight (June 24). It quickly evolves into a large wave train, the number of waves increasing during the propagation, the distance between two successive waves decreasing from the first to the last. The leading wave speed (relative to bottom) evolves from 1.7 to 2.1 m.s^{-1}.

Fig. 8 shows an example of the images obtained along axes 15 and 16 on June 24, between 8H 20 to 8H 51 UT. The central part of the Strait is missing, because the swath of the radar (10 km) does not allow us to sample the entire Strait width. This picture is obtained around HW + 5. The principal wave train, generated at the Camarinal sill, occupies a large part of the Tarifa Narrows. Nine waves are visible, the resolution of the radar is 10 m. This figure shows clearly how complex is the wave system in the Strait of Gibraltar. Indeed, besides the propagating wave train, in the Tarifa Narrows, a strong signal appears, West of Tarifa (Figure 8, image 15.4), composed of two curved lines oriented NW-SE, joining a structure parallel to the Spanish coast continental shelf. Figure 5 (stick diagram for mooring 2B on June 24, on which an arrow in SAR II indicates the time of Figure 8) shows that, just after HW, the current is SW at all depths (except 128 m) until HW + 2.30 H, then turns NE below 180 m and later (HW + 4) at 100 m, and exhibits sudden changes of direction (with SE components), and oscillations of the temperatures (Fig. 6).

The structure appearing west of Tarifa could be related to lee waves formed along the continental shelf during the south phase of the current, evolving eastward and northward till HW + 6.

This figure reveals also, in front of the principal wave train and perpendicular to it, in the southern part of the Strait, another wave system, which could have been reflected by the Moroccan coast. This will be developed later, in a future paper.

In conclusion, we argue that synthetic aperture radar technique is a valuable tool for studying internal oceanic processes. Its fine resolution clearly reveals a lot of information about internal dynamics. On board an aircraft, surveying a zone of interest, the time evolution dimension is added, giving the possibility of calculating the speed of the wave train, and following the evolution of the internal structure in great detail.

3. SPOT Satellite Imagery:

Launched in February 1986, the French satellite SPOT-1 became operational in May 1986. It is an earth observation system equipped with two High Resolution Visible (HRV) imaging instruments, measuring in one or more spectral bands the solar radiation the solar radiation reflected by the surface. The system can be used in two spectral modes: (1) A panchromatic mode (spectral band: 0.51 μm - 0.73 μm) with a 10 m resolution; (2) A multispectral mode: with green (0.5 - 0.59 μm), red (0.61 - 0.68 μm) and near infra-red (0.79 - 0.89 μm) bands and a 20 m resolution. The most specific characteristics of the HRV imaging sensor is its oblique-viewing capacities. As pointed out by Wadsworth and Piau (1987), these capabilities make SPOT imagery, when acquired under special viewing conditions, as useful as radar imagery for the detection of oceanographic features like internal waves. Unlike this, the spectral mode has no real importance and both panchromatic and multispectral SPOT imagery may be used for oceanographic purposes.

We present here an image (Fig. 9) acquired above the Strait of Gibraltar, on June 18, 1988 at 11H 11 UT. Relative to tide at Tarifa, this image is at HW + 7 (C = 66). It has been processed (selection in the histogram of the image of radiometric values corresponding to the ocean, and enlargement of that part of the histogram) in order to enhance oceanic features, so the land and coast details are not clearly visible.

453

Figure 9: SPOT-1 Image of the Strait of Gibraltar on June 18, 1988 at 11 H 11 UT. Panchromatic mode was used for this image (resolution 10 meters). The image has been processed to enhance oceanographic features to the detriment of land details. Relative to the tide in Tarifa, this image is at HW (4 H 17 UT, C = 66) + 7.

This image will be described in more detail in another paper. In the Tarifa Narrows, a well-developed wave train system, where more than 15 waves are counted, is followed by a second weaker system, of 4 to 5 waves, which could be the signature of the 2nd mode travelling internal bore (Armi and Farmer, 1988). It is preceded, in the northeastern part of the Strait, just south of the Rock of Gibraltar, by a 4 to 5 wave train. A large number of secondary systems, propagating perpendicular to and towards the Spanish continental shelf, can be noted. In particular, we see waves entering the Algeciras Bay, already signaled by G.Watson (1989). This image has to be compared with the SAR image, Figure 8, which is quite similar in some of its details.

We notice also the strong signal, west of Tarifa, which probably has the same origin as the one we signaled on the SAR image, Figure 8. This structure will propagate in the northwestern part of the Strait.

We arrive at the conclusion that this new satellite technique is quite interesting for studying internal ocean dynamics in detail. Its high resolution gives a fine synoptic view of the internal processes of a large zone (60 X 60 km). Its disadvantage, excepting the fact that it is quite expensive, is that it requires cloud-free conditions and that the satellite surveys the same zone only each 26 days, although varying angle of view allows the possibility of obtaining data for a particular zone between J - 5/10 to J + 5/10 (J being the day when the satellite passes directly over the zone of interest). Combining these different techniques and in situ data will obviously give the best results, and we have this opportunity for studying the Strait of Gibraltar.

Acknowledgments: This work has been supported by DRET (Contracts 86-040 and 87-180). The SAR experiment over the Strait of Gibraltar was supported by CNES, DRET, GDTA and IGN. We are grateful to Alain Wadsworth and IFP for their help in the realization of the SPOT images. We thank R. Pillsbury and the University of Oregon for giving us access to the mooring data. Jean Distrophe worked with patience and efficiency to the mounting of the SAR images. We thank SPOT-IMAGE for their help. NOAA-AVHRR images were produced on the SPHINX image processing system at the L.O.A in Lille (France). Thanks also to Larry Pratt and Barbara Gaffron for their careful correction of the manuscript.

4. References:

ARMI, L. and D.M. FARMER (1988).- The flow of Mediterranean Water through the Strait of Gibraltar.- Progress in Oceanography, 21, 1, 105 pp.

DORMAN, C., R. BEARDSLEY, R. LIMEBURNER, J. TAPIA CONTRERAS and B. ABDELAZIZ (1988).- Surface wind and Marine Boundary Layer measurements in the Gibraltar Experiment.- in Seminario sobre la Oceanografia fisica del Estrecho de Gibraltar, Madrid, 24-28 Octubre 1988, Almazan, Bryden, Kinder and Parilla, Eds., SECEG publ., 567 pp., 21-25.

FARMER, D.M. and L. ARMI (1988).- The Flow of Atlantic Water through the Strait of Gibraltar.- Progress in Oceanography, 21, 1, 105 pp.

LACOMBE, H. and C. RICHEZ (1982).- The regime of the Strait of Gibraltar.- Hydrodynamics of semi-enclosed seas, J.C.J. Nihoul, Ed., Elsevier, Amsterdam, 13-73.

LA VIOLETTE, P.E. and H. LACOMBE (1988).- Tidal-induced pulses in the flow through the Strait of Gibraltar.- Oceanologica Acta, N° SP., 13-27.

PILLSBURY, R.D., D. BARSTOW, J.S. BOTTERO, C. MILLEIRO, B. MOORE, G. PITTOCK, D.C. ROOT, J. SIMPKINS III, R.E. STILL and H. BRYDEN (1987).-

Gibraltar Experiment current measurements in the Strait of Gibraltar, October 1985 - October 1986.- Oregon State University, Data Rpt 87-29, Ref 139, December 1987.

RICHEZ, C., D. VAILLANT, A. WADSWORTH (1987).- Airborne SAR Experiment in the Strait of Gibraltar. GIBEX: 22-24 June 1986.- LODYC-CNES-GDTA Int. Rept, LODYC/87/01, April 1987.

RICHEZ, C. (1988).- Airborne Synthetic Aperture Radar observations of internal waves propagation in the Strait of Gibraltar.- in Seminario sobre la Oceanografia fisica del Estrecho de Gibraltar, Madrid, 24-28 Octubre 1988, Almazan, Bryden, Kinder and Parilla, Eds., SECEG publ., 567 pp., 482-486.

WADSWORTH, A. and P. PIAU (1987).- SPOT, a Satellite for Oceanography?.- Proc. IGARSS'87, Ann Arbor.

WATSON, G. (1989).- Internal waves in the Strait of Gibraltar: A study using radar imagery.- Ph.D. Thesis, Department of Oceanography, University of Southampton

THE BAROTROPIC TIDE IN THE STRAIT OF GIBRALTAR

JULIO CANDELA
Center for Coastal Studies, A-009
Scripps Institution of Oceanography
La Jolla, California 92093

ABSTRACT. The Strait of Gibraltar represents the transition between two very different tidal regimes, that in the North Atlantic where the tidal ranges exceed 2 m and that in the Mediterranean where the ranges are less than 1 m. The surface tide within the strait is principally forced by the Atlantic tide and has a strongly semidiurnal character, i.e. ~96% of the pressure and ~74% of the current variability observed are contained within this band. The cotidal chart for the principal semidiurnal constituent M_2 presents a complicated diffraction pattern within the strait. Corange lines (lines of equal amplitude) are aligned perpendicular to the axis of the strait with a >50% magnitude decrease between Atlantic and Mediterranean sides. Cotidal lines (lines of equal phase) show a southwestern propagation emanating from a point source around Tarifa, implying a ~40 min time delay between maximum elevation at the north and south shores around the longitude of Tarifa. Using simultaneous pressure and current observations, it is shown that this apparent complicated behavior of the surface tide is the manifestation of simple, first order, along- and across-strait momentum balances.

1. Introduction

The Strait of Gibraltar connects the Mediterranean Sea with the North Atlantic ocean. It has a length of ~60 km and a minimum width (Pt. Cires section) of ~14 km. The main sill (Camarinal sill) is located between Pt. Paloma (Spain) and Pt. Altares (Morocco) with a maximum depth of ~300 m (Figure 1). East of the sill the strait deepens rapidly to ~600 m south of Tarifa and to 900 m at the Gibraltar-Ceuta section. West of the sill the depths are ~400 m north of Tangier, but shallowed to 350 m at a secondary sill north of Cape Spartel, to then slope gently towards the deep North Atlantic.

As first suggested by Lacombe and Richez [1982], a useful conceptual way to classify the flow variability through the strait is into three distinct types: Long-term, subinertial and tidal. Long-term flows relate to the two-way baroclinic exchange between upper layer inflowing Atlantic and lower layer outflowing Mediterranean waters. Since the variability of these long-term exchange flows are of very low frequency, i.e. seasonal to interannual periodicities, most of the steady or quasi-steady hydraulic two-layer flow theories that are dicussed in this volume are applicable to this type of flows. Caution should be exercised, however, to make sure that the strong time variability imposed at subinertial and tidal frequencies does not violate any of the main assumptions of the theory applied. Bryden and Stommel [1984] and Bormans et al. [1986] discuss the forcing mechanisms for these flows, which have an order of magnitude of 1 Sv in each layer.

L. J. Pratt (ed.), The Physical Oceanography of Sea Straits, 457–475.
© 1990 *Kluwer Academic Publishers.*

STRAIT of GIBRALTAR

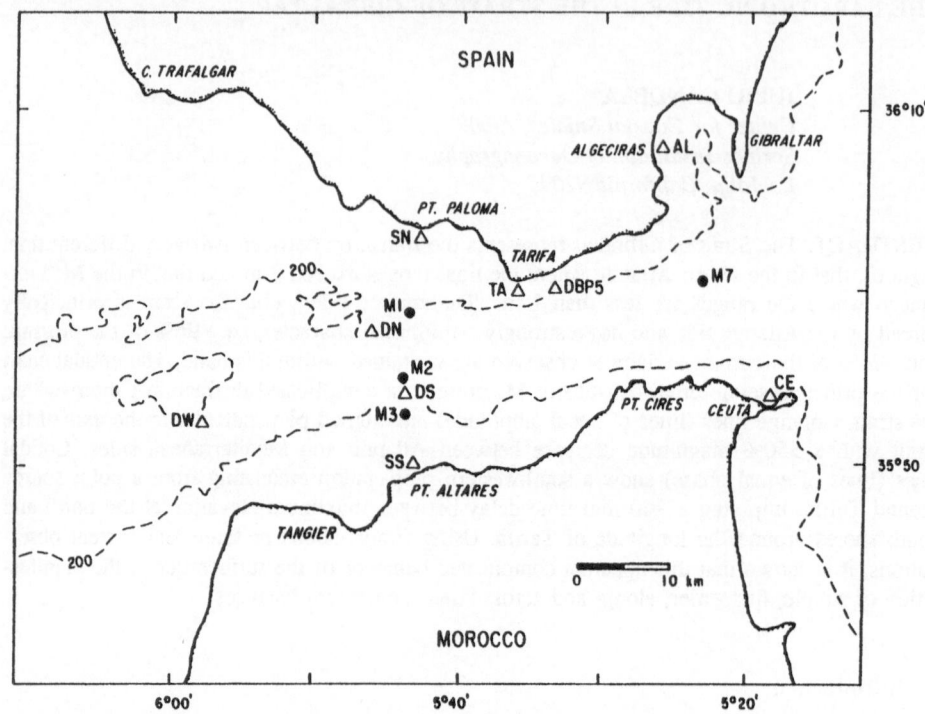

Figure 1. Map of the Strait of Gibraltar showing the locations of places referred to in the text. Also shown are the locations of the bottom pressure sensors (Δ) and the current meter moorings (•) used in this paper.

Subinertial flows, with periods from days to a few months, are mainly barotropic with an rms value of 0.4 Sv and have been shown to be forced principally by atmospheric pressure fluctuations over the Mediterranean sea (Candela et al. [1989a]).

Tidal flows have a dominant barotropic behavior in the strait, with an rms value of 2.1 Sv and and are forced principally by the North Atlantic tide (Candela et al. [1989b]).

Although this flow classification is helpful to guide our criterion about the flow characteristics in the strait, it is too simplified in the sense that it overlooks the appreciable interactions that are observed among the distinct flow types (Bryden et al. [1988], Candela et al. [1989a,b]).

This paper concentrates on describing the main kinematic and dynamic behavior of the surface tide in the strait, as it is observed through the analysis of pressure and current data gathered during the recent Gibraltar Experiment (Kinder and Bryden [1987]).

2. Data and Methods used

During the Gibraltar Experiment (October 85- October 86), pressure, currents, salinity and temperature time series were recorded at several locations in the strait. Here data from nine subsurface, bottom-mounted pressure instruments, and currents from moorings M1, M2 and M3 at the sill and M7 at the eastern end (Figure 1), will be looked at in detail. Tables 1 and 2 give deployment information for the pressure and current meter data, respectively.

Station	Depth (m)	Latitude North	Longitude West	Start	Stop	Type of Instrument
DW	260	35°53'	05°58'	October 25, 1985	March 17, 1986	BP.SIO
SN	11	36°03'	05°43'	October 22, 1985	January 26, 1986	BP.SIO
				March 4, 1986	August 6, 1986	
DN	210	35°58'	05°46'	October 26, 1985	March 3, 1986	BP.SIO
DS	210	35°54'	05°44'	October 27, 1985	March 21, 1986	BP.SIO
SS	12	35°50'	05°43'	April 17, 1986	August 20, 1986	BP.SIO
DP5	160	36°00'	05°34'	October 19, 1985	March 19, 1986	BP.UNH
TA	5	36°01'	05°36'	October 10, 1985	November 14, 1985	AA.IHM
				December 16, 1985	February 20, 1986	
				March 11, 1986	July 16, 1986	
AL	5	36°08'	05°26'	October 10, 1985	May 1, 1986	AA.IHM
				May 12, 1986	June 26, 1986	
CE	5	35°53'	05°18'	October 18, 1985	May 9, 1986	AA.IHM
				January 1, 1986	August 7, 1986	STG.IHM

ABBREVIATIONS: DW - deep west, SN - shallow north, DN - deep north, DS - deep south, SS - shallow south, DP5 - Doppler acoustic logger #5, TA - Tarifa, AL - Algeciras, CE - Ceuta, BP.SIO - bottom pressure from Scripps, BP.UNH - bottom pressure from the University of New Hampshire, AA.IHM - Aanderaa bottom pressure from Instituto Hidrografico de la Marina, STG.IHM - Standbar tide gauge from IHM.

TABLE 1. Information for the bottom pressure measurements obtained during the Gibraltar Experiment. Station name, depth, location, start and stop times of observation and instrument type are listed.

Station	Water Depth (m)	Latitude North	Longitude West	Start	Stop	Depth of Instrument (m)
M1	222	35°58.26'	05°44.62'	October 22, 1985	May 4, 1986	143
				October 22, 1985	May 4, 1986	156
				October 22, 1985	May 4, 1986	167
				October 22, 1985	May 4, 1986	215
M2	321	35°54.79'	05°44.41'	October 22, 1985	November 23, 1985	123
				October 22, 1985	November 23, 1985	143
				October 22, 1985	November 23, 1985	153
				October 22, 1985	November 23, 1985	191*
				October 22, 1985	November 23, 1985	254
				October 22, 1985	November 23, 1985	306
M3	190	35°53.42'	05°44.20'	October 21, 1985	April 21, 1986	110
				October 21, 1985	April 21, 1986	140
				October 21, 1985	April 21, 1986	180
M7	916	35°59.98'	05°22.75'	October 19, 1985	March 27, 1986	54
				October 19, 1985	March 27, 1986	193

* This instrument had a malfunctioning rotor and did not record speed, but the direction, pressure, temperature, and conductivity records are of good quality.

TABLE 2. Information for some of the current meter measurements obtained during the Gibraltar Experiment. Mooring name, water depth, location, start and stop times of observation and depth of instrument are indicated. These data were provided by Harry Bryden from WHOI and are described in Pillsbury et al. [1987].

Standard methods for time series analysis were used to obtain basic statistics (Bendat and Piersol [1986], Godin [1978]). On average, power spectra of the pressure data indicate that 96.1% of the variance is contained in the semidiurnal band (2 cpd), while low frequencies (<.5cpd) account for 1.6%, and the diurnal band 1%. The variance of the current field, obtained by vector power spectra (Table 3), indicates that 78% is at semidiurnal, 11.6% at diurnal and 6.6% at low (<.5cpd) frequencies.

To look more closely into the structure within the tidal bands, harmonic analyses (Godin [1972]), verified by admittance calculations (Godin [1976] and Godin [1978]) using Cadiz, Deep North or Ceuta pressure as input, were performed on the pressure and current data. Table 4 gives the amplitude and phase for the two principal constituents in the semidiurnal band, i.e. M_2 and S_2, and Table 5 gives the constituent ellipse of the current series for these two frequencies.

M_2 is the principal lunar semidiurnal constituent at a frequency of 28.98 degrees per hour. In the strait it represents more than 70% of the energy content in the semidiurnal band and can be taken as the best single frequency representative of the tidal motions. Even though M_2 accounts for most of the tidal variability observed, restricting a description of the tides to it would imply leaving out subtleties of the tidal behavior, due to interactions between

Sta.	Depth (m)	Length of series (hrs)	Mean Amp. (cm/s)	Mean Orien.	Subinertial M (cm/s)	Subinertial m	Subinertial θ	Diurnal M (cm/s)	Diurnal m	Diurnal θ	Semidiurnal M (cm/s)	Semidiurnal m	Semidiurnal θ
M1	143	4657	16	−144°	31	1	−5°	35	−1	−11°	92	−6	−8°
	156	4157	19	−164°	28	1	−18°	32	−1	−26°	86	−6	−20°
	167	4649	20	−150°	27	1	−8°	31	−2	−16°	83	−6	−9°
	215	4657	12	168°	14	1	−21°	16	0	−17°	48	−2	−16°
M2	123	766	11	−117°	46	−1	13°	49	0	11°	130	−4	12°
	143	766	17	−150°	33	1	5°	47	0	13°	116	−5	12°
	153	766	28	−147°	35	1	10°	51	−2	13°	127	−8	15°
	254	766	53	−150°	18	0	25°	35	1	26°	88	1	30°
	306	766	38	−134°	18	−1	18°	29	1	41°	65	2	43°
M3	110	4358	13	51°	43	0	−9°	48	3	6°	123	−3	9°
	140	4358	6	138°	38	1	−2°	40	1	6°	111	−5	12°
	180	4358	25	−157°	27	−1	19°	26	0	20°	71	−4	22°
M7	54	3808	54	7°	55	2	10°	30	−6	9°	36	−5	14°
	193	3808	22	−152°	15	−1	35°	16	1	25°	39	2	26°

TABLE 3. Mean current and vector power spectral calculations for the current meter measurements used in the paper. Spectral estimates are for a frequency band resolution of 1cpd and are expressed by the ellipse parameters of the first 3 bands resolved, i.e. subinertial, diurnal and semidiurnal. M is the semimajor axis, m the semiminor axis and θ the orientation of M in degrees with respect to east. The sign of m indicates the mean sense of rotation of the currents within the band, positive values implying a counterclockwise mean rotation.

constituents, which are likely to occur, and be appreciable, in places with complicated topography and large tidal amplitudes, such as Gibraltar. To avoid this common bias of just looking at the structure of the predominant constituents, it was also decided to isolate the complete tidal signal in the pressure and currents, by applying the PL64 filter (Flagg et al. [1976]), and using the resulting series to investigate the dynamics of the tidal motions in the strait.

When inter-comparing pressure and/or current observations, as when evaluating the terms in the momentum equations, it is helpful to discern that part of the variability which is common to all simultaneous measurements within the study area, from the part which is local or due to individual noise. One method for attaining this goal is the Empirical Orthogonal Function analysis (EOF) (Kundu et al. [1975]). In the next section EOF analysis is used to obtain the principal modes constituting the subsurface tidal pressure field in the strait (when applied to the 9 simultaneous high-pass (>.5cpd) pressure observations), and the principal tidal current structure modes across the sill section (when applied to the 12 available high-pass principal axis simultaneous currents from moorings M1, M2 and M3 at the sill). See Candela et al. [1989b] for details of the analysis procedure.

Station (Length of Series, Hrs.)	M$_2$		S$_2$	
	Amp.	Phase	Amp.	Phase
DW (3414)	78.5	56.1°$_+$	29.0	82.2°$_{DN}$
SN (3464)	52.3	47.6°$_+$	18.5	73.4°$_{DN}$
DN* (3257)	60.1	51.8°$_+$	22.5	73.8°$_+$
DS* (3459)	54.0	61.8°$_+$	21.1	83.3°$_{DN}$
SS (2991)	57.1	66.8°$_+$	20.6	92.3°$_{DN}$
DP5 (3406)	44.4	47.6°$_+$	16.1	73.9°$_{DN}$
TA (3043)	41.2	41.2°$_+$	14.7	67.9°$_{DN}$
AL (4857)	31.0	48.0°$_+$	11.1	73.9°$_{DN}$
CE (8760)	29.7	50.3°$_+$	11.4	75.6°$_+$

Amplitude in millibars ($\approx$ cm), phase in degrees with respect to GMT.

$_+$ Direct harmonic analysis, with inference to separate K$_2$ from S$_2$ using their relation at Cadiz.

$_{DN}$ Admittance with deep north predictions.

* Corrected for internal contribution.

TABLE 4. Principal semidiurnal tidal constituents for the observed bottom pressure records. The specific type of analysis used to obtain the tidal constituents is indicated in each case and explained in Candela et al. [1989b]. Amplitudes are in mb (~cm) and phases are in degrees with respect to GMT.

Station	Depth of Instrument (m)	M_2 M (cm/s)	m (cm/s)	θ	g	S_2 M (cm/s)	m (cm/s)	θ	g
M1	143	85	-5	353°	133°	31	-2	353°	159°
	156	81	-6	340°	133°	29	-2	340°	159°
	167	77	-6	351°	132°	28	-2	351°	158°
	215	44	-2	344°	124°	16	-1	344°	150°
M2	123	112	-3	11°	154°	40	-1	11°	180°
	143	106	-6	14°	154°	38	-2	14°	180°
	153	113	-9	17°	153°	41	-3	17°	179°
	254	79	1	30°	133°	28	0	30°	159°
	306	58	3	43°	133°	21	1	43°	159°
M3	110	115	-3	9°	146°	41	-1	9°	172°
	140	104	-5	12°	148°	37	-2	12°	174°
	180	66	-3	12°	145°	24	-1	12°	171°
M7	54	21	-6	20°	208°	8	-2	20°	234°
	193	37	2	26°	139°	13	1	26°	165°

TABLE 5. Principal semidiurnal tidal ellipses for the current meter observations listed in Table 2. Admittance calculations with Deep North pressure station predictions have been used to separate K_2 from S_2. M, m and θ are as in Table 3. The phases (g), which indicate the time of maximum flood current (into the Mediterranean), are in degrees with respect to GMT.

3. Results

Based on the values listed on Table 4, the cotidal chart for M_2 can be constructed (Figure 2). There are two main features to be noted on this chart. The M_2 amplitude decreases more than 50% from the Atlantic to the Mediterranean sides, but is otherwise uniform across the strait. In contrast, the lines of constant phase (cotidal lines) are oriented east-west along the channel, implying ~40 minute time difference between the time of maximum amplitude at the north and south shores around the Tarifa section. According to the theoretical work of Rocha and Clarke [1987] on the tidal behavior in a strait connecting to semi-infinite oceans, some of their conclusions are: i) If the strait is narrow enough (i.e. width (W)<<Rossby radius (Ro)), sea level constants change linearly from one end to the other, with steep gradients within the strait in the case of differing tides in each ocean. ii) For a short strait (i.e. length/W~1), because of end effects, strong gradients in sea level occur across the strait even when W<<Ro. iii) The deep sea tide "sees" the strait as a point source at a distance greater than 6W from the center point of the entrance to the strait. For Gibraltar Ro~1000 km, length/W~5 and the tides at both sides differ appreciably, so that the points above seem to apply and give some theoretical explanation to the observed M_2 pattern.

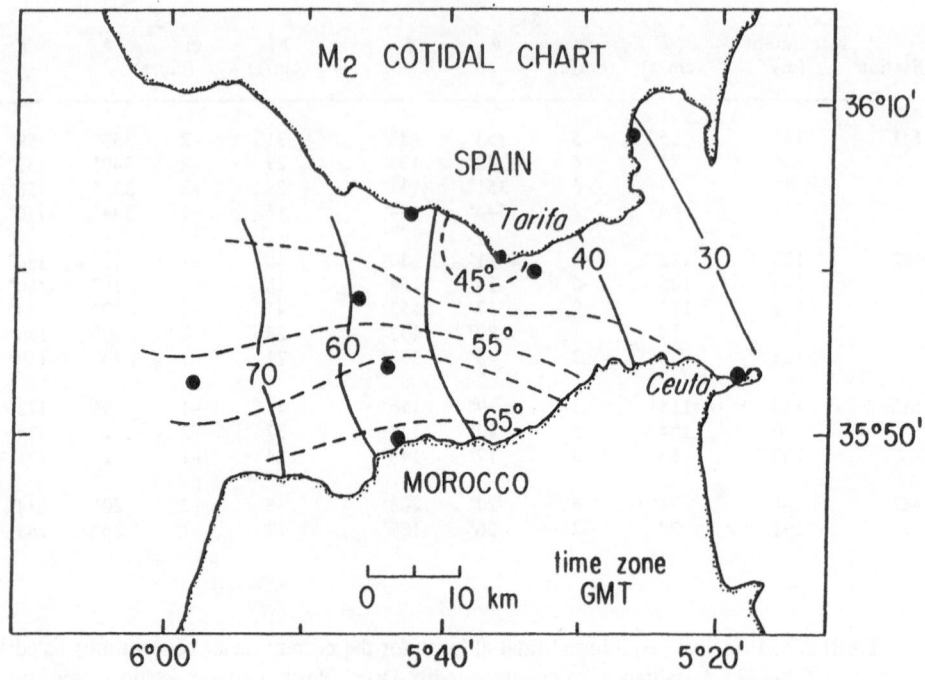

Figure 2. Cotidal chart for M_2 (frequency = 28.98°/h), principal lunar semidiurnal tidal constituent. Deduced from analysis of 9 bottom pressure stations (•). Amplitude lines (continuous) are in mb. Phase lines (dashed) are in degrees with respect to GMT.

Our present concern is to compare the pressure and current observations in the strait, so as to discern the principal dynamical balances that manifest themselves in such a sea surface pattern. Disregarding frictional and non-linear effects, the along-strait pressure gradient should be balanced by the local acceleration of the flow and the across-strait pressure gradient by the flow itself, through geostrophy. If these two balances hold, it may be useful to decompose the pressure field into orthogonal modes, since u and $\partial u/\partial t$ are orthogonal. Empirical orthogonal function analysis (EOF), which decomposes a field of observations into orthogonal components, is thus applied to the simultaneous highpassed (frequency $\sigma > .5$cpd) pressure data.

The first mode that results from the EOF decomposition (Figure 3a), represents 97.6% of the tidal variance and consists of a standing wave pattern that has an along-strait amplitude gradient decaying toward the Mediterranean. The spatial weights contoured in Figure 3a represent the rms value of the fluctuations of the tide associated with this mode. The second mode (Figure 3b) explains only 1.8% of the total variance, and its spatial weights describe an across-strait standing wave with a zero crossing running through the center of the strait. This EOF analysis of the pressure field greatly simplifies the description of the tides in the strait,

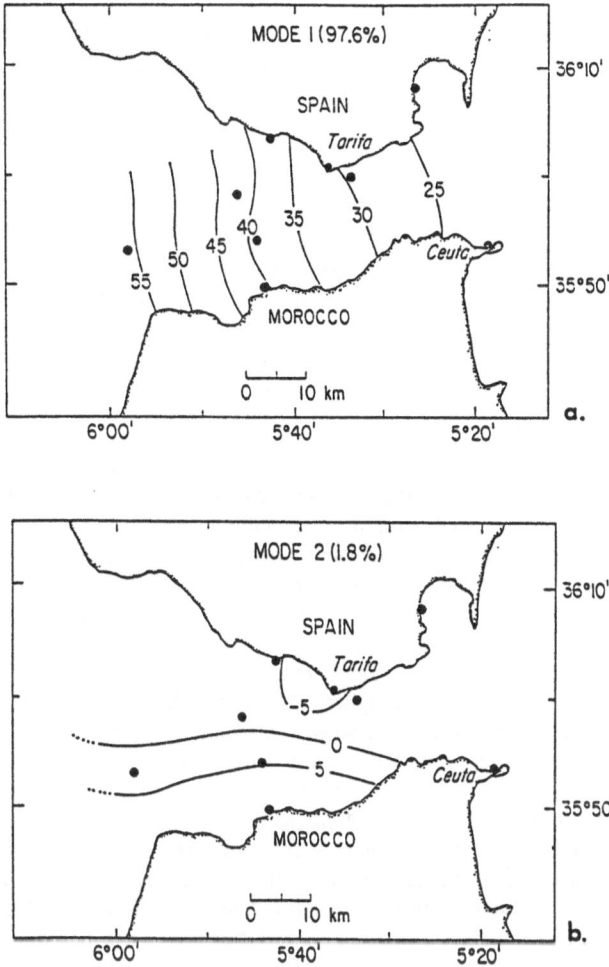

Figure 3. Empirical Orthogonal Function (EOF) analysis of the tidal pressure field observed in the strait. a) Contours of the ~patial structure of Mode 1 representing
97.6% of the total tidal variance. b) Spatial structure of Mode 2 representing 1.8%
of the variance. Contour values for both modes are in mb and represent the standard deviation of the pressure field associated with each mode. Nine simultaneous
pressure stations (•), covering a period of ~2 months (1447h) were used in the
analysis.

and with the use of these two modes it is possible to describe the behavior of any of the principal tidal constituents.

A harmonic analysis of the time evolution of the two modes (Figure 4) provides the amplitude and phase of the M_2 constituent (or any other principal constituent) associated with each
mode. Mode 1 has an M_2 amplitude of 1.37 and a phase of 53.8° and when multiplied by
the weights of Figure 3a, gives the main M_2 signal in the strait. The M_2 signal thus obtained

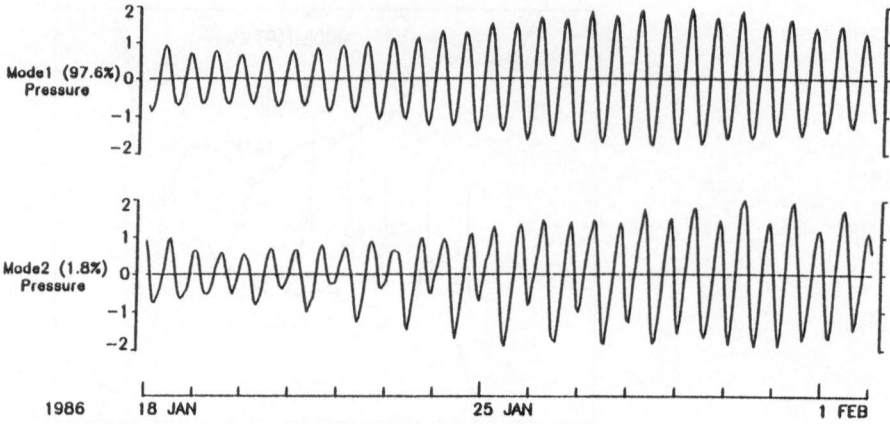

Figure 4. Series of the time coefficients for the two EOF modes of the pressure field shown in Figure 3. The coefficients are chosen to be dimensionless, but when multiplied with the appropriate spatial weights, reproduce the dimensional time series of the pressure fluctuations due to each mode. Although two months of simultaneous observations were used in the analysis, only a 15 day period is shown for clarity.

shows nearly the same along-strait amplitude gradient as that of Figure 2, but indicates a simultaneous M_2 tide throughout the strait with a phase of 53.8°. The addition of the contribution due to mode 2, with an M_2 amplitude of 1.27 and phase of 142.6°, only slightly alters the M_2 amplitude, but provides the across-strait phase gradient observed, reproducing the M_2 cotidal chart of Figure 2. This reconstruction of M_2, based on the two principal empirical modes, indicates that at least the spatial structure of these modes could have been deduced directly from the cotidal chart of Figure 2. If instead of plotting the amplitude and phase of M_2 at each location, plots of its real and imaginary parts are made, nearly identical patterns to those of Figures 3a and 3b are obtained. Both forms of decomposition set the principal tidal field as being composed of two standing wave patterns orthogonal to each other: one with an amplitude gradient along the strait and the other with a cross-strait gradient with a zero crossing in the middle.

To determine how these pressure modes are coupled dynamically with the current field, observations from M1, M2 and M3 at the sill section, and from M7 at the eastern end, are considered. Low-frequency fluctuations, with periods longer than a day, are eliminated by subtracting their contribution from the raw observations of the u-east and v-north current components. The principal axis components of the high-pass time series are then computed and listed in Table 6.

Instead of comparing the pressure field to individual point measurements, it is more appropriate to find covarying modes of fluctuation for the tidal sill currents using EOFs. The first mode explains 92.3% of the variance and its weights, which are the rms value of the currents due to this mode at each location, are listed in Table 6 and shown in Figure 5a. This figure also illustrates the cross section of the sill and indicates the location of the current

Mooring	Depth of Instrument (m)	Orientation	Principal Axis Component Semi Major Axis (cm/s)	Semi Minor Axis (cm/)s	EOF Mode 1	Mode 2
M1	143	-8.5°	71.0	12.1	73.5	-14.2
	156	-21.4°	66.8	11.6	68.6	-20.7
	167	-10.4°	64.1	11.9	64.8	-21.9
	215	-16.2°	37.2	8.4	34.0	-16.4
M2	123	12.2°	98.9	14.7	96.4	19.8
	143	12.1°	89.1	15.5	86.4	17.5
	153	15.3°	97.4	16.0	94.8	15.8
	254	29.1°	68.8	13.6	65.9	-15.8
	306	41.4°	52.4	12.8	48.9	-14.6
M3	110	8.7°	92.9	13.2	90.9	3.9
	140	10.9°	83.6	14.3	82.5	6.0
	180	21.2°	55.8	14.1	49.3	8.5
					92.3%	4.1%
M7	54	10.7°	31.0	14.4		
	193	25.9°	29.1	6.7		

TABLE 6. Main statistics of the tidal (highpassed σ>.5cpd) currents measured at Gibraltar's main sill (M1, M2 and M3) and eastern end (M7). The orientation of the principal axis tidal currents is measured counterclockwise from the east. The semimajor and semiminor axis are denoted by the rms value of the current (cm/s) along that axis. The spatial weights for the principal axis tidal current EOF modes are shown and the variance represented by each mode is also indicated.

meters on each mooring. This mode, where tidal currents fluctuate everywhere in phase, with some amplitude variation, is the closest approximation to a "barotropic" tidal current mode at the sill section. A direct average of the weights in the section gives 0.71 ms^{-1}, which when multiplied by the cross-sectional area at the sill (2.951 10^{6}m^2) gives an estimate of 2.1 Sv for the rms value of the tidal transport through the strait. The time coefficients of this first mode are shown in Figure 6. The second mode of the EOF analyses accounts for 4.1%. Its spatial structure (Figure 5b and Table 6) resembles a first baroclinic mode with a node coinciding with the approximate mean location of the interface between Atlantic and Mediterranean waters. The time evolution of this mode is quite erratic (Figure 6), which might indicate that the available data is not adequate for completely resolving its structure or time behavior. However, this second mode is the manifestation of the internal tide at the sill section and the EOF analysis has provided a quantitative estimate of its contribution to the currents sensed at the section.

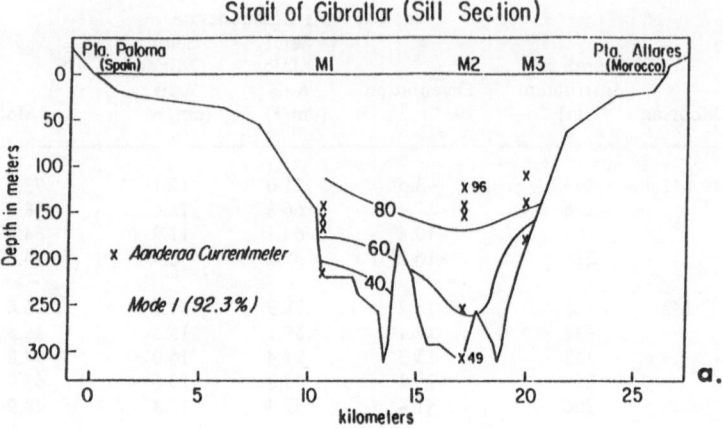

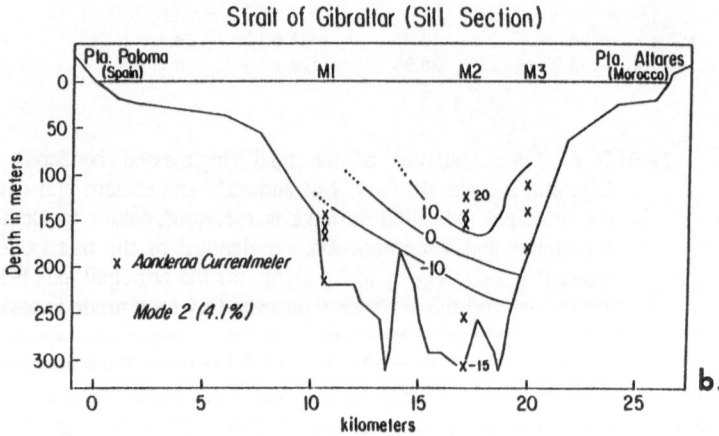

Figure 5. Spatial structure of the EOF analysis modes of the principal axis tidal currents at Gibraltar's main sill section. a) Mode 1 explaining 92.3% of the variance at these frequencies ($\sigma > .5$cpd). b) Mode 2, which explains 4.1% of the total variance contained in the tidal current field. Contours are in cm/s and relate to the standard deviation of the tidal currents associated with each mode.

The mode 1 illustrated in Figure 5a is taken to represent the spatial structure of the main tidal current signal at the sill and its time behavior as that shown on Figure 6. This current mode will be used in conjunction with the pressure field modes previously described, to verify the simple dynamical balances assumed for the tides in the strait. This mode of the

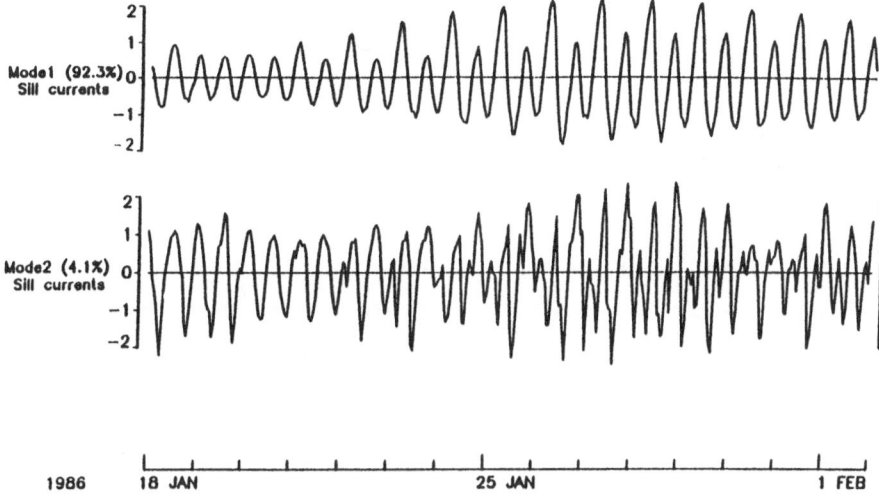

Figure 6. Time coefficients of the two EOF modes of the principal axis tidal currents shown in Figure 5, chosen to be dimensionless. Only a 15 day period is shown for clarity.

currents at the sill has an M_2 phase of 140.2° and an amplitude of 1.24. The phase difference of nearly 90° between the main current and pressure modes at the sill supports the idea of a standing wave behavior for the M_2 constituent in the strait. It is interesting to notice that in the strait, ebb current (toward the Atlantic) precedes high water, behaving as if a node for each of the principal semidiurnal constituents was present inside the Mediterranean. The 90° phase relation is also indicative of a nearly zero tidal energy flux through the strait.

To investigate the along-strait momentum balance, the along-strait pressure gradient is estimated by fitting a plane of the form: $P = a_0 + a_1 x + a_2 y$ (where P is pressure, x and y the east and north Cartesian coordinates of the stations location, and the a's fitting coefficients), to the weights of mode 1 corresponding to the four pressure stations located over the sill area, i.e., SN, DN, DS and SS. The results give $a_1 = \partial P/\partial x = -0.1067\ 10^{-2}$ mb/m and $a_2 = \partial P/\partial y = -0.9251\ 10^{-4}$ mb/m, so the local maximum pressure gradient has magnitude of $1.071\ 10^{-3}$ mb/m and is oriented -175° with respect to east. If this is taken as representative of the local along-strait pressure gradient, then the term $-1/\rho\ \partial P/\partial x_0$ (where $\rho = 1028\ Kg/m^3$ and x_0 is the along-strait coordinate) has an rms value of $10.4\ 10^{-5}\ m/s^2$, and when multiplied by the mode 1 time coefficients (Figure 4) gives a time series for this term. To estimate $\partial u/\partial t$ the mean value of mode 1 of the sill currents (71.32 cm/s) Figure 5a, is multiplied by its time coefficients, and a time derivative of the resulting series is calculated using splines [Forsythe et al., 1977]. The comparison between the two terms (Figure 7a) is remarkable; the correlation coefficient between the two series is 0.94. Considering that the current and pressure fields being compared are independent measurements, the agreement underlines the quality of both sets of observations.

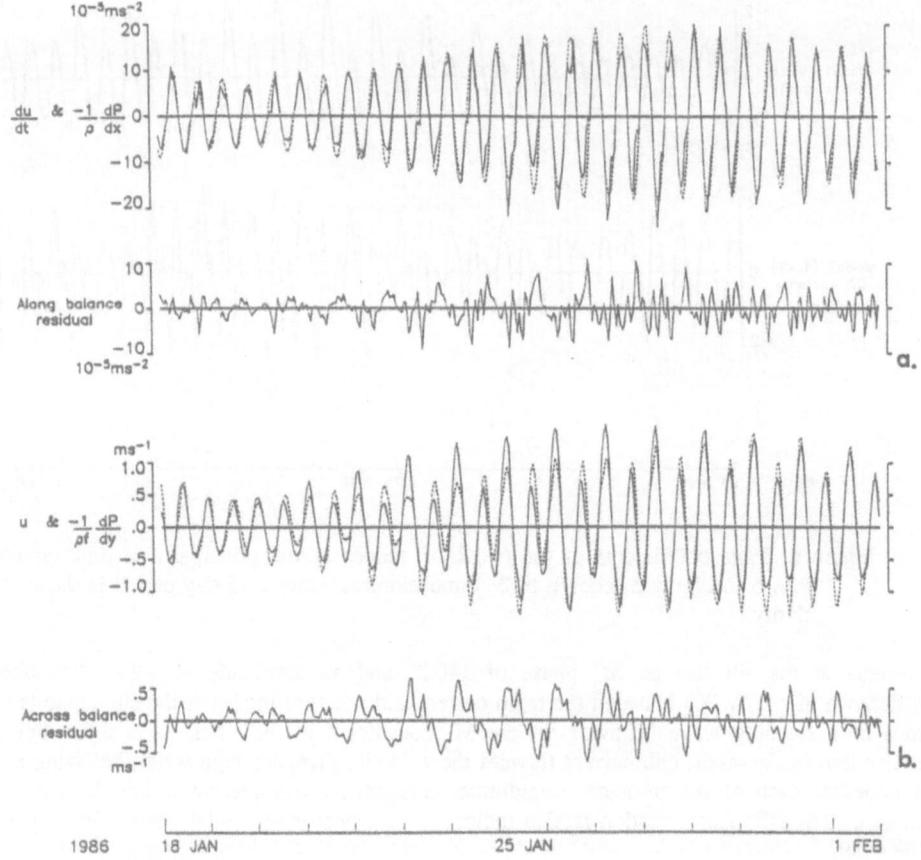

Figure 7. Series of the principal terms in the along- and across-strait tidal momentum balances at the main sill. a) Principal along-strait momentum balance terms, upper plot shows $\partial u/\partial t$ (continuous line) and $-\rho^{-1}\partial P/\partial x$ (broken line). The lower plot shows the difference between the two terms plotted above. b) Principal terms in the across-strait tidal momentum balance at Gibraltar's main sill. The upper plot shows the observed tidal current u (continuous line) and the across-strait surface pressure gradient $-(\rho f)^{-1} \partial P/\partial y$ (broken line). The lower time series (across balance residual) is the difference of the two terms plotted above.

The residual between the two terms (lower plot Figure 7a) is noisy and its variance is about an order of magnitude smaller than any of the two terms considered. It is interesting to note that friction seems to play a minor role in the balance. If present, it should be at least an order of magnitude smaller than either of the other two terms, since its contribution would have to be extracted from the residual. However a friction term is expected to be in phase and correlated with the current u, which is not well supported by these observations. Still, if it is assumed that at least the variance contained in the residual, with an rms of $3.5 \ 10^{-5} \ ms^{-2}$, could be explained by a linear frictional term of the form λu, then $\lambda \sim 5 \ 10^{-5}s^{-1}$, since the rms of the tidal u is $0.71 \ ms^{-1}$. This λ implies an e-folding decay time scale (5.6 h) of about half of the semidiurnal tidal period, a result in agreement with that obtained for the subinertial flows through the strait by Candela et al. [1989a]. The value of $\lambda = 5 \ 10^{-5}s^{-1}$, as obtained above, implies having a perfect correlation coefficient ($r = -1.$) between u and the along-strait momentum balance residual. The available series give $r = -0.36$, so the data only supports a value of $\lambda = 1.84 \ 10^{-5}s^{-1}$ (15.4 h). This last value agrees well with the accepted relationship of $\lambda = CU/H$, where C is a dimensionless drag coefficient ($\sim 3. \ 10^{-3}$), U an rms current velocity ($0.71 \ ms^{-1}$) and H the hydraulic depth (cross-sectional area/width $\sim$ 120 m), which gives $\lambda = 1.75 \ 10^{-5}s^{-1}$. Even though the data points to this smaller value for the linear friction coefficient, it is believed that this represents a lower limit to the possible magnitude of the effects of friction in the strait. The noise content in the along-strait momentum balance residual does not permit a proper evaluation of the effects of friction, but based on the available data λ should have a value between $1.8 - 5. \ 10^{-5}s^{-1}$ in the strait. If friction is instead parameterized by a quadratic relationship, the value for the dimensionless drag coefficient should then fall between 0.003 and 0.008.

To verify if the simple along-strait balance found at the sill is also present at the eastern end, the first mode of pressure is compared with current measurements from mooring M7. For the eastern end, the along-strait pressure gradient implied by mode 1 (Figure 3a) is calculated by a plane fit to the weights from stations CE, AL, DP5 and TA. The magnitude of this gradient is about half of that found at the sill, with a value of $4.816 \ 10^{-4}$ mb/m, oriented $-157°$ with respect to east. This gives an rms value for the term $-\rho^{-1} \ \partial P/\partial x_0$ at the eastern end of $4.69 \ 10^{-5} m/s^2$. Mooring M7 has only two instruments, i.e., M7-54 m and M7-193 m. The upper layer currents (M7-54 m) present a drastically different variance distribution in comparison with currents measured elsewhere in the strait (Table 3), even with respect to the currents in the layer below (M7-193 m). This weakening of the upper layer tidal currents at the eastern end is partially explained by comparing upper and lower layer transports between the sill and the eastern sections and accounting for the particular behavior of the interface separating the two layers, as dicussed by Candela et al. [1989b]. In any event the observations at M7-54 m do not comply with a simple balance between local acceleration and the along-strait pressure gradient. The lower layer (M7-193 m), however, does comply with this simple along-strait momentum balance (Figure 8). It is encouraging to find such a good agreement (correlation coefficient 0.91), even when comparing the pressure field with the acceleration term deduced from the currents measured at only one point in the lower layer. The correlation between the residual in Figure 8 and the u in the lower layer (M7-193 m) is only -0.28, but the slope value of the regression surprisingly gives again $\lambda = 1.85 \ 10^{-5}s^{-1}$, confirming the lower limit estimate for the effects of friction obtained at the sill previously.

The across-strait momentum balance is verified by comparing mode 2 of the pressure field (Figure 3b) with the first mode of currents at the sill (Figure 5a). Again fitting a plane to the four sill stations for mode 2 (Figure 3b) a north-south pressure gradient of $-0.63 \ 10^{-3}$ mb/m is obtained. This gives an rms value for the term $-(\rho f)^{-1} \ \partial P/\partial y$ (where $f = 8.51 \ 10^{-5}s^{-1}$ is the Coriolis parameter) of 0.72 m/s, which when multiplied by the time coefficients of mode 2 (lower plot Figure 4) gives a time series for this term. To see how geostrophic the tidal flow is at the sill section, $-(\rho f)^{-1} \ \partial P/\partial y$ is compared to the first mode of currents at the sill. The agreement is reasonably good (Figure 7b), the correlation coefficient is 0.92, but there are appreciable discrepancies mainly in amplitude.

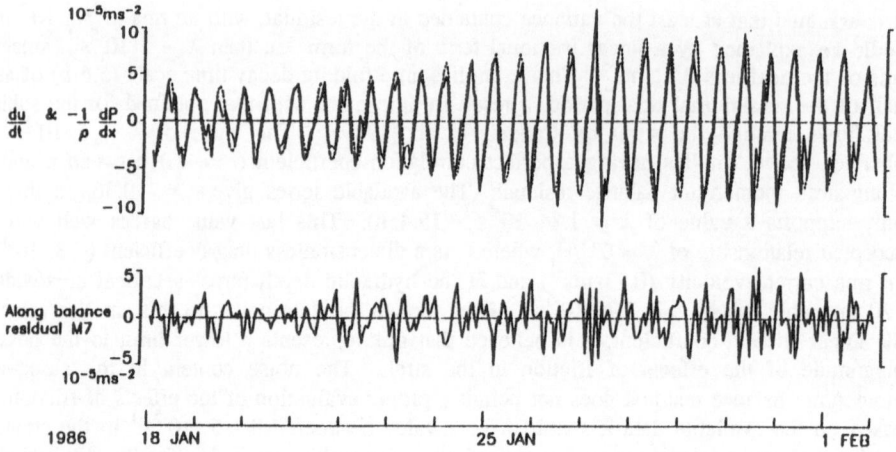

Figure 8. Series of the principal terms in the along-strait tidal momentum balance for the
lower layer at the eastern end of the strait, i.e. Algeciras-Ceuta section. The upper
plot shows $\partial u/\partial t$ (continuos line) computed from the principal axis tidal current
observed at M7-193m and $-\rho^{-1}\partial P/\partial x$ (broken line) obtained from the pressure data
around the section. The lower curve gives the difference of the two terms plotted
above. Note the difference in vertical scales when comparing with Figure 7.

In contrast to $\partial P/\partial x_o$, $\partial P/\partial y$ is expected to vary considerably with depth. With the pressure data at hand and due to the corrections[1] performed on the deep sill pressure measurements, $\partial P/\partial y$ is measuring the across-strait pressure gradient at the surface and so relates to the surface currents. On the other hand, the mode 1 current (Figure 5a) used in Figure 7b refers to a "barotropic" current. In principle the residual of the across-strait balance (Figure 7b) should be related to the baroclinic current. A good correlation could be expected between this residual and mode 2 of the sill currents (Figure 6 lower plot), which has an obvious baroclinic structure, but the data does not show it. Unfortunately there are no current measurements available near the surface and the current mode decomposition used is mostly based on observations from the lower layer. The internal tide at the sill would have to be very stable and have appropriate vertical symmetry, for the second mode (Figure 5b) to also account for the upper layer baroclinic variability. Another consideration is that $\partial P/\partial y$ is not likely to be constant across the strait. It is a well known fact, usually pointed out to mariners on charts of the strait, that tidal currents inshore turn earlier than offshore in the strait. Then, at given times within a tidal cycle, the shape of $\partial P/\partial y$ should differ substantially from a straight line as is implicit in the plane fit used to evaluate $\partial P/\partial y$. It may be that most of the observed residual could be explained by this discrepancy alone.

---

1. The correction referred to consisted of eliminating the baroclinic pressure contribution from the deep pressure measurements at the sill, i.e. DN and DS. Density and depth series from current meters in nearby moorings were used in the calculation (see Candela et al. [1989b] for details).

At the eastern end, the tidal currents in the upper layer do not seem to agree with a geostrophic balance. This is at first surprising when at subinertial frequencies it is observed that the pressure difference between Ceuta and Algeciras (or Gibraltar) is in excellent agreement with a geostrophic across-strait balance. In fact, it is very reasonable to take the across-strait pressure difference as representative of the subinertial fluctuations of the upper layer inflow into the Mediterranean [Kinder and Bryden, 1987].

At tidal frequencies though, the small magnitude of the tidal current at M7-54 m, coupled to the nearly same amplitude and phase for the tidal pressure signal observed at the north (AL) and south (CE) shores, makes the corroboration of a tidal geostrophic balance very dependent on small errors in the data. Across-strait pressure differences relate to an average current across the strait, along a line joining the two pressure instruments at either side. If there is across-strait variability of the along-strait tidal current, the pressure difference and the currents measured at a point in between need not be well correlated. In any case, the complicated behavior of the upper layer currents at the eastern end can very well imply that other terms need to be included to describe appropriately the across-strait momentum balance there. The lower layer tidal current, however, might still show a simple geostrophic balance across the strait. To verify this, deep across-strait pressure measurements are needed and are not available at present.

Apart from the complicated tidal behavior observed on the surface layer at the eastern end, the following interpretation of the observed tides in the strait can be given as a conclusion to this section: The strait has to match two very different tidal regimes (Atlantic and Mediterranean), mainly by an along-strait amplitude gradient. Since this amplitude gradient has very little cross-strait structure, it cannot accommodate a geostrophically balanced tidal flow. As a consequence, a gradient in phase is needed across to produce the pressure difference between the north and south shores to balance the Coriolis acceleration associated with the tidal currents.

4. Discussion and Conclusions

It is interesting to observe that even in a place like Gibraltar, with complicated topography and strong flows (O(.5m/s)) at lower frequencies (<.5cpd), the tides in the strait present simple dynamical balances which are the same as those governing the tide in the deep ocean. The best wave model for the deep ocean tides is Kelvin waves (Hendershott [1973]). A Kelvin wave traveling along a boundary has a geostrophic balance in the across-boundary direction, while in the along-boundary direction the acceleration of the flow is balanced by the pressure gradient and some dissipation or friction term. In a strait, these same balances seem to dominate the dynamics of the barotropic tide. This tidal behavior has been verified in other straits, i.e. Garrett and Petrie [1981] and Godin et al. [1981]. For example, in the Strait of Juan de Fuca, south of Vancouver Island on the west coast of Canada, the M_2 cotidal chart has a configuration completely reversed from that in the Strait of Gibraltar. There, cotidal lines are somewhat perpendicular to the axis of the strait, increasing in magnitude towards the interior. Corange lines are also perpendicular to the axis in the western part of the strait, but become parallel to the axis towards the eastern end increasing in magnitude southward. Still, tidal currents and sea level observations comply with the dynamic balances of a Kelvin wave. Actually this cotidal pattern can be shown to be the manifestation of a local degenerated amphidromic system, whose node has been forced to fall outside the channel due to the effects of friction. Interestingly, a linear friction coefficient of $\lambda = 4 \ 10^{-5} \ s^{-1}$ is required to best fit the M_2 tidal observations in Juan de Fuca, which is within the range of values deduced from the Gibraltar data discussed previously.

Acknowledgments

The work described here is a part of the Gibraltar Experiment. The author wishes to express his appreciation to his Spanish and Moroccan colleagues and the various field groups who made the experiment so successful. In particular we wish to thank Dennis Conlon and Tom Kinder whose continued support and enthusiasm made the program possible. Special thanks are due to: Harry Bryden, from WHOI, for providing his current meter data, Captain Antonio Ruiz, from Instituto Hidrografico de la Marina in Cadiz, for furnishing pressure data from the Spanish Ports of Tarifa, Algeciras and Ceuta, and Neal Petigrew, from UNH, for supplying pressure data from his Tarifa South (DP5) station. Tangier's Port Captain, Commandant El Affaqui, is greatly thanked for his invaluable help during the experiment. This work was supported by the Office of Naval Research under contract No. N00014-85-C-0223.

References

Candela, J., C. D. Winant, and H. L. Bryden, Meteorologically forced subinertial flows through the Strait of Gibraltar. *J. Geophys. Res.*, in press, 1989a.

Candela, J., C. D. Winant, and A. Ruiz, Tides in the Strait of Gibraltar. *J. Geophys. Res.*, submitted, 1989b.

Bendat, J. S. and A. G. Piersol, *Random data, Analysis and Measurement Procedures*, 2nd. ed., 566 pp., Wiley-Interscience, New York, 1986.

Bormans, M., C. Garrett, and K. R. Thompson, Seasonal variability of the surface inflow through the Strait of Gibraltar, *Oceanol. Acta 9*, 403-414, 1986.

Bryden, H. L., and H. M. Stommel, Limiting processes that determine basic features of the circulation in the Mediterranean Sea, *Oceanol. Acta 7*, 289-296, 1984.

Bryden, H. L., E. C. Brady, and R. D. Pillsbury, Flow through the Strait of Gibraltar, *The Strait of Gibraltar Symposium*, Madrid, October 24-29, 1988.

Flagg, C. N., J. A. Vermersch, and R. C. Bearsdley, M.I.T. New England shelf dynamics experiment (March, 1974) data report, II, the Moored Array, *Rep. 76-1, Dept. of Meteorol., Mass. Inst. of Technol.*, Cambridge, 1976.

Forsythe, G. E., M. A. Malcom, and C. B. Moler, *Computer Methods for Mathematical Computations*, Prentice Hall, Inc., Englewood Cliffs, New Jersey, xi + 259 p., 1977.

Garrett, C. and B. Petrie, Dynamical aspects of the flow through the Strait of Belle Isle, *J. Phys. Oceanogr., Vol.11, No 3*, 376-393, 1981.

Godin, G., *The Analysis of Tides*, Univ. Toronto Press, xxi + 264 pp., 1972.

Godin, G., The use of the admittance function for the reduction and interpretation of tidal records, *Mar. Sci. Dir. MS. Rep. 42*, 45 pp., 1976.

Godin, G., L'analyse des donnees de courants: theorie et practique, *Mar. Sci. Dir. MS. Rep. 49*, 91 pp., 1978.

Godin, G., J. Candela, and R. de la Paz, An analysis and interpretation of the current data collected in the Strait of Juan de Fuca, *Marine Geodesy, 5* (3), 273-302, 1981.

Kundu, P. K., J. S. Allen, and R. L. Smith, Modal decomposition of the velocity field near the Oregon coast, *J. Phys. Oceanogr., 5*, 683-704, 1975.

Hendershott, M. C., Ocean Tides, *EOS Trans. AGU, 54*, 76-86, 1973.

Kinder, T. H., and H. L. Bryden, The 1985-1986 Gibraltar Experiment: Data collection and preliminary results, *EOS Trans. AGU, 68*, 786, 1987.

Lacombe, H., and C. Richez, The regime of the Strait of Gibraltar, in *Hydrodynamics of Semi-enclosed Seas*, edited by J.C.J. Nihoul, pp. 13-74 pp., 1982.

Pillsbury, R. D., D. Barstow, J. S. Bottero, C. Milleiro, B. Moore, G. Pittock, D. C. Root, J. Simpkins III, R. E. Still, and H. L. Bryden, Gibraltar Experiment, current measure-·ments in the Strait of Gibraltar, October 1985 - October 1986, College of Oceanography, *OSU Data Report 87-29*, reference 139, 284 pp., 1987.

Rocha, C., and A. J. Clarke, Interaction of ocean tides through a narrow single and narrow multiple strait, *J. Phys. Oceanogr., 17* (12), 2203-2218, 1987.

GENERATION AND KINEMATICS OF THE INTERNAL TIDE IN THE STRAIT OF GIBRALTAR

N. A. BRAY[1]
C. D. WINANT[1]
T. H. KINDER[2]
J. CANDELA[1]

[1] Center for Coastal Studies, Scripps Institution of Oceanography, La Jolla, CA 92093, USA

[2] Office of Naval Research, Code 1121CS, 800 N. Quincy St., Arlington, VA 22217, USA

ABSTRACT. Observations from moorings and hydrographic surveys during the Gibraltar Experiment are used to describe the structure in space and time of the internal tide in the strait. By internal tide is meant those fluctuations of the interface (separating inflowing Atlantic from outflowing Mediterranean waters) that have a persistent and demonstrable phase relationship to the barotropic tide. Two modes of fluctuation dominate the variance at semi-diurnal frequency: rising and falling of the interface throughout the strait, which explains about half of the total variance, and fluctuations in the cross-strait slope of the interface, which explains about 25% of the variance. The largest amplitude fluctuations are found at the Camarinal sill. The interface rises and falls approximately in quadrature with the barotropic current, and a kinematic argument for why that should occur is presented. The cross-strait slope of the interface changes approximately in phase with the barotropic tidal currents, corresponding to changes in the vertical shear, in what may be an incomplete geostrophic adjustment in the cross-strait direction.

1. Introduction

Locally generated internal fluctuations, occurring at tidal frequencies within and near the Strait of Gibraltar, contribute several times as much energy to the system as does the mean buoyancy-driven exchange. Interaction of the barotropic tide with the Camarinal sill generates large amplitude fluctuations of the interface between Atlantic and Mediterranean waters. Of these fluctuations, some are resolved into traveling solitary wave packets, some overturn and mix the water column, and some are found to have consistent phase relationships throughout the strait. It is the latter we examine here, and will refer to as the internal tide.

There are two types of interface fluctuations associated with the internal tide: the interface depth increases or decreases, and the cross-strait slope of the interface varies. We will argue that the depth of the interface varies as a result of kinematic requirements associated with the large amplitude barotropic tidal currents, and that the cross-strait slope varies as a result of fluctuations in the vertical shear at tidal frequencies.

L. J. Pratt (ed.), The Physical Oceanography of Sea Straits, 477–491.
© 1990 Kluwer Academic Publishers.

2. Description of the Internal Tide from Observations

As no one type of observation is ideal for constructing an adequate description of the internal tide in the strait, we have combined moored time series observations (current, temperature, salinity, and bottom pressure) with hydrographic survey measurements wherever we can demonstrate that the different approaches are consistent. In the Gibraltar Experiment data set, the primary limitation of the moored observations is lack of resolution: both horizontally and within the upper inflow layer. Use of hydrographic observations in the strait is limited principally by aliasing of the semi-diurnal tide. In the present work, we have utilized CTD time series (yoyo) casts, maps constructed from stations taken at the same phase of the semi-diurnal tide and rapid cross-strait surveys to avoid, to some extent, these aliasing problems.

2.1 MEAN INTERFACE GEOGRAPHY

It has long been recognized that there exists a 'mean' structure to the interface: shallower east of the sill, deep west of the sill and with a tilt downward to the south to accommodate the geostrophically balanced buoyancy-driven exchange through the strait (Lacombe and Richez, 1982). This average interface structure is apparent, despite the large amplitude fluctuations in the depth of the interface, particularly near the Camarinal sill, as illustrated in Figure 1, a map constructed from all of the hydrographic data taken in the strait during the Gibraltar Experiment (cruises in November 1985, March, June and September of 1986; [Bray, 1986; Shull and Bray, 1989; Kinder et al. 1986, 1987]).

The average distribution of interface depth was constructed in the following way: the average depth of the salinity interval 37 to 38 was first determined for each station. (The structure of the interface is relatively insensitive to this definition: for instance, defining the interface as 36.9 to 37.1 gives the same result as 37 to 38 for all topics discussed here.) Second, the stations were grouped according to time relative to high water at Tarifa (hereafter HW), independent of season or spring/neap cycle. We chose this differentiation because the largest variances in interface depth, by far, are associated with semi-diurnal fluctuations; spring/neap and seasonal variations, though not negligible for some calculations, are less important in determining interface depth. Also, there are insufficient data to further subdivide the categories without losing significant horizontal resolution. Third, each group of stations corresponding to observations taken within a given hour relative to HW was objectively mapped onto a horizontal grid, using standard techniques and an exponentially decaying correlation scale. (Parameters used were 40 km decay scale and an error variance of 0.03). Finally, all 12 grids were averaged, point by point, to create the map shown in Figure 1.

As a comparison, the data were also divided into rectangular areas within the strait and simply averaged. The patterns derived from the two techniques are indistinguishable, except in the northwest corner where there are very few data. The advantage of the objective mapping technique is that it provides gridded data for each of the hourly maps, facilitating calculations we present later.

2.2 SEMI-DIURNAL FLUCTUATIONS OF THE INTERFACE

Given the average structure of the interface, we can then proceed to examine fluctuations about that average. We will present three different analyses of those fluctuations: first by simply subtracting the average field from each of the hourly maps (Figure 2), second by using an empirical orthogonal function (EOF) decomposition of the CTD-derived hourly maps (Figure 3), and third through an harmonic analysis of the interface fluctuations inferred from moored measurements of temperature and salinity (Figure 4). In general, we will not make a distinction between semi-diurnal and M_2, although we recognize that the latter is only one of several constituents contributing to the semi-diurnal band. Thus, the CTD observations will include all fluctuations occurring within that band, but harmonic analyses will be related specifically to the M_2 constituent; the error associated with neglecting the other components in the strait is small compared to sampling errors and noise from other processes, such as internal waves.

The time sequence of fluctuations illustrated in Figure 2 has several interesting characteristics in common with the harmonic analysis of Figure 4:
- the largest fluctuations are found at the sill;
- fluctuations are larger on the south side of the sill than in the north;

- there is a sense of phase propagation from west to east (e.g., the steady eastward movement of the zero line from -30 km at high water HW to +20 km at HW +5);
- there is a tendency for phase propagation from north to south at the sill (e.g., the intrusion of positive anomaly begins on the northern side of the sill at LW and rapidly moves south). The amplitude of fluctuations overall is larger in the CTD-derived data than for the mooring-derived interface; this may result from inadequate resolution of the upper layer by the moored instruments.

A more compact analysis of the CTD-derived interface fluctuations is shown in the EOF results of Figure 3. In the top panels of Figure 3 we have reproduced the average interface depth map, together with a map of standard deviations of interface depth calculated from the hourly grids. The center panels illustrate the spatial structure of the first and second EOF modes, while the time structure of those modes appears in the lower left panel. For comparison, Figure 4 is reproduced to the same scale in the lower right panel.

The temporal structure of both modes is nearly sinusoidal, with semi-diurnal period. However, there is roughly 90° phase difference between the two. In the strait, the barotropic tidal current lags the surface elevation by roughly 90° (Candela, Winant and Ruiz, 1989), with maximum flood tide occurring 3 hours after HW. The first two EOF modes then correspond approximately to maximum surface elevation (mode 1) and maximum tidal currents (mode 2). This sinusoidal time dependence suggests that the EOF decomposition is usefully describing a geophysical process. The first EOF mode (44% of the variance) has its maximum amplitude in time one hour before high water and its minimum one hour before low water. The spatial structure associated with this mode corresponds to the interface either rising or falling throughout the strait, with maximum amplitude on the southern side of the sill. The second mode (27% of the variance) has its maximum one hour before maximum flood current and its minimum one hour after maximum ebb current relative to Tarifa.

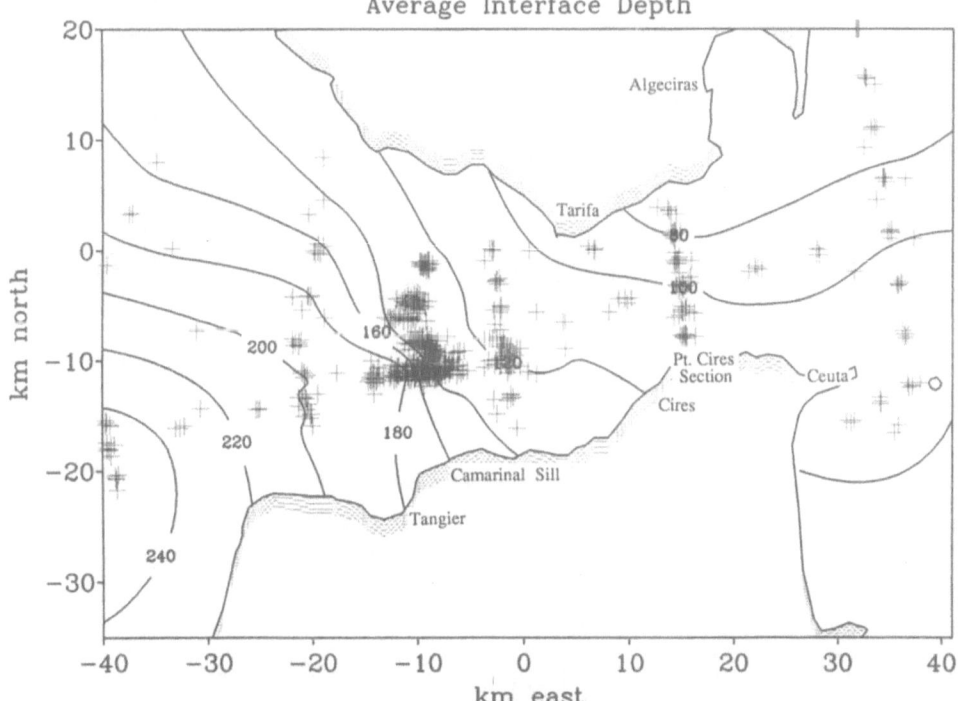

Figure 1.   The average interface depth, in meters, throughout the Strait of Gibraltar, based on hydrographic data. Observations used include those from four Gibraltar Experiment cruises. Geographic locations referred to in the text are noted; crosses denote station locations.

INTERFACE DEPTH ANOMALY

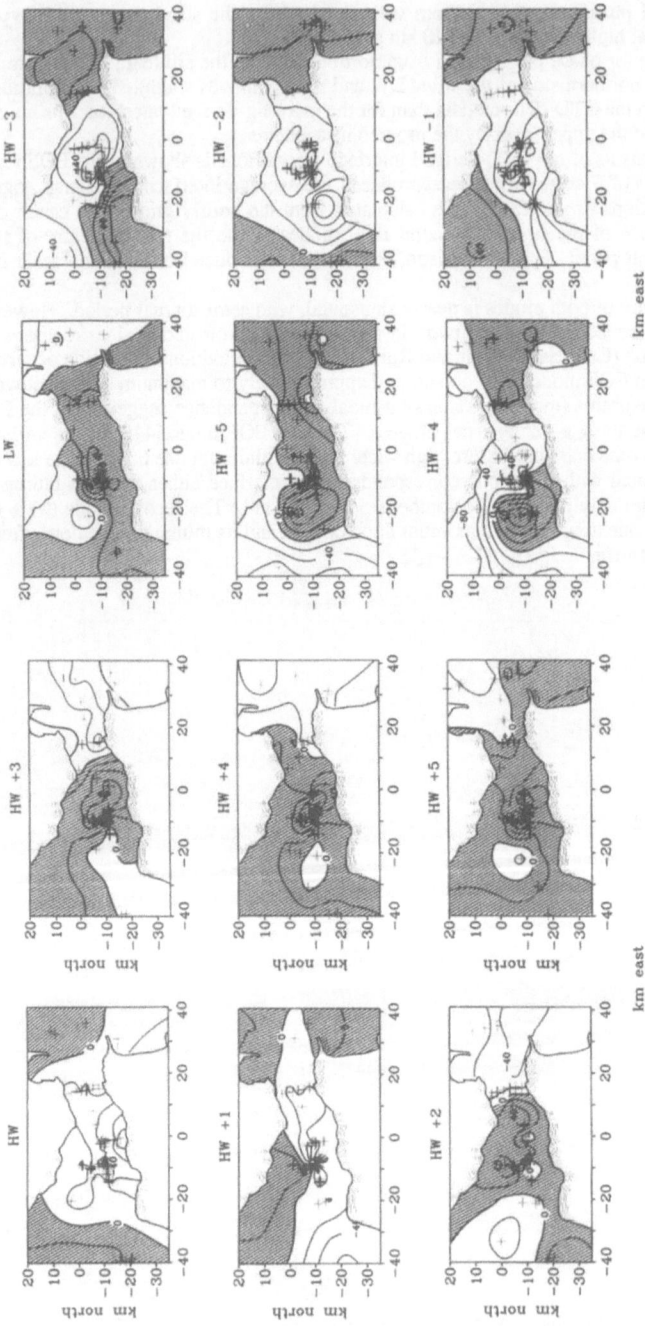

Figure 2.   Interface depth anomaly from the average shown in Figure 1, as a function of hours from high water at Tarifa. Depths in meters. Shaded areas represent regions where the interface is deeper than the mean value.

INTERFACE DEPTH (37 - 38‰)

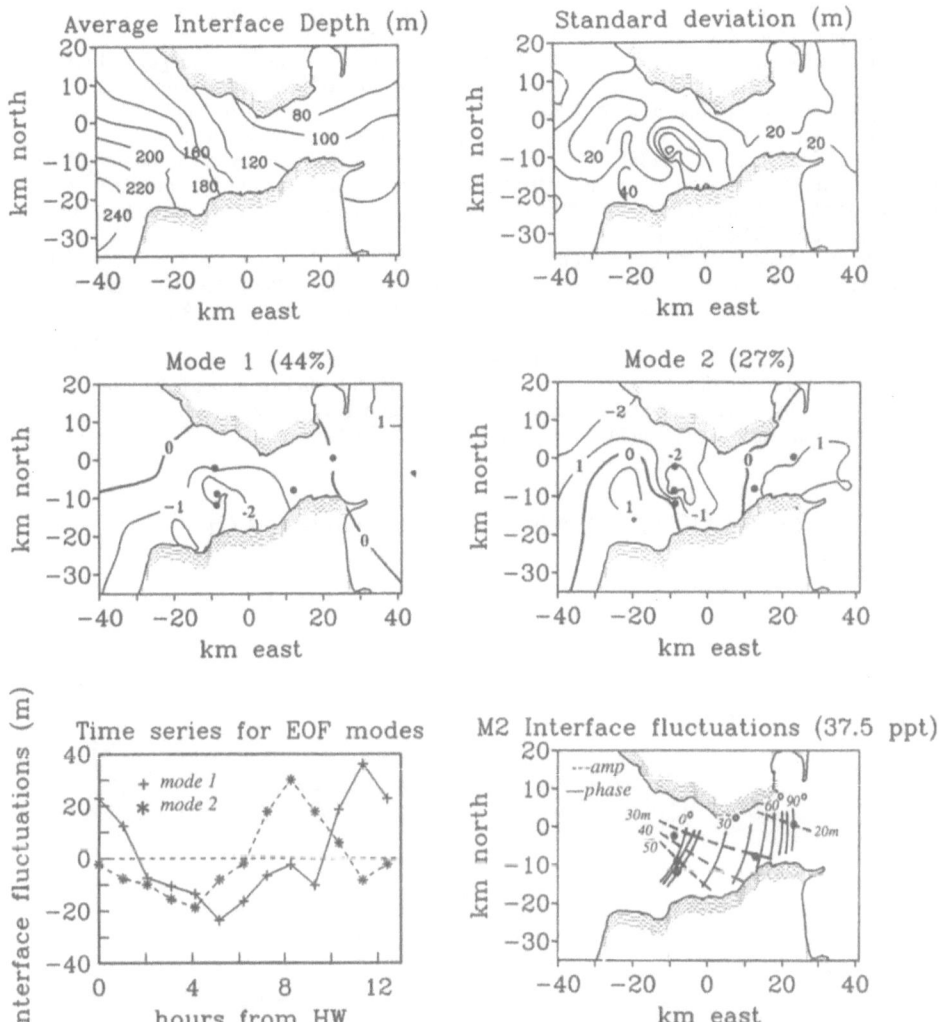

Figure 3. Empirical orthogonal analysis of fluctuations in interface depth as derived from CTD observations. Top panels are average and standard deviation of 12 maps, each corresponding to the horizontal structure of the interface at a given hour relative to HW. Center panels are the spatial structures for modes 1 and 2. Lower left panel shows the time behavior of the two modes, and lower right panel the harmonic analysis of interface depth from Figure 4, reduced to the same scale for comparison with the EOF analysis.

482

Historically, the surface tide at Tarifa has been used as reference for phase considerations of oceanographic phenomena in the strait. A more general reference would be Greenwich, as is typically used in tidal analyses. For the M_2 surface tide constituent, the phase of Tarifa relative to Greenwich is 41°, and a phase change of 1 hour for M_2 corresponds to 28.9°. Thus, the EOF decomposition implies phases (Greenwich) of 12° for mode 1 and 99° for mode 2.

The traditional, and very useful, description of tides in terms of amplitude (co-range) and phase (co-tide) isopleths is a result of harmonic analysis utilizing a number of observation points. Although this approach has been most often used to describe surface elevations, it is equally applicable to interface elevations. From moored time series of temperature and conductivity at five locations in the strait (Figure 4), the depth of the interface as a function of time was inferred by assuming a vertical salinity profile and interpolating (or extrapolating when the interface was shallower than the top instrument) between the measured points. Here the interface was defined to be the 37.5 salinity surface, or the middle of the range used to obtain the CTD-derived interface.

Despite the limited horizontal resolution, a fairly clear pattern of phase propagation from west to east emerges in Figure 4. There is also a phase propagation from north to south over the sill, with an average value of about 0° (Greenwich), consistent with the first EOF mode of Figure 3. Also consistent is the increase in amplitude from north to south across the sill. (Recall, however, that these two analyses are not directly comparable, in that phase information is treated differently: phase information in the EOF analysis is limited to the time series behavior, so that real spatial variations in phase may be obscured.) Given the 90° (3 hour) phase lag at the eastern end of the strait, one might expect to see more variance in EOF mode 2 at the eastern end than in mode 1. This appears to be the case.

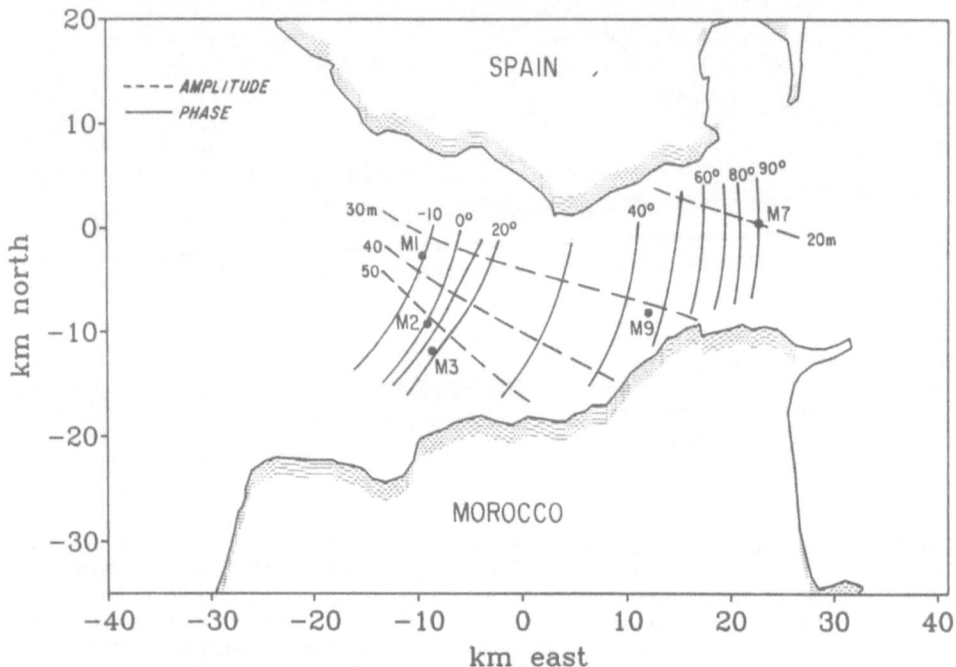

Figure 4.   Harmonic analysis (co-tide and co-range lines) for the M_2 constituent of the interface fluctuations as determined from moorings (located by dots).

3. Cross-Strait Structure and Fluctuations

While some useful information about the cross-strait structure of the internal tide can be gleaned from the horizontal maps of the previous section, sufficient observations were taken during the Gibraltar Experiment to allow us to examine cross-strait sections at the main constricting sill, Camarinal, and in the eastern end of the strait near Pt. Cires (see Figure 1 for locations).

3.1 CAMARINAL SILL

Most of the moored observations taken during the experiment were concentrated in the area of the Camarinal sill, the principal vertical constriction for exchange through the strait, with a maximum depth of 320 m. The Tarifa Narrows, however, is the narrowest section, located 15 km east of the sill, and with a depth of about 700 m. Three moorings were located across the sill, on the north (M1), in the center (M2) and at the south (M3) (Figure 5). Mooring M2 was lost after about a month, but is still useful for tidal period analyses. In addition to the moored observations, a large number of CTD time series (yoyo) casts were occupied across and near the sill. Cross-strait sections constructed from these casts are described in detail in Bray and Lacombe (1989) and will not be discussed here.

An EOF decomposition of the cross-strait structure of the observed tidal currents is shown in Figure 5 (spatial structure only for modes 1 and 2). Most of the variance (92.3%) is contained in the first mode, which is barotropic, with some decrease in speed near the bottom. The structure of mode 2 reflects the fluctuations in the vertical shear at tidal frequencies. The time behavior of these two modes is illustrated in Figure 6, along with time series of interface height at M1 and M3 for a 2.5 day period in April of 1986. The top two panels of Figure 6 are the observed interface depths on the south and north sides of the sill. The third panel is the bottom pressure (approximately the surface elevation) measured on the north side of the sill, and is included as an indication of the phase of the surface tide.

In the bottom two panels of Figure 6 the EOF modes of the tidal currents are compared with the cross-strait interface fluctuations: (panel 4) the average depth of the interface (simply the arithmetic average of the interface depths on the north and south sides of the strait) vs the time series of the barotropic tidal current (mode 1); and (panel 5) the cross-strait slope of the interface (calculated as the difference in interface depth from south to north) vs the vertical shear (mode 2). The correspondence between the two time series is remarkably close for both modes. The time lag between the average interface depth and the mode 1 currents is about 3 hours, whereas the interface slope appears to be in phase with the mode 2 currents.

3.2 PT. CIRES SECTION

In November of 1985 a series of 13 hydrographic sections was occupied across the eastern end of the strait, near Pt. Cires (see Figure 1 for section location). The sections were done in two groups: one at spring tides and one at neap tides. Succeeding sections were occupied as quickly as possible, alternating direction of transit each time. One transit, comprising 6 stations, took from 2 to 4 hours to accomplish. We have taken these 13 transits and compiled them into a time series, with 'time' taken to be the midpoint of the interval required by a given transit. Some aliasing of the tide undoubtedly occurs with this approach, but the resultant EOF decomposition is fairly convincing that tidal frequency fluctuations are resolved by these observations.

The average salinity and its standard deviation as determined from the 13 surveys across the Cires section are shown in Figure 7. Note that the interface is shallower to the north (to the right) in the average and that there are regions on either side of the strait where the standard deviation is high, reflecting variations in the slope of the interface. The first two EOF modes calculated for this series of sections have generally sinusoidal, semi-diurnal time behavior. Like the EOF modes shown in Figure 3 for the interface depth as a function of horizontal position in the strait, the first mode in the Cires analysis corresponds to salinity uniformly increasing or decreasing across the strait (in the depth range occupied by the interface). The maximum change occurs at the average depth of the interface, and on the south side of the strait. This salinity fluctuation is equivalent to the interface rising and falling. The time behavior is also similar to that of Figure 3: nearly in phase with the surface elevation, plotted as a dotted line in the time series plot

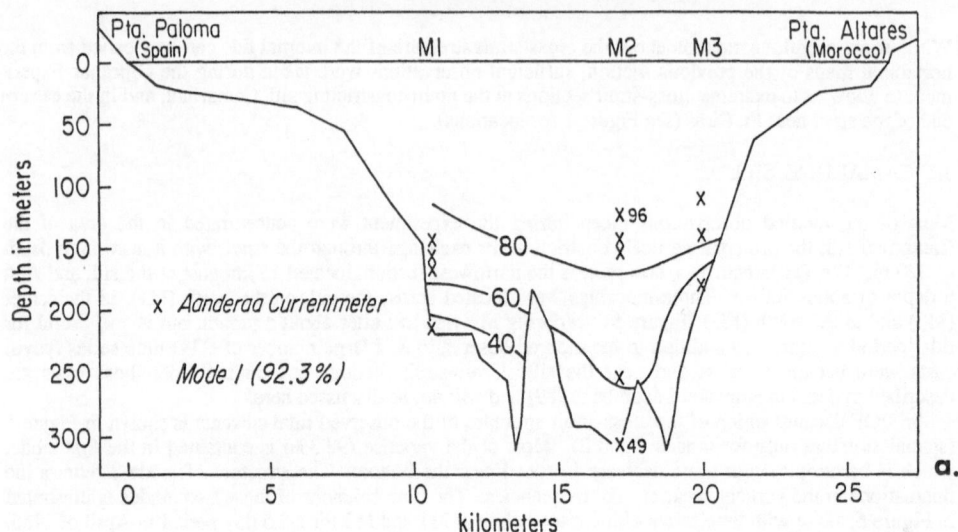

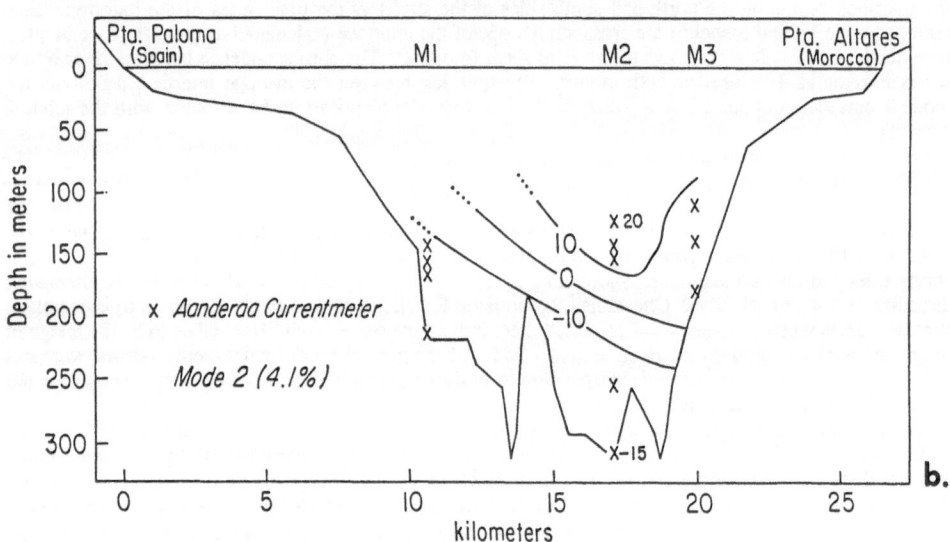

Figure 5.   Spatial structure of the first two modes of the tidal currents ($\sigma > 0.5$ cpd) from a section across the sill (north is on the left). Contours are in cm sec^{-1}. (Redrawn from Candela, Winant, and Ruiz, 1989).

for mode 1. The phase difference between surface elevation and interface depth has changed, however, and the surface here leads the interface by about 2 hours, whereas at the sill the interface leads the surface by 1 hour. This spatial phase shift is consistent with the harmonic analysis shown in Figure 4.

Mode 2 of the Cires section is less sinusoidal in time, but is at least partly correlated with maximum tidal current (indicated here by phase-shifting the tidal elevation by 3 hours, the observed phase difference between elevation and current in the strait). The spatial structure of mode 2 corresponds to changing the slope of the interface: for example, when salinity increases on the north side, it decreases on the south, reflecting an increased tilt.

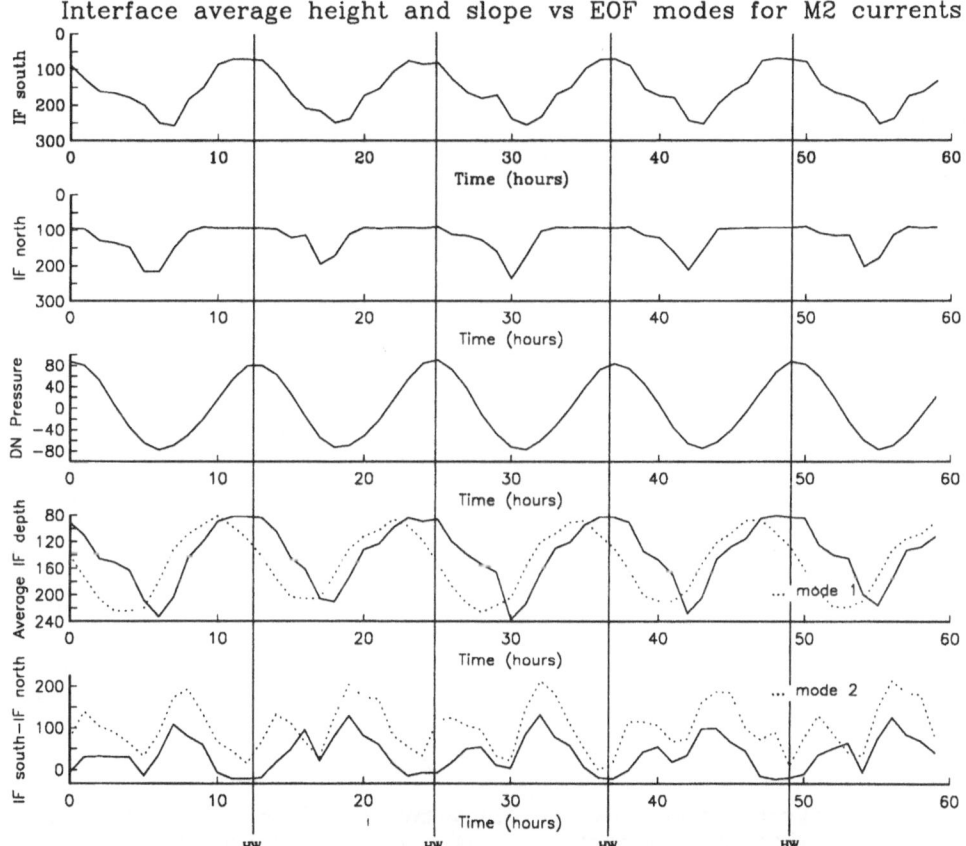

Figure 6. Interface depths in meters from moored data on the south (top panel) and north (second panel); bottom pressure from the sensor at DN, corresponding to the surface elevation (third panel). In the lower two panels: comparison of the average interface depth (solid line) and the time series of mode 1 from Figure 5 (dotted line); and comparison of the interface difference (solid line) and the time series of mode 2 from Figure 5 (dotted line).

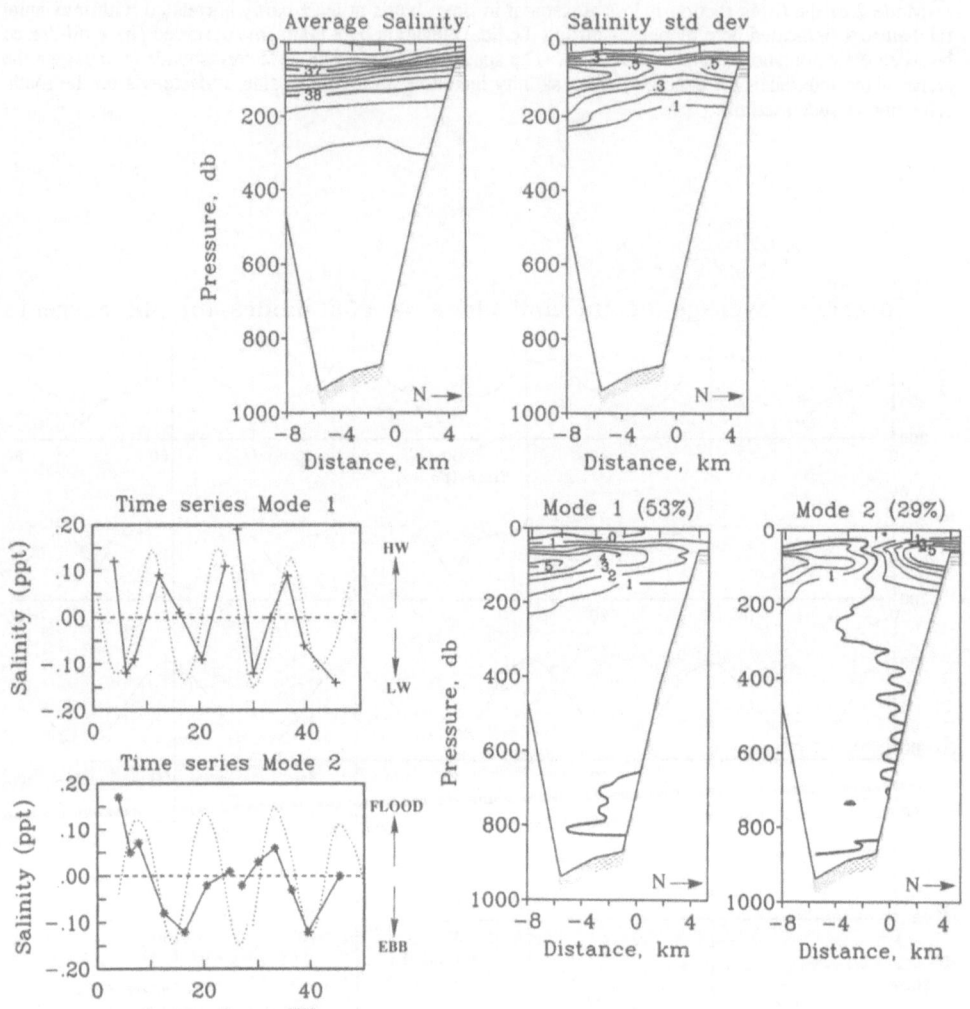

Figure 7.   Empirical orthogonal analysis of the Cires hydrographic sections. Top panels are average and standard deviations of 13 cross-strait sections occupied in rapid succession. Center panels illustrate the temporal and spatial behavior of the first two modes. For comparison in the time series plots, surface elevation is plotted as a dotted line with mode 1, and surface elevation shifted by 90°, to represent maximum current, is plotted as a dotted line with mode 2.

4. Discussion

The overall structure of the internal tide in the strait can be summarized as follows. The surface elevation of the M_2 tide is nearly in phase throughout the strait. The barotropic tidal currents are also nearly in phase throughout the strait (except in the upper layer at the eastern end of the strait), and 90° out of phase with the surface elevation, with maximum flood current following high water. The elevation of the interface has fixed phase relative to either maximum current or surface elevation at the Camarinal sill and the Pt. Cires section, though with different lags in the two locations. At the sill, the interface leads the surface elevation by about an hour, while at Cires the interface lags the surface elevation by about 2 hours. In addition to rising and falling, the interface slope also changes in phase with the tidal frequency vertical shear fluctuations at the sill, and in phase with the tidal current at the Cires section. Based on the EOF analyses of CTD data throughout the strait and across the Pt. Cires section, roughly 50% of the total variance is associated with the rising and falling of the interface, while another 25% is associated with the change in cross-strait interface slope.

The 3-hour phase difference observed between the sill and the eastern end of the strait in the interface fluctuations corresponds to a propagation speed of about 3 m sec^{-1}, 2 to 3 times the expected phase speed of an internal wave travelling along the interface. It is likely, therefore, that the response of the interface in the eastern strait results from a combination of local response to local tidal currents and westward propagation of some type of adjustment occurring east of the sill.

Why should the interface rise and fall in phase with the surface elevation? The simplest argument we can construct is kinematic, involving continuity of mass in two layers of different depths and significant along-strait depth variation, when a large barotropic current is imposed, as illustrated schematically in Figure 8. In the limit of the lower layer depth going to zero west of the strait and the upper layer vanishing east of the strait, and assuming no flow through the interface, the interface must act as deformable surface to accommodate the additional mass flux associated with the barotropic tidal currents. The rate of change of the depth of the interface will be maximum when the tidal current is maximum, and the interface will reach its deepest and shallowest points at slack tides (low water and high water, respectively). Because the surface elevation and tidal currents in the strait are in quadrature, the depth of the interface is, coincidentally, in phase with the surface elevation. However, in this argument, it is the tidal current that drives fluctuations of the interface. In addition to the vertical constraint imposed by the sill, there is a horizontal constraint imposed by the Tarifa Narrows, that will have a significant effect on the continuity argument. It is therefore necessary to consider the convergence and divergence of the transport rather than simply the velocity field in constructing the governing equation here. Let $W(x)$ be the width of the strait, $h(x, t)$ the height of either layer, $u(t)$ the barotropic tidal current, and x and t along-strait distance and time. The continuity equation is then:

$$\frac{\partial Wh}{\partial t} = \frac{\partial (Whu)}{\partial x} \tag{1}$$

Candela, Winant and Ruiz (1989) estimated from moored observations at the sill and at the eastern end of the strait the time rate of change of the lower layer volume enclosed by those moorings and the difference in transports in the lower layer through the two sections. This corresponds to evaluating the two terms of equation (1), integrated in x, as a function of time:

$$\int_{x_1}^{x_2} W \frac{\partial h}{\partial t} \, dx = Whu \mid_{x_1} - Whu \mid_{x_2} \tag{2}$$

The two time series were found to be significantly correlated at the 99% confidence level, and the comparison is shown in Figure 9. The agreement between these time series suggests that the kinematic argument that the interface rises and falls in order to compensate the additional mass introduced into the system by the tidal currents is sound.

KINEMATICS OF TIDAL FLUCTUATIONS OF THE INTERFACE

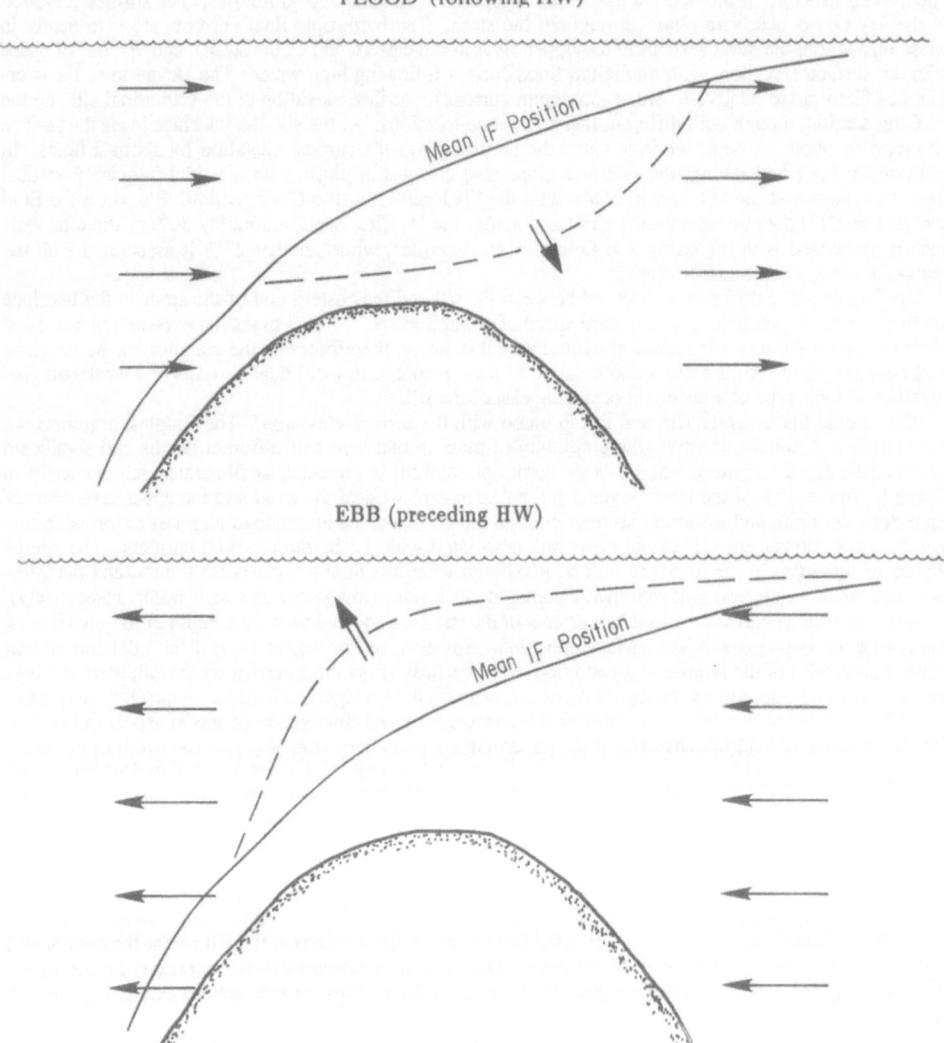

Figure 8.   Schematic of the effect of a uniform barotropic current acting upon a two layer system with very different layer depths east and west of the sill, and significant change in layer depth moving along the strait.

Why should fluctuations in the slope of the interface be associated with vertical shear at tidal frequencies? The answer to this question at low frequencies is fairly clear: a geostrophic balance is established, and the cross-strait slope of the interface accommodates a known vertical shear. In the case of an oscillatory flow, at higher frequency, there may not be sufficient time for the mass field to adjust to a geostrophic balance before the flow field changes, or even reverses. The time required for adjustment is the travel time for an internal wave to cross the strait: $\sqrt{g'h}$, or about 3 hours for the Camarinal sill. This time period corresponds to one-quarter of a semi-diurnal period, so that it is unlikely that full adjustment ever occurs. However, there does appear to be a partial adjustment, and, as the vertical shear produced at tidal frequencies is not negligible, the variance associated with changing cross-strait slope of the interface is a significant contribution to the total.

Throughout this paper we have limited the discussion to the semi-diurnal frequency band. However, Candela, Winant and Bryden (1989) have shown that important low frequency fluctuations in transport through the strait are both barotropic and associated with in-phase fluctuations of the interface. It is of interest, then, to examine the similarities between the high frequency barotropic flows discussed here and lower frequency phenomena. Although a thorough discussion is beyond the scope of this paper, we offer a summary figure. The left panel of Figure 10 illustrates the coherence between the first current EOF modes at the sill, calculated using all frequency bands, not just the M_2, and the average interface depth; and the right panel illustrates the coherence between the second mode and the interface slope, or difference, across the strait. The spatial structures of these two modes, though not shown, are similar to the tidal modes seen in Figure 5: mode 1 is associated with barotropic current fluctuations, and mode 2 with vertical shear.

The average interface depth and the mode 1 currents are coherent at the 95% confidence level from the lowest frequencies, corresponding to about 40 days, through the diurnal band, and with significant peaks at semi-diurnal and higher frequencies. The phase changes from approximately zero at low frequencies to 90° at the semi-diurnal and higher frequencies. There is less coherence at low frequency between mode 2 and interface slope, with an important exception at the fortnightly period (14 days). Coherence is high again at diurnal and higher frequencies. The phase relationship, unlike the mode 1 case, does not appear to change with frequency, but remains at zero for all frequencies where the coherence is significantly different from zero.

From this summary figure, we may surmise that low frequency barotropic fluctuations do not behave in the same way as tidal barotropic fluctuations: the phase relationship between current and interface depth is zero, rather than 90°, and there is no apparent tendency for the cross-strait slope of the interface to change, or equivalently, for vertical shear to be generated. Fortnightly fluctuations also behave differently than do those in the semi-diurnal band, in that there is again zero phase lag between the current and the interface depth fluctuations, as well as significant coherence between cross-strait interface slope and shear.

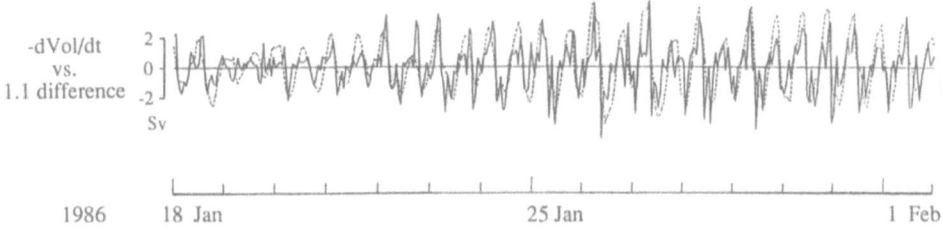

Figure 9.   Comparison of the change in volume with time (solid line) with the difference in transport in the lower layer at either end of the volume (dotted line). The volume is defined by the Camarinal sill on the west and mooring M7 at the eastern end of the strait (see Figure 4 for mooring locations), the (time-varying) interface depth and the bottom of the strait. (Redrawn from Candela, Winant, and Ruiz, 1989).

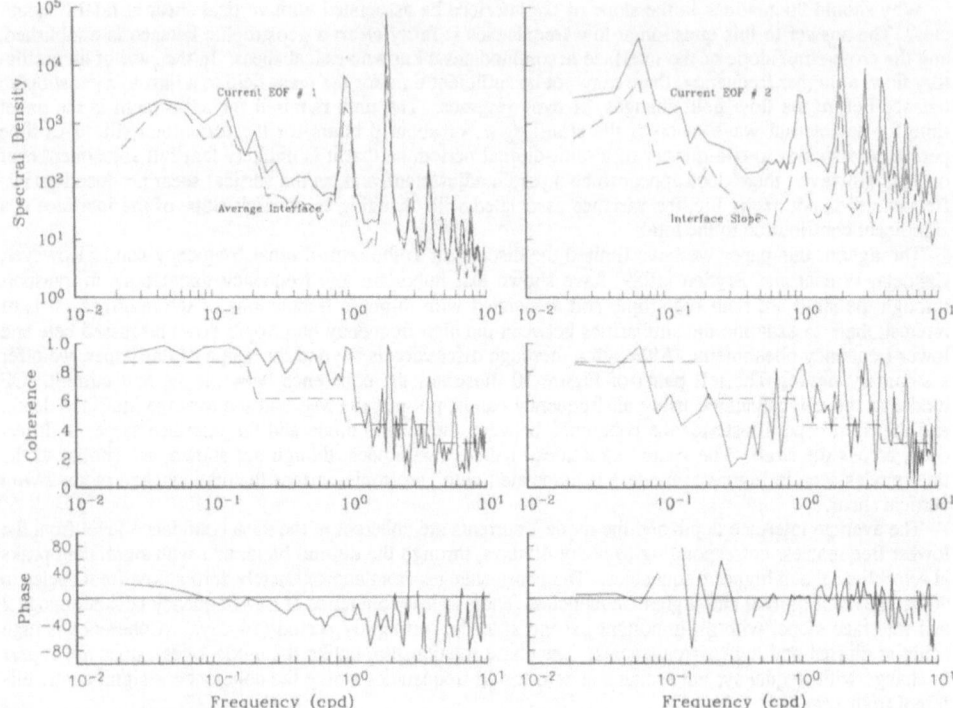

Figure 10.   Spectra and coherence and phase estimates for (left panels) current EOF mode 1 and cross-strait average interface depth, and (right panels) current EOF mode 2 and cross-strait interface slope. Dashed lines are 95% confidence limits.

5. Acknowledgements

The observations discussed in this paper were collected during the Gibraltar Experiment, under the auspices of the United States Office of Naval Research. We are grateful to Harry Bryden and Dale Pillsbury for permission to use the moored current observations they collected during the experiment. NAB was supported under ONR contract #N00014-85-C-0407 during the acquisition and analysis of observations and the preparation of this manuscript. THK received support from ONR under Code 422CS program element 61153N, and CDW and JC from ONR contract #N00014-85-C-0223. Special thanks go to Joan Semler and Michael Clark for their painstaking preparation of the camera-ready final version of the paper. As always, we are indebted to the many people who made it possible to acquire observations at sea, especially in as challenging an environment as the Strait of Gibraltar.

6. References

Bray, N. A. (1986) Gibraltar experiment CTD data report, March-April 1986, USNS *Lynch, SIO Ref. 86-21*, 212 pp., Scripps Inst. of Oceanogr., La Jolla, Calif., USA.

Bray, N. A., and H. Lacombe (1989) Tidal modulation of water mass exchange in the Strait of Gibraltar. In preparation.

Candela, J., C. D. Winant, and H. L. Bryden (1989) Meteorologically forced subinertial flows through the Strait of Gibraltar, accepted by *J. Geophys. Res.*

Candela, J., C. D. Winant, and A. Ruiz (submitted 1989) Tides in the Strait of Gibraltar, *J. Geophys. Res.*

Kinder, T. H., D. A. Burns, and R. D. Broome (1986) Hydrographic measurements in the Strait of Gibraltar, November 1985, *Naval Ocean Research and Development Activity Report 141*, 332 pp.

Kinder, T. H., D. A. Burns, and M. R. Wilcox (1987) Hydrographic measurements in the Strait of Gibraltar, June 1986, *NORDA Tech. Note 378-1* (Appendix), 355 pp.

Lacombe, H., and C. Richez (1982) The regime of the Strait of Gibraltar, in *Hydrodynamics of Semienclosed Seas*, edited by J. C. J. Nihoul, pp. 13-74, Elsevier, Amsterdam.

Shull, S., and N. A. Bray (1989) Gibraltar Experiment CTD data report: September-October 1986, USNS *Lynch*. In preparation.

THE STRUCTURE OF THE INTERNAL BORE IN THE STRAIT OF GIBRALTAR AND ITS INFLUENCE ON THE ATLANTIC INFLOW

N. R. PETTIGREW and R. A. HYDE
Institute for the Study of Earth, Oceans and Space
University of New Hampshire
Durham, New Hampshire USA 03824

ABSTRACT. As part of the Gibraltar Experiment, acoustic Doppler current profilers were deployed in the Tarifa Narrows section of the Strait of Gibraltar. Analysis of data from a Doppler profiler deployed near midstrait has shown that the passage of large amplitude nonlinear internal waves exerts profound influence on both the transient and mean flow structures in this section of the strait. Direct current measurements in the Tarifa Narrows yield an inflow transport estimate of one Sverdrup in the Atlantic layer. Of this total approximately 60% is associated with the internal bore, and only 40% is due to the steady circulation.

1. Introduction

The geometry of the Strait of Gibraltar has strong influence upon both the mean and high-frequency flow fields. The most apparent flow constrictions within the strait are the Camarinal sill, at which the bottom shoals to less than 300 m, and the narrowest section of the strait just east of Tarifa. In the latter location, which we refer to as the Tarifa Narrows, the strait is only 15 km wide at the surface (Figure 1).

The Strait of Gibraltar is a region well known for extreme physical oceanographic conditions. Perhaps the most striking of these characteristic features is the periodic and rapid deepening of the density interface between waters of Atlantic and Mediterranean origin, and the associated sudden eastward flow acceleration in the upper (Atlantic) layer. This phenomenon is often referred to as the passage of an internal bore or the arrival of a "velocity front." The feature is believed to be generated near the sill around high water, and to propagate eastward at a speed of 3-4 knots (Lacombe and Richez, 1982).

The passage of the internal bore apparently has its most pronounced effect on the southern side of the strait in the Tarifa Narrows. Data from repeated hydrographic stations by Lacombe and Richez (1982) sometimes showed the interface to plunge more than 200 m in one hour. At the same time, shipboard current measurements showed the near-surface currents to increase by roughly 1 m/sec. Lacombe and Richez also report that the deepening of the interface and the increase of velocity in the Tarifa Narrows both coincide with the passage of a series of surface "slicks" propagating eastward. Recently the propagation of these surface patterns has been investigated by both land-based (Watson, 1989) and airborne (Richez, 1988) radar systems.

Among the scientific goals of our participation in the Gibraltar Experiment were the characterization of the vertical and temporal structure of the internal bore, and the assessment of the bore's contribution (if any) to the mean inflow of Atlantic water into the

493

L. J. Pratt (ed.), The Physical Oceanography of Sea Straits, 493–508.
© 1990 *Kluwer Academic Publishers.*

Mediterranean basin. The discussion of these issues is the principal subject of this brief manuscript.

1.1 EXPERIMENTAL METHODS AND SAMPLING

As part of the Gibraltar Experiment, Doppler Acoustic Profiling Current Meters (DAPCMs) were moored in the Tarifa Narrows section of the Strait of Gibraltar. Figure 1 shows the location of a DAPCM mounted in the upper flotation sphere of a subsurface mooring in the deep central region of the Narrows. The Doppler instrument was moored approximately 400 m above the bottom in order to bring it within profiling range of the near-surface layer.

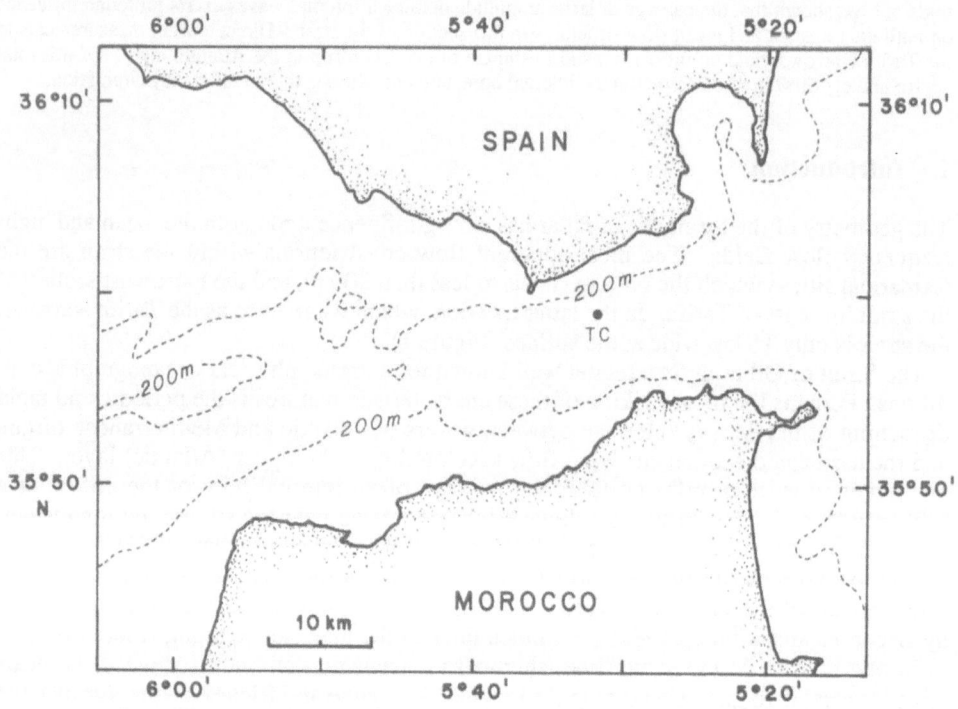

Figure 1. Chart showing the location (TC) of the moored Doppler Acoustic Profiling Current Meter (DAPCM) used in this study.

Doppler current profiles with 10 m vertical resolution, as well as pressure, temperature, and conductivity (all at float depth), were recorded every 5 minutes during the period from March 17- April 23, 1986. Our characterization of the bore will be based on these data that alone have sufficient vertical and temporal resolution to determine the detailed bore structure.

In order to determine the detailed structure of the bore, it is necessary to isolate it from the other strong signals present in the strait. The task of separating currents associated with the tidally-generated bore from the engendering tidal currents is particularly difficult. The main problem is that the nearly-constant phase relationship between the bore arrival and the tide confounds attempts to use traditional methods of tidal analysis to discriminate accurately between tidal currents and the tidally-generated bore. Accordingly, modifications were made to the response method of tidal analysis (Munk and Cartwright, 1966). These modifications enable one to remove the contaminating influence of phase-locked events on tidal analysis and prediction. This method of tidal analysis will be briefly discussed in a later section.

2. Results

2.1 VELOCITY PROFILES

Figure 2 shows a short (40 hour) series of eastward velocity profiles during the neap tide at the beginning of the experimental record. The profiles have been subsampled at 10 minute intervals in order to reduce overcrowding of the figure. The profiles are characterized by a "slab-like" flow structure in the upper and lower layers separated by a region of very high vertical shear.

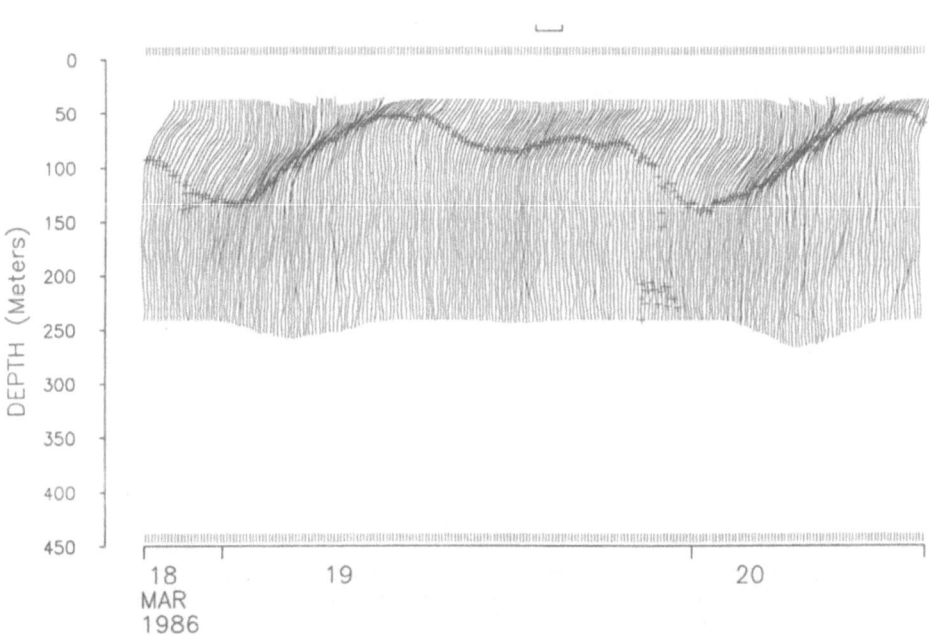

Figure 2. Vertical profiles of the eastward velocity component measured during neap tide. The profiles show slablike upper and lower layer currents separated by a high-shear transition layer. The plus signs indicate the zero crossings so that currents above the plus sign are eastward and those below are westward.

The plus signs that appear in each profile mark the location of a zero crossing of the eastward velocity with water deeper than the plus sign flowing westward and water shallower flowing eastward into the Mediterranean. It is interesting to note that the zero crossings during this neap tidal period occur very close to the bottom of the shear or transition layer, and that the depth of the base of the shear layer undergoes diurnal vertical oscillations of roughly 50 m. Since, as we verify later, the maximum shear layer is associated with the hydrographic transition layer between the inflowing Atlantic water and the outflowing Mediterranean water, these data show that during the neap tide most of the water in the transition layer is being advected eastward into the Mediterranean. The lack of data above 35 m depth is the result of acoustic contamination due to side-lobe reflections from the sea surface (see Pettigrew et al., 1987).

Figure 3 shows a series of vertical profiles of eastward velocity during a spring tide. In comparison with the conditions during neap tide several important differences may be noted. The high-shear layer is thicker, its vertical oscillation is greater; and it occurs twice daily (although there is still some diurnal inequality). The tidal currents during the spring tides are stronger relative to the mean circulation, so that during the incoming tide, the flow in the Atlantic, transition, and Mediterranean layers are all eastward.

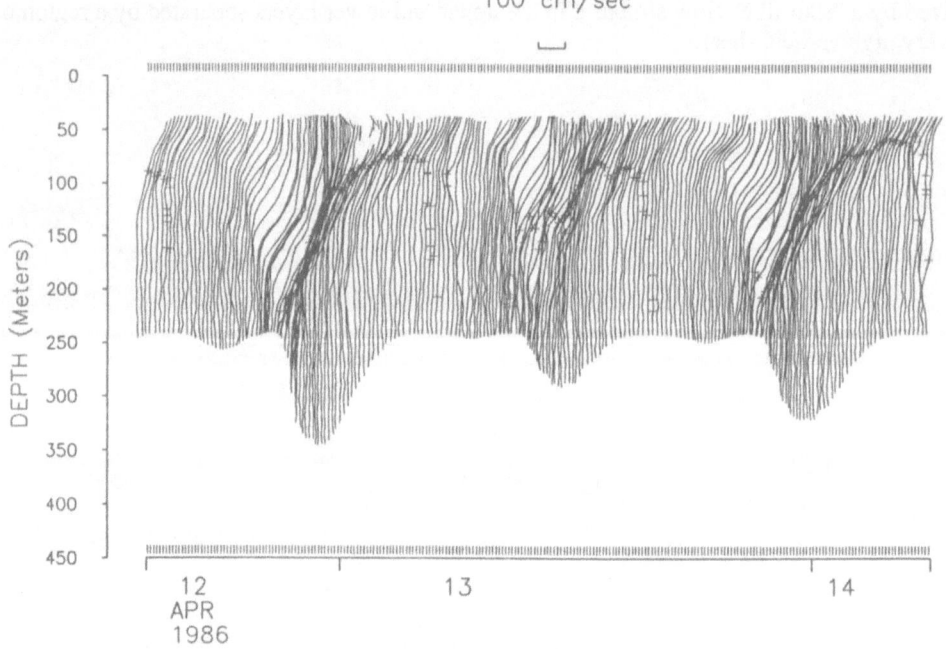

Figure 3. Vertical profiles of the eastward velocity component measured during spring tide. The profiles show a more complicated structure than was evident during neap tide, with the entire water column moving eastward during the incoming tide.

The lower ends of the profiles shown in Figure 3 trace the vertical excursions experienced by the mooring in response to the strong currents. In the example shown, the mooring plunged 100 m late on April 12 as the tidal currents at 250 m (and deeper) changed from weak eastward to strong westward. As demonstrated in this plot, the remote sensing capability of the DAPCM allowed us to maintain current measurements up to within 36 m of the surface despite the pronounced vertical excursions of the mooring and the Doppler instrument.

2.2 CURRENT TIME SERIES

Time series plots of eastward currents are presented in Figure 4. The data are shown at 40 m depth intervals (rather than at 10 m as measured) for fixed depths between 40 and 240 m. While some shallow data are missing, especially during the very strong spring tides of late March, when the instrument occasionally plunged out of range of the near-surface, nearly continuous records were achieved at fixed depths even in the face of the extreme mooring motion. The ability to maintain current measurements at fixed depths in these strongly varying flows proved crucial in the determination of both the vertical and temporal structures of the internal bore.

The first things noted in these current records are that the flow is dominated by fluctuations of tidal frequency and that the currents are very strong. The peak flows exceed 2 m/sec to the east in the upper layer and 2 m/sec toward the west in the lower layer. Mean currents were found to be approximately 0.3 m/sec toward the west at 240 m depth, 0.9 m/sec toward the east at 40 m depth, and zero at approximately 115 m.

The time series current records also clearly demonstrate the important influence of the spring-neap tidal cycle on both the vertical and temporal flow structure in the Strait of Gibraltar. While the diurnal inequality is quite marked during the neap tides at 80 m, at depths of 200 m or greater, neap tidal currents are characterized by semidiurnal fluctuations. During the spring tides the currents are more semidiurnal in nature throughout the water column. It is also noteworthy that at 120 m depth the sign of the tidally-averaged currents changes between the spring and neap phases with currents westward during the neaps and eastward during the springs. This occurrence suggests spring/neap variations in the transport and in the depth of the interface between the Atlantic and Mediterranean waters.

Especially evident during the spring tide, are large eastward spikes superimposed upon a rather regular tidal variation. A more detailed view of these events is shown in Figure 5 showing time series eastward currents from April 8-10. During these events, the eastward current in the upper 100 m increases by more than 1 m/sec over a period of 5 or 10 minutes. This signature is the velocity front referred to by Lacombe and Richez (1982). Thus, Figure 5 shows a detailed vertical and temporal velocity record of the passage of the internal bore. Closer examination of the data reveals that the abrupt eastward acceleration occurs on the falling tide, that the signature in successively deeper layers is successively lagged in time, and is of diminished amplitude at depths exceeding 160 m.

The significance of the internal bore to the mean flow field and to the transport through the Strait of Gibraltar is clearly evident in Figure 5. As an example, at 120 m the mean eastward flow during this 3-day period is due entirely to the passage of the bore. Therefore, we see that a substantial fraction of the mean inflow velocity in the upper layer is actually due to the passage of large nonlinear internal waves.

498

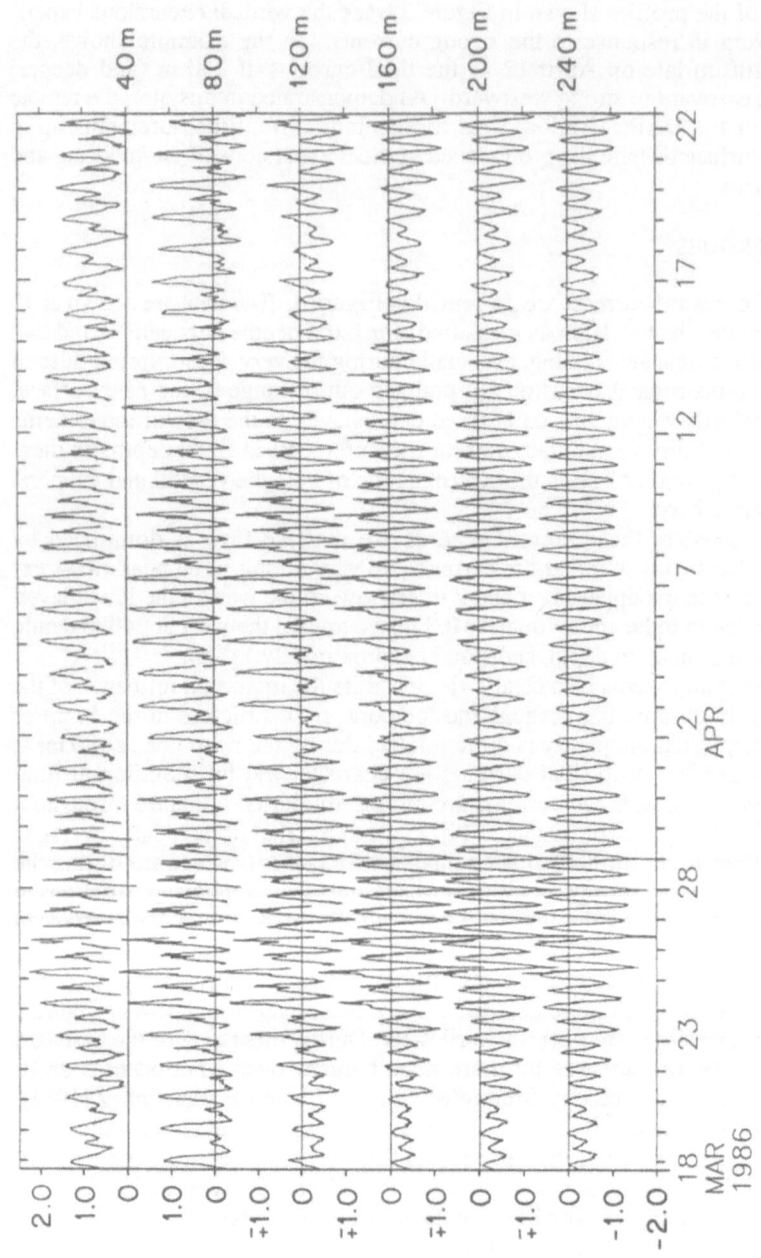

Figure 4. Time series plots of the eastward flow component at fixed depths in 40 m increments. Data gaps generally occurred when strong tidal currents pulled the mooring down out of profiling range of the near-surface region.

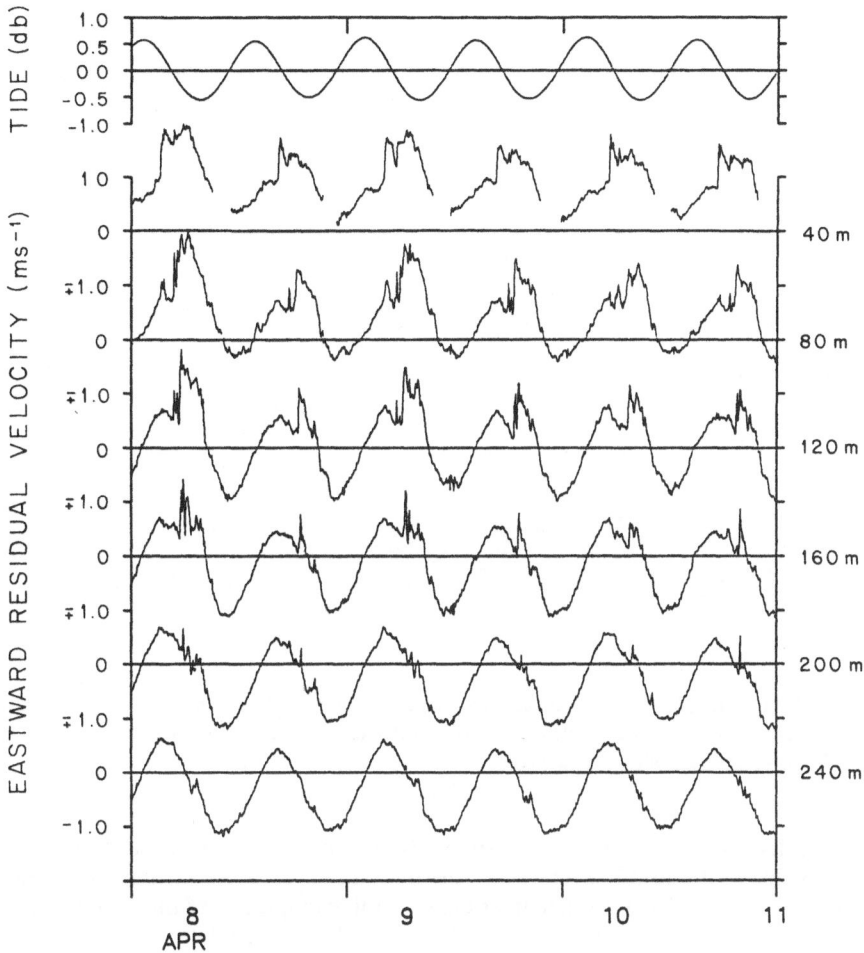

Figure 5. Detailed view of current structure during spring tide. The dominant influence of the bore is clearly evident.

2.3 INTERFACE OSCILLATIONS

As mentioned earlier, Lacombe and Richez (1982) report that the passage of the velocity front was accompanied by the rapid deepening of the hydrographic interface between the Atlantic and Mediterranean layers. This observation is supported by Figure 6 showing calculated tidal elevations along with time-series measurements of pressure, salinity, and temperature at the depth of the buoy. During the spring tide periods, warm fresh spikes appear in the records coincident with the occurrence of the eastward velocity pulse (that is, slightly lagging mean tide while going from high to low). These data confirm that the interface has been displaced downward. In fact, the TS values of the spikes are representative of water properties commonly associated with the bottom of the transition layer

500

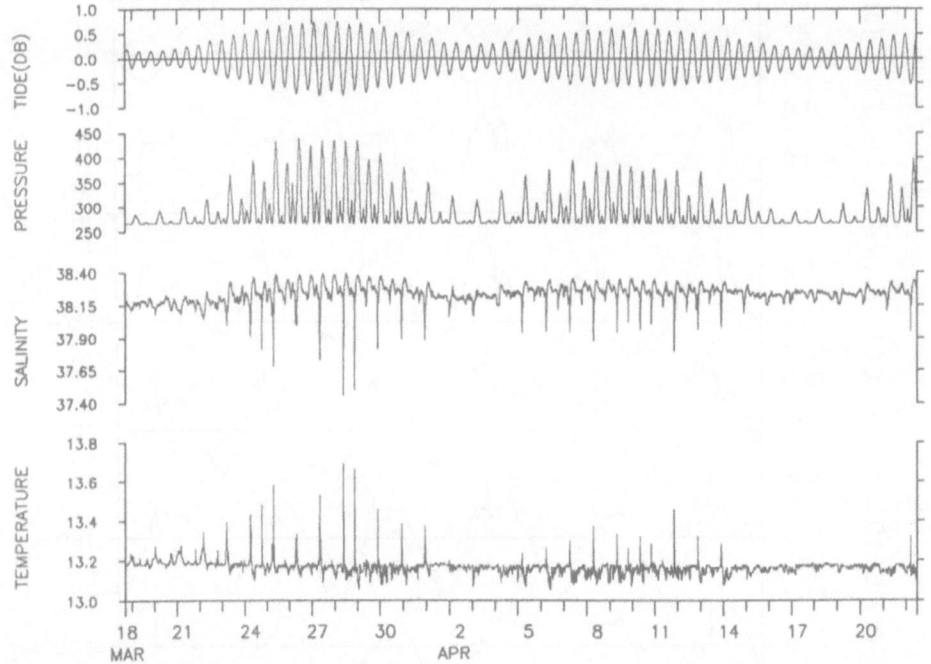

Figure 6. Time series plots of pressure, salinity, and temperature at the depth of the Doppler instrument. The warm fresh spikes show that the density interface between the Atlantic and Mediterranean layers is depressed during the passage of the internal bore.

(Lacombe and Richez, 1982). Unfortunately, the increased drag associated with the change of tide in the lower layer causes the buoy to descend 100-200 m and reenter the Mediterranean layer. Therefore, neither the maximum amplitude of the TS change nor the duration of the interface depression are well measured by the TS data.

The Doppler profiles themselves provide a record of the vertical excursions of the interface in some detail. Figures 7 and 8 are plots of vertical shear profiles corresponding to the velocity profiles shown in Figures 2 and 3. This presentation of the data provides a remarkably effective imaging of the high-shear layer. Figure 7, especially when viewed on end, shows a "ridge" of high shear that daily undergoes vertical excursions of approximately 50 m. The thickness of the high shear layer is approximately 30-50 m. Figure 8 shows an example of the shear over a 40 hour period during a spring tide. In this case the high shear layer, which our TS measurements have identified as the hydrographic interface, descends 200 m with a predominantly semidiurnal periodicity. In this figure we can clearly see the interface descend to the depth of the Doppler instrument just prior to the change in tide and the resultant descent of the mooring. These occurrences coincide with TS spikes noted in Figure 6.

It is also interesting to note that the shear layer appears broader and more diffuse on the trailing edge of the bore (that is, as the interface begins to rise after reaching its maximum depth) as is consistent with the occurrence of mixing. In some cases, such as the third

5 cm/sec/meter

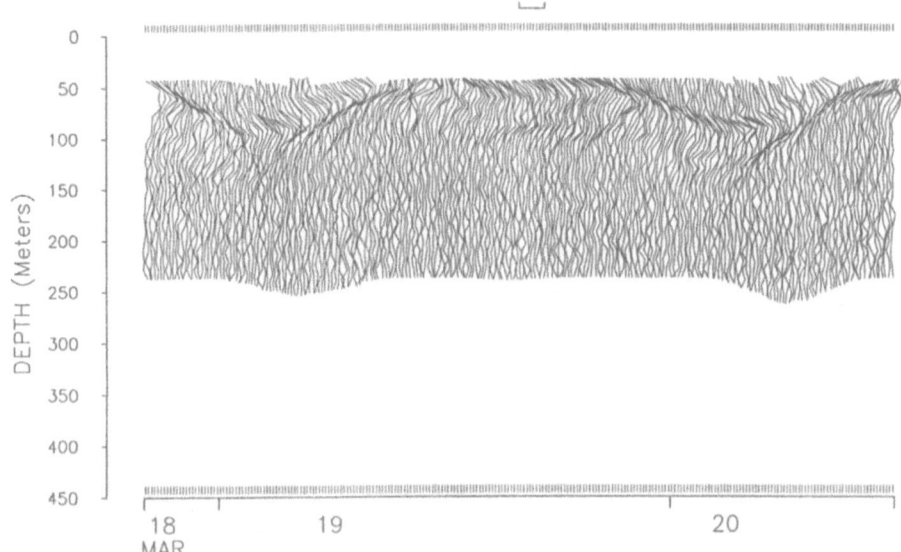

Figure 7. Profiles of vertical shear corresponding to the current profiles in Figure 2. The ridge of high shear, corresponding to the interface, is displaced downward by approximately 50 m during the passage of the internal bore.

event in Figure 8, the high-shear layer on the trailing edge of the bore appears to contain multiple shear maxima. This occurrence is suggestive of the action of mixing associated with shear instabilities. These interpretations are supported by the work of Wesson and Gregg (1988) who observed dissipation rates in the trailing edge of the bore to be 10^{-5} to 10^{-4} W kg^{-1} as compared to 10^{-6} to 10^{-9} W kg^{-1} prior to the bore's arrival.

Although the depth of the maximum vertical shear is somewhat noisy, especially during spring tides, we believe that it represents a good estimate of the depth of the interface in the strait. Evidence in support of this assertion comes not only from our time series TS measurements (Figure 6), but also from the hydrographic data of Kinder et al. (1986) and Bray (1986). The average depth of the 27.7 potential density surface (chosen as the approximate middle of the transition layer) calculated from 40 hydrographic stations within 5 km of the Doppler deployment site is approximately 75 m. By comparison, the average depth of the maximum shear is calculated to be 80 m. Evidence of the instantaneous correlation between the depths of the interface and maximum shear may be found in the data of Armi and Farmer (1989) and/or Farmer and Armi (1989).

2.4 TIDAL ANALYSIS

In order to study the structure of the tidal bore and to estimate its contribution to the mean flow in the Strait of Gibraltar, it is necessary to separate the bore currents from the engen-

502

5 cm/sec/meter

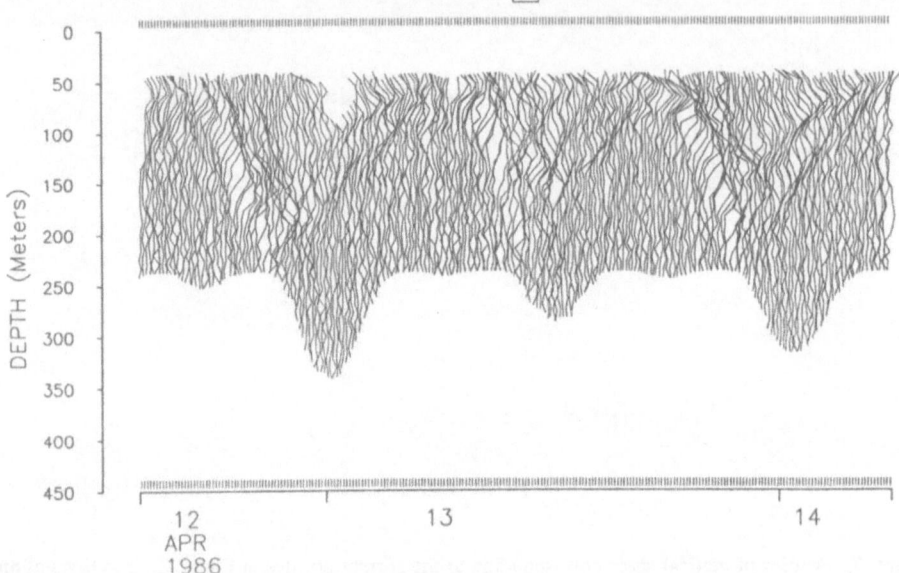

Figure 8. Profiles of vertical shear corresponding to the current profiles in Figure 3. The ridge of high shear, corresponding to the interface, is displaced downward by approximately 200 m during the passage of the internal bore.

dering tidal currents. Since the tides in the strait are so strong, errors involved in removing the tides could easily disguise the true character of the residual (bore).

The common procedure for analyzing short (less than a year) oceanographic records for tides is the response method of tidal analysis introduced by Munk and Cartwright (1966), and operationally defined by Cartwright et al. (1966), and Zetler and Munk (1975). In this procedure a linear causal relation is hypothesized between input (noise-free tidal series synthesized from known tidal coefficients) and output (observations) time series. For this linear system, the input and output functions can be related by a set of complex weights that correspond to the impulse response characterizing the system. In practice, the weights are calculated using a least-squares fit of the demeaned observations based on the reference (input) series. Once these response weights are known, the tidal component of the observations may be calculated (predicted) for any time period over which reference series may be constructed. Since long records (and accurate tidal constants) are often available for coastal tidal elevations, noise-free reference series are usually constructed from elevation tidal coefficients when analyzing current records. The reader unfamiliar with this technique is referred to the above references for details.

While the response method has proved a very useful method for analyzing and/or removing tides from records, it may be confounded by either data gaps or large amplitude events that occur at approximately constant intervals with respect to the tide. Both of these situations occur in the Doppler current data collected in the Gibraltar Experiment. Data gaps occurred in the upper layers when the Doppler mooring occasionally dipped to depths exceeding its working range from the surface and, as shown earlier, the tidally-generated internal bore produced current spikes of very large amplitude. The contaminating influence of the tidal bore on tidal predictions is illustrated in the two upper panels of Figure 9, which show a short section of the spring-tide current record from the Gibraltar Experiment and a tidal prediction using traditional response analysis. Under these conditions the method results in an overprediction of the amplitude and a shift in phase of the underlying tidal currents. This error results from including the bore currents in the least squares fit between currents and the noise-free tidal reference series. While under some circumstances these discrepancies may not be considered serious, when the intent is to study the

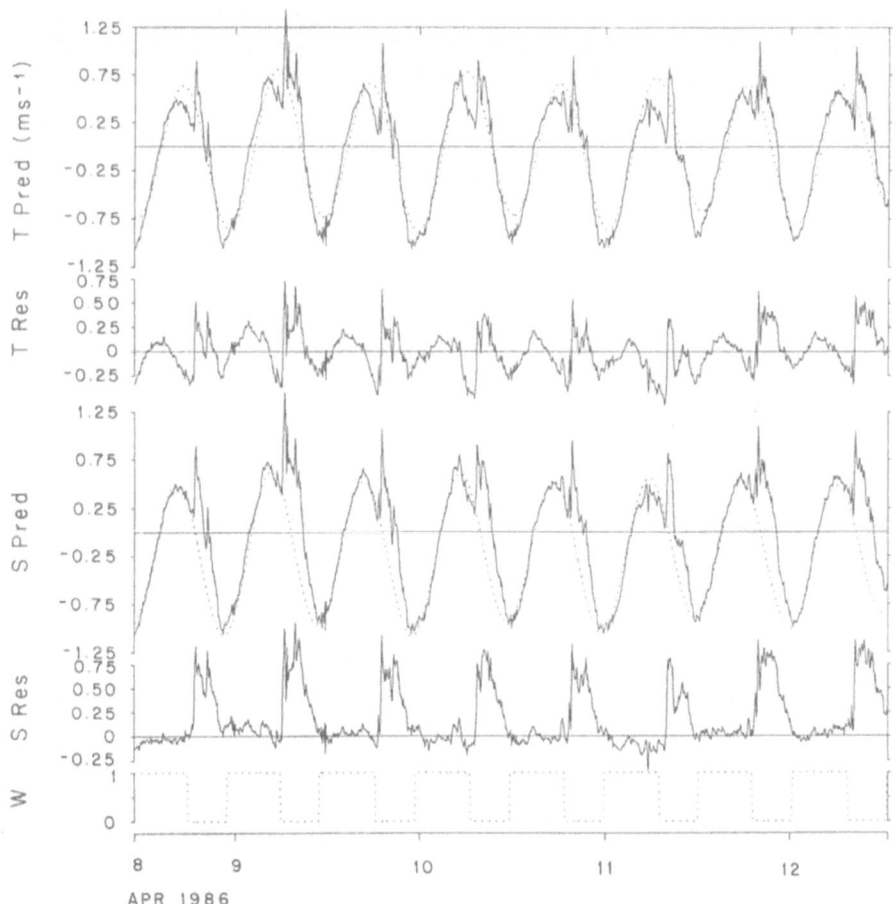

Figure 9. A comparison of tidal predictions and residuals using the response method and the modified response method of tidal analysis. The top two panels show the effect of the bore on the tidal prediction and residuals. The third and fourth panels show the improved performance using the modified response method. The bottom panel shows the weighting function used to remove the bore from the tidal analysis.

residual the procedure is clearly inadequate. The second panel in Figure 9 would indicate erroneously that the fundamental periodicity in the bore is quarter diurnal rather than semi-diurnal. We have produced the same erroneous signature using synthetic examples.

The elimination of this contamination is quite easily achieved. The principal change is to perform the least-squares fit over that subset of data points for which the bore, gaps, and other contaminating features are absent. This modification is achieved by zero-weighting the offending points in the least-squares minimization, in effect removing their influence entirely. The other procedural change is to calculate the mean or steady part of the series as the first weight of the least-squares fit, thereby releasing the mean value from the influence of the bore. The effect of using this modified, or selective, response analysis is shown in the third and fourth panels of Figure 9. Using this procedure the underlying tides can now be accurately predicted and the nature of the residual can be investigated. The removal of data points is effective and without risk as long as the data so treated do not represent large portions of each tidal cycle. The bottom panel of Figure 9 shows the sections of the data given zero weight in this example.

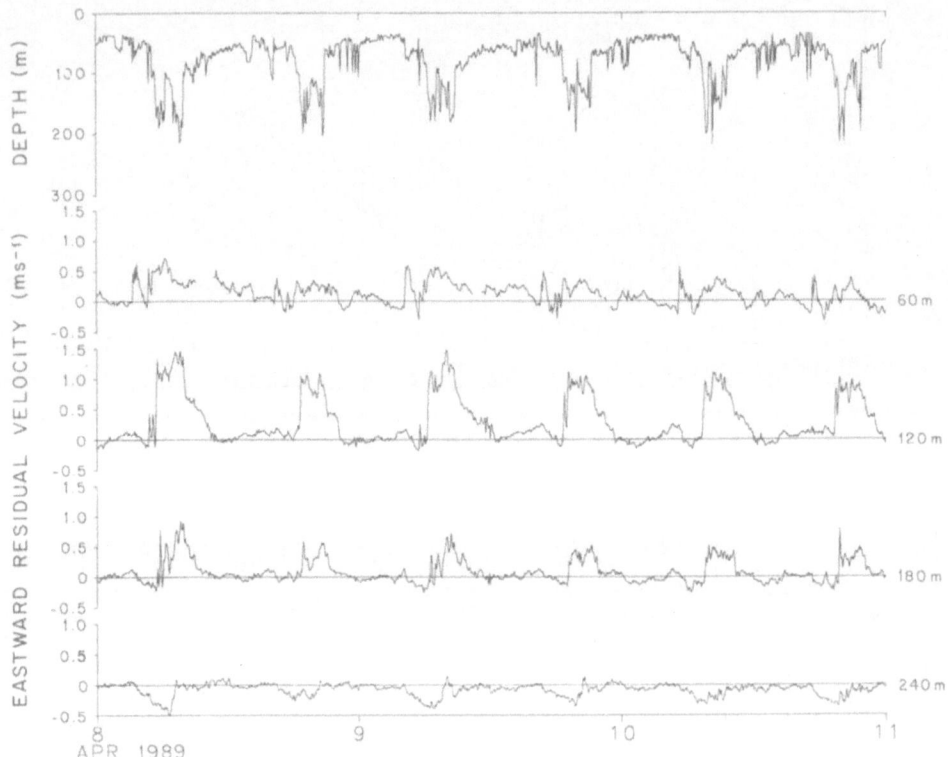

Figure 10. A time series plot characterizing the internal bore during the spring tide. The upper panel shows the large interface displacements, and the lower four panels show the tidal residual at four depths. The strength of the bore current is strongest at depths within the high shear layer (120 m), and has a primarily westward contribution at the 240 m level. At mid-depths the contribution of the bore to currents is eastward and apparent only after the interface descends to that level.

Figure 10 shows a short section (during spring tide) of residual currents at four depths after the tides have been removed using the selective response method. Also shown is the depth of the interface as estimated by the depth of the maximum vertical shear. Inspection of these records reveals several interesting and consistent features. The correlation between interface depth and current residuals shows clearly that the bore completely dominates the residual flow. The most obvious characteristics of the bore are that its amplitude is largest at depths within the high-shear transition layer, and that the leading edge of the current front is more abrupt than its trailing edge.

By inspection of Figure 10 several revealing characteristics of the bore may be established. First, although the interface depth has significant high frequency content, there is a clear depression of the order of 100-150 m that has a characteristic time scale of roughly 3 hours. There is also a tendency for the interface to rise partway through the event and then re-descend (this interpretation is consistent with, but not suggested by the velocity records). The second characteristic of principal interest is that at 240 m, normally well below the depth of the interface, the primary velocity signature of the bore is westward rather than eastward. This feature is characteristic of "first mode" internal solitary waves or bores. The fact that the passage of the bore enhances eastward flow above the interface and westward flow below, suggests that it is an agent for increasing the two-layer exchange between the Atlantic and Mediterranean basins.

2.5 INFLOW TRANSPORT

A time series estimate of transport in the Atlantic layer in the Tarifa Narrows can be made using the Doppler data set. The volume transport (Q) may be defined as

$$Q(t) = U(t)A(t) \tag{1}$$

where $U(t)$ is the velocity averaged over the upper layer and $A(t)$ represents the cross sectional area of the channel above the interface. In the present calculation $U(t)$ was taken to be the vertical average of upper-layer currents at site TC (Figure 1), and $A(t)$ was calculated from the topography and the time series measurements of the depth of the maximum shear. In order to estimate the contribution of the bore to the mean transport in the upper layer, the following decomposition was performed.

$$U = U_s + U_T + U_b + U' \tag{2}$$

where U_s is the "steady" flow component, U_T is a tidal (zero-mean) component, U_b is the velocity signature of the bore, and U' represents a zero-mean deviation of nontidal origin. Both U_s and U_T are provided by the modified (selective) response method tidal fit. The upper layer cross sectional area, $A(t)$, can be similarly decomposed to give:

$$A = A_s + A_T + A_b + A'. \tag{3}$$

Substituting equations (2) and (3) in equation (1) leads to

$$Q = (U_s + U_T + U_b + U') (A_s + A_T + A_b + A'). \tag{4}$$

Since the tidal analyses show that $U_b >> U'$ (see Figure 6) and $A_b >> A'$, we may approximate

$$U_b + U' \cong U_b \; ; \; A_b + A' \cong A_b. \tag{5}$$

Substituting and expanding, we arrive at the expression

$$Q = U_s A_s + U_s A_T + U_s A_b + U_T A_s + $$
$$ + U_T A_T + U_T A_b + U_b A_s + U_b A_T + U_b A_b. \tag{6}$$

Taking the time-average, and dropping terms that are zero by definition, we have

$$\overline{Q} = U_s A_s + U_s \overline{A_b} + \overline{U_b} A_s + \overline{U_T A_T} + \overline{U_T A_b} + \overline{U_b A_T} + \overline{U_b A_b} \tag{7}$$

$$1.0 = \quad 0.4 \qquad 0.2 \qquad 0.2 \qquad 0.0 \qquad 0.0 \qquad 0.0 \qquad 0.2$$

where the overbar notation represents the time average operator over one spring neap cycle, and the numerical values are given to a tenth of a Sverdrup (1 Sv = 10^6 m^{3}s^{-1}). Based on sensitivity studies, we estimate confidence limits on these calculations to be $\pm$ 15%.

These calculations show several results of general interest. Our direct transport estimates are significantly lower than most previous direct and indirect estimates of inflow transport through the strait. Lacombe and Richez (1982) estimated the inflow transport as 1.26 Sv, while older indirect estimates have been as high as 1.87 Sv (see Hopkins, 1978). The lower estimate found in the present study is probably due, at least in part, to the fact that the data were collected during the low period of the seasonal cycle (Bormans et al., 1986). In addition, our estimates are based upon data with significantly more spatial and temporal resolution than has previously been used, and being direct transport estimates, do not rely upon any assumed salt or water balances.

One of the most significant outcomes of this analysis is our finding that the steady component of the strait circulation actually carries only 40% of the mean Atlantic inflow in the Tarifa Narrows. As shown above, the majority (60%) of the inflow transport is associated with the internal bore. The contributions of the internal bore are split roughly equally between the mean deepening of the upper layer, the mean increase in upper layer flow speed, and the correlation between the increased layer thickness and increased flow that characterize the bore. It is noteworthy, although not surprising, that the linear tidal fluctuations make no contribution to the net transport.

3. Summary

Moored acoustic Doppler current measurements have provided a uniquely detailed view of the energetic flow in the Tarifa Narrows. These measurements make clear the important role of internal oscillations in determining the vertical and temporal structure of the currents in the upper several hundred meters. In particular, the passage of a large internal solitary wave, or bore, has been shown to dominate the current records at depths to 160 m.

Data analysis has shown that the passage of the bore enhances the two-way exchange between the Atlantic and Mediterranean basins. In fact, decomposition of the transport calculations suggest that the internal bore is the principal agent of Atlantic-layer mass transport in the Tarifa Narrows. Direct transport estimates show approximately one Sverdrup of Atlantic water flowing into the Mediterranean of which approximately 0.4 Sv is due to the steady component of the thermohaline circulation pattern. The remaining 0.6 Sv transport is accounted for by the passage through the strait of the internal bore. The average inflow transport was found to be significantly less than classical estimates of the long-term inflow. The difference is probably due to the combination of low seasonal flow during the experimental period and more accurate current measurements than have previously been used.

ACKNOWLEDGEMENTS

The authors would like to thank Dr. J.D. Irish for his insights into the response method of tidal analysis, and for keeping us from going too far in our modifications. We would also like to thank Drs. Bray and Kinder who kindly provided CTD data with which to compare our shear measurements, and Drs. H. Bryden and D. Conlon for their unflagging enthusiasm when the going got tough. This work was supported by grants and contracts from the Office of Naval Research, Coastal Sciences Division.

REFERENCES:

Armi, L. and D.M. Farmer, 1989. The flow of Mediterranean water through the Strait of Gibraltar, *Progress in Oceanography,* **21,** 1-105.

Bormans, M.C., C. Garrett, and K.R. Thompson, 1986. Seasonal variability of the surface inflow through the Strait of Gibraltar, *Oceanologica Acta,* **9**: 403-414.

Bray, N.A., 1986. *Gibraltar Experiment, CTD Data Report,* SIO Ref. Series #86-21, 212 pp.

Cartwright, D., W. Munk, and B. Zetler, 1969. Pelagic tidal measurements, *EOS,* **50,** 472-477.

Farmer, D.M. and L. Armi, 1989. The flow of Atlantic water through the Strait of Gibraltar, *Progress in Oceanography,* **21,** 1-105.

Hopkins, T.S., 1978. Physical processes in the Mediterranean basins, In *Estuarine Transport Processes,* B. Kjerfve, Ed., Univ. South Carolina Press, Columbia, SC, 269-310.

Kinder, T.H., D.A. Burns and R.D. Broome, 1986. *Hydrographic Measurements in the Strait of Gibraltar,* Naval Ocean Research and Development Activity Report 141, pp.355.

Lacombe, H. and C. Richez, 1982. The regime of the Strait of Gibraltar, in *Hydrodynamics of Semi-Enclosed Seas,* J.C.J. Nihoul, ed., Elsevier, Amsterdam, 13-74.

Munk, W. and D. Cartwright, 1966. Tidal spectroscopy and prediction. *Proc.Roy. Soc. London,* A259, 533-581.

508

Pettigrew, N.R., J.D. Wood, E.H. Pape, G.J. Needell and J.D. Irish, 1987. Acoustic Doppler current profiling from moored subsurface floats, *Proceedings of Oceans '87*, 110-116.

Richez, C., 1988. Airborne synthetic aperture radar observations of internal wave propagation in the Strait of Gibraltar, *Taller Sobre La Oceanografia Fisica Del Estrecho De Gibraltar*, Madrid, 24-28 Octubre.

Watson, G., 1989. A study of internal wave propagation in the Strait of Gibraltar using shore-based marine radar images, *Journal of Physical Oceanography*, (in press).

Wesson, J.C. and M.C. Gregg, 1988. Turbulent dissipation in the Strait of Gibraltar and associated mixing, in *Small-Scale Turbulence and Mixing in the Ocean*, ed., J. Nihoul, Elsevier, Amsterdam.

Zetler, B.D. and W.H. Munk, 1975. The optimum wiggliness of tidal admittances, *Journal of Marine Research*, Supp., 33, 1-13.

TRANSIENTS IN THE NONLINEAR ADJUSTMENT TO GEOSTROPHY

James O'Donnell
Department of Marine Sciences
The University of Connecticut
Groton, Connecticut 06340
U.S.A.

ABSTRACT. The consequences of nonlinearity on the character of the waves generated during the adjustment to geostrophic equilibrium of a layer of incompressible fluid are investigated by comparing the results of a linear analysis to the approximate solution to the full nonlinear problem obtained numerically. The calculations show that when nonlinear advection and rotation are both important to the dynamics, the evolution of the flow is quite different from that found in the linear, rotating problem of Cahn and the nonlinear, nonrotating problem of Stoker. Solutions presented demonstrate that the adjustment is accomplished by the formation of a large amplitude jump in layer depth which propagates in the opposite direction to that of the initial discontinuity, i.e. towards high pressure. Comparison of this solution to that for the linear problem suggests that the jump is formed by the steepening of Poincare waves and that nonlinear effects have little influence on the rate of approach to the final geostrophic state.

1. INTRODUCTION

Surface layers of buoyant fluid are ubiquitous in the ocean but are particularly important in sea straits where there is an exchange of fluids of differing density, and at the mouths of estuaries where there are localized sources of brackish water. At the mouth of the Chesapeake Bay, for example, a large plume of brackish water has been observed by Boicourt et al. (1987) which then forms a coastal current directed to the right of the source, and Aure and Saetre (1981) have observed the formation of a northward flowing coastal current along the coast of Norway by the outflow from the Skagerrak. The dynamics in the region where the initial alongshore turning takes place has been shown by Garvine (1987) and Chao (1988) to be principally inertial, but becomes semi-geostrophic further downstream in the coastal current. These dynamics are therefore closely related to the strait exchange flows discussed by Whitehead (1986), Garrett and Toulany (1982) and Dalziel (1988), among others.

This simple momentum balance is, in natural circumstances, periodically disrupted by the intrinsic instabilities of the flow itself, by changes in the source buoyancy flux and by wind stress. Between these perturbations however, the buoyant layer must adjust to a state in which the horizontal pressure gradient and the Coriolis acceleration are in equilibrium and, since the timescale and nature of the adjustment are important facets of real coastal currents and strait flows, the behavior of buoyant layers during the adjustment process is the focus of attention here. In particular, we solve the simplest problem that retains the effects of stratification and advection and ignore the influences of geometry, complicated initial conditions, surface fronts, and all frictional effects.

In the following section, prior work on related problems will be summarized and the problem addressed in the remainder of the paper will be presented mathematically in section

509

L. J. Pratt (ed.), The Physical Oceanography of Sea Straits, 509–516.
© 1990 *Kluwer Academic Publishers.*

510

3. In section 4, the numerical solution technique will be briefly outlined and the solutions presented and discussed. The main results and conclusions will be summarized in the closing section.

2. PREVIOUS WORK

The evolution of a pressure discontinuity in a layer of fluid in a nonrotating environment is well known. Stoker (1957) found exact solutions to the fully nonlinear "breaking dam" problem by employing the method of characteristics and his solution for layer depth at unit time intervals is shown in figure 1. Immediately after the release of the discontinuity, a bore, surge, or hydraulic jump forms and propagates towards lower pressure and a rarefaction travels in the opposite direction. In the limit of large time, the pressure field becomes uniform and the fluid stationary with all of the potential energy of the initial state transported out of the domain.

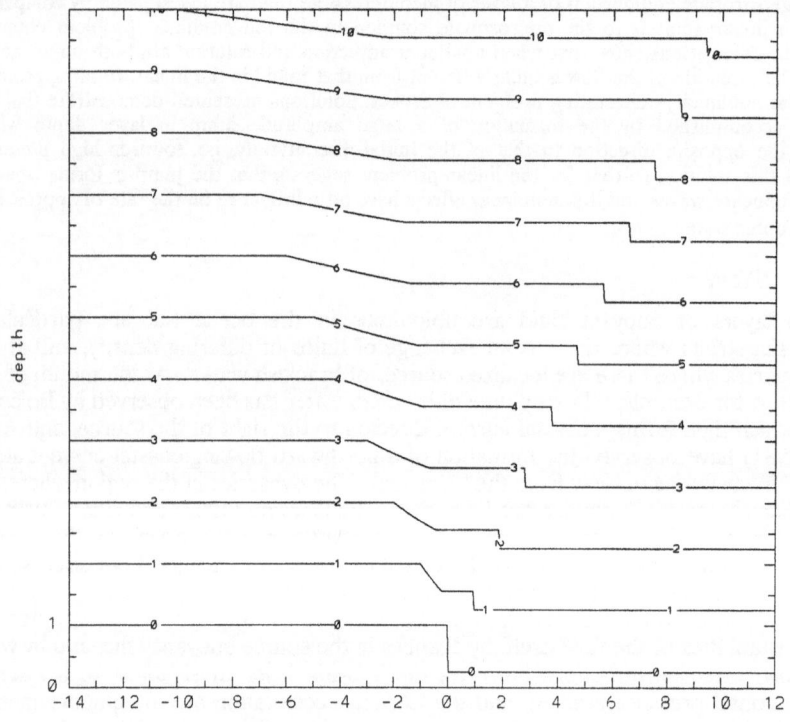

Figure 1. Stoker's analytic solution for layer thickness in the breaking dam problem at unit intervals of time. Note that D is plotted against x for the values of t indicated on the curves and offset by unity.

Some of the influences of rotation were demonstrated in a pair of landmark papers by Rossby (1937, 1938) in which he obtained the final state of a nonlinear adjustment using the principal of conservation of potential vorticity. Though the transient component of the motion was not determined, the natural length scale of the adjustment became clear as the distance a small amplitude interfacial wave would propagate in a pendulum day, or the Rossby deformation radius. In the presence of rotation, the spread of the pressure anomaly and the release of potential energy is inhibited so that in the final state, the pressure field is in geostrophic balance with a steady motion normal to the plane of the initial discontinuity. The total energy of the final state is then nonzero, though less than that of the initial state with the "missing" energy assumed to be contained in the waves generated during the adjustment.

This conjecture was subsequently confirmed when the character of the transition was determined by Cahn (1945) who found analytic solutions for the evolution of infinitesimal discontinuities. This, and related work, is thoroughly discussed in the monograph of Gill (1982). Figure 2 shows Cahn's solution for the interface displacement at unit time intervals and the final geostrophic state. Initially, the flow evolves as in the linear, nonrotating problem, see Gill (1982, p110), with infinitesimal discontinuities moving off in each direction at the shallow water phase velocity of the layer. As is apparent in the figure, an increase in pressure propagates to the right (a compression) and a decrease (a rarefaction) to the left. After $t = O(1)$, rotational effects become important and cause the appearance of a wake of dispersive Poincare waves behind each of the fronts. This solution will be discussed further after the presentation of the approximate solution to the full nonlinear problem.

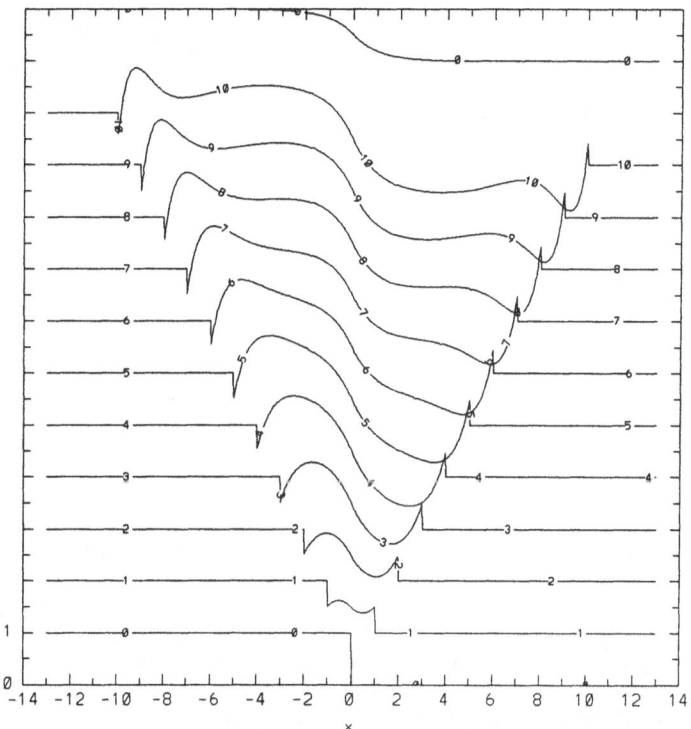

Figure 2. The analytic solution for layer depth perturbation in the linearized adjustment problem of Cahn (1945).

3. THE PROBLEM

The particular Rossby problem we consider here, is the evolution of an initially stationary layer of buoyant fluid which is bounded at $x = 0$ by a temporary "dam", and which overlies a much thicker layer of slightly denser fluid. Using D to indicate the layer depth, the initial conditions may be specified as,

$$D(x,y,t<0) = \begin{cases} \delta & x>0 \\ 1 & x\leq 0 \end{cases} \tag{1}$$

with

$$u(x,y,t<0) = v(x,y,t<0) = 0 \tag{2}$$

where u and v are the velocity components in the x and y directions respectively. These conditions are identical to those employed by Stoker (1957) and are shown at the bottom of figure 1.

For a short time after the removal of the dam at $t = 0$, the motion is complicated, but subsequently, the long wave equations (see, for example, Gill (1982), section 7.10) provide a good model of the motion. Using the local inertial period and the internal Rossby deformation radius as the time and length scales, these may be stated in dimensionless variables as,

$$\frac{\partial D}{\partial t} + \frac{\partial uD}{\partial x} = 0 \tag{3}$$

$$\frac{\partial u}{\partial t} + u\frac{\partial u}{\partial x} + \frac{\partial D}{\partial x} - v = 0 \tag{4}$$

$$\frac{\partial v}{\partial t} + u\frac{\partial v}{\partial x} + u = 0 \tag{5}$$

where y derivatives are omitted since there is no dependence of the initial conditions on the y coordinate and boundaries normal to that direction are very far from the region of interest.

4. THE SOLUTION

The numerical solution of the adjustment problem posed in (1)-(5) is complicated by the discontinuity in the initial conditions and, as will become apparent, the formation of other discontinuities during the adjustment process. Recently, several methods which resolve and track such discontinuities have been proposed, c.f. O'Donnell and Garvine (1983), O'Donnell (1988), and Bennett and Cummins (1988), but in this paper the simpler and well known Lax-Wendroff method which "smears" the discontinuities over a few grid points is employed. The main advantages of this scheme are that it is simple to program and allows jumps to develop in the interior of the domain without the complicated control logic of the sophisticated schemes. But, on the other hand, it is weakly dissipative and overshoot and oscillation of the solution in the neighborhood of discontinuities is common. Details of the procedure are provided by Houghton (1969) and therefore omitted here. Note, however; all calculations presented were performed with a spatial increment $\Delta x = 0.025$, and time step $\Delta t = C_N \Delta x / \max |u \pm c|$, where c is the shallow water phase speed and the Courant number, $C_N = 0.65$; and that the effects of high wavenumber oscillations were minimized by initiating the numerical scheme with the solution of Stoker (1957) evaluated at $t = 0.05$.

In figure 3, the computed solution for the evolution of the layer depth during the adjustment to geostrophy when $\delta = 0.25$ is presented. The initial evolution is essentially the same as in the non-rotating case described by Stoker (1957), i.e. the propagation of the discontinuity to the right and a rarefaction to the left, and the aforementioned high wavenumber oscillations at in the neighborhood of the jump and the original location of the discontinuity are evident.

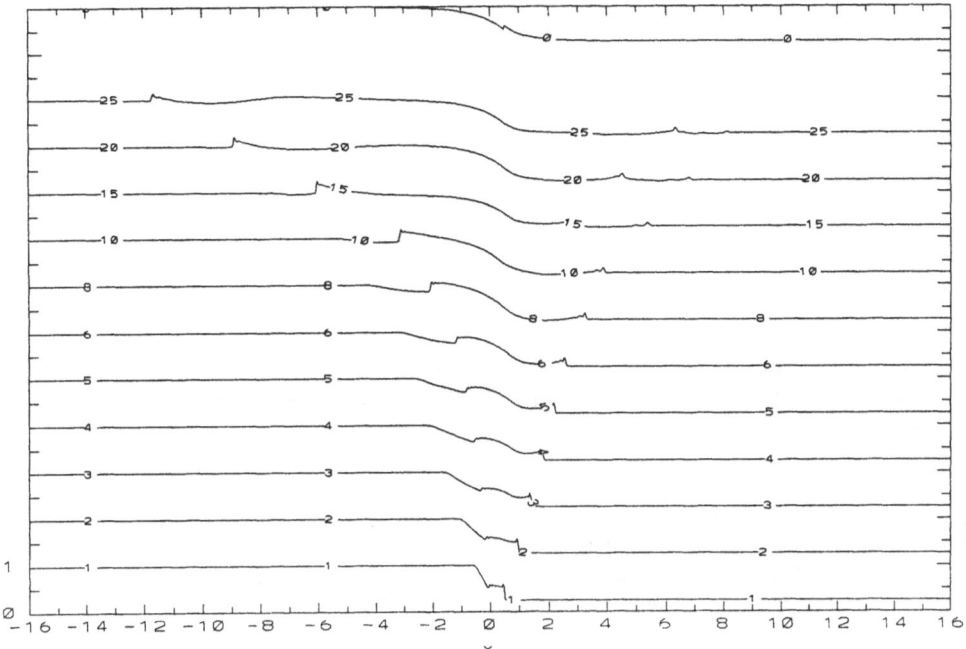

Figure 3. The numerical solution for the layer depth in the geostrophic adjustment problem with $\delta=0.25$. In the plot, solutions are offset by unity and the solution time is indicated in the line breaks. The analytic solution for large time is shown at the top of the figure.

The interesting and novel features of the solution, however, are the rapid decay of the amplitude of the rightward propagating shock, and the formation and growth of the leftward facing jump in the neighborhood of $x=0$ (see the solution at $t=3$). It is evident in the figure that at $t=10$ only a small ripple remains of the initial discontinuity which continues to propagate away to the right, closely followed by a second small amplitude jump that is barely resolved in the calculation. By this time, the leftward propagating jump has grown and accelerated, moving up the pressure gradient in the opposite direction to the particle motion. Its amplitude has decayed somewhat by $t=25$, though it remains the main transient of the flow. The solution for both the layer depth and x direction velocity in the neighborhood of this jump is presented in greater detail in figure 4 and shows that $u\,\partial u/\partial x$ remains a significant term in the momentum balance in the neighborhood of the propagating jump even at large time.

The origin of the leftward propagating jump can be understood by a comparison of figures 2 and 3, the linear and nonlinear solutions respectively. Since the leftward propagating

514

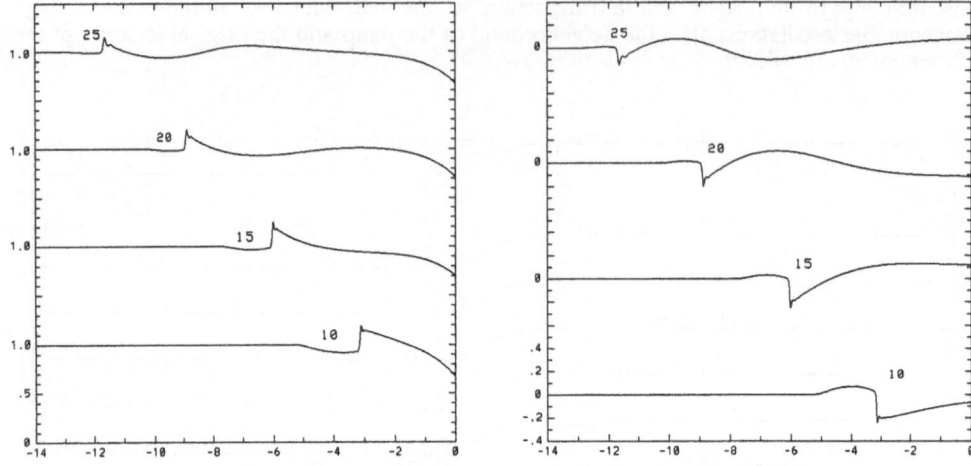

Figure 4. A detailed view of the leftward propagating jump shown in figure 3. The layer depth is shown on the left and the x component of velocity on the right.

rarefaction is followed in the linear solution by a trailing wake of Poincare waves which increase the pressure after the passage of the front, and since it is well known that a compression wave will steepen in a nonlinear long wave model, it is reasonable to conclude that the secondary jump is caused by the nonlinear steepening of the Poincare wake. The calculation of the trajectories of characteristic lines (not shown here) support this conclusion since the u-c family of wave fronts show a tendency to converge.

The decay of the rightward propagating jump can also be understood by comparison of figures 2 and 3. In the linear solution, the compression front is followed by a rarefaction of Poincare waves which should be expected to expand, thereby reducing the amplitude of the leading jump.

The complete adjustment to geostrophy in the neighborhood of the initial position of the discontinuity is only achieved when all the wave energy has propagated to large $|x|$ which, as Gill (1976) has pointed out, takes a long time in the linear model since the group velocity of long Poincare waves is small. The influence of nonlinearities on the rate of approach to equilibrium is demonstrated in figure 5, which compares $u(x=0,t)$ in the linear and nonlinear solutions. Clearly, nonlinear effects have no significant influence on the rate of approach to the final geostrophic state.

5. SUMMARY

The results of the calculation presented above demonstrate that the adjustment to geostrophy of a large amplitude pressure anomaly in a rotating layer of fluid is accomplished by the formation of a jump in layer depth that propagates away from the position of the initial discontinuity in the direction of higher pressure, in contrast to the behavior in the absence of rotation. A comparison of the solutions to the linear and nonlinear problems and consideration of the trajectories of characteristic lines leads to the conclusion that the jump is caused by the nonlinear steepening of short wavelength Poincare waves.

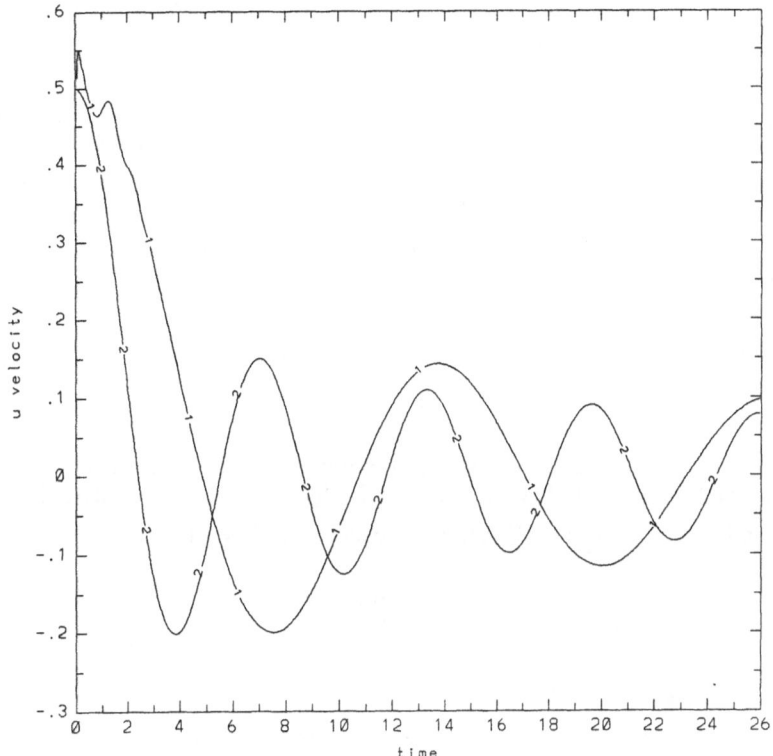

Figure 5. The evolution of the x direction velocity at $x=0$ in the nonlinear (1), the linear (2) problems.

The problem studied here is much too simple to be considered a model of the real ocean or atmosphere, of course, but does highlight the need for caution in the oceanic application of unsteady layer models which ignore rotation. It also prompts the development and analysis of a model with two active layers and bottom topography which would allow the investigation of transient behavior in highly stratified flows through straits.

Acknowledgements
This work was supported by grant OCE-8816566 from the U.S. National Science Foundation and facilitated by equipment provided by The University of Connecticut.

6. REFERENCES

Aure, J. and R. Saetre (1981) Wind effects on the Skagerrak outflow. In "The Norwegian Coastal Current", R. Saetre and M. Mork edts. Univ. of Bergen.

Bennett, A.F. and P.F. Cummins. (1988) Tracking fronts in solutions to the shallow water equations. J. Geophys. Res., 93, 1293-1301.

Boicourt, W.C., S.-Y. Chao, H.W. Ducklow, P.M. Glibert, T.C. Malone, M.R. Roman, L.P. Sanford, J.A. Fuhrman, C. Garside and R.W. Garvine (1987) Physics and microbial ecology of a buoyant plume on the continental shelf. Eos, v68, n31, 666-668.

Cahn, A. (1945) An investigation of the free oscillation of a simple current system. J. Meteorol., 2, 113-119.

Chao, S.-Y. (1988) River forced estuarine plumes. J. Phys. Oceanogr., 18, 72-88.

Dalziel, S.B. (1988) Two-layer hydraulics: maximal exchange flows. Ph.D. dissertation, Dept. of Applied Mathematics and Theoretical Physics, University of Cambridge. England.

Garrett, C.J.R. and B. Toulany (1982) Sea level variability due to meteorological forcing in the northeast Gulf of St. Lawrence. J. Geophys. Res., 87, 1968-1978.

Garvine, R.W. (1987) Estuary plumes and fronts in shelf waters: A layer model. J. Phys. Oceanogr., 17, 1877-1896.

Gill, A.E. (1976) Adjustment under gravity in a rotating channel. J. Fluid Mech., 77, 603-621.

Gill, A.E. (1982) Ocean-Atmosphere Dynamics. Academic Press, New York, N.Y. 662pp.

Houghton, D.D. (1969) Effects of rotation on the formation of hydraulic jumps. J. Geophys. Res., 74, 1351-1360.

O'Donnell, J. (1988) A numerical technique to incorporate frontal boundaries in two dimensional layer models of ocean dynamics. J. Phys. Oceanogr., 18, 1584-1600.

O'Donnell, J. and R.W. Garvine (1983) A time dependent, two-layer model of buoyant plume dynamics. Tellus, 35A, 73-80.

Rossby, C.-G. (1937) On the mutual adjustment of pressure and velocity distributions in certain current systems, 1. J. Mar. Res., 1, 15-28.

Rossby, C.-G. (1938) On the mutual adjustment of pressure and velocity distributions in certain current systems, 2. J. Mar. Res., 2, 239-263.

Stoker,J.J. (1957) Water Waves. Interscience, New York, N.Y., 567pp.

Whitehead, J.A. (1986) Flows of a homogeneous rotating fluid through straits. Geophys. Astrophys. Fluid Dynamics, 36, 187-205.

IV. Outflows, Turbulence and Mixing

CAN MIXING IN EXCHANGE FLOWS BE PREDICTED USING INTERNAL HYDRAULICS?

Gregory A. LAWRENCE
Department of Civil Engineering
University of British Columbia
Vancouver, British Columbia, V6T 1W5
Canada

ABSTRACT. The gravitational exchange of two fluids of slightly different density through a constriction is analysed using internal hydraulic theory. The results show that Long's (1956) criterion for the stability of long internal waves is violated. However, it is the stability of short wave instabilities, most notably the Kelvin-Helmholtz instability, that is most relevant. These instabilities cause mixing between the layers. An upper bound on the vertical extent of this mixing can, to a first approximation, be predicted from internal hydraulic theory.

1. Introduction

The gravitational exchange of fluid through a strait connecting two seas of slightly different density is a long-standing problem of considerable oceanographic interest. Marsili (1681) related the exchange of fluid caused by the removal of a barrier separating fresh and salt water in a tank to the flow in the Bosporus. Defant (1961) discusses several other straits (including Gibraltar, Bab el Mandel, Hormuz and Messina), where significant exchange flows occur. It is the exchange flow through the Strait of Gibraltar that has attracted perhaps the greatest interest, see Kinder & Bryden (1987), both because of its strategic location and the tremendous volume of water (about 10^6 m^3/s in each direction), that is transported through it. Armi & Farmer (1989) have presented a detailed model of exchange through the Strait of Gibraltar, based on extensive field measurements, and the application of the hydraulics of layered flows as presented in Armi (1986), Armi & Farmer (1986), and Farmer & Armi (1986).

The degree to which exchange flows affect oceanographic conditions on either side of a strait is dependent, at least in part, on the amount of mixing that occurs within the strait. There is always some mixing between the two layers, but it is generally far from complete. The present paper reviews the theory of the hydraulics of two-layer flows, and then extends it to investigate the possibility of predicting the extent of mixing in exchange flows.

L. J. Pratt (ed.), The Physical Oceanography of Sea Straits, 519–536.

2. Review of the Hydraulics of Two-Layer Flows

Four assumptions, subsequently referred to as the hydraulic assumptions, are used in the study of layered flows. They are: (i) the fluids are inviscid; (ii) the pressure is hydrostatic; (iii) the density and velocity are constant except at the interfaces between the layers, which are characterised by discontinuities in density and/or velocity; and (iv) the flow is non-rotating. Although these assumptions are frequently violated, and can, under certain circumstances, be relaxed, the one-dimensional equations that arise from them are basic to an understanding of layered flows. Houghton & Isaacson (1970), Baines (1984), and Armi (1986) have expressed these one-dimensional equations in the general form:

$$v_t + C\, v_x = D\, f_x \tag{1}$$

where x is the flow direction and t is time. If the flow has m layers: v and f are vectors with 2m and 2 elements respectively; C is a 2m x 2m matrix; and D is an 2 x 2m matrix. Armi (1986) gives the coefficients of v, f, C, and D for both single- and two-layer flow.

As a consequence of the hydrostatic assumption (1) has only long wave solutions. The characteristic velocities (celerities) of these long waves are specified by the eigenvalue equation:

$$\mathrm{Det}\,(\,C - \lambda\, I\,) = 0 \tag{2}$$

where I is the identity matrix. In general, the characteristic velocities are given by:

$$\lambda^\pm = u^* \pm c \tag{3}$$

The ratio of the convective velocity, u^*, to the phase speed, c, is traditionally known as the Froude number; i.e., $Fr = u^*/c$. If $Fr = 1$ the flow is said to be critical.

For steady flow the non-dimensional equations (1) may be rewritten as:

$$v_x = \frac{R}{\mathrm{Det}\,C} \tag{4}$$

where $R = (\mathrm{Adj}\,C)\,D\,f_x$. Solutions exist if the matrix C is non-singular or, if at locations where the singularity condition, $\mathrm{Det}\,C = 0$, is satisfied, the regularity condition, $R = 0$, is also satisfied. The term "control", is traditionally used to describe locations where $\mathrm{Det}\,C = 0$. At control locations the values of the dependent variables, represented by v, can be determined using the regularity and singularity conditions. In single-layer flow :

$$\mathrm{Det}\,C = gy\,(\,F^2 - 1\,) \tag{5a}$$

with the single-layer Froude number:

$$F^2 = \frac{u^2}{gy} \tag{5b}$$

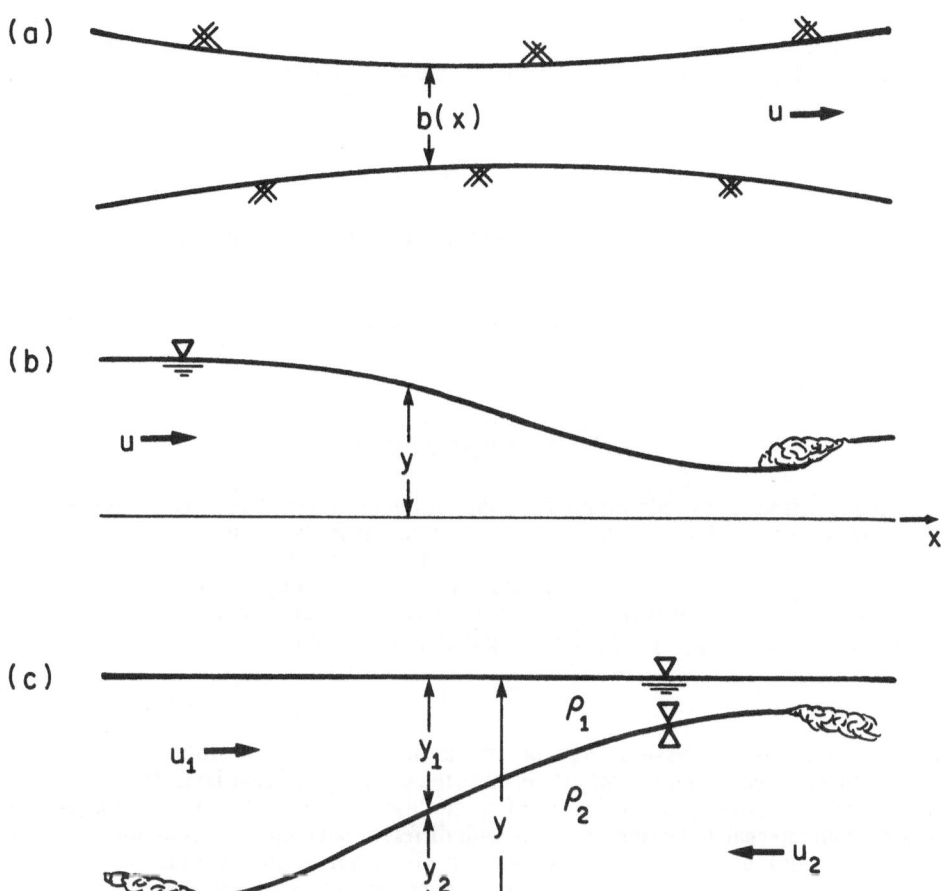

Fig. 1   (a) Plan view for all flows and side view for (b) single-layer flow, and (c) two-layer exchange flow.

where u is the cross-sectionally averaged flow velocity, y is the depth of flow, and g is the gravitational acceleration. The flow is said to be subcritical, critical, or supercritical depending on whether F^2 is greater than, equal to, or less than unity. It is customary to use F^2 rather than F, since it is F^2 that appears in the solutions of the hydraulic equations, and since, depending on the convention used, u may be negative. Definition diagrams for single and two-layer flow through a constriction are given in Fig. 1.

For single layer flow the singularity condition becomes:

$$F^2 = 1 \tag{6}$$

and we see that a single layer flow is always critical at a control. A little more care must be taken when considering two-layer flows where:

$$\text{Det } C = \varepsilon\, g^2\, y_1\, y_2\, (1 - G^2) \tag{7}$$

with the composite Froude number:

$$G^2 = F_1^2 + F_2^2 - \varepsilon\, F_1^2\, F_2^2 \tag{8}$$

where the densimetric Froude numbers of the individual layers $F_n^2 = u_n^2 / \varepsilon g y_n$, i = 1,2; ρ_n, u_n, and y_n are the density, velocity and thickness of layer n, subscript 1 refers to the upper layer, and subscript 2 refers to the lower layer. Note than in two-layer exchange flow (sec Fig. 1c) the velocity of the layer flowing left-to-right is positive by convention, and the velocity of the layer flowing right-to-left is negative. The relative density difference between the layers $\varepsilon = (\rho_2 - \rho_1)/\rho_2$. The singularity condition is:

$$G^2 = 1 \tag{9}$$

Comparison of this expression with the singularity condition for single layer flow (6) suggests that the parameter G^2 may determine the criticality of two-layer flow just as the parameter F^2 determines the criticality of single-layer flow. However, a two-layer flow supports both internal and external waves with different characteristic velocities. It is not immediately obvious how the single parameter, G^2, can determine the criticality of two-layer flows, since it cannot be the ratio of convective velocity to phase speed for both internal and external waves. Further discussion of this problem is best deferred until after a brief discussion of the solution of Schijf & Schönfeld (1953) for flows satisfying the Boussinesq approximation, $\varepsilon \ll 1$. This assumption is made throughout the remainder of this paper. Those interested in the non-Boussinesq case are referred to Lawrence (1990).

2.1. BOUSSINESQ TWO-LAYER FLOWS

In a two-layer system the eigenvalue equation (2) has solutions of the form:

$$\lambda_E^{\pm} = u_E^* \pm c_E \tag{10a}$$

$$\lambda_I^{\pm} = u_I^* \pm c_I \qquad (10b)$$

The external characteristic velocities, $\lambda_E^{\pm}$, correspond to free surface wave motions; and the internal characteristic velocities, $\lambda_I^{\pm}$, correspond to internal (interfacial) wave motions. The convective velocities are denoted, u_E^* and u_I^*, and the phase velocities, c_E and c_I. Using the results of Schijf and Schönfeld (1953), for the Boussinesq case, we obtain the external Froude number:

$$F_E = \frac{\tilde{u}}{\sqrt{gy}} \qquad (11)$$

where the flow weighted mean velocity $\tilde{u} = (u_1y_1 + u_2y_2)/y$, and $y = y_1 + y_2$. So for Boussinesq two-layer flows the external Froude number is, as we would expect, the same as the single layer Froude number, see (5b). The internal Froude number :

$$F_I = \frac{u_1y_2 + u_2y_1}{\sqrt{g'yy_1y_2(1 - F_\Delta^2)}} \qquad (12)$$

where $g' = \varepsilon g$. The stability Froude number:

$$F_\Delta^2 \equiv \frac{\Delta u^2}{g'y} \qquad (13)$$

where $\Delta u = u_2 - u_1$. There are two important points that we can immediately make with respect to the stability Froude number. Firstly, from (12), we see that Long's (1956) stability criterion for long internal waves, i.e. $F_\Delta^2 \leq 1$, must be satisfied for the internal Froude number to have real values. Secondly, the stability Froude number can be regarded as an inverse bulk Richardson number. The significance of these points will be discussed below.

2.2. FUNDAMENTAL RELATIONSHIP BETWEEN FROUDE NUMBERS

The above review has identified four Froude numbers of importance to the study of two-layer flows: the external Froude number, F_E; the composite Froude number, G; the internal Froude number, F_I; and the stability Froude number, F_Δ. Lawrence (1985, 1990) uses the fact that the determinant of a matrix is the product of its eigenvalues, i.e., Det $C = \lambda_E^+ \lambda_I^+ \lambda_I^- \lambda_E^-$, together with the expressions for the characteristic velocities (10) and the expression for Det C (7) to obtain:

$$(1 - G^2) = (1 - F_\Delta^2) (1 - F_E^2) (1 - F_I^2) \qquad (14)$$

Equation (14) is the fundamental relationship between the Froude numbers relevant to two-layer flows. For the remainder of this paper the free surface deflection, and the external Froude number, will be assumed negligible. In which case:

$$(1 - G^2) = (1 - F_\Delta^2)(1 - F_I^2)$$ (15)

An important point to note is that the composite Froude number does not distinguish between stable and unstable flows. From (15) we see that a value of $G^2 > 1$ may apply to either, a supercritical ($F_I^2 > 1$), stable ($F_\Delta^2 < 1$) flow, or an unstable flow (i.e., $F_\Delta^2 > 1$, and $F_I^2 < 0$). Since the composite Froude number has been used successfully in many studies of two-layer flow we must first ask whether this point is of practical relevance. Do flows that violate Long's stability criterion occur? This question will be addressed for Boussinesq two-layer flows in general in the following section and for exchange flows through a constriction in §4. The possibility of using these results to predict the extent of interfacial mixing will be investigated in §5.

3. The Stability Froude Number

In this section general expressions will be derived for the stability Froude number, for both exchange and unidirectional flows. The first step is to rewrite the definition of the stability Froude number (13) in terms of dimensionless variables:

$$F_\Delta^2 = (q_2' / b' y_1' y_2')^2 (q_r y_2' \pm y_1')^2$$ (16)

where $y_n' = y_n/y$, $q_2' = q_2 / \sqrt{g' b_o^2 y^3}$, q_2 is the volumetric flow rate in the lower layer, and the relative width of the channel, $b' = b/b_o$, where b_o is the width at the narrowest point (throat) of the constriction. The negative sign applies to unidirectional flow, and the positive sign applies to exchange flow. Flows from left-to-right are positive, so for the example given in Fig. 1c, q_2 and u_2 are negative. The flow rate ratio $q_r = |q_1| / |q_2|$.

Defining $u_n' = u_n/\sqrt{g'h}$ the continuity equation becomes:

$$q_n' = y_n' u_n' b' \qquad n = 1,2$$ (17)

The variation of F_Δ^2 is best illustrated using the Froude number plane, see Armi (1986). This involves expressing F_Δ^2 in terms of F_1^2 and F_2^2. The assumption of negligible free surface deflection gives:

$$y_1' + y_2' = 1$$ (18)

which together with the definition of the densimetric Froude numbers, yields:

$$y_1' = \frac{q_r^{2/3} F_2^{2/3}}{F_1^{2/3} + q_r^{2/3} F_2^{2/3}}$$ (19a)

$$y_2' = \frac{F_1^{2/3}}{F_1^{2/3} + q_r^{2/3} F_2^{2/3}}$$ (19b)

and $$\left\{ \frac{q_2'}{b'} \right\}^2 = \frac{F_1^2 F_2^2}{\{F_1^{2/3} + q_r^{2/3} F_2^{2/3}\}^3}$$ (19c)

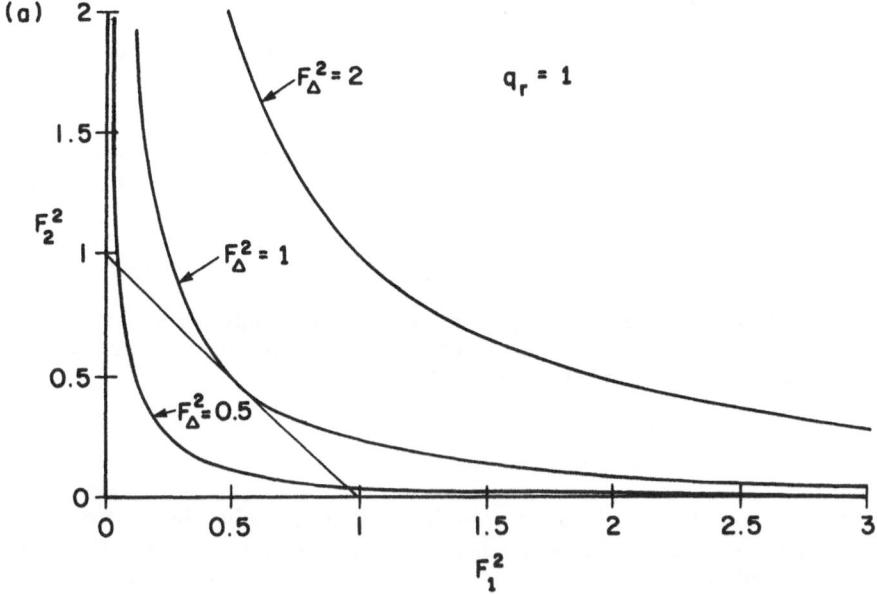

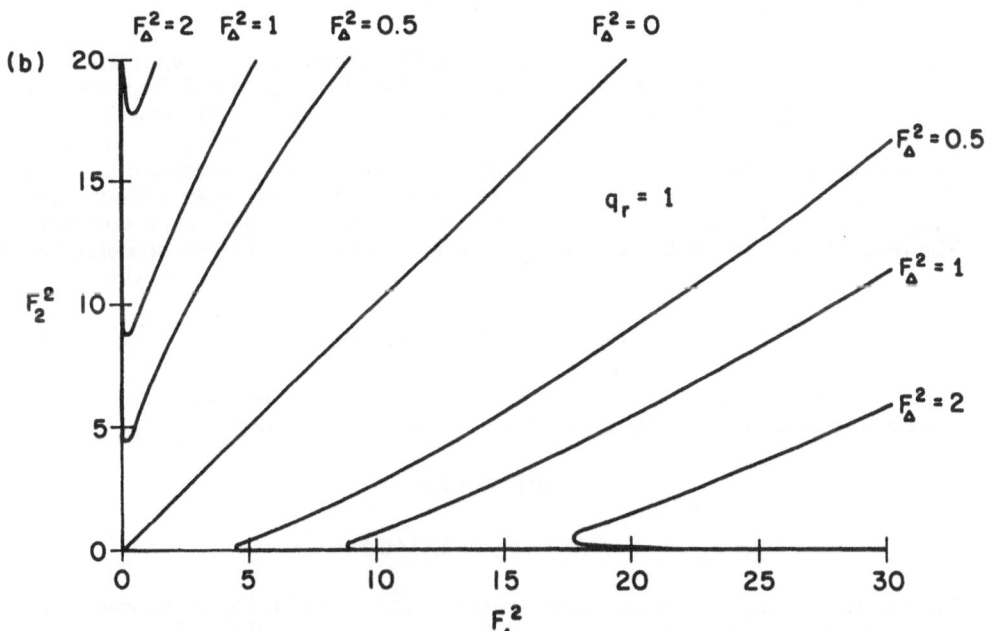

Fig. 2 Contours of F_Δ^2 in the Froude number plane for Boussinesq flows with $q_r = 1$.
(a) Exchange flow. (b) Unidirectional flow.

Substituting (19) into (16) gives:

$$F_{\Delta}^2 = \frac{F_1^{2/3} F_2^{2/3} (q_r^{1/3} F_1^{2/3} \pm F_2^{2/3})^2}{F_1^{2/3} + q_r^{2/3} F_2^{2/3}} \tag{20}$$

Contours of F_{Δ}^2 are plotted on the Froude number plane for $q_r = 1$, for both exchange (Fig. 2a), and unidirectional flows (Fig. 2b). Two aspects of these plots need to be emphasized:

(i) For exchange flows, and to a lesser extent, unidirectional flows, there are significant portions of the Froude number plane where Long's stability criterion is violated. The accessibility of these regions will be investigated below.

(ii) The fact that, for $q_r = 1$, Long's stability criterion is only violated if the flow is supercritical, is shown in Fig. 2a. This result is seen to apply for all values of q_r by expanding F_{Δ}^2 and G^2 in terms of q_r, y_1 and y_2 to obtain:

$$\frac{F_{\Delta}^2}{G^2} = 1 - \frac{(q_r^2 y_2'^2 \pm y_1'^2)^2}{q_r^2 y_2'^3 + y_1'^3} \tag{21}$$

Here the positive sign indicates unidirectional flow, and the negative sign exchange flow. Equation (21) in conjunction with the fundamental relationship between G^2, F_{Δ}^2, & F_I^2 (15) requires that all internally subcritical flows satisfy Long's stability criterion.

4. Exchange Flow Through a Constriction

The variation of the stability Froude number and other parameters of interest will now be derived for exchange flow through a constriction. This flow has already been studied in some detail by Armi & Farmer (1986), subsequently referred to as A&F. Only the case in which both layers are flowing, which A&F call moderately barotropic flow, will be considered. A&F showed that for all values of q_r Long's stability criterion is satisfied at the throat of a constriction, but they did not present results for any other sections. A&F's analysis will now be extended to evaluate the variation of F_{Δ}^2 throughout the constriction. The momentum equation for two-layer flow can be re-expressed as the conservation of what A&F call the dimensionless energy difference between the layers, $\Delta H'$, where

$$\Delta H' = y_2' + (u_2'^2 - u_1'^2)/2 \tag{22}$$

A&F show that if q_r is respectively less than, or greater than, unity then there is a virtual control downstream, or upstream, of the throat. At the virtual control

$$u_{1v}' = -u_{2v}' \tag{23}$$

and

$$\Delta H_v' = y_{2v}' = 1/(1+q_r) \tag{24}$$

The subscript v is used to denote conditions at the virtual control. The term virtual control is used to describe any location, other than the throat, where the flow is critical. As indicated in Fig. 1c, internal hydraulic jumps may occur, $\Delta H'$ will not be conserved in

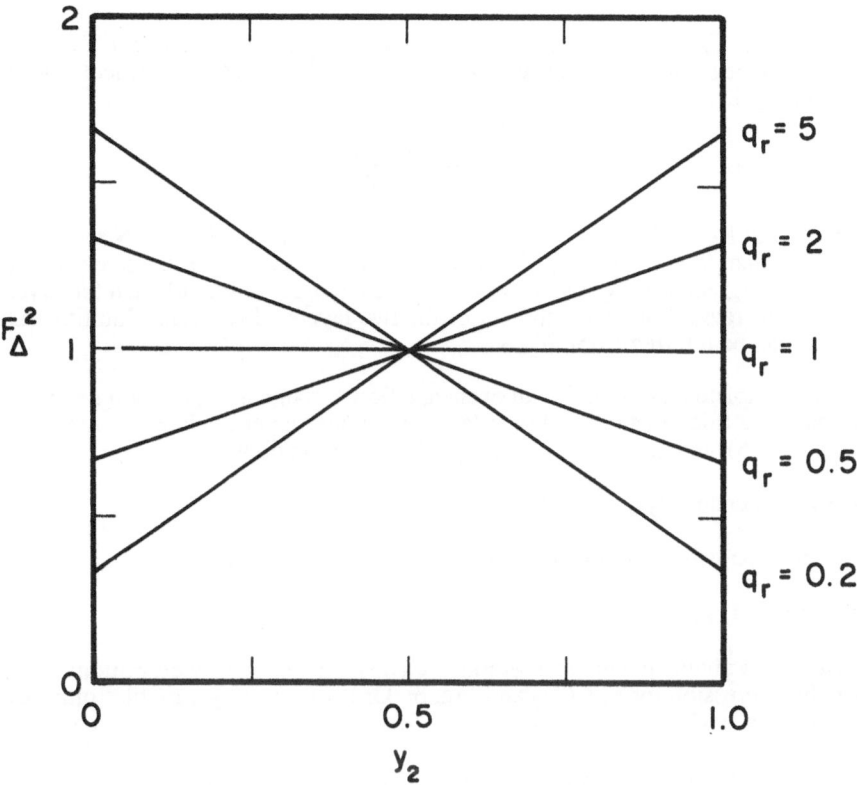

Fig. 3　Variation of F_Δ^2 with the dimensionless interface height y_2 for exchange flow through a constriction (from Lawrence 1990).

these jumps, but for the present analysis they can be assumed to occur outside the region of interest. If $\Delta H'$ is conserved (22) - (24) yield

$$\{ \frac{q_2}{b} \}^2 = \frac{2 (y_1 y_2)^2}{(1 + q_r) (y_1 + q_r y_2)} \tag{25}$$

The primes used to denote dimensionless quantities will now be dropped.

The variation of F_Δ^2 is obtained from the dimensionless expression for F_Δ^2 (16), the assumption of negligible free surface deflection (18), and the requirement that $\Delta H'$ is conserved (25), i.e.:

$$F_\Delta^2 = \frac{2 \{ 1 - (1 - q_r) y_2 \}}{1 + q_r} \tag{26}$$

The variation of F_Δ^2 with y_2 for various values of q_r is plotted in Fig. 3. Note that, if $q_r = 1$ the flow is marginally stable ($F_\Delta^2 = 1$) throughout. If $q_r \neq 1$ a moderately barotropic flow attains marginal stability when $y_2 = 0.5$, and becomes unstable when the layer with the higher flow rate becomes the thinner of the two layers. The overall limiting value of $F_\Delta^2 = 2$, as can be inferred from Fig. 3.

For a more complete understanding of exchange flow through a constriction it is important to determine the thickness of the lower layer (the upper layer thickness is then obtained from equation 18), and the width of the channel at three locations:

(i) at the virtual control (y_{2v} and b_v);

(ii) at the position where the flow is marginally stable (y_{2m} and b_m);

(iii) at the throat (y_{2o} and b_o).

Only two of the above heights and widths are constants: $b_o = 1$ by definition; and $y_{2m} = 0.5$, see the expression for F_Δ^2 (26) and Fig. 3. Of the remaining variables only one, $y_{2v} = 1/(1+ q_r)$, is trivial to determine. Two others, y_{2o} and b_v, have been evaluated numerically, see Fig. 5 in A&F . However, it is not necessary to resort to numerical solutions, and the derivation of algebraic expressions for y_{2o}, and b_v will be presented below. In addition, a new expression for the width of the channel at the point of marginal stability is derived. The height of the interface at the throat is determined by using the fact that b has a minimum value at the throat, in conjunction with the expression for q_2/b (25), to obtain:

$$3 (q_r - 1) y_{2o}^2 + (5 - q_r) y_{2o} - 2 = 0 \tag{27}$$

Solving (27) for $q_r = 1$ gives $y_{2o} = 0.5$, recall that in addition if $q_r = 1$, then $y_{2v} = y_{2m} = 0.5$, so that the flow is both critical and marginally stable at the throat, as noted by A&F. Evaluating the expression for q_2/b (25) at the throat (where $b = 1$), gives $q_2 = 1/4$. Substituting back into (25) gives the variation of layer thickness with channel width; i.e.,

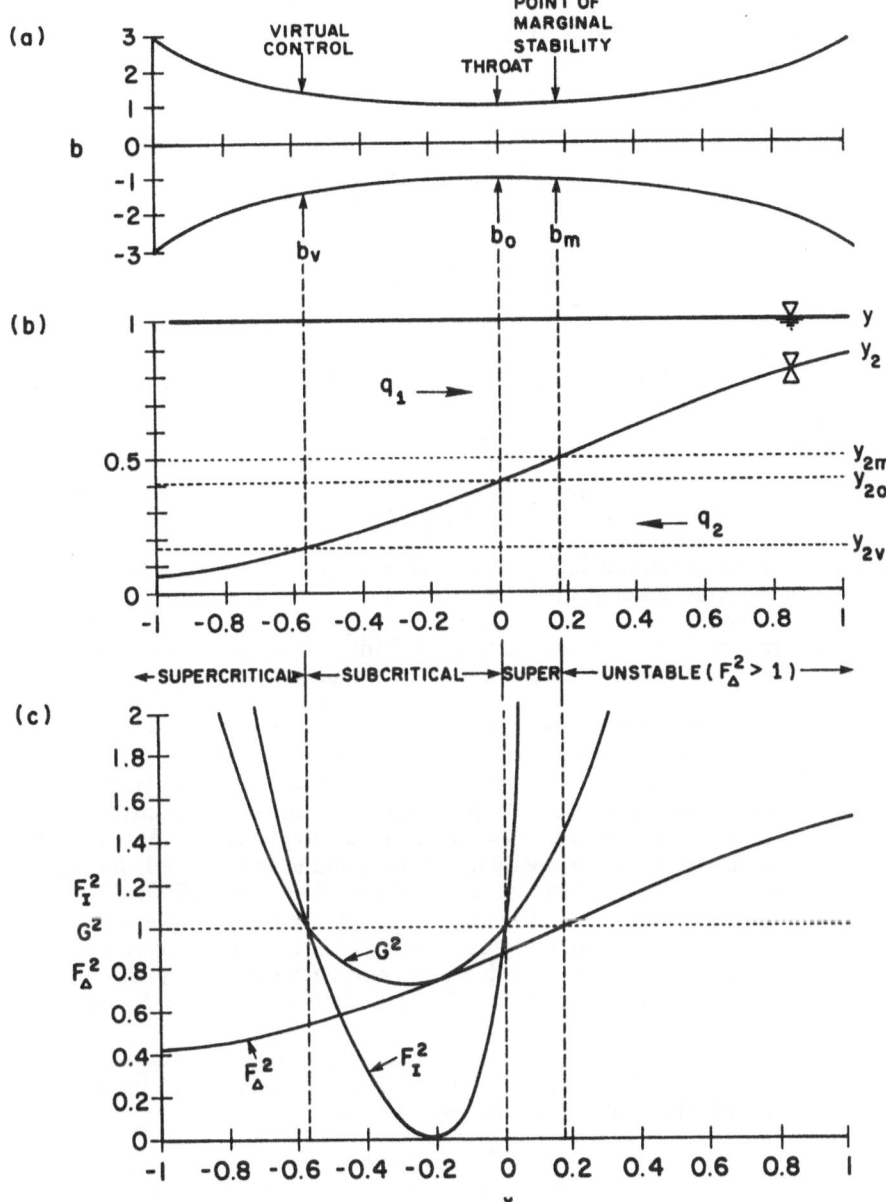

Fig. 4   Exchange flow through a constriction with $q_r = 5$ and $b' = \exp(x^2)$. (a) Plan view of the constriction. (b) Plot of the variation of interface height. (c) Plot of the variations of F_I^2, F_Δ^2, and G^2 (from Lawrence 1990).

$$y_2 = \begin{cases} \frac{1}{2}\left\{ 1 - \sqrt{1 - \frac{1}{b}} \right\} & x < 0 \\ \frac{1}{2}\left\{ 1 + \sqrt{1 - \frac{1}{b}} \right\} & x > 0 \end{cases} \tag{28}$$

Solving (27) for $q_r \neq 1$ gives:

$$y_{2o} = \frac{q_r - 5 + \sqrt{q_r^2 + 14\,q_r + 1}}{6\,(\,q_r - 1\,)} \tag{29}$$

and

$$q_2 = \sqrt{\frac{2\,\gamma}{1 + q_r}} \tag{30}$$

where $\gamma = \dfrac{(\,(1 - y_{2o})\,y_{2o}\,)^2}{1 - (1 - q_r)\,y_{2o}}$. The general expression for the width of the constriction, obtained from (25) and (29), is:

$$b^2 = \gamma\,\frac{y_1 + q_r\,y_2}{(\,y_1\,y_2\,)^2} \tag{31}$$

The above results are illustrated in Fig. 4 for an exchange flow with $q_r = 5$ and $b = \exp(x^2)$. A plan of the constriction is given in Fig. 4a, and an elevation showing the variation of interface height in Fig. 4b. The three heights y_{2v}, y_{2o}, and y_{2m} are plotted showing that the virtual control and the point of marginal stability occur on opposite sides of the constriction. The virtual control is on the side where the slower moving layer is thinner.

The variations of F_Δ^2, G^2, and F_I^2 plotted in Fig. 4c satisfy the fundamental relationship between them (15). To the left of the virtual control the flow is supercritical with a passive upper layer. Between the virtual control and the throat the flow is subcritical. Note that the internal Froude number drops to zero at the point where the internal convective velocity, $u_c = u_1\,y_2 + u_2\,y_1$, equals zero, which for $q_r = 5$ corresponds to $y_2 = 1/(1+\sqrt{5})$. To the left of this point the internal convective velocity is negative, and to the right it is positive. For further discussion of the internal convective velocity see Lawrence (1990). Between the throat and the point of marginal stability the flow is supercritical, and the lower layer is passive. To the right of the point of marginal stability Long's stability criterion is violated and the validity of hydraulic analysis in this region must be questioned. In this region $F_I^2 < 0$, corresponding to imaginary convective velocities. Some of the above features are also represented on the Froude number plane given in Fig. 5.

Substituting y_{2v} and y_{2m} into the expression for b (31) gives expressions for the width at the virtual control and at the point of marginal stability; i.e.:

$$b_v^2 = 2\,\gamma\,\frac{(1+q_r)^3}{q_r} \tag{32}$$

$$b_m^2 = 8\,\gamma\,(1 + q_r) \tag{33}$$

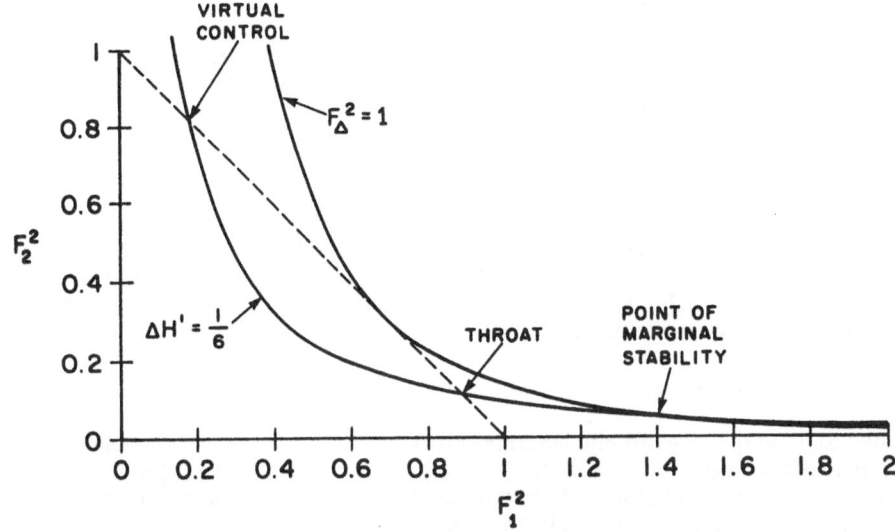

Fig. 5   Exchange flow through a constriction with $q_r = 5$ and b' = $\exp(x^2)$ represented on the Froude number plane. The flow follows the line of constant dimensionless energy difference between the layers, $\Delta H' = 1/(1+q_r) = 1/6$ (from Lawrence 1990).

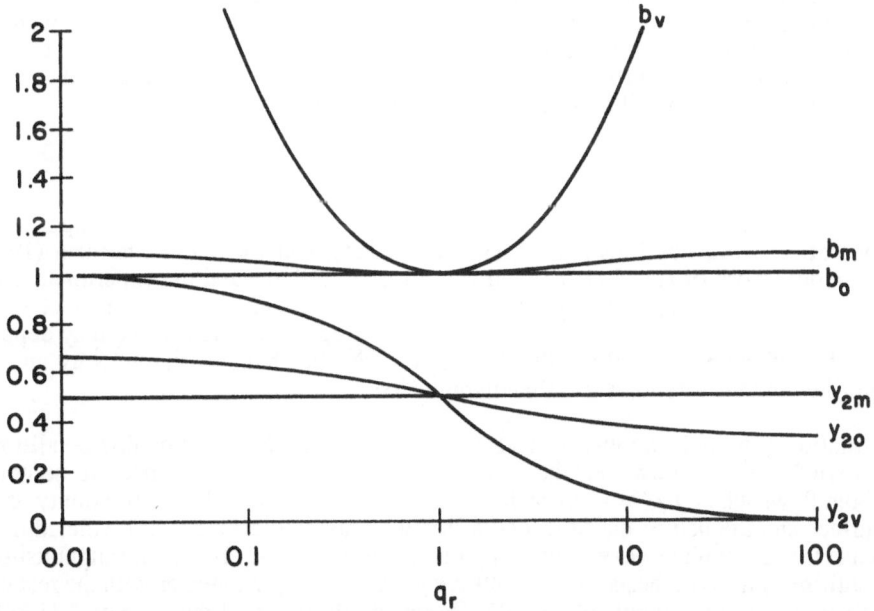

Fig. 6   Plot of the variation with q_r of the widths and interface heights at the virtual control, the point of marginal stability, and the throat (from Lawrence 1990).

These variations are plotted on Fig. 6 together with the variations of y_{2v}, y_{2o}, and y_{2m}. The most important new result is the fact that $b_m \leq b_v$ which ensures the presence of a region of unstable flow unless the constriction is highly asymmetric. Lawrence (1990) shows that this relationship also holds for non-Boussinesq flows. For very large and very small values of q_r the width at the virtual control approaches infinity whereas the width at the point of marginal stability approaches a finite value:

$$(b_m)_{max} = \sqrt{\tfrac{32}{27}} \tag{34}$$

So the width of the contraction at the point of marginal stability is always less than approximately 9% greater than at the throat. Therefore significant regions of unstable flow can be expected in exchange flows through constrictions. An examination of the consequences of the violation of Long's stability criterion is given in the next section.

5. Interfacial Mixing in Two-Layer Flows

For flows satisfying the hydraulic assumptions (i.e. layered, inviscid, non-rotating flows with a hydrostatic pressure distribution), Long's stability criterion applies. However, Long (1956), himself, notes that: "If we abandon the hydrostatic assumption momentarily, we find that sufficiently short infinitesimal waves are unstable for any shear." Even at values of $F_\Delta^2 \ll 1$ the interface is unstable to a number of short wave instabilities (Thorpe, 1987; Lawrence et al. 1990), the most notable being the Kelvin - Helmholtz instability. If the initial thickness of the density interface is sufficiently small Kelvin-Helmholtz billows will form on the interface. These billows may become unstable to a subharmonic wave that causes alternate billows to pair. As this pairing proceeds smaller scale three-dimensional instabilities result in mixing at a molecular level. In a free shear layer pairing will continue until the billows become so large that the shear is no longer strong enough to overcome the buoyancy forces resisting pairing. Koop & Browand (1979) observed that smaller scale three-dimensional disturbances effectively mix most, but not all, of the fluid entrained by the Kelvin-Helmholtz billows, resulting in a final mixing layer thickness

$$\delta = J_{max} F_\Delta^2 y \tag{35}$$

where the bulk Richardson number, $J = g' \delta / \Delta u^2$. This result is supported by the theoretical work of Miles (1961), Howard (1961), and Corcos & Sherman (1976); the numerical work of Hazel (1972) and Patnaik et al. (1976); and the experimental work of Thorpe (1973), Koop & Browand (1979), Lawrence (1985), and Lawrence et al. (1990). Estimates of J_{max} vary between about 0.25 and 0.32. The actual value may depend on a number of factors including: the definition of δ; the shape of the velocity and density profiles; and the Reynolds and Prandtl numbers.

Equation (35) only applies to free shear layers, i.e., $\delta \ll y$. For this condition to be satisfied $F_\Delta^2 < 1$; however, the present study has shown that we need to be concerned about flows where $F_\Delta^2 > 1$. In such flows pairing is restrained by the proximity of the free surface and the bed of the flow, and a smaller density interface than predicted by (35) is expected. Preliminary experiments by Lawrence, Guez, and Browand (unpublished) have confirmed this hypothesis. The results are plotted on Fig. 7 together with the results of the mixing layer experiments of Koop & Browand (1979), and Lawrence et al (1990). Also

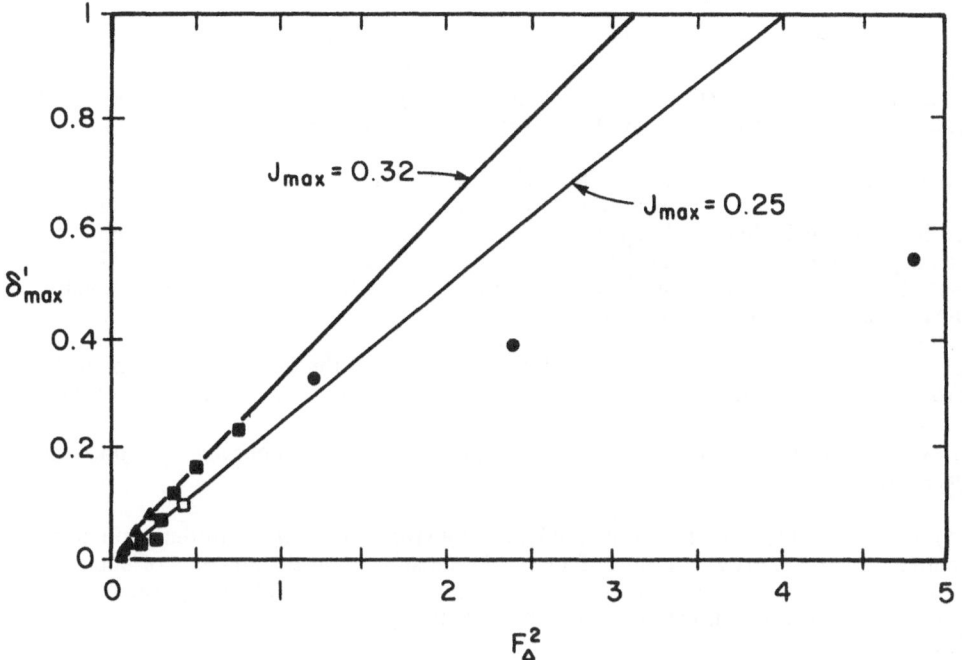

Fig. 7   Variation of dimensionless maximum interface thickness $\delta' = \delta/y$, with the stability Froude number, $F_\Delta{}^2$:

- ■ Koop and Browand (1979)
- □ Lawrence (1985)
- ▲ Lawrence, Browand & Redekopp (1990)
- ● Lawrence, Guez & Browand (unpublished)

included in Fig. 7 is a measurement of Lawrence (1985) obtained in a study of two-layer flow over an obstacle.

Another possible limitation on the thickness of a mixed layer is the fact that the process of pairing takes a finite length of time. It may be that an instability is advected out of a region of high shear before a significant amount of mixing has occurred. Therefore the value given by (35) should be considered an upper bound on the mixing layer thickness.

Field data of from the Strait of Gibraltar are also consistent with the above results; for example, Fig. 13.1 of Armi & Farmer (1989) shows Kelvin-Helmholtz instabilities at 1pm on 14 April 1986 when $\delta \approx 0.2$ y and $F_\Delta^2 \approx 0.6$. Although rotation causes the interface to be deeper on the south side of the strait than the north, the site from which the above data was obtained was located near the center of the channel and thought to be representative of the average of conditions across the strait.

The above results suggest that the following account of the importance of the stability Froude number is of more relevance than Long's stability criterion. Consider three cases:

(i) If $F_\Delta^2 \ll 1$, then hydraulic analysis applies and the maximum mixing layer thickness is given by (35).

(ii) If $F_\Delta^2 = O(1)$, then hydraulic analysis may still be relevant with the application of momentum and energy correction factors like those used in the analysis of single-layer flows (see Henderson, 1966), together with the introduction of an analogous density correction factor. In this case the relationship between δ and F_Δ^2 outlined in Fig. 7 would have to be used.

(iii) If $F_\Delta^2 \gg 1$, then the flow approaches that of a bounded, homogeneous mixing layer. Almost complete mixing occurs and hydraulic analysis cannot be used.

6. Summary

The stability Froude number F_Δ^2 is an important parameter in the analysis of exchange flows. It value can be computed using internal hydraulic theory. If $F_\Delta^2 \leq 1$ the the flow is said to satisfy Long's stability criterion for long internal waves and we can place confidence on the predictions of internal hydraulic analysis even though there may be some mixing at the interface caused by short wave instabilities. In addition, if $F_\Delta^2 \leq 1$, an upper bound on the thickness of the mixed layer can be estimated from past studies.

In past studies of the internal hydraulics of two-layer flows it has generally been assumed that $F_\Delta^2 \leq 1$; however, the present study has shown that significant regions of flow with $F_\Delta^2 > 1$ can be expected in exchange flow through a constriction. On the other hand in all cases $F_\Delta^2 < 2$, and it appears likely that internal hydraulic analysis will prove to be an aid in the prediction of mixing in exchange flow through a constriction.

Acknowledgements

The author gratefully acknowledges the financial support of the US Office of Naval Research's Fluid Mechanics and Oceanography Programs, and the Canadian Natural Sciences and Engineering Research Council. Many thanks to Larry Armi, Fred Browand and Laurent Guez for their advice, encouragement and assistance.

References:

Armi, L. 1986 The hydraulics of two flowing layers of different densities. *J. Fluid Mech.* **163**, 27.

Armi, L. & Farmer, D. M. 1986 Maximal two-layer exchange flow through a contraction with barotropic net flow. *J. Fluid Mech.* **164**, 27.

Armi, L. & Farmer, D. M. 1989 The flow of Mediterranean water through the Strait of Gibraltar. *Progress in Oceanography* **21**, No. 1.

Baines, P. G. 1984 A unified description of two-layer flow over topography. *J. Fluid Mech.* **146**, 127.

Corcos, G. M. & Sherman, F. S. 1976 Vorticity concentration and the dynamics of unstable free shear layers. *J. Fluid Mech.* **73**, 241.

Defant, A. 1961 *Physical Oceanography, Vol. 1*, Pergamon Press.

Farmer, D. M. & Armi, L. 1986 Maximal two-layer exchange over a sill and through the combination of a sill and contraction with barotropic flow. *J. Fluid Mech.* **164**, 53.

Farmer, D. M. & Armi, L. 1989 The flow of Atlantic water through the Strait of Gibraltar. *Progress in Oceanography* **21**, No. 1.

Hazel, P. 1972 Numerical studies of the stability of inviscid stratified shear flows. *J. Fluid Mech.* **51**, 39.

Henderson, F. M. 1966 *Open Channel Flow*. MacMillan.

Houghton, D. D. & Isaacson, E. 1970 Mountain winds. *Stud. in Num. Analysis* **2**, 21.

Howard, L. N. 1961 Note on a paper of John W. Miles. *J. Fluid Mech.* **10**, 509.

Kinder, T. H., & Bryden, H. L. 1987 The 1985-86 Gibraltar experiment: data collection and preliminary results. *EOS,* **68**, No. 40, 786.

Koop, C. G. & Browand, F. K. 1979 Instability and turbulence in a stratified fluid with shear. *J. Fluid Mech.* **93**, 135.

Lawrence, G. A. 1985 *The hydraulics and mixing of two-layer flow over an obstacle*. Ph. D. thesis, University of California, Berkeley. 122pp.

Lawrence, G. A. 1990 The hydraulics of Boussinesq and non-Boussinesq two-layer flows. *J. Fluid Mech.* (in press).

Lawrence, G. A., Browand, F. K., & Redekopp, L. G. 1990 The stability of a sheared density interface. Accepted for publication in *Physics of Fluids A: Fluid Dynamics*.

Long, R. R. 1956 Long waves in a two-fluid system. *J. Met.* **13**, 70.

Marsili L. F. 1681 Osservazioni intorno al Bosforo Tracio o vero Canale di Constantinopoli, rappresentate in lettera alla Sacra Real Maesta Cristina Regina di Svezia, Roma, reprinted in *Boll. Pesca, Piscic. Idrobiol.*, **11**, 734-758 (1935).

Miles J. W. 1961 On the stability of heterogeneous shear flows. *J. Fluid Mech.* **10**, 496.

Patnaik, P. C., Sherman, F.S., and Corcos, G. M. 1976 A numerical simulation of Kelvin-Helmholtz waves of finite amplitude, *J. Fluid Mech.* **73**, 215-240.

Schijf, J. B. & Schönfeld, J. C. 1953 Theoretical considerations on the motion of salt and fresh water. In *Proc. of the Minn. Int. Hyd. Conv. Joint meeting of the IAHR and Hyd. Div., ASCE.* p 321.

Thorpe, S. A. 1973 Turbulence in stably stratified fluids: a review of laboratory experiments. *Boundary Layer Meteorol.* **5**, 95.

Thorpe, S. A. 1987 Transitional phenomena of turbulence in stratified fluids. *J. Geophys. Res.* **92**, 5231.

A SIMPLE MODEL OF THE DESCENDING MEDITERRANEAN OUTFLOW PLUME

MOLLY O. BARINGER AND JAMES F. PRICE
Woods Hole Oceanographic Institution
Woods Hole, Massachusetts 02543
USA

ABSTRACT. Observations show that the outflow of Mediterranean water from the Strait of Gibraltar forms a distinct plume that flows along the northern continental slope in the Gulf of Cadiz. In a model with Froude number dependent mixing, the plume mixes with overlying fresher waters in two stages as it descends to an equilibrium depth of about 1100 m. The plume cools from 13.5 C at the western end of the strait to 11.8 C at Cape St. Vincent, and the salinity freshens from 38.2 ppt to 36.5 ppt. The controlling external parameter is the bottom slope, which is nearly invariant in our sensitivity experiments because the path of the plume is constrained by rotation. Thus the end product of mixing remains insensitive to reasonable changes in the internal and external parameters.

1. Introduction

Evaporation and cooling in closed basins produces some of the densest waters in the world. The Mediterranean water flows out through the Strait of Gibraltar, as a dense plume cascading down the bottom slope and mixing with the overlying water. The Mediterranean plume slowly loses its high salinity as it mixes with the fresh North Atlantic water and flows northwest toward Cape St. Vincent where it becomes neutrally buoyant. We are interested in understanding this mixing process and the dynamics which control the plume. We have developed a simple model to investigate the sensitivity of the outflow characteristics to variations in the boundary conditions at the strait.

2. The Plume Model

Our model consists of a homogeneous bottom layer representing the plume, overlain by a continuously stratified, inactive upper layer (figure 1). The width, as a function of alongstream distance, is specified *a priori* from field measurements, while the height is then determined from continuity. Our model includes rotation, mixing into the plume, and quadratic bottom friction. The entrainment stress is of the form given by Ellison and Turner (1959) and is parameterized by a Froude number (Price 1979). The model equations (figure 2) are integrated along the stream axis to yield unique velocity, temperature and salinity for each model realization.

The physical parameters that influence the plume's descent are the density anomaly, the bottom slope, bottom friction, and the ambient stratification. Strong buoyancy forcing accelerates the plume down the bottom slope. This increases the Froude number and causes the plume to mix with the overlying North Atlantic water.

L. J. Pratt (ed.), The Physical Oceanography of Sea Straits, 537–544.
© 1990 *Kluwer Academic Publishers*.

538

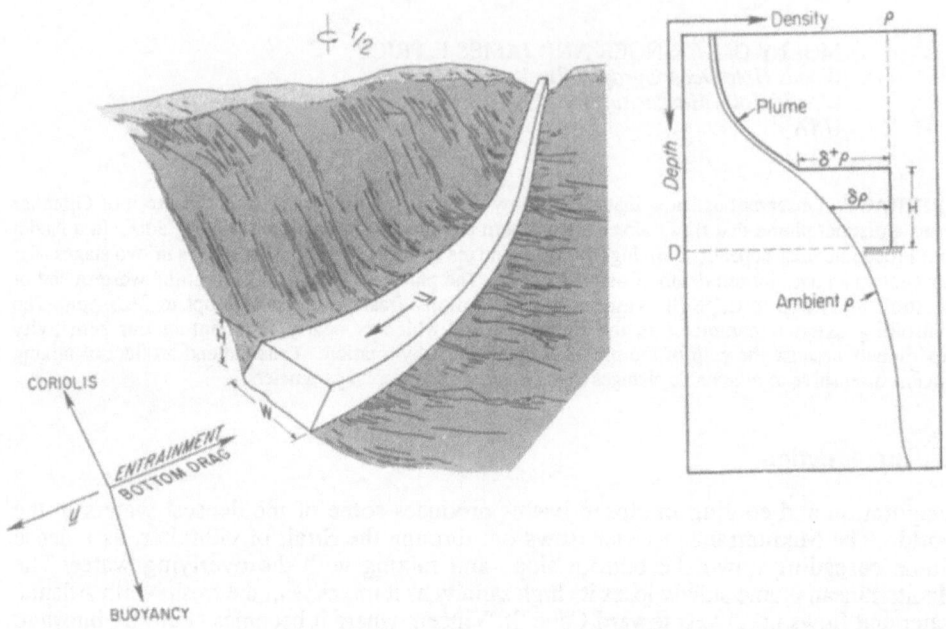

Figure 1. Model geometry and density profile. The plume is idealized as a single layer of height H and width W. It is allowed to flow freely down the continental slope beginning at the western end of the Stait of Gibraltar.

DESCENDING PLUME MODEL

$$\mathbf{U}\cdot\nabla\mathbf{U} + \mathbf{f}\mathbf{x}\mathbf{U} = g\delta\rho\nabla D/\rho_o - \nabla\hat{P}/\rho_o - C|U|U/H - E(\mathbf{U} - \hat{\mathbf{U}})/H$$

$$\left.\begin{array}{l} \mathbf{U}\cdot\nabla T = -E\delta^+ T/H \\[2mm] \mathbf{U}\cdot\nabla S = -E\delta^+ S/H \end{array}\right\} \ \rho(T, S, z)$$

$$\mathbf{U}\cdot\nabla H = E - H\mathbf{U}\cdot\nabla W/W - H\nabla\cdot\mathbf{U}$$

$$E = |U|\cdot 5\times 10^{-4}\cdot R_v^{-4}, \quad R_v = \frac{g\delta^+\rho H}{\rho_o(\mathbf{U} - \hat{\mathbf{U}})^2}$$

$$\mathbf{x}(t) = \int_{t_o}^{t} \mathbf{U}dt$$

$\mathbf{U}, H, W, T, S$ are given as the IC.

$\hat{P}$, $\hat{T}(z)$, $\hat{S}(z)$, $D(\mathbf{x})$, and f are specified.

$W(\mathbf{x})$ is specified ("channel aprox.")

$C = 3\times 10^{-3}$ is specified for a rough bottom.

Figure 2. Model equations. The plume is assumed to have downstream dependence only (i.e. it is assumed steady). There are two features of the present model that are less than fully satisfying: 1) neglect of the pressure gradient term due to the plume itself, and 2) the specification of the plume width as a constant fraction of the downstream distance as if the plume were constrained by channel walls (which it is in some regions only).

Initial experiments suggested that bottom slope is the dominant parameter. The actual Gulf of Cadiz bottom topography is very rough with many canyons, and cannot be well represented by a flat plane. Accordingly, we use a digitized five minute (8 km) resolution topography of the Gulf of Cadiz, somewhat smoothed, as shown in figure 3. The bottom friction coefficient used was 0.003, a value often cited for rough bottoms.

3. Results

Figure 4 shows the current and density anomaly as a function of along-stream distance. The overall trend of the model solution follows historical observations reasonably well (see figure). The plume becomes neutrally buoyant and separates from the bottom off the coast of Portugal. The plume equilibrates at a mean depth of about 1100 m with salinity freshening from 38.2 ppt to 36.5 ppt and temperature cooling from 13.5 deg C to 11.8 deg C.

3.1. MIXING

The model demonstrates a two stage mixing process. The most vigorous mixing occurs in the first 50 km after the plume exits the strait (figure 3 shows the plume path over the topography with each 25 km marked). The large initial density anomaly together with the bottom slope forces the plume to accelerate sufficiently to drive the Froude number above one, which initiates strong mixing. The initial mixing stage is shut off by an inertial oscillation that decelerates the plume by forcing it up the slope.

The second mixing stage occurs approximately 190 km downstream and is less dramatic, with only a barely perceptible entrainment stress (figure 5). Mixing occurs after acceleration of the plume around a bump. After the second mixing event the plume is nearly neutrally buoyant. It continues down the slope with very little mixing until it reaches equilibrium depth. It becomes neutrally buoyant and separates from the bottom in a region of strong slopes.

This two stage mixing has also been observed in data presented by Heezen and Johnson (1969). The strongest mixing occurs close to the straits where the velocity and density anomaly are the greatest. This supports our hypothesis that the mixing can be parameterized through a bulk Froude (or Richardson) number.

3.2. DYNAMICS

The momentum balances in the along-stream and cross-stream directions are shown in figure 5. The stresses are up to 5 Pa, demonstrating the massive power of these underwater waterfalls. Initially the entrainment stress and bottom drag are very important in controlling the plume dynamics. Farther downstream, the plume becomes a damped geostrophic current. Bottom stress is not necessary to balance the buoyancy in a rotating system. Rotation can thus inhibit vertical motion even in the presence of strong slopes.

3.3. PARAMETER DEPENDENCE

The plume width is an externally imposed function based on data from Smith (1975). We would prefer to determine the width from dynamics, but find that changing the specified plume width by fifty percent does not appreciably change the plume salinity, temperature or equilibrium depth.

GULF OF CADIZ

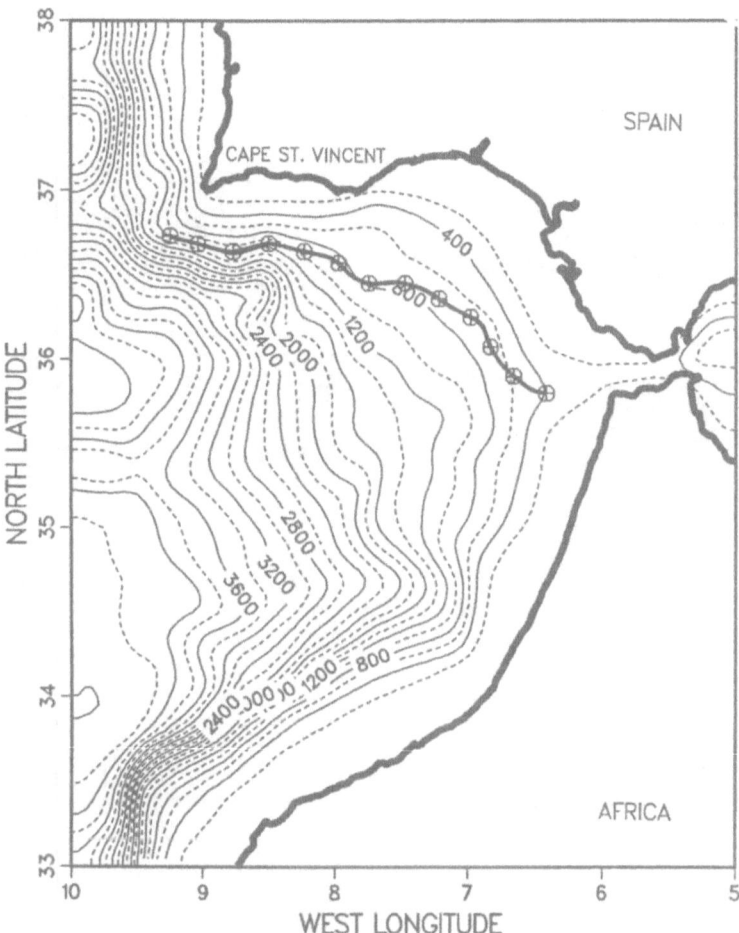

Figure 3. Gulf of Cadiz bathymetry and modeled plume path.
Markers are placed every 25 km along the plume's path.

CURRENT SPEED

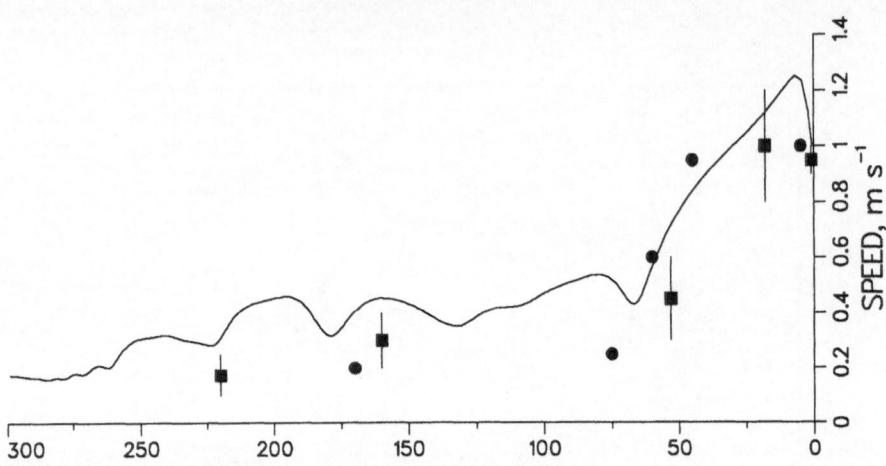

DENSITY ANOMALY

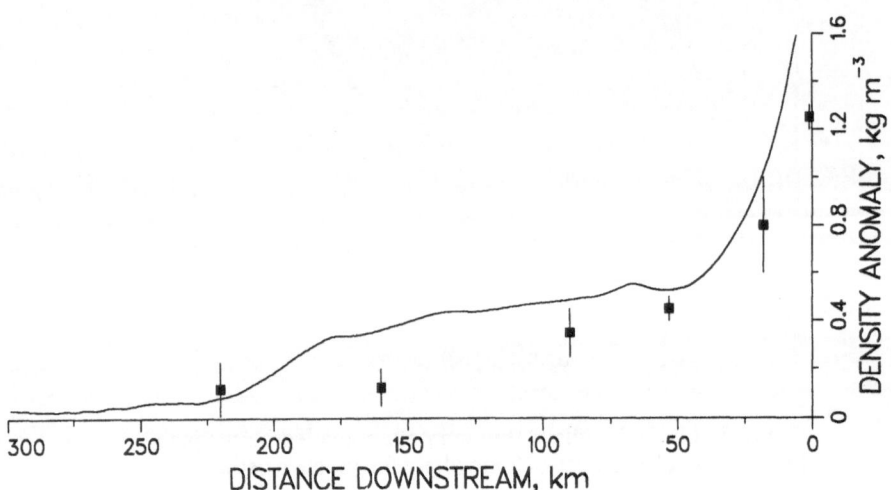

DISTANCE DOWNSTREAM, km

Figure 4. Current speed and density anomaly along the axis of the plume. Density anomaly is the density difference between the plume and the surrounding waters. Data with error bars is from Smith (1975), data without error bars is from Heezen and Johnson (1969).

CROSS–PLUME

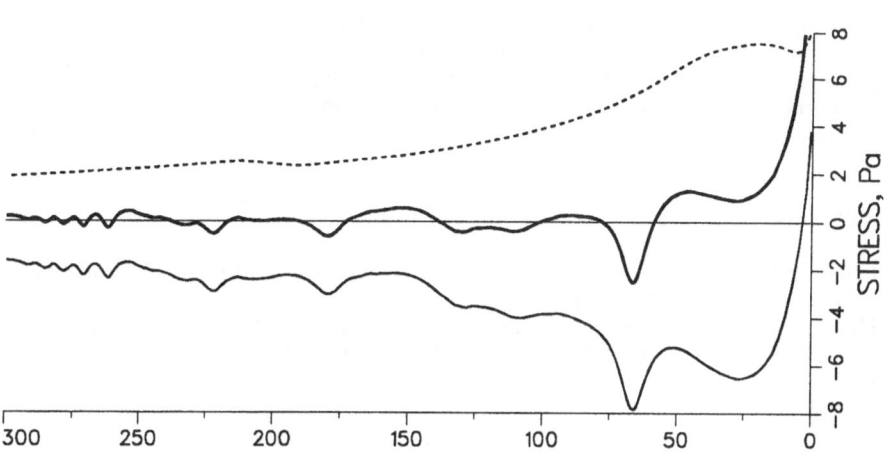

ALONG–PLUME

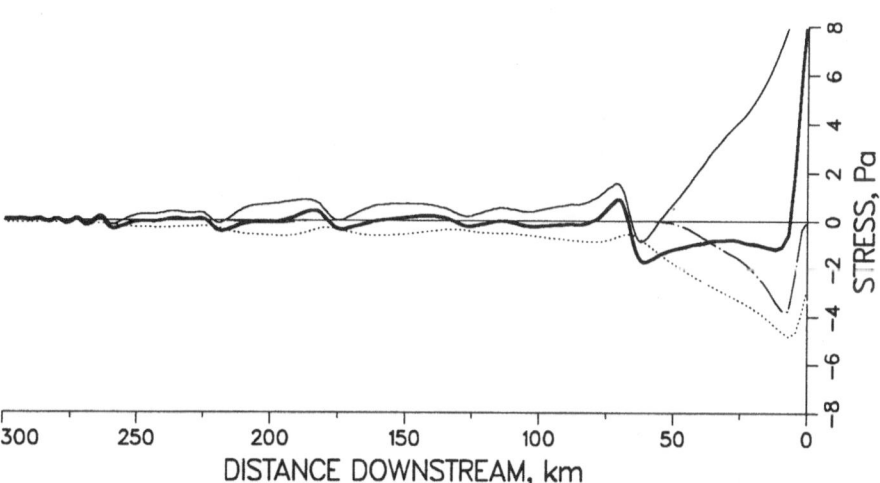

DISTANCE DOWNSTREAM, km

Figure 5. Momentum balance in the cross-plume and along-plume directions as a function of downstream distance. The dashed line is the Coriolis force; the chain-dotted line is the entrainment stress; the dotted line is the bottom drag; the light solid line is the buoyancy, and the heavy line is the sum of all the above (i.e. the resulting force on the plume).

Several model experiments have been conducted to determine the sensitivity to variations in initial conditions and internal parameters such as the bottom friction. Surprisingly, the model shows no substantial changes with reasonable variations in the drag coefficient, outflow rate or initial density anomaly. By far, the controlling external parameter is the bottom slope. The path of the plume remains virtually invariant in all experiments (constrained by rotation), and hence the equilibrium state is not sensitive to external parameters.

References

Ellison, T. H., and J. S. Turner (1959) 'Turbulent entrainment in stratified flows', J. Fluid Mechanics, 6, 423-448.

Heezen, B. C., and L. Johnson (1969) 'Mediterranean undercurrent and micro-physiography west of Gibraltar', Bull. Inst. Oceanographique, 67, 1-96.

Price, J. F. (1979) 'On the scaling of stress-driven entrainment experiments', J. Fluid Mechanics, 90, 509-529

Smith, P. C. (1975) 'A streamtube model for bottom boundary currents in the ocean', Deep Sea Research, 22, 853-873.

FRICTION IN A SHALLOW TWO-LAYER FLOW IN A ROTATING OCEAN

FL. BO PEDERSEN
Institute of Hydrodynamics
and Hydraulic Engineering
Technical University of Denmark
Building 115
DK 2800 Lyngby
Denmark

ABSTRACT. A brief summary is given on friction in homogeneous as well as in two-layer stratified flow. The importance of friction on flows in a rotating ocean is demonstrated by a simple case, namely a surface layer flow from a two-layer ocean into a strait. Characteristics of the flow are outlined for non-frictional and frictional flow, respectively. While the classical, non-frictional theory may describe the model test in distorted scale very well, it is unable to explain observed field measurements. The major reason for the discrepancy is shown to be the omission of friction.

1. THE FRICTION FACTOR

1.1. Definition

The friction factor f is by definition related to the shear stress τ by the formula

$$\tau = (f/2)\rho V^2 \qquad (1)$$

where V is a reference velocity, and ρ is the density.
     From Reynolds' modelling law it is known that the friction factor depends on

$$f=f \text{ (Re, Cross-sectional shape,} \\ \text{sand-roughness} = k) \qquad (2)$$

where Re is the Reynolds number.

1.2. Homogeneous flow

In ordinary hydraulics, for fixed boundaries, the friction factor is given by

L. J. Pratt (ed.), The Physical Oceanography of Sea Straits, 545–557.
© 1990 Kluwer Academic Publishers.

$$f = f \ (\text{Re}, \ k/R) \qquad\qquad (3)$$

i.e. it is only dependent on the Reynolds number Re and on the relative roughness, which means that the shape factor can be incorporated into the hydraulic radius R.

Based on a large amount of experimentally determined data, Colebrook and White [1939] suggested a formula for turbulent flow which in a slightly modified version reads

$$\sqrt{\frac{2}{f}} = 6.4 - 2.45 \ \ln \ \left[\frac{k}{R} + \frac{3.3}{\text{Re}}\sqrt{\frac{2}{f}}\right] \qquad\qquad (4)$$

1.3 Two-layer stratified flow

In Bo Pedersen [1980] the interfacial shear stress τ_i was related to the friction factor f_i by

$$\frac{\tau_i}{\rho} = \frac{f_i}{2} \ (U_m - u_i)^2 \qquad\qquad (5)$$

where U_m = maximum velocity
u_i = interfacial velocity

which means that the reference velocity used is the maximum velocity difference within the flow influenced by the interface. Using the associated layer thickness y_o in forming the Reynolds' number Re_i

$$\text{Re}_i = \frac{(U_m - u_i)(y - y_o)}{\nu} \qquad\qquad (6)$$

an empirical equation for the friction factor was established, based on carefully treated experimental data from the literature. The result is reproduced in Fig. 1. It shall be emphasized that the results summarized above may only be used in a fully developed flow which has a pronounced two--layer character.

2. ON FRICTION IN A TWO-LAYER ROTATING OCEAN

As the main purpose is to demonstrate the importance of friction in nature, we may choose a hydrodynamically simple case for this illustration. To this end we look at the inflow to a long narrow channel of upper layer water from a two-layer ocean with zero upstream potential vorticity. To avoid problems with free contraction the entrance has guide-walls following the streamlines of a vena contracta.

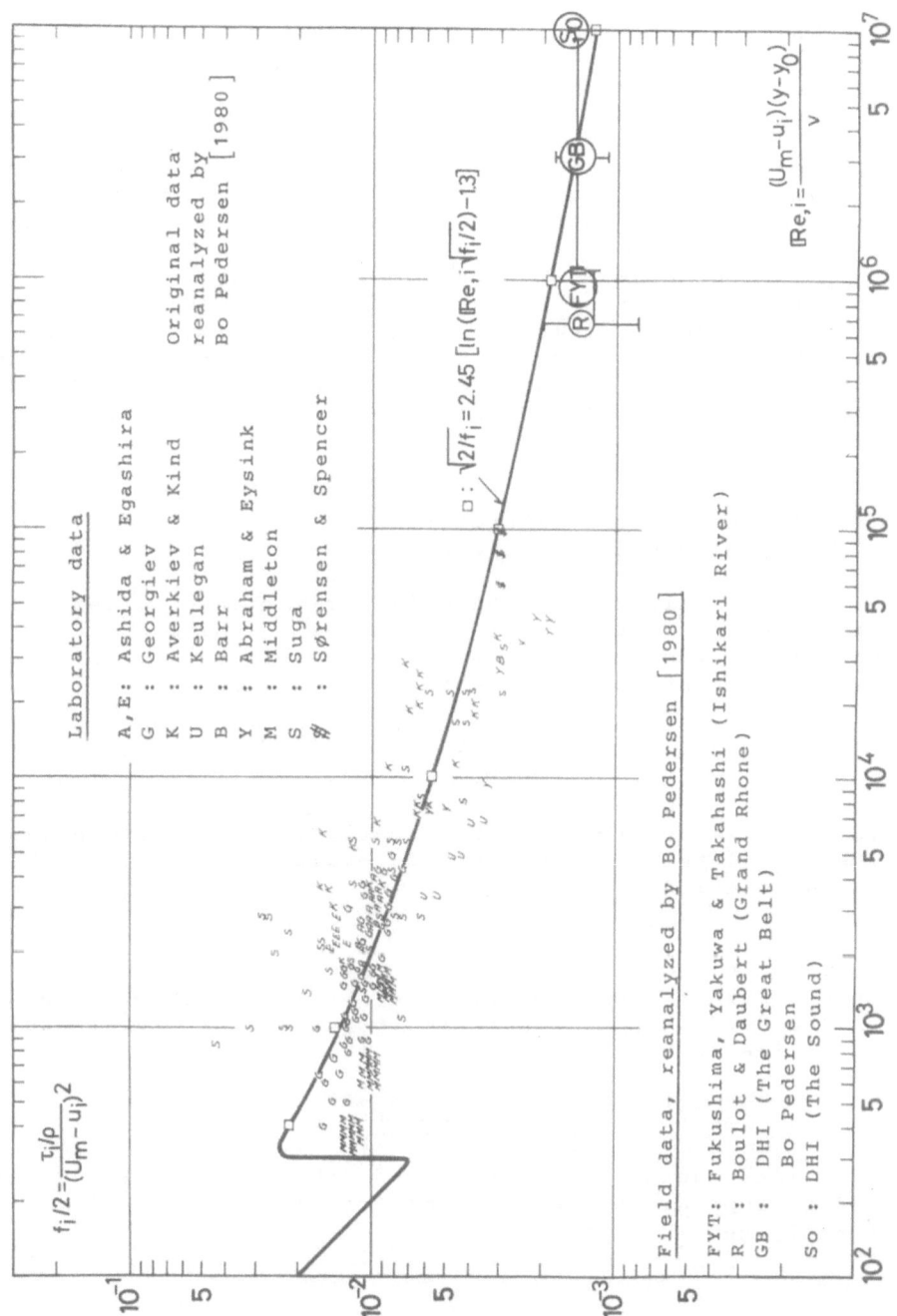

Fig. 1. The interfacial friction factor for a gradually
varying two-layer stratified fluid. (U_m = max
velocity, u_i = interfacial velocity, $y-y_0$ =
distance between U_m and u_i). For data see
Bo Pedersen [1980].

2.1. Scaling considerations for model tests

As the length-to-depth ratio in geophysical flows often is
of the order of magnitude of, say, 1000, one has to use a
highly distorted scale in the laboratory to model any geophysical two-layer flow. For the sake of argument, we choose
a distortion scaling factor of 1000 in the following, which
means that the length-to-depth ratio in the present example
is simply one.

In the above mentioned entrance flow we have a length L
for the zone of flow establishment, by which we mean the
distance from the inlet to the cross-section where the geostrophic balance is established. This length L is, in the
model as well as in the field, of the order of magnitude of
the width of the flow. Hence, as typical values for the
length of the flow to the depth of the flow, we obtain the
ratios

L/y = 1 in the laboratory
L/y = 1000 in the field

The implication of this is a remarkable difference in
the importance of the shear stresses in the laboratory and
the field, respectively. We may quantify this difference by
recalling that the shear stress amounts to

$$\tau_i = \frac{f_i}{2} \rho (U_m - u_i)^2 \sim \frac{f_i}{2} \rho \frac{1}{3} v^2 \qquad (7)$$

where the reference velocity has been related to the mean
velocity in accordance with experience, Bo Pedersen [1980].
The associated head loss can be evaluated by use of the well
known relation

$$\tau = \rho g y I \qquad (8)$$

where I is the energy gradient, i.e. the head loss ΔH per
unit distance. For a distance of L, the order of magnitude
of the head loss (or approximately the water level drop) can
be estimated to be

$$\Delta H \approx IL = \frac{\tau_i}{\rho g y} L = \frac{fi/2 \rho v^2}{3 \rho g y} L = \frac{fi/2}{3} \frac{L}{y} \left(\frac{v^2}{2g}\right) \qquad (9)$$

Introducing the length-to-depth ratios outlined above
and appropriate values for the friction factor (Fig. 1) we
finally get

$$\Delta H \approx \frac{10^{-2}}{3} \; 1 \; \frac{v^2}{2g} \approx 0.003 \; \frac{v^2}{2g} \quad \text{in the laboratory} \quad (10a)$$

$$\Delta H \approx \frac{3 \cdot 10^{-3}}{3} \; 10^3 \; \frac{v^2}{2g} \approx \frac{v^2}{2g} \quad \text{in the field} \quad (10b)$$

From these crude order of magnitude calculations we may conclude that the head loss in a labratory experiment with distorted scale only amounts to less than 1% of the kinetic energy involved in the flow, and hence may be treated theoretically as non-frictional.

In the geophysical example, on the contrary, the head loss is of the same order of magnitude as the kinetic energy, which means that the use of Bernoulli's equation, the potential vorticity conservation equation, and similar frictionless approximations are meaningless in geophysical flows with a typical length-to-depth ratio of 1000 or more.

As shall be demonstrated below, this difference in the importance of the friction has a great bearing on the velocity distribution in the established flow.

2.2. Governing equations for surface flow

We consider a simple two-layer flow in a long straight channel, where the upper layer is flowing, while the lower layer is at rest, see Fig. 2.

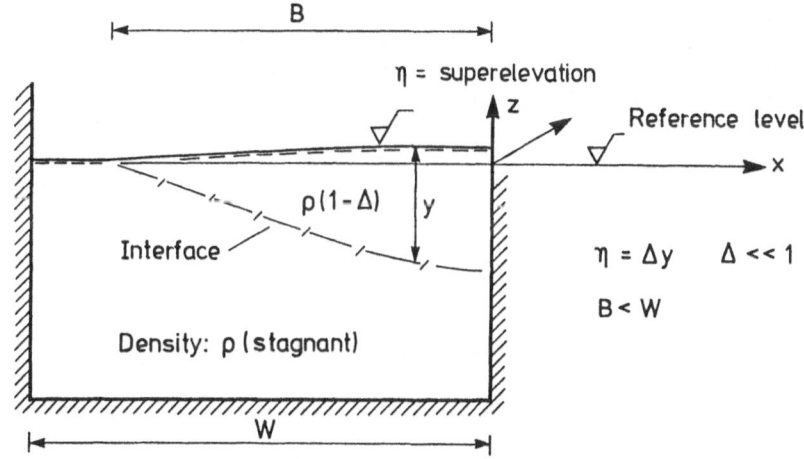

Fig. 2. The simple two-layer flow considered in the text.

When the flow is fully established, it is in the lateral direction governed by a force balance between the lateral component of the pressure gradient and the Coriolis force. This so-called geostrophic balance is valid independently of frictional or non-frictional flow conditions, simply because the cross flow components are secondary currents. Hence, the geostrophic balance applies for the conditions in the laboratory (in a rotating setup) as well as in the field.

In the flow direction the current is governed by a local balance between the flow direction component of the acceleration of gravity (or the horizontal pressure gradient) and the friction. Of course this condition can only be applied when dealing with frictional flow. For the non-frictional flow, the established flow is governed by the boundary conditions - in the present case the upstream located ocean - and the Bernoulli equation.

2.2.1. Non-frictional flow

The solution to the non-frictional surface flow is given in Fig. 3. For convenience, we have chosen a width of the strait which is slightly larger than the width of the coastal jet. In the frictionless calculations it is assumed that the water level drop $\bar{\eta} - \eta$ in Fig. 3 is converted into kinetic energy ($V^2/2g$). As demonstrated in chapter 2.1. above the actual head loss in the laboratory experiment only amounts to less than 1% of the available kinetic energy, and hence one may expect that the results shown in Fig. 3 agree very well with laboratory experiments. Similar experiments have been performed by Whitehead [1980], which confirm the agreement.

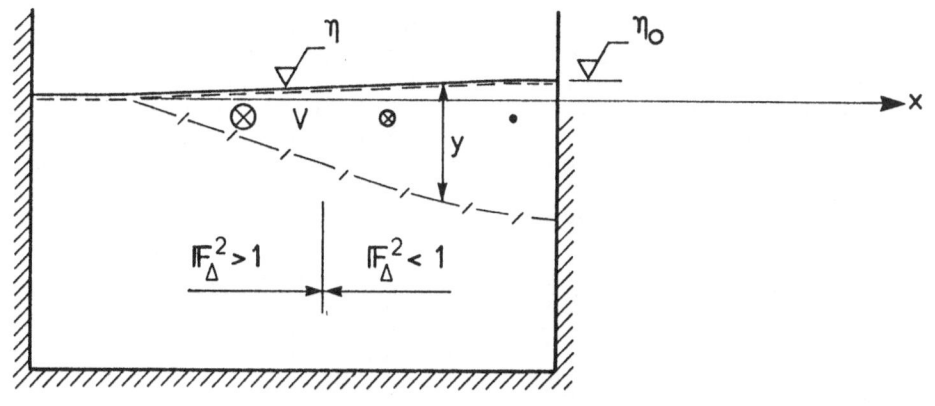

$$y = \frac{1}{2} \frac{f_c^2 B^2}{\Delta g} \left[1 - (\frac{x}{B})^2\right]$$

$$\eta + \frac{V^2}{2g} = \eta_o$$

$$V = (f_c B)(\frac{x}{B})$$

$$\mathbb{F}_\Delta^2 = 2 \frac{(\frac{x}{B})^2}{\left[1 - (\frac{x}{B})^2\right]}$$

Fig. 3. The non-frictional solution to the coastal jet in
geostrophic balance.

2.2.2. Frictional flow

It would have been nice to present a tableau for the
established frictional flow situation, similar to Fig. 3,
but there is a minor problem, namely that no solution exists
for a stationary coastal current surrounded by stagnant
water! But why is that so?

A stationary coastal current demands a longitudinal
gradient of the water level to compensate for the frictional
head loss. Outside the front the water is stagnant and
therefore without any gradient. Hence, the frontline is
horizontal. If the water level along the coast is falling,
and the frontline is horizontal, the flow cannot maintain
its geostrophic balance.

In the real world there are many elements which may overcome the dilemma presented, such as a frontal jet, a wind field, entrainment, detrainment etc. Instead of philosophizing over all these possible extra conditions, we will turn directly to a set of actual field measurements and extract some valuable conclusions out of the observations.

2.3. Case story. The Sound

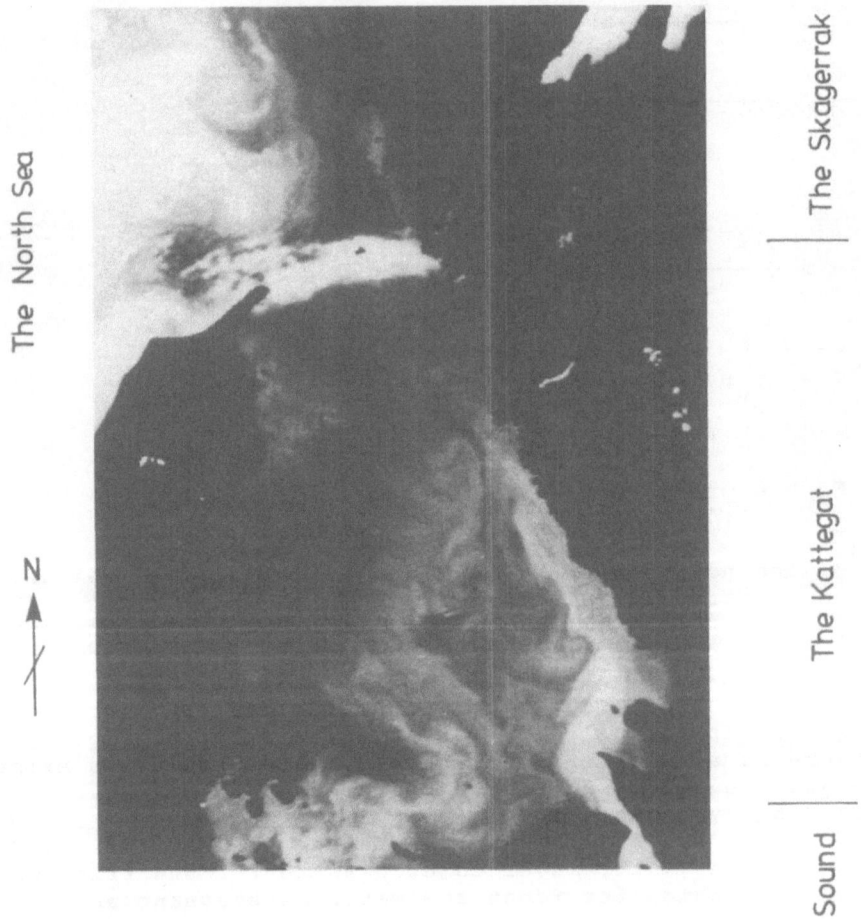

Fig. 4. Satellite image showing an outflow of cold (light grey), brackish water from the Baltic Sea. 1987-05-25. From Fonselius [1987].

The Sound is the second largest link between the Baltic Sea and the Kattegat, physically separating Denmark and Sweden. In Fig. 4 is shown a satellite image of the northernmost part of the Sound, the Kattegat, and part of the Skagerrak, which is in free connection with the North Sea. The outflowing cold (light gray) brackish water is visible along the Swedish coast all the way to Skagerrak. The narrowest cross-section in the Sound is approximately 5 km wide, which is slightly less than the internal Rossby radius of deformation. With a northgoing current, this section normally has critical flow conditions.

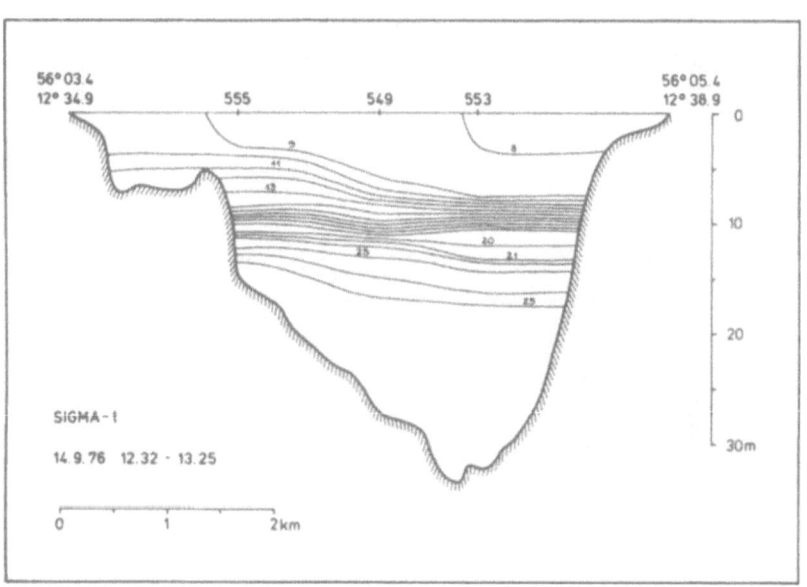

Fig. 5. The sigma-t distribution just north to the most narrow cross-section in the sound. 1976-09-14, 12:32 to 13:25. From T.S. Jacobsen and P. Bo Nielsen [1978].

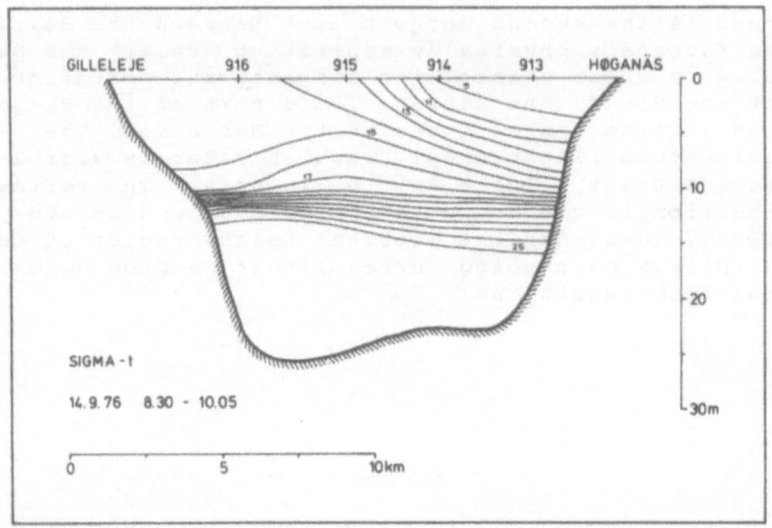

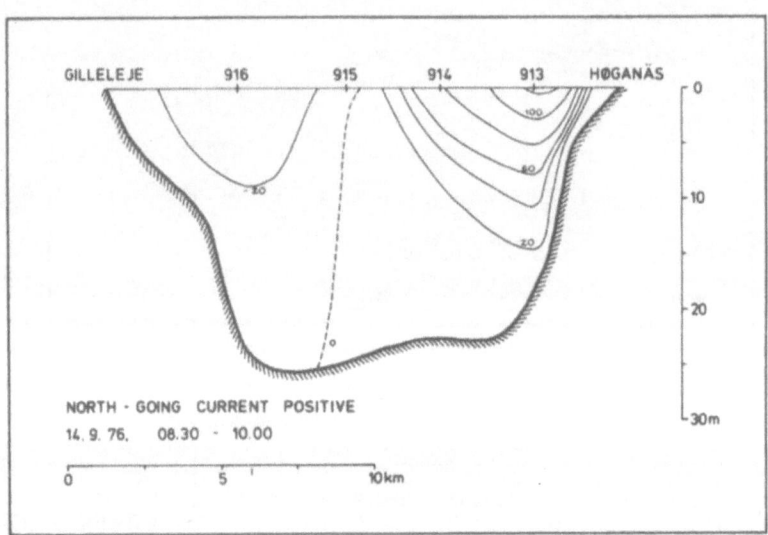

Fig. 6. a) The sigma-t distribution some 10 km north to the
            section shown in Fig. 5. Notice the changed
            length scale.

         b) The velocity distribution some 10 km north to
            the section shown in Fig. 5.

         Both measured 1976-09-14, 8:30-10:05.

         From T.S. Jacobsen and P. Bo Nielsen [1978].

In Fig. 5 is shown the observed sigma-t distribution in this cross-section during a persistent outflow condition. Despite the very strong current (will be shown below) there is almost no sign in the density (pressure) field of the Coriolis force, because the flow has not run the distance necessary to adjust itself to this. Some 10 km further downstream, the flow is fully established as shown in Fig. 6. In this cross-section we have observations of the density field and the velocity field.

Concerning the density distribution, three things are striking. First, it may be noticed that the density structure indicates a sort of three-layer structure, which is caused by the Kattegat being a two-layer system with a density of the upper layer much higher than the density of the outflowing Baltic water, while the density of the nearly stagnant lower layer water is the same in the Sound and in the Kattegat. Second, it is observed that the density of the flowing upper layer water has increased as a result of the entrainment during the passage from the narrow section to the present section. Third, it is noticed that the density distribution over the flowing part of the upper layer is nearly linear in the vertical as well as in the lateral direction, a feature often observed in geophysical jets.

The velocity distribution is interesting in many ways. First, we may compare it with the theoretically determined distribution for non-frictional flow, see Fig. 3. The non--frictional solution has its maximum velocity to the left, i.e. an anticyclonic circulation, while the real velocity distribution yields a cyclonic circulation (except for the shallow coastal region, where the current has bottom contact and hence a depth dependent velocity variation). We shall elaborate on this in section 2.4. below. Second, it is observed that in the region where an interface exists (i.e. outside the shallow coastal region), the velocity distribution is nearly linear horizontally as well as vertically, as was the case for the density distribution. This implies that the gradient Richardson number is a constant across the flow. Third, it may be noticed that the higher friction factor for the coast compared to the interface is reflected in a larger velocity gradient along the coast. Finally, the compensating flow for the entrainment can be observed in the left half of the upper layer. This special feature is caused by the flow taking place in a slightly diverging strait with a width a little greater than the internal Rossby radius.

2.4. Discussion

By comparing the theoretically obtained results - assuming

556

frictionless flow - with field observations many discrepancies were observed, which may all be explained by including
friction in the theory.

Let us start with the gradient Richardson number, which
is inversely proportional to the internal Froude number
squared. In the field this was observed to be independent of
the position in the flow. This is reasonable for two reasons.
First, the energy gradient I (see above) is directly proportional to the internal densimetric Froude number squared.
Hence the constant Froude number implies that the energy
gradient, or alternatively the streamwise water surface
gradient, is the same for all the streamlines. If the
streamwise water surface gradient is independent on the
lateral position in the jet, then the lateral water surface
gradient is unchanged from one cross-section to another
which means that the geostrophic balance is maintained along
the path of the coastal jet, contrary to the fictitious
steady flow treated in chapter 2.2.2. Second, it is ques-
tionable if a quasi-steady flow could exist with a Froude
number varying across the flow section. Take for instance a
flow with a Froude number in the vicinity of the critical
value. A cross-sectionally varying Froude number would for
this flow imply that the flow was subcritical in one part of
the section and supercritical in the other, say half of the
flow field. Any small disturbance in the subcritical part
would be transferred upstream as well as downstream to
adjust for the small change. In the supercritical part, on
the contrary, the same disturbance would only be able to go
downstream for adjustments. Hence the geostrophic balance
would be destroyed upstream.

Why is it possible for the non-frictional solution to
have a densimetric Froude number which varies from zero to
infinity across the flow? Answer: Because an ideal fluid
does not have to obey any law of common sense. But why is
it possible to do experiments of this sort of flows, which
confirm the non-frictional theory? Answer: because the
experiments do not extend beyond the zone of flow
establishment.

The frictionless theory implied that the water level
drop was converted into kinetic energy, confer with section
2.2.1. In chapter 2.1. it was calculated that the normal
head loss associated with a distance of the order of
magnitude of the zone of flow establishment was equal to the
total kinetic energy available in the real geophysical
flows. It is therefore obvious that the frictionless theory
does not apply to real flows. Besides the difference in the

densimetric Froude number distribution, the most remarkable
difference in the non-frictional and the frictional flow
seems to be the non-frictional anticyclonic and the frictional cyclonic circulation, respectively. As most of the
classical, oceanographic theories are developed on the
assumption of ideal fluids, this finding may give rise to
some considerations concerning other types of flow in a
rotating ocean.

2.5. References

Bo Pedersen, Fl. [1980] "A Monograph on Turbulent
   Entrainment and Friction in Two-Layer Stratified
   Flow". Technical University of Denmark, Institute
   of Hydrodynamics and Hydraulic Engineering, Series
   Paper No. 25. Doctoral dissertation.
Colebrook, C.F. [1939] "Turbulent flow in pipes with
   particular reference to the transition region between
   the smooth and rough pipe laws". Journ. Institution
   Civil Engineers, 1939; see also Engineering Hydraulics,
   ed. by H. Rouse, Chap. VI, Steady flow in pipes and
   conduits by V.L. Streeter, New York 1950.
Fonselius, S. [1987] "Kattegatt - havet i väster"
   ("The Kattegat - the Western Sea") SMHI Oceanografi,
   No. 18. (In Swedish).
Jacobsen, T. and Nielsen, P. Bo [1978] "Hydrographical
   Observations in Øresund, September 1976. A data
   report". University of Copenhagen, Institute of
   Physical Oceanography, Report No. 37.
Whitehead, John A. [1980] "Rotating Critical Flows in
   the Ocean". Second International Symposium on
   Stratified Flows, pp. 72-80. The Norwegian Institute
   of Technology, Trondheim, Norway, 24-27 June.

THE BREAKUP OF OUTFLOWS AND THE FORMATION OF 'MEDDIES'

DORON NOF
Department of Oceanography and the
Geophysical Fluid Dynamics Institute
The Florida State University
Tallahassee, Florida 32306-3048, U.S.A.

ABSTRACT. It is suggested that the formation of "Meddies" (i.e., mid-Atlantic eddies containing Mediterranean water) is related to intermittencies in the transport of the Mediterranean outflow.

In this note, a new nonlinear mechanism for the generation of anticyclonic lens-like eddies from boundary currents is proposed. In contrast to the familiar generation processes that rely on flow instabilities or vortex shedding due to the geometry of the boundary, the present mechanism is related to intermittency in the current's mass transport. The essence of the new mechanism is that intermittencies in the transport [such as those in the Mediterranean outflow (see Fig. 1)] lead to unbalanced patches of fluid near the boundary. These unbalanced patches break up into a discrete set of eddies that interact with the boundary.

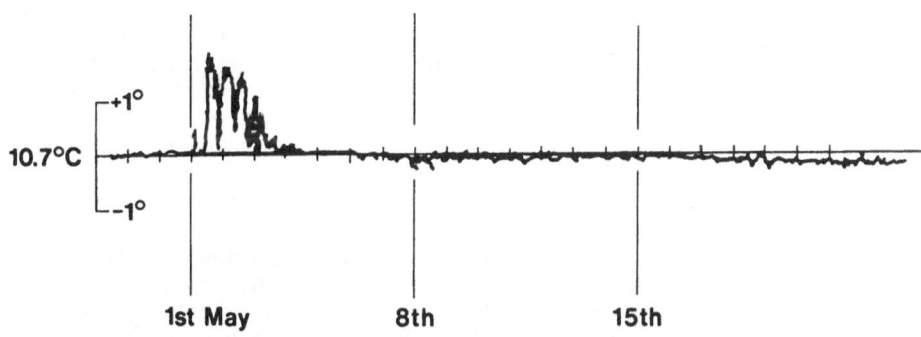

Figure 1a.   The temperature record of the Mediterranean outflow at approximately 500 meters from the bottom during spring 1971 (adapted from Gründlich 1981). The location of the mooring is marked on Fig. 1b. Note that from the 1st to the 4th of May an abrupt increase in the temperature (and presumably the transport) took place.

L. J. Pratt (ed.), The Physical Oceanography of Sea Straits, 559–566.
© 1990 Kluwer Academic Publishers.

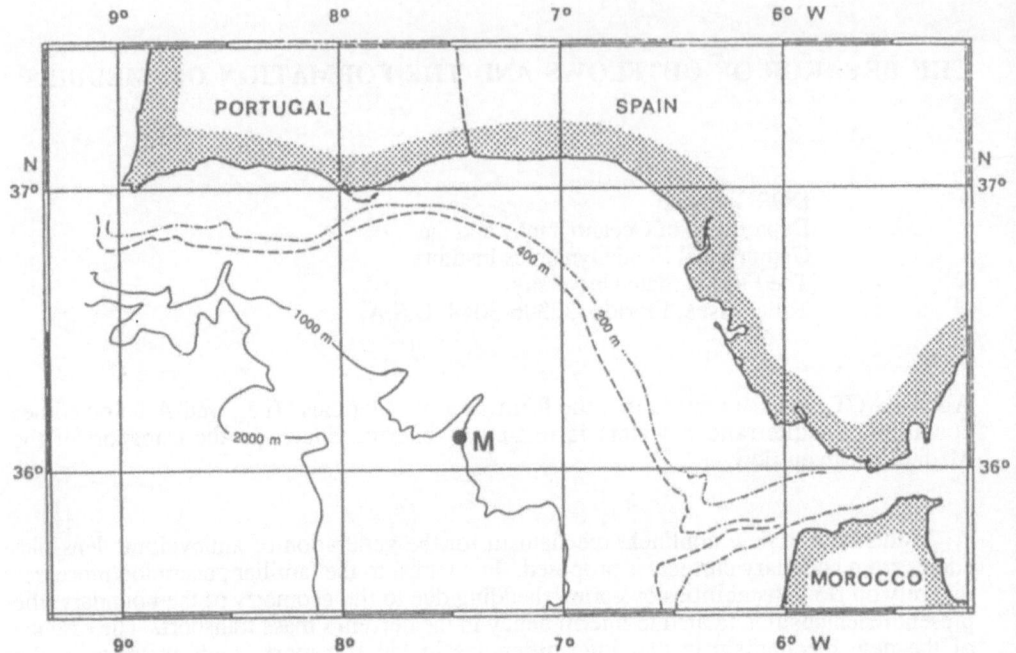

Figure 1b.   The approximate location (M) of the instruments used by Gründlich (1981) to measure the temperature of the Mediterranean outflow.

The above process is modeled as follows. We begin with a rectangular box containing the motionless (light) fluid near the boundary (Fig. 2). At, say, t = 0, the conceptual box is removed and the unbalanced fluid undergoes two main processes. The first involves the establishment of a set of eddies via geostrophic adjustment whereas the second is associated with the interaction of the set with the wall. These two processes are examined independently even though in reality the processes are, obviously, taking place at the same time.

To examine the first process we consider the nonlinear collapse of a (light) rectangular box in the open ocean away from the boundary (Fig. 3). The general structure of the resulting final chain of eddies can be computed analytically by using the usual connecting principles, the conservation of potential vorticity and mass. It turns out, however, that the number of eddies and their detailed structure cannot be computed unless one invokes an additional constraint. To resolve this difficulty, the integrated angular momentum constraint, which is rarely used in oceanographic modeling, has been applied. Namely, with the aid of the torque constraint − which, in contrast to common knowledge, does not usually coincide with the conservation of potential vorticity − the problem can be closed.

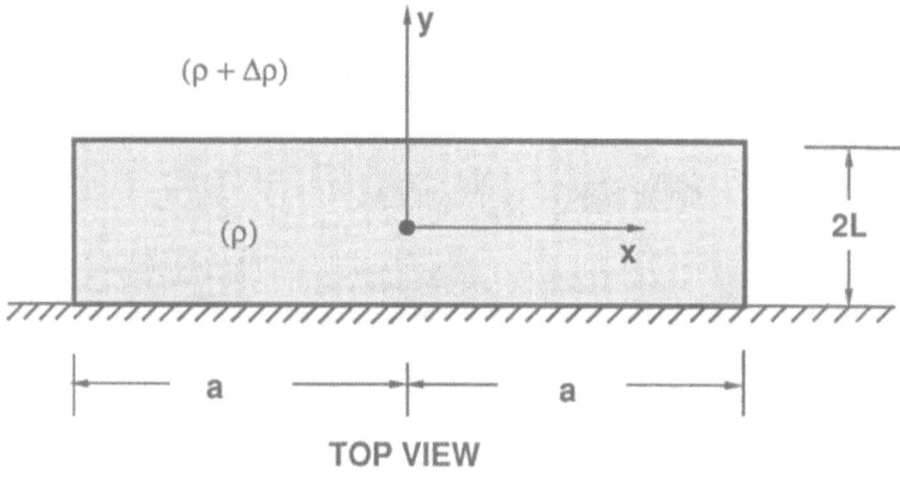

TOP VIEW

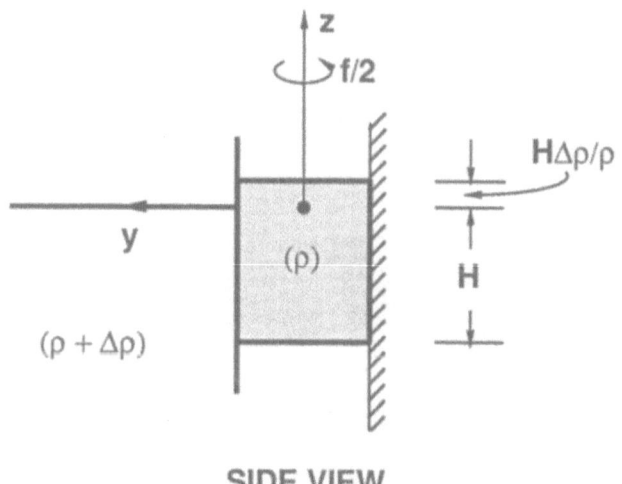

SIDE VIEW

Figure 2. Schematic diagram of the model under study. Initially, the light fluid (shaded) is contained within a rectangular box that is later removed. After withdrawal the light fluid undergoes a collapse and an adjustment process that involves the generation of eddies and their interaction with the wall. Note that, from a mathematical point of view, the problem is identical to that of a collapse sandwiched between two infinitely deep layers. Such a situation corresponds to the Mediterranean outflow.

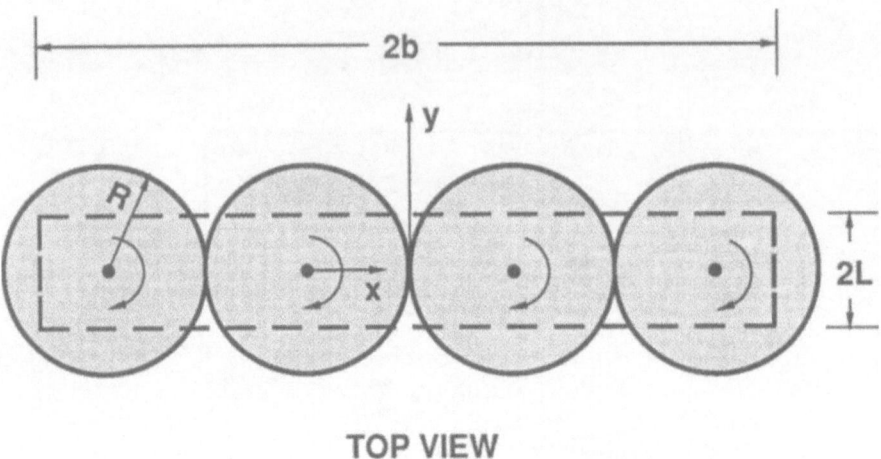

TOP VIEW

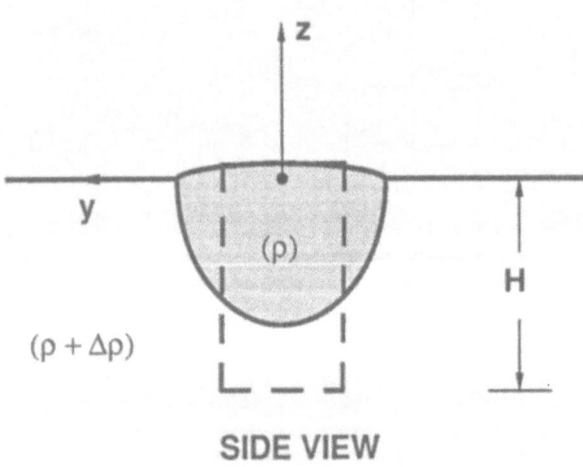

SIDE VIEW

Figure 3. A schematic diagram of our first collapse process. In contrast to Fig. 2, where the collapse is taking place in the vicinity of the wall, we consider here the collapse of a rectangular box (dashed line) in the open ocean away from boundaries. The collapse produces a chain of eddies whose detailed structure is computed analytically.

The details of the second process (i.e., chain-wall interaction) are examined with the aid of the (above) results for the chained off-shore eddies and the single eddy-wall interaction analysis of Nof (1988) (see Fig. 4). A combination of these two studies shows that a chain of lens-like eddies forced against a wall would leak fluid (Fig. 5) until all the eddies in the chain are merely "kissing" the wall.

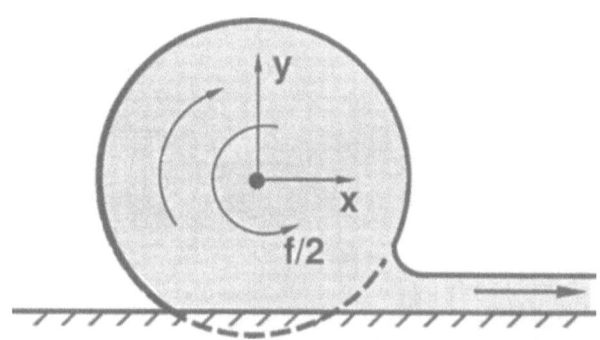

Figure 4. A sketch of the leakage associated with a single eddy interacting with a wall (Nof, 1988).

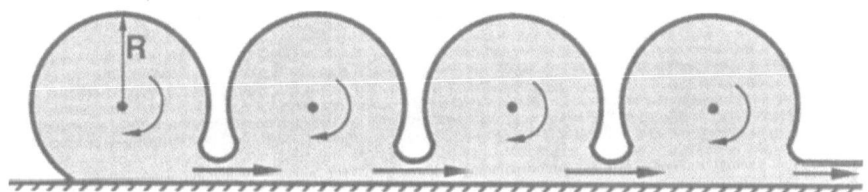

Figure 5. The expected multi-leakage of a chain of eddies interacting with a wall. Ultimately, the fluid which is in direct contact with the wall will be entirely removed and we will be left with a discrete set of eddies that are "kissing" the wall and are separated from each other.

Simple qualitative laboratory experiments on a rotating table support the conclusion that intermittencies in the current's transport lead to a chain of eddies that leak along the wall (Nof, 1989). Ultimately, the leaked fluid would presumably be entirely removed so that a discrete set of eddies would result.

This indicates that the observed mid-Atlantic eddies resulting from the Mediterranean outflow (Meddies) might be formed by a mechanism similar to our newly described process.

564

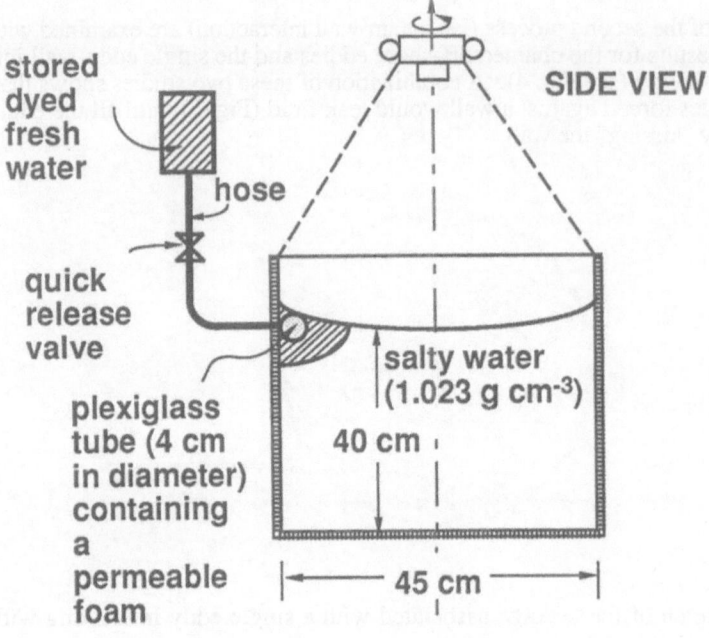

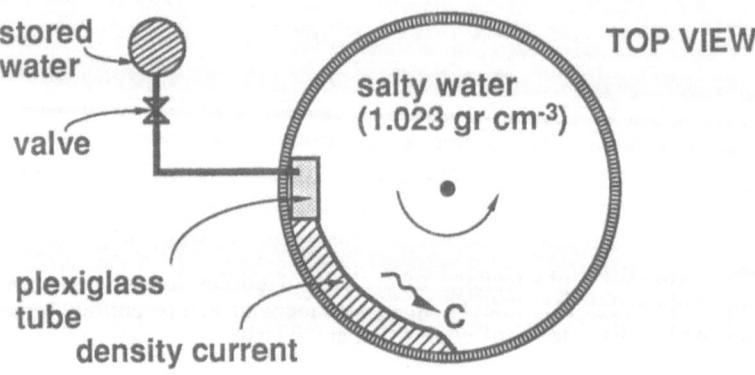

Figure 6. A sketch of the experimental apparatus. The intermittent density current is established by releasing the stored light fluid for a finite amount of time (approximately 20 seconds). The "wiggly" arrow indicates propagation of the density current.

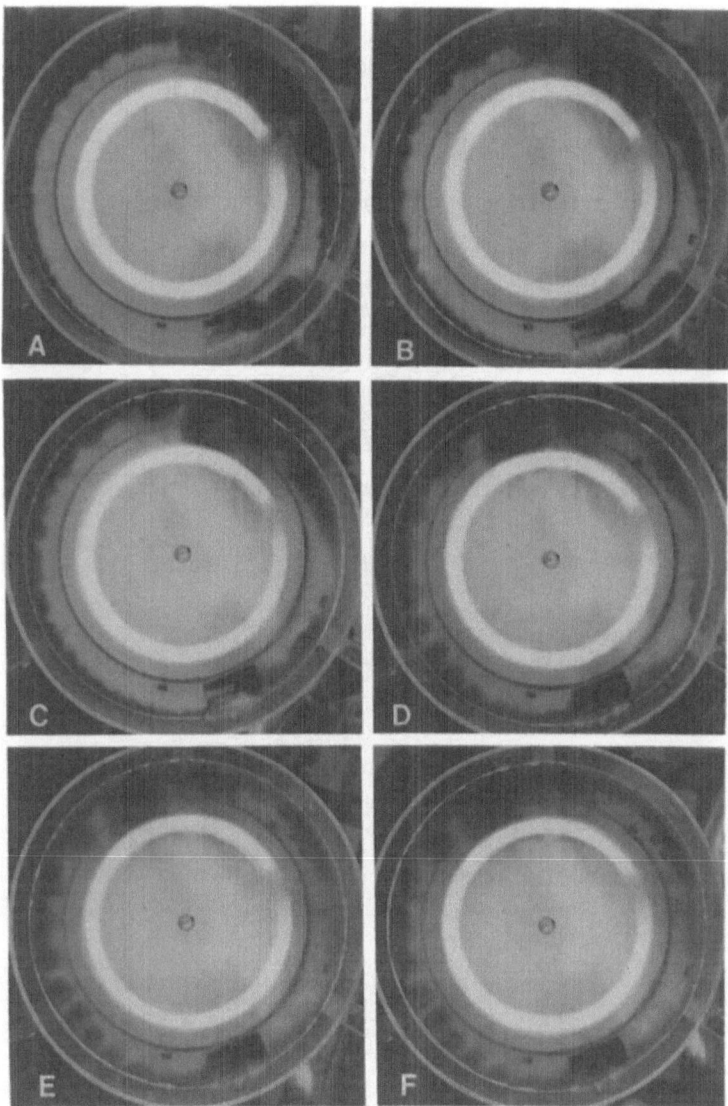

Figure 7. Subsequent photographs of an intermittent density current. The sequence shows the structure of the current and the formation of eddies during various stages of the breakup. Physical constants: f = 4.0 sec^{-1}; $\Delta\rho/\rho$ = 0.023; the mass flux of the initial current was a few cubic centimeters per second. Photograph (a) corresponds to the stage where the mass flux has already been terminated and the eddies are beginning to form. Note that the leakage at the front continues at all times during the breakup. The displayed process (a-f) lasted for about 30 seconds. The eddies at the front are smaller than those in the rear due to the initial wedge-like shape of the unbroken current.

REFERENCES

Gründlich, M. L. (1981) 'On the observation of a solitary event in the Mediterranean Outflow west of Gibraltar', *"Meteor" Forsch.-Ergebn., 23*, 3-46.

Nof, D. (1988) 'Eddy-wall interactions', *J. Mar. Res., 46*, 527-555.

Nof, D. (1989): 'Lenses generated by intermittent currents', submitted to *Deep-Sea Res.*

The Dynamics of Two Dimensional Turbulence

T. Maxworthy
Departments of Mechanical & Aerospace Engineering
University of Southern California
Los Angeles, California 90089-1453

ABSTRACT. The diffusive and spectral character of a decaying quasi two-dimensional turbulent field has been determined experimentally. The spectral slopes found are steeper than those expected from classical theories but comparable to those found in numerical simulation. The geometry of the resulting vorticity fields has been explored using the techniques of fractal geometry. A fractal description has been found to be a useful adjunct to the more conventional measures of turbulent activity.

1 Introduction

Most fluid dynamics problems of geophysical interest take place in a fluid which is both stratified and rotating. Elementary scaling arguments show that for small scale features the effects of rotation can be ignored to a good approximation. In particular mixing events due to any number of mechanisms; internal wave interaction and breaking, shear layer instability, boundary mixing etc. can be considered, in their early evolution, to be of a three dimensional (3D) character. A number of mechanistic arguments suggest that this 3D state will persist unaffected by the local stratification until the local Froude number of the turbulence $Fr = U'/\ell N$ is of order one. Where U' is the r.m.s. turbulent velocity, ℓ the integral length scale and N the intrinsic frequency of the basic stratification $\left(\frac{g}{\rho}\frac{\partial \rho(z)}{\partial z}\right)^{1/2}$, here $\rho(z)$ is the density distribution of the ambient fluid and g the acceleration due to gravity. Thus Fr can be expressed as a ratio of two length scales ℓ_0/ℓ where $\ell_0 = (U'^3/\ell N^3)^{1/2}$, often referred to as the Osmidov or Richardson length scale. Experiments, e.g., Browand & Hopfinger (1985), Liu et al. (1987), suggest that when Fr, or alternatively Nt, is $O(1)$ the turbulence starts to feel the constraint of its stratified surroundings. Here t is the time since the generation of the 3D turbulence. As Nt increases, scales somewhat larger than ℓ_0 begin to dominate and vertical fluctuations are severely but not completely suppressed. By the time Nt is around thirty all turbulent vertical motion has been suppressed and the flow consists entirely of horizontally interacting eddies of various sizes plus weak internal waves. These motions then constitute a quasi-2D turbulent flow which should then evolve according to the classical ideas of such motions, e.g., Kraichnan (1967) and Batchelor (1969), hereafter K-B. In this and similar

L. J. Pratt (ed.), The Physical Oceanography of Sea Straits, 567–574.
© 1990 Kluwer Academic Publishers.

568

theories an initial concentration of kinetic energy at wave-number k_0 evolves to produce a k^{-3} enstrophy cascade range, while the center-of-gravity of the distribution moves to smaller wave-numbers, i.e. there is a transfer of energy to larger scales. The evolution to a k^{-3} spectrum is not borne out by most numerical simulations of the processes, e.g., McWilliams (1984), find evolution to a k^{-5} spectrum and the appearance of isolated vortices at large times. Recent work by Santangelo et al. (1989) suggests that many of the discrepancies between various numerical simulations are intimately tied to the initital conditions imposed on the system. They find that the spectrum usually has two distinct slopes over a high and low wave number range, that the slopes vary with time and that for long times are steeper than a k^{-3} spectrum. In 1985 we[1] decided to attempt an experimental study of the questions raised by both the k^{-3} theory and the early numerical simulations. One of the experiments is described below.

2 The Experiment

This experiment was performed in a two-layer, stratified tank 2.4m square. A grid of vertical bars was towed across the tank and the flow visualized by photographing neutrally buoyant particles suspended in the interface between the two fluid layers. The resulting particle streaks were digitized and their coordinates placed in computer memory. Velocity vectors were computed and the velocity field (u) interpolated onto a regular 64×64 grid. From these data the vorticity and kinetic energy (u^2) fields could be calculated. It was also possible to calculate the diffusive properties at each time step by calculating the structure-function-diffusivity $[\delta D/\delta t]^2$ and quadratic $|\delta D^2/\delta t|$ as functions of particle pair displacement D. Here δD is the change in D that takes place during the time interval δt. It follows that if one assumes that the diffusion of particle distance D apart is due to vortex structures of order D in diameter (i.e., a local approximation) that:

$$\left[\frac{\delta D}{\delta t}\right]^2 \approx D^{n-1} \text{ and } \left|\frac{\delta D^2}{\delta t}\right| \approx D^{\frac{n+1}{2}} \tag{1}$$

if the energy spectrum has the form $E \approx k^{-n}$, where k is the wave-number.

In what follows we calculated the quantities (1) as well as the energy spectrum found by performing a 2D Fourier transform on the U^2 field and then averaging along arcs of constant k in order to recover a 1D spectrum. We consider the flow at times during the decay process when the flow is essentially two-dimensional, i.e. all 3D and internal wave motion have disappeared.

3 Results

Figure 1 shows the contours of constant vorticity (ω) for three different times during the evolution of the turbulent field; $Nt = 170$, 310 and 475. The appearance of fewer, larger vortices as time progresses is particularly noteworthy as is the observation that they are

---

[1]The work reported here has been done in collaboration with Ph. Caperan and G. R. Spedding.

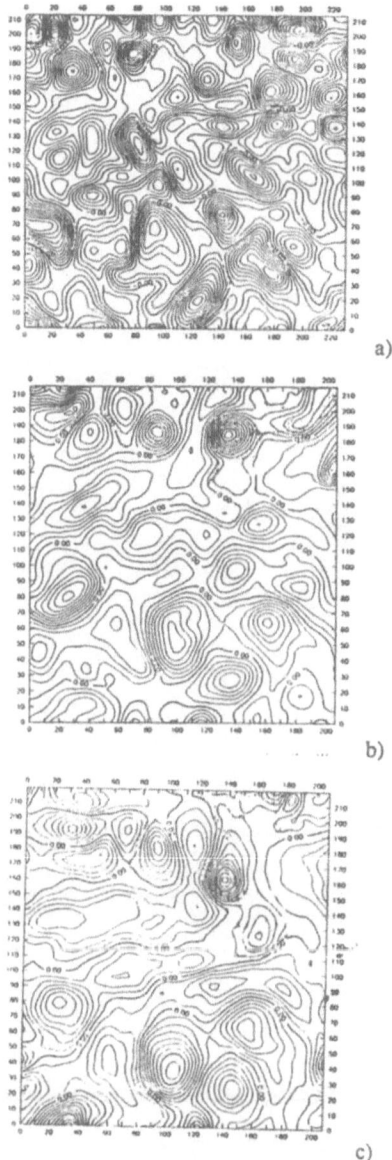

Figure 1. Vorticity distributions $|\omega|$ for the three times $Nt = 170$, 310 and 475 from top to bottom

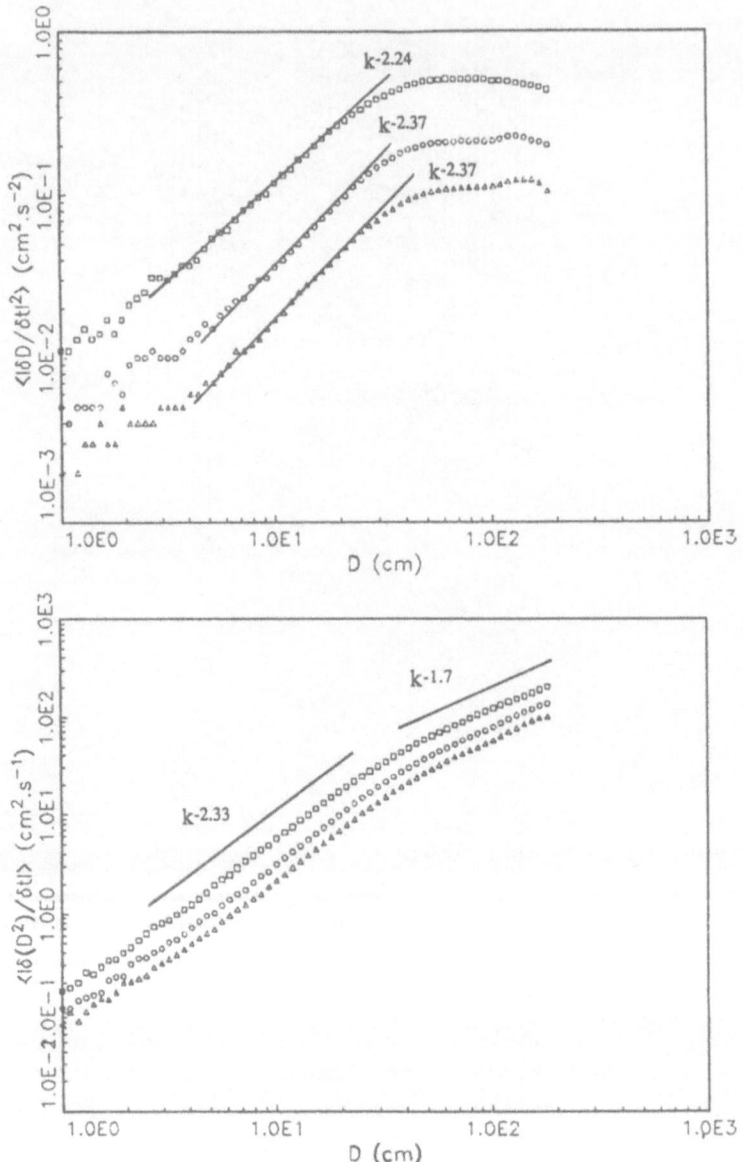

Figure 2. Structure function and quadratic diffusivity as a function of particle pair spacing. The inferred slopes of the energy spectra from equation 1 are superimposed. For the times $Nt = 170$ (□), 310 (O), and 475 (△).

closely packed rather than sparsely distributed (see comments in the Introduction). Also, one can pick out, upon close inspection strong evidence of vortex pairing, the main mechanism by which larger vortices appear predominant at later times.

When one subjects the data to the analyses discussed briefly in §2, one finds results like figs. 2 and 3. In the former we have marked the inferred slopes of the energy spectra using equation (1), and note that they are consistently less than the value of -3 expected from the classical theory. They are, in fact, in close agreement with values found in other experiments using the same techniques (e.g., Griffiths & Hopfinger, 1986). On the other hand, slopes found by direct Fourier Transform of the U^2 field (fig. 3) give values, at small wavenumbers, which are consistently higher than the classical results and which are close to those found in numerical simulations. The reasons for these differences are of some interest since they must ultimately limit the usefulness of (1) when applied to natural data. Since the spectral slope given by the direct calculation must, reasonably, be assumed to be correct, and since calculations using velocity correlation data given the same result, it seems that the assumption that the diffusion is due to local processes is incorrect, and that motions of particles a distance D apart are in fact due to larger scale motions. This point has been made previously by Babiano et al. (1985) but, as far as we are aware, this is the first experimental confirmation.

A second interesting point can be made by observing that while our spectral distributions are virtually identical to those of McWilliams (1984) the two vorticity distributions are quite dissimilar. This leads to the natural question as to how one can *quantitatively* distinguish between the two fields. Of course the problem has arisen because all phase information has been eliminated from the problem in constructing the energy spectrum. How can such information be reintroduced into our description? Or put in a slightly different way, how can one describe the geometry of the vorticity or U^2 fields in a quantitative way?

4 A Fractal Description Of The Vorticity Field

As a result of much recent work on the subject of fractal geometry, e.g., Mandelbrodt (1982), Peitgen & Saupe (1988), we can apply standard techniques in order to determine the distribution of fractal dimensions of our fields. The idea is to apply the box-counting method (Peitgen & Saupe, 1988, pg. 60) to the various level contours of absolute vorticity $|\omega|$ and kinetic energy. Thus on plots like figure 1 we generate contours of vorticity for various values between the maximum and minimum. Apply the box-counting algorithm to these various contour maps results in plots like those shown in figure 4 for $Nt = 170$. Here N is the number of boxes of side L required to cover the regions of vorticity within the contours. The various lines represent the several levels of vorticity chosen $\left[T = \frac{\omega - \omega_{min}}{\omega_{max} - \omega_{min}}\right]$ where ω_{max} and ω_{min} are the maximum and minimum values of $|\omega|$ respectively. The final result is the form shown in figure 5 where the fractal dimensions found from the slope of the lines in figure 4 are plotted as a function of level, T, for the three times $Nt = 170$, 310, and 475. We have also sketched in the type of distribution of $D^{(2)}$ one would expect from the vorticity fields found by McWilliams (1984), i.e., a very rapid change in D once the vorticity maps

---

[2]Not to be confused with the particle separation of §§2 & 3.

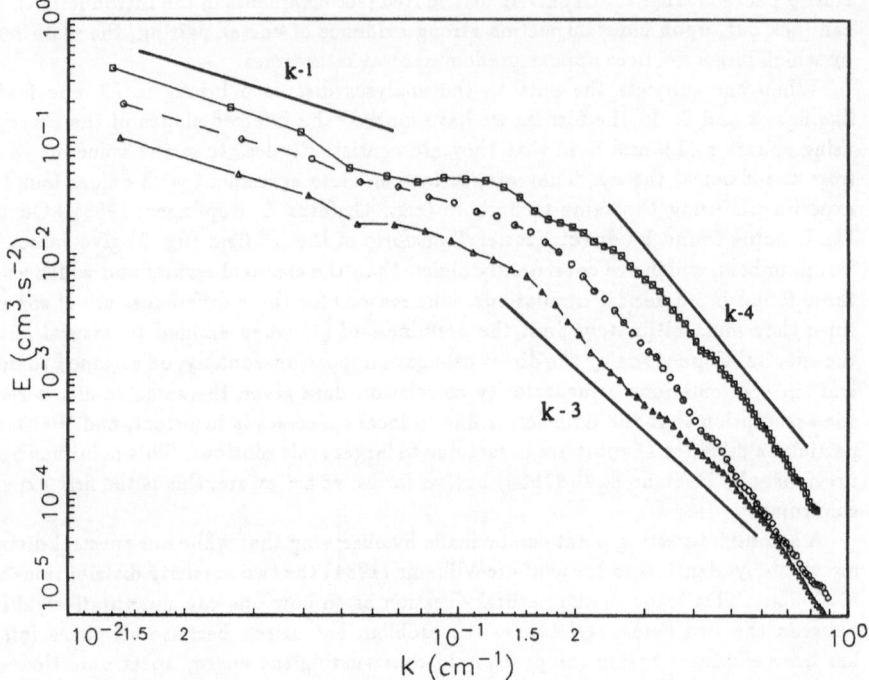

Figure 3. Energy spectrum calculated directly from the U^2 field with the same symbols as figure 2.

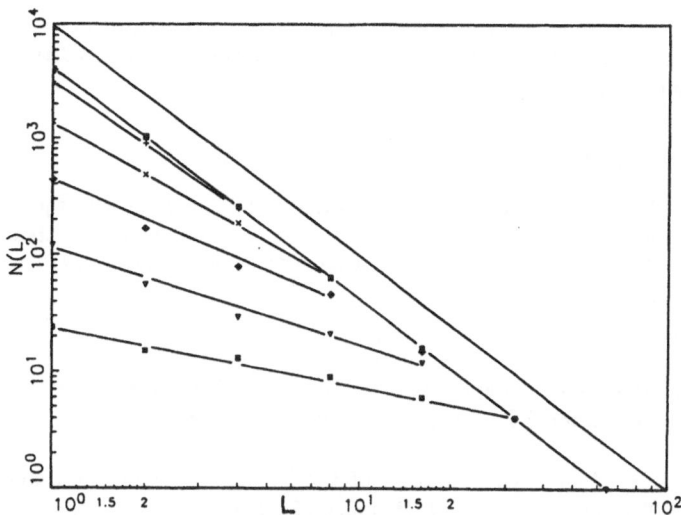

Figure 4. Results of the box-counting algorithm performed on the $|\omega|$ field at $Nt = 170$. Number of boxes N of side L required to cover the $|\omega|$ field at various levels of $|\omega|$, i.e., T.

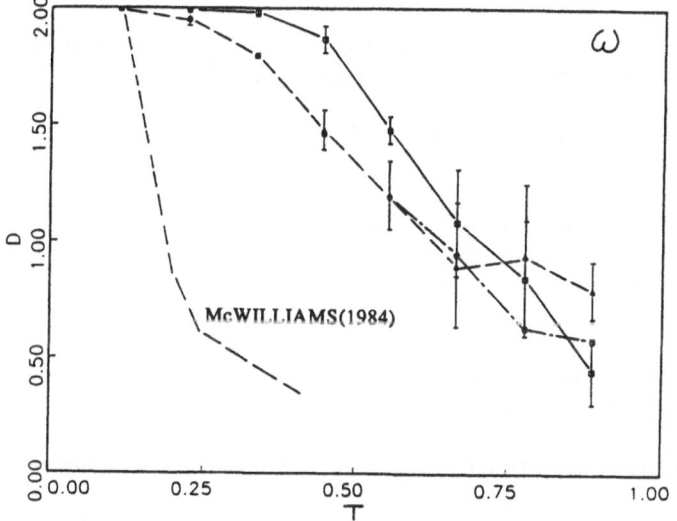

Figure 5. Distribution of fractal dimension, D, as a function of level, T, for the three times and symbols of figure 2.

highlight only the isolated vortices and not the low level background vorticity. It appears that this technique holds some promise for describing the geometry and turbulent fields and distinguishing between them in a quantitative fashion.

5 Acknowledgments

This contribution is a brief résumé of some of the work carried out with the critical assistance of Drs. G. Spedding and Ph. Caperan, their input is gratefully appreciated. The work was supported by ONR Contract No. N-00014-89-J-1400 to the University of Southern California.

6 References

Babiano, A., Basdevant, C. and Sadourny, R. 1985, "Structure Functions and Dispersion Laws in Two-Dimensional Turbulence", *J. Atmos. Sci.*, **42**, 941-49.

Batchelor, G. K. 1969, "Computation of the Energy Spectrum in Homogeneous, Two-Dimensional Turbulence", *Phys. Fl.* (Suppl.), **12**, II233.

Browand, F. K. and Hopfinger, E. J. 1985, "The Inhibition of Vertical Turbulence Scales by Stable Stratification". In *Turbulence and Diffusion in Stable Environments*, Clarendon Press, Oxford.

Griffiths, R. W. and Hopfinger, E. J. 1986, "The Structure of Mesoscale Turbulence and Horizontal Spreading at Ocean Fronts", *Deep-Sea Research*, **31**, 245-69.

Kraichnan, R. H. 1967, "Inertial Ranges in Two-dimensional Turbulence", *Phys. Fl.*, **10**, 1417-23.

Liu, Y. N., Maxworthy, T. and Spedding, G. R. 1987, "Collapse of a Turbulent Front in a Stratified Fluid", *J. Geophs. Res.*, **92**, 5427-33.

Mandelbrodt, B. B. 1982, *The Fractal Geometry of Nature*, W. H. Freeman & Co., New York.

McWilliams, J. C. 1984, "The Emergence of Isolated Coherent Vortices in a Turbulent Flow", *J. Fluid Mechs.*, **146**, 21-43.

Peitgen, H-O. and Saupe, D. 1988 *The Science of Fractal Images*, Springer-Verlag, New York.

Santangelo, P., Benzi, R. and Legras, B. 1989, "The Generation of Vortices in High-Resolution,, Two-Dimensional Decaying Turbulence and the Influence of Initial Conditions on the Breaking of Self-Similarity", *Phys. Fl.*, **A1**, 1027-34.

V. CURRENT RESEARCH PROBLEMS

CURRENT RESEARCH PROBLEMS

LAWRENCE J. PRATT and KARL R. HELFRICH
Department of Physical Oceanography
Woods Hole Oceanographic Institution
Woods Hole, Massachusetts, 02543, U.S.A.

Below is a list of research problems and questions that the participants wrote down towards the end of the meeting. The suggestions were made in response to the question "What problem would you personally like to see solved within the next 5–10 years?" As the reader will see, there is a great deal of interest in the hydraulics of time-dependent strait and sill flows. Related issues include internal bore propagation, rectification, and the applicability of hydraulic control concepts. Other major topics of interest include the determination of maximal *vs.* submaximal exchange, the prediction of interfacial mixing and its consequences, and a host of issues related to flow with multiple density layers or continuous stratification. The following is not a comprehensive list of important issues but rather a statement of personal interests.

Arisoy: Modern technology has provided us with new instruments for oceanographic sampling. With these, it is possible to make measurements at very small time intervals, on the order of minutes, say. In the long range, such practices lead to an enormous amount of data to be processed and evaluated. I believe we have to evaluate the data needs of particular problems versus actual data collection procedures.

Baines: How are singularities which arise in linear coastally-trapped wave theory resolved? Reference, Gill *et al.* (1986).

Baringer: What governs potential vorticity evolution in descending plumes? Can potential vorticity conservation be used as a constraint to determine the interface structure of the Mediterranean outflow? How important is frontal dynamics in plumes? What effects do mixing and friction have on the solution and its stability?

Borenäs: What determines the potential vorticity in recirculation regions in straits?

Bryden: What are the effects of temporal variability (especially on tidal time scales) in the hydraulic control problem? Does the amplitude of the tide influence the amount of exchange? The amount of mixing and dissipation? What is the role of the bore generated by the barotropic tidal interaction with the sill?

L. J. Pratt (ed.), The Physical Oceanography of Sea Straits, 577–580.
© 1990 Kluwer Academic Publishers.

Candela: Can we quantify mixing and friction mechanisms and their relation to the low frequency tidal modulation of the baroclinic exchange flows through straits.

Cannon: To what extent does the longitudinal pressure gradient and its variations affect flow through a strait? What causes the variations?

Farmer: How does a (a) contraction, (b) sill, (c) combination, act to control classical three-layer estuarine flow?

Garrett: Is the Gibraltar exchange maximal or submaximal?

Geyer: What is the influence of tides on net exchange, mean transport and drag?

Ottesen-Hansen: How do outflowing surface plumes break up?

Helfrich: To what extent can time-dependent flows be hydraulically controlled? Can control be exerted in some average sense? Also, how does hydraulic control arise in flows with continuous stratification?

Hunkins: We would like to know how, from first principles, to predict the exchange of water and its properties through a strait of given geography between basins of given hydrography. One of the most serious obstacles appears to be the exchange and recirculation which occurs locally within the strait. How are we able to predict and account for the mixing? Note that this mixing may occur through vertical entrainment as in Gibraltar or horizontally as in the lateral recirculation within Fram Strait.

Lawrence: How do we predict mixing due to interfacial instability?

McClimans: We need to develop a model (laboratory or computer) which simulates coupled exchange, tide and mixing processes. We also need to develop a reliable, precision T–S chain for monitoring strait hydrography in several cross sections. Does the Aanderaa system give sufficient performance? Also, what determines the fate of mixed water in the strait? Does it get entrained to one or both of the layers locally? Does it have its own dynamics and leak out like internal coastal currents (with small deformation radius) along the sides of the channel?

Møller: With references to the talks by Fleming Bo Pedersen (enhanced effects on bulk flow properties) and Armi's and Imberger's on mixing, I would like to suggest the inclusion of friction and mixing mechanisms into three-dimensional numerical models. There still seems to be a wide gap between "physicists" and "modellers" on this subject. Further, I would like to suggest that the assimilation of field data into models and rational vertification procedures for models be developed further.

Murray: The Red Sea with an areal extent of over 450,000 square kilometers (Morcos, 1970) is one of the great marginal sea basins in the world ocean. Extreme evaporation rates in the northern Red Sea produce the unique Red Sea Deep Water (RSDW) which enters the Indian Ocean through the Bab el Mandab Strait and spreads out over the Indian Ocean (Wyrtki, 1971), reaching eastward as far as Sumatra and southward as far as South Africa.

Despite its obvious economic and geopolitical importance, our knowledge of the oceanographic processes operating in the Strait of Bab el Mandab remains primitive. The most recent observations reported by Maillard and Soliman (1986) provide a range of temperature, salinity, oxygen, and velocity variations in the Strait. A pronounced sill of about 100 km length extends northward from the narrowest section (20 km width) at Perim Island to the shallowest section (130 m depth) at the Hanish archipelago, providing an application of double control point theory (Armi and Farmer, 1988). A two-layer system with a dense (high salinity) lower layer outflow into the Gulf of Aden is present in the winter season. Strong southward-directed winds in summer, however, reverse a 30–50 m thick surface layer and set up a distinct three-layer system. Vertical differences on the sill of 4 ppt in salinity and 4°C in temperature in combination with strong currents reaching 100 cm/sec provide the setting for a rich variety of as-yet-unstudied internal hydraulic phenomena. Intensive vertical mixing between layers is indicated by the CTD data available. Outflow of the dense RSDW into the Gulf of Aden also offers a variety of dynamical problems. We can conjecture, for example, the presence of meso-scale cyclones, "Reddies," in the outflow, as in the Mediterranean Sea case. Additional problems associated with sub-tidal fluctuations in the mean flows and the tidal dynamics in the Strait all suggest the Bab el Mandab area as a logical focus of a major international observational effort in the next decade.

Nof: Why should there be a maximal exchange?

O'Donnell: What is the spatial structure of the internal bore in the Strait of Gibraltar and how does it evolve? Perhaps observations with a ship-mounted ADP and a tow-yoed CTD would yield the answers.

Bo Pedersen: How does friction and/or mixing influence the hydraulic control in a gradually varying contraction in a shallow two-layer rotating strait, where the width of the contraction zone is about the internal Rossby Radius?

Pettigrew: (1) What is the role of bore transport in establishing the equilibrium density difference between an ocean basin and a marginal sea that are connected by a strait? Given an imbalance between evaporation, precipitation, and runoff in the marginal sea, does the occurrence of bore transport make equilibrium possible at a lower density difference than would occur under steady conditions? (2) What is the influence of rotation on the structure and propagation of internal bores in straits?

Pratt: We need to develop a shock joining theory for rotating jumps and bores. The missing link is an understanding of the importance of the potential vorticity change across a shock. We also need to develop models which link sill processes, specifically critical control, to the circulation in the upstream basin.

Renouard: We need to learn more about transient (i.e. non-steady) hydraulics in the Strait of Gibraltar including bore propagation. What is the solution to the initial value problem for a nonlinear wave in a rotating fluid?

Salusti: There are many treatments of internal solitary waves in two-layer and rotating systems and in systems with vertical shear. I would be very pleased to see a theoretical model that can easily be compared with field and tank experiments. A stochastic version of these deterministic treatments that are usually applied to infer the velocity field in marine straits would be very stimulating. Also, there is a lot of theoretical work on hydraulic control in baroclinic strait flows, but I would like to see a simple model: (a) in a rotating system, (b) with very weak viscosity, and (c) taking the tides into account.

Sugimoto: My present interests include (1) the year-to-year variation of seasonality of the Tsushima–Tsugaru Warm Current focusing on the effect of air–sea interaction in the East China Sea shelf, the intrusion of the Kuroshio water, and the ring–ring interaction in the outflow area of the Tsugaru warm water. (2) The controlling physics of the volume transport through the Japanese straits, focusing on barotropic inflow / baroclinic outflow adjustment. (3) The effect of straits and gaps on meso- and small-scale structures and their short-term fluctuations (using moored instruments, ferry boats and satellite SST) and their effect on residence time (using drifting buoys).

Ünlülata: There is a need to learn more about time-dependent hydraulics and submaximal exchange flow as affected by neighboring basins.

References

Armi, L. and D. M. Farmer, 1988. The flow of Mediterranean Water through the Strait of Gibraltar. *Progr. Oceanogr.*, **21**(1), 1–105.

Gill, A. E., M. K. Davey, E. R. Johnson and P. F. Linden, 1986. Rossby adjustment over a step. *J. Mar. Res.*, **44**(4), 713–728.

Maillard, C. and G. Soliman, 1986. Hydrography of the Red Sea and exchange with the Indian Ocean in summer. *Oceanol. Acta*, **9**(3), 249–269.

Morcos, S. A., 1970. Physical and chemical oceanography of the Red Sea. *Oceanogr. Mar. Biol. Ann. Rev.*, **8**, 73–202.

Wyrtki, K., 1971. *Oceanographic Atlas of the International Indian Ocean Expedition*, U.S. National Science Foundation, Washington, D.C.; Amerind Publishing Company, New Delhi, 1988, republication.

Index